Revolutionary Automobile Production Systems for Optimal Quality, Efficiency, and Cost

Kakuro Amasaka
Aoyama Gakuin University, Japan

A volume in the Advances in Logistics, Operations, and Management Science (ALOMS) Book Series

Published in the United States of America by
    IGI Global
    Business Science Reference (an imprint of IGI Global)
    701 E. Chocolate Avenue
    Hershey PA, USA 17033
    Tel: 717-533-8845
    Fax: 717-533-8661
    E-mail: cust@igi-global.com
    Web site: http://www.igi-global.com

Library of Congress Cataloging-in-Publication Data

CIP DATA PROCESSING

Revolutionary Automobile Production Systems for Optimal Quality, Efficiency, and Cost
 Kakuro Amasaka
  2024 Business Science Reference

ISBN: 9781668483015(hc) I ISBN: 9781668483022(sc) I eISBN: 9781668483039

This book is published in the IGI Global book series Advances in Logistics, Operations, and Management Science (ALOMS) (ISSN: 2327-350X; eISSN: 2327-3518)

British Cataloguing in Publication Data
A Cataloguing in Publication record for this book is available from the British Library.

For electronic access to this publication, please contact: eresources@igi-global.com.

Advances in Logistics, Operations, and Management Science (ALOMS) Book Series

John Wang
Montclair State University, USA

ISSN:2327-350X
EISSN:2327-3518

MISSION

Operations research and management science continue to influence business processes, administration, and management information systems, particularly in covering the application methods for decision-making processes. New case studies and applications on management science, operations management, social sciences, and other behavioral sciences have been incorporated into business and organizations real-world objectives.

The **Advances in Logistics, Operations, and Management Science** (ALOMS) Book Series provides a collection of reference publications on the current trends, applications, theories, and practices in the management science field. Providing relevant and current research, this series and its individual publications would be useful for academics, researchers, scholars, and practitioners interested in improving decision making models and business functions.

COVERAGE

- Finance
- Marketing engineering
- Computing and information technologies
- Production Management
- Risk Management
- Networks
- Decision analysis and decision support
- Operations Management
- Information Management
- Services management

IGI Global is currently accepting manuscripts for publication within this series. To submit a proposal for a volume in this series, please contact our Acquisition Editors at acquisitions@igi-global.com or visit: https://www.igi-global.com/publish/.

Titles in this Series

For a list of additional titles in this series, please visit:
www.igi-global.com/book-series/advances-logistics-operations-management-science/37170

Novel Six Sigma DMAIC Approaches to Project Risk Assessment and Management
Vojo Bubevski (Independent Researche, UK)
Business Science Reference • © 2024 • 259pp • H/C (ISBN: 9798369328187) • US $295.00

Digital Transformation for Improved Industry and Supply Chain Performance
Muhammad Rahies Khan (Bahria University, Karachi, Pakistan) Naveed R. Khan (UCSI University, Malaysia) and Noor Zaman Jhanjhi (Taylor's University, Malysia)
Business Science Reference • © 2024 • 421pp • H/C (ISBN: 9798369353752) • US $325.00

Theories and Practices for Sustainable Urban Logistics
Olcay Polat (Pamukkale University, Turkey) and Görkem Gülhan (Pamukkale University, Turkey)
Business Science Reference • © 2024 • 297pp • H/C (ISBN: 9798369314470) • US $290.00

Generative AI and Multifactor Productivity in Business
Festus Fatai Adedoyin (Bournemouth University, UK) and Bryan Christiansen (Southern New Hampshire University, USA)
Business Science Reference • © 2024 • 284pp • H/C (ISBN: 9798369311981) • US $285.00

Global Cargo Industry Resilience of Asia-Pacific Shipping Industries
Bivek Datta (Mangalayatan University, Jabalpur, India) and Harshada Rajeev Satghare (Vishwakarma University, Pune, India)
Business Science Reference • © 2024 • 320pp • H/C (ISBN: 9798369316023) • US $290.00

Robo-Advisors in Management
Swati Gupta (Chitkara University, India) Sanjay Taneja (Graphic Era University (Deemed), India) Vijay Kumar (Chaudhary Ranbir Singh University, India) Ercan Ozen (Usak University, Turkey) and Luan Vardari (University "Ukshin Hoti" Prizren, Kosovo)
Business Science Reference • © 2024 • 396pp • H/C (ISBN: 9798369328491) • US $325.00

Neuroleadership Development and Effective Communication in Modern Business
Jyoti Kukreja (Jagannath International Management School, India) Shefali Saluja (Chitkara Business School, Chitkara University, India) and Sandhir Sharma (Chitkara Business School, Chitkara University, India)
Business Science Reference • © 2024 • 360pp • H/C (ISBN: 9798369343500) • US $350.00

IGI Global
PUBLISHER of TIMELY KNOWLEDGE

701 East Chocolate Avenue, Hershey, PA 17033, USA
Tel: 717-533-8845 x100 • Fax: 717-533-8661
E-Mail: cust@igi-global.com • www.igi-global.com

Table of Contents

Table of Contents

Detailed Table of Contents

Chapter 1

In this chapter, the author describes the demand for advancing management technology in global production to succeed in corporate management. Recently, advanced companies in countries all over the world, including Japan, are shifting to global production. Particularly, today, the pressing management issue for Japanese manufacturers to survive in the global market is the "uniform quality worldwide and production at optimum locations." Specifically, therefore, to reconstruct world-leading, the urgent mission for Japanese manufacturers is the establishment of "advanced principles of management technology" which realizes the "simultaneous achievement of QCD (quality, cost and delivery)" for the market value creation.

Chapter 2

In this chapter, the author describes the foundation of automobile production systems for customer value creation through the development of the automobile management technology in corporate strategy. Specifically, to realize this, the author has researched (i) What is the main concern of top management and the Top management and manager class?, (ii) Future management technology aiming the high reliability of company, organization, and human resources, (iii) Creation of the scientific quality management employing customer science principle, and (iv) Developing science TQM to strengthen of automobile management technology using science SQC. Concretely, as the typical example, the author illustrates Toyota's management technology which has come to represent Japanese manufacturing through the (i) Outline of Toyota's corporate management system in Japan, and (2) Implementation of a dual core fundamental "TPS and TQM".

Chapter 3

In this chapter, the author describes the fundamentals of TPS using basic JIT as a dual core principle "TQM and TPS" in Toyota. Specifically, as the basic principle of manufacturing via the TPS, the author describes the simultaneous realization of quality and productivity via a lean system and application examples of fundamentals of TPS. As actual examples using basic JIT, the author focuses on the process control and process improvement using TPS fundamentals: Manufacturing methods and costs; Daily control activities at the production site; Daily improvement activities at the production site; and Innovation of the production process.

Chapter 4

In this chapter, the author describes the foundation and effectiveness of applications of TPS for realizing QCD studies by developing JIT strategy based on the chapter, "Fundamentals of Toyota Production System (TPS) Using Basic JIT." Specifically, the author illustrates the innovation of the production in automobile rear axle unit assembly line by having cooperation between on-site white-collar engineers and supervisors and workers with affiliated and non-affiliated suppliers by the author's patens. Actually, to do the solution of bottleneck problems of manufacturing technology, the author has verified the validity of "Kaizen" through the whole of production process in welding, machining and assembling.

Chapter 5

In this chapter, the author discusses on inter-enterprise link and partnering chains that concretely carries out the platform functions of TPS and TQM activities of Toyota and Toyota group companies while depending on a framework that recognizes collaborating relation between automaker (vehicle assembler) and parts manufacturers (supplier), which is indispensable in producing good products. This study approaches the TPS and TQM activities for excellent QCD studies from a new angle of view of partnering to strengthen Japan supply system (JSS) using the strategic quality management—performance measurement model (SQM-PMM). It also touches on actual study of new development that is looked upon with expectation in near future. Specifically, this chapter studies on potential development by Toyota's Total Task Management Team (TTMT) activity among Toyota Group, which draws attention as a new development in the quality management, by quoting the typical case of the brake pad quality assurance and others.

Chapter 6

To be successful in the future, a global marketer must develop an excellent quality management system that can impress consumers and continuously provide excellent quality products in a timely manner through corporate management. Then, the author has established the new JIT, new management technology principle surpassing JIT for manufacturing in the 21st century. Specifically, new JIT contains a dual hardware and software system, as the next generation technical principle for the customer value creation.

Chapter 7

In recent years, customers have been selecting products that fit their lifestyles and their set of personal values. For this reason, manufacturers' success or failure in global marketing will depend on whether or not they are able to grasp precisely the customers' preferences and are then able to advance their manufacturing to adequately respond to the demands of the times. In this chapter, therefore, the author describes a new automobile product development design model (NA-PDDM) using a dual corporate engineering strategy for innovation of automobile product engineering fundamental employing TDS. Specifically, NA-PDDM contains both the Exterior design engineering strategy and Driving performance design engineering strategy by using customer science principle (CSP) developing advanced TDS. Concretely, the foundation of NA-PDDM consists of the automobile exterior design model with three core methods (AEDM-3CM), automobile optimal product development design model (AOP-DDM), and CSP-customer information analysis and navigation system (CSP-CIANS).

Chapter 8

To advance the Japan manufacturing foundation, the author describes the new japan production management model (NJ-PMM) called advanced TPS that surpasses conventional JIT practices in order to re-construct world-leading management technologies. Specifically, the author mentions the strategic development of a dual global engineering model (DGEM) by possessing the new japan global production model (NJ-GPM) and new japan global manufacturing model (NJ-GMM) surpassing JIT. The effectiveness of DGEM was verified through the actual applications to auto-manufacturing in Toyota and suppliers.

Chapter 9

In this chapter, the author describes the new automobile sales marketing model (NA-SMM) using four core elements based on a dual corporate marketing strategy for innovation of auto-dealers' sales. To realize this, the author develops both the customer science principle (CSp) and science SQC, new quality control principle. Specifically, foundation of NA-SMM consists of the scientific customer creative model (SCCM), networking of customer science principle application system (NCSp-AM), video unites customer behavior and maker's designing intentions (VUCMIN) and scientific mixed media model (SMMM). The validity of NA-SMM is then verified through each model of actual applications to customer creation in Toyota.

Chapter 10

In this chapter, the author introduces the application studies of driving force in new JIT strategy that contributes to the evolution of Japanese automobile production. The author believes that the key to successful global production is the excellent QCD studies by total task management team (TTMT) activities between the assembly-maker and suppliers. Specifically, the author has established the strategic stratified task team model (SSTTM) using partnering performance measurement model (PPMM) for global SCM strategy. To actualize this, the author introduces typical case studies of how this model improved the bottleneck problems of auto-manufactures in the world for realizing simultaneous QCD fulfilment in Toyota and suppliers.

Chapter 11

In this chapter, to strengthen the New JIT strategy more, the author has created a strategic patent value appraisal model (SPVAM) that contributes to corporate management employing strategic enhancing corporate reliability model (SECRM) based on the developing TJS (total job quality management system).

Specifically, to strengthen management technology in corporate innovation, improvement of patent value signifies engineers' value creation at work (invention by white-collar workers). This model consists of several elements each for inventive technique and patent right and has verified the validity of the model at Toyota and other leading corporations. Furthermore, standardization has been carried out in order to spread the effectiveness of SPVAM, and Amasaka new JIT laboratory - patent performance model (A-PPM) has been created and its effectiveness will be investigated through applications at Toyota and others.

Chapter 12

In this chapter, to develop new JIT strategy, the author develops a scientific approach to identifying customers' tastes employing automobile exterior design model (AEDM) for customer value creation. To realize this, the author uses the customer science principle (CSp) aiming to achieve the intelligence design concept method (CSp-IDCM). AEDM improves the design business process so that implicit knowledge on customer is turned into explicit knowledge. To strengthen automobile exterior design, the author has developed the Automobile Exterior Design Model with 3 Core Methods (AEDM-3CM) as follows; (A) Improvement of design business process methods for automobile profile design, (B) Creation of automobile profile design using psychographics approach methods, and (C) Actual studies on automobile profile design, form and color matching support methods (APFC-MSM) employing automobile profile design, form and color optimal matching model (APFC-OMM). The validity of AEDM was verified through case studies of the actual application examples.

Chapter 13

In previous chapters, the author has developed the AEDM (automobile exterior design model) for realizing profile design, form, and color matching. In this chapter, furthermore, the author has created the automobile exterior and interior color matching model using design SQC (AEDM-EICMM). Specifically, then, the author has developed the Design SQC to Toyota vehicle Passo using CSp-IDCM (intelligence design concept method using customer science principle) as follows: (1) Preference surveys were conducted to determine which "exterior colors, interior colors, and front panel colors" suit the preferences of women in their twenties (20's); (2) Color combinations were created based on the data obtained from the preference surveys; (3) Effectiveness of the AEDM-EICMM was confirmed by conducting surveys to determine whether the color combinations created were suitable for women in their 20's, and (4) Similar applications in Toyota and others.

Chapter 14

In this chapter, the author has established the highly-accurate CAE analysis model for bolt-nut loosening solution as a developing new JIT strategy. To enable the high quality assurance - based research aimed at the innovation of product design processes, it is necessary to initiate the transition from the conventional prototype testing method to predictive evaluation method by the combination of various experiments and CAE. This model's validity is verified with application to study on loosening mechanism of bolt-nut tightening as the worldwide auto-manufacturers. bottleneck technology. Specifically, moreover, the author has created an excellent "New design nut" for the prevention of bolt-nut looseness in excellent cost performance as the viewpoint of bottleneck solution in the vehicle market claim "looseness of bolt-nut tightening".

Chapter 15

To strengthen corporate management technology, the author has recently worked out the epoch-making innovation in work qualityy for auto-global production strategy. Specifically, to realize the high-productivity in New JIT strategy, the author has organized the strategic global production activity named AWD-6P/J (aging working development six projects) for evolution of the work environment, and has verified its validity at an advanced car manufacturer Toyota. While many vehicle assembly shops depend on a young male and female workforce, innovation in optimizing an aging workforce is a necessary prerequisite of TPS (JIT). Elements necessary for enhancing work value and motivation, and work energy, including working conditions and work environment (amenities and ergonomics), were investigated through objective survey, and analyzed from labor science perspectives.

Chapter 16

In this chapter, the author describes the new global partnering production model (NGP-PM) for Japan's expanding overseas manufacturing strategy. Recently, leading companies aimed to succeed in localizing production as a global production strategy; the key to this is success in global production in order to get ahead in the worldwide quality competition. Specifically, to improve quality at leading manufacturers' overseas production bases from the perspective of global production, the new global partnering production model (NGP-PM)—the strategic development of the new global production model (NJ-PM) for Toyota's expanding overseas manufacturing—is established. In the enforcement stage, to development of New JIT strategy, strategic global deployment of created strategic productivity improvement model (SPIM) has proved especially the validity of these models.

Chapter 17

Faced with a sluggish economy, car sales have been disappointing in recent years. Given this situation, it is critical that auto-dealerships shift the focus of their sales and marketing activities from attracting new customers to keeping the customers they already have. To develop advanced TMS, this move can be expected not only to reduce sales costs, but also contribute to healthy profits. To strengthen auto-dealerships, therefore, the author proceeds with researching customer satisfaction (CS) and customer loyalty (CL) as a way of boosting marketing effectiveness by using covariance structure analysis/structural equation modeling (SEM), clarifying the key factors that comprise CL, and help improve the marketing strategy. Then, as the application examples, the author focuses on the development and effectiveness for CS and CL by employing video unites customer behavior and maker's designing intentions (VUCMIN) and total direct mail management model (TDMMM).

Chapter 18

In this chapter, to strengthen New JIT strategy, the author has established the scientific mixed media model (SMMM) for boosting automobile dealer visits by developing advanced TMS, strategic customer creation model based on the TMS (total marketing system) named new Japan marketing management model (NJ-MMM) in order to realize the automobile market creation. Specifically, SMMM develops and validates the effectiveness of putting together four core elements: (1) Video that unites customer behavior and manufacturer design intentions (VUCMIN), (2) Customer motion picture–flyer design method (CMP-FDM), (3) Attention-Grabbing train car advertisements (AGTCA), and (4) Practical method using optimization and statistics for direct mail (PMOS-DM) into new strategic advertisement methods designed to enhance marketing and the desire in the automotive industry. At present, SMMM was applied to a dealership representing an advanced car manufacturer Toyota, where its effectiveness was verified.

Preface

To be successful in the future a global marketer must develop an excellent corporate management technology that can impress consumers (customers) and continuously provide high value products in a timely manner through corporate management for global manufacturing in the 21st century. Therefore, to realize global manufacturing that places top priority on customers with an excellent product quality "QCD" (Quality, Cost and Delivery) and in a rapidly changing technical environment, it is essential to create a "new corporate management technology principle" capable of changing the work process quality of all divisions for reforming the super-short-term development production.

Today, one of the greatest contributions that Japan made to the world is "Just in Time (JIT)". JIT is a production system that enables provision of what customers desire when they desire it. The Japanese-style production system represented by the current "Toyota Production System (TPS)", so-called JIT, is a production system which has been developed by Toyota Motor Corporation. TPS (JIT) is also introduced the number of enterprises in the United States and Europe" as a key management technology. However, Toyota's production system "TPS", which is representing the Japanese-style production system today, has already been developed as an internationally shared system, known as the "Lean System" and is no longer an exclusive technology of Toyota.

The top priority issue of the industrial field today is the "new deployment of global marketing" for surviving the era of global quality competition. The pressing management issue particularly for Japanese manufacturers to survive in the global market is the uniform quality worldwide and production at optimum locations which is the prerequisite for successful global production. To overcome this issue, it is essential to renovate not only Toyota Production System, which is the core principle of the production process, but also establish core principles for marketing, design and development, production and manufacturing, and other departments.

Recently, because of that realization, the author has created the "New JIT, new management technology principle: Surpassing JIT" which realizes the simultaneous achievement of QCD into effective management strategy. New JIT contains both of the "hardware and software systems", as the "next generation management technology principle", for transforming management technology into a management strategy in the world.

The hardware system consists of the 5 core elements: "Total Development System (TDS), Total Production System (TPS), Total Marketing System (TMS), Total Business Intelligence Management System (TIS) and Total Job Quality Management System (TJS)". The software system consists of the 3 core elements: "Science TQM, new quality management principle using Science SQC, new quality control principle", "Strategic Stratified Task Team Model (SSTTM)" and "Strategic Patent Value Appraisal Model (SPVAM)".

Aims and Structure

The Japan production technology principle that contributed most to the world in the latter half of the 20th century was the Japanese-style production system typified by "Toyota Production System" (TPS) based on the "Just in Time" (JIT).

In recent years, however, the TPS so-called "Traditional TPS" is introduced in a number of enterprises in the "advanced countries" (United States & Europe) and "developing countries" as the key management technologies, and it is no longer Japanese exclusively (or Toyota's) technology.

Therefore, to get over the "world quality competition", a future successful global marketer must develop an "excellent quality management system" as a "next-generation corporate strategy" that impresses users and continuously provides excellent quality products in a timely manner through corporate management for reforming the super-short-term development production in "customer satisfaction (CS), customer loyalty (CL)".

To realize this, it is important to develop a "new production technology principle" and establish a "new business process management principles" to enable global production. Moreover, "new systematic marketing activities" independent from past experience are required for sales and service divisions to achieve firmer relationships with customers. In addition, a "new quality management technology principle" linked with overall activities for higher work process quality in all divisions is necessary for an enterprise to survive.

In aims and structure, then, the author focuses on the progress of "Toyota Production System" (TPS) through "From JIT to New JIT" based on the Total Quality Management (TQM).

Specifically, as the "theory and practice of TPS", the author introduces the demonstration of "from traditional TPS to advancing TPS", and its effectiveness under the "worldwide corporate management technology" where digital engineering develops rapidly in global production development.

To realize this, it is essential to renovate not only "conventional TPS", which is the core principle of the auto-production process as the "improvement of productivity by *Kaizen*" based on the VE (value engineering) activities", but also establish the "new core principles" for marketing, development and design, production and other departments as the "innovation of corporate management technology" by the total linkage of the "worldwide business process in Japan and overseas".

A future successful global marketer must develop an "excellent quality management system that impresses users, and continuously provides excellent quality products in a timely manner through strategic corporate management. Specifically, to realize this, the author shows the "Section 1: Foundation of Automobile Production Systems: Developing Toyota Production System (TPS), Section 2: Revolutionary Automobile Production Systems: Innovation of Toyota management technology, and Section 3: Optimal Quality, Efficiency, and Cost: Realizing Simultaneous QCD Fulfilment" through the Chapter 1-20.

Kakuro Amasaka
Aoyama Gakuin University, Japan

Introduction

The increasing sophistication and diversification of customers' wants are needed for the development of global production, which acts in concert with the overseas deployment of production bases, a pressing management issue. To succeed in global production, achieving worldwide uniform quality and simultaneous launch (production at optimal locations) is an urgent task. The simultaneous achievement of QCD requirements that reinforce the product appeal is required to realize this global production system (Amasaka, 2015, 2017a, 2022a,b, 2023a).

Then, to transforming Japanese corporate management strategy in automobile global production, the author describes the "Revolutionary Automobile Production Systems for Optimal Quality, Efficiency, and Cost." through Sections 1, 2 and 3 in this book. Specifically, first, Section 1 is the "Foundation of Automobile Production Systems: Developing Toyota Production System (TPS)". Second, Section 2 is the Revolutionary Automobile Production Systems: Innovation of Toyota management technology. Third, Section 3 is "Optimal Quality, Efficiency, and Cost: Realizing Simultaneous QCD Fulfilment".

In Section 1, the Japanese management technology that made the biggest impact on the world in the latter half of the 20th century was the "Toyota Production System" (TPS). It is often also referred to as to "Just-in-Time" (JIT) or "Total Quality Management" (TQM) through the Chapter 2 to 6 (Ohno, 1977; Amasaka, 1988, 2002, 2015, 2017; Toyota Motor Cor6., 1996). As TPS became practiced as Lean System around the world and was further developed, it lost its status as a Japanese production system. In recent years, the superior quality of Japanese products has rapidly lost ground (Goto, 1999; Taylor and Brunt. 2001; Amasaka, 2002, 2013; Amasaka, Ed.; 2007a, 2012).

In Section 2, to be successful in the near future, a global marketer must develop an excellent management technology that can impress customers and continuously provide high value products in a timely manner (See to Chapter 2, 3 and 6 in Section 1). Therefore, to transforming Japanese corporate management strategy based on the "Customer Science principle" (CSp), the author has created the "New JIT, new management technology principle: surpassing JIT" using the 5 core systems as follows;

The "Total Development System" (TDS), "Total Production System" (TPS), "Total Marketing System" (TMS), "Total Business Intelligence Management System" (TIS) and "Total Job Quality Management System" (TJS) named "New Manufacturing Theory" (NMT) in Japanese manufactures through the Chapter 7 to 12 with typical proof examples (Amasaka, 2002, 2005, 2014, 2015, 2017a, 2022a,b, 2023a; Amasaka et al., 2008; Amasaka, Ed., 2019) (See to Chapter 2 and 7). Specifically, as a high-linkage business process of product design, production and sales marketing, New JIT develops the "Science TQM, new quality management principle" with "Science SQC, new quality control principle" (Amasaka, 1999, 2004a,b, 2008a, 2013, Amasaka, Ed., 2007a, 2012) (See to Chapter 2 and 3).

Concretely, to realize strengthening of "New JIT" strategy for global marketing creation, the author has created the "Advanced TDS, TPS, TMS, TIS & TJS" by employing "Strategic Stratified Task Team Model" (SSTTM) and "Strategic Patent Value Appraisal Model" (SPVAM) as the "Driving force in New JIT strategy" for the "simultaneous achievement of QCD" (Amasaka, 2000, 2007a,b, 2008b, 2009a,b, 2011, 2015, 2017b, 2018a, 2019a, 2020a,b, 2021, 2022a,b,c,d, 2023a,b; Amasaka et al., 2008; Amasaka, Ed., 2019) (See to Chapter 7-12).

In Section 3, to expand of New JIT development, the author has verified the validity of the "Revolutionary Automobile Production Systems for Optimal Quality, Efficiency, and Cost" through the application examples as follows; (Amasaka et al., 2008; Amasaka, Ed., 2007a,b, 2012, 2019 ; Amasaka, 2015, 2017, 2022a,b, 2023a).

In Chapter 13 to 15, firstly, the author has demonstrated the typical case studies of the "product development and design: (i) "Developing Automobile Exterior Design Model for customer value creation", (ii) "Automobile Exterior and Interior Color Matching Model using Design SQC" and (iii) "Highly-Accurate CAE Analysis Model for bolt-nut loosening solution" (Amasaka et al., 1999, 2012; Amasaka and Nagaya, 2002; Amasaka, 2007a,b,c,d,e, 2008c, 2009a,b, 2010, 2012a,b, 2017b, 2018b, 2019b; Sakai and Amasaka, 2008; Amasaka et al., 2012; Asakura et al., 2011; Koizumi et al., 2014; Toyoda et al., 2015).

In Chapter 16 to 17, secondly, the author has demonstrated the typical case studies of the "production engineering, manufacturing and SCM": (iv) "Epoch-making innovation in work quality for auto-global production" and (v) "New Global Partnering Production Model for overseas manufacturing" (Ebioka et al., 2007; Yamaji and Amasaka, 2009; Amasaka and Sakai, 2010, 2011). In Chapter 18 to 19, thirdly, the author has demonstrated the typical case studies of the "sales marketing": (vi) "CS and CL to boost marketing effectiveness in auto-dealerships" and (vii) "Scientific Mixed Media Model for boosting auto-dealer visits" (Yamaji et al., 2010; Ishiguro and Amasaka, 2010; Amasaka, 2011; 2023b; Okutomi and Amasaka, 2013; Amasaka et al., 2013).

Kakuro Amasaka
Aoyama Gakuin University, Japan

REFERENCES

Amasaka, K. (1988). *Concept and progress of Toyota Production System (Plenary Lecture), Co-sponsorship: The Japan Society of Precision Engineering and others*. Hachinohe, Aomori-ken.

Amasaka, K. (1999). A study on "Science SQC" by utilizing "Management SQC": A demonstrative study on a new SQC concept and procedure in the manufacturing industry. *International Journal of Production Economics, 60-61*, 591–598. doi:10.1016/S0925-5273(98)00143-1

Amasaka, K. (2000). Partnering chains as the platform for quality management in Toyota. *Proceedings of the 1st World Conference on Production and Operations Management, Sevilla, Spain*.

Amasaka, K. (2002). New JIT, a New Management Technology Principle at Toyota. *International Journal of Production Economics, 80*(2), 135–144. doi:10.1016/S0925-5273(02)00313-4

Amasaka, K. (2004a). Development of *"Science TQM"*, a new principle of quality management: Effectiveness of Strategic Stratified Task Team at Toyota. *International Journal of Production Research*, *42*(17), 3691–3706. doi:10.1080/0020754042000203867

Amasaka, K. (2004b). *Science SQC, new quality control principle: The quality control principle: The quality strategy of Toyota.* Springer-Verlag Tokyo. doi:10.1007/978-4-431-53969-8

Amasaka, K. (2005). Constructing a *Customer Science* Application System *"CS-CIANS"* - Development of a global strategic vehicle *"Lexus"* utilizing *New JIT -. WSEAS Transactions on Business and Economics*, *2*(3), 135–142.

Amasaka, K. (2007a). *New Japan Production Model*, an advanced production management principle: Key to strategic implementation of *New JIT. The International Business & Economics Research Journal*, *6*(7), 67–79.

Amasaka, K. (Ed.). (2007a). New Japan Model: Science TQM-Theory and practice of strategic quality management. Study group on the ideal situation on the quality management on the manufacturing, Maruzen, Tokyo.

Amasaka, K. (2007b). High linkage model *"Advanced TDS, TPS & TMS"*: Strategic development of *"New JIT"* at Toyota. *International Journal of Operations and Quantitative Management*, *13*(3), 101–121.

Amasaka, K. (Ed.). (2007b). Establishment of a needed design quality assurance framework for numerical simulation in automobile production. Working Group No. 4 Studies in JSQC, Study group on simulation and SQC, Tokyo.

Amasaka, K. (2007c). The validity of *"TDS-DTM"*, a strategic methodology of merchandise: Development of *New JIT*, Key to the excellence design *"LEXUS". The International Business & Economics Research Journal*, *6*(11), 105–115.

Amasaka, K. (2007d). Highly Reliable CAE Model: The key to strategic development of *New JIT. Journal of Advanced Manufacturing Systems*, *6*(2), 159–176. doi:10.1142/S0219686707000930

Amasaka, K. (2007e). The validity of *Advanced TMS,* a strategic development marketing system using *New JIT. The International Business & Economics Research Journal*, *6*(8), 35–42.

Amasaka, K. (2008a). Science TQM, a new quality management principle: The quality management strategy of Toyota. *The Journal of Management & Engineering Integration*, *1*(1), 7–22.

Amasaka, K. (2008b). Strategic QCD studies with affiliated and non-affiliated suppliers utilizing New JIT. Encyclopedia of Networked and Virtual Organizations, III(PU-Z), 1516-1527.

Amasaka, K. (2008c). An Integrated Intelligence Development Design CAE Model utilizing New JIT: Application to automotive high reliability assurance. *Journal of Advanced Manufacturing Systems*, *7*(2), 221–241. doi:10.1142/S0219686708001589

Amasaka, K. (2009, June). (2009a). The foundation for advancing the Toyota Production System utilizing New JIT. *Journal of Advanced Manufacturing Systems*, *80*(1), 5–26. doi:10.1142/S0219686709001614

Amasaka, K. (2009b). Proposal and validity of Patent Value Appraisal Model "TJS-PVAM": Development of "Science TQM" in the corporate strategy. *The China-USA Business Review*, 8(7), 45–56.

Amasaka, K. (2010). Proposal and effectiveness of a High-Quality Assurance CAE Analysis Model: Innovation of design and development in automotive industry, *Current Development in Theory and Applications of Computer Science. Engineering and Technology*, 2(1/2), 23–48.

Amasaka, K. (2011). Changes in marketing process management employing TMS: Establishment of Toyota Sales Marketing System, *China & USA. Business Review (Federal Reserve Bank of Philadelphia)*, 10(7), 539–550.

Amasaka, K. (Ed.). (2012). Science TQM, New Quality Management Principle: The quality management strategy of Toyota. Bentham Science Publishers.

Amasaka, K. (2012a). Constructing Optimal Design Approach Model: Application on the Advanced TDS. *Journal of Communication and Computer*, 9(7), 774–786.

Amasaka, K. (2012b). *Prevention of the automobile development design, precaution and prevention.* Japanese standards Association Group, Tokyo.

Amasaka, K. (2013). The development of a Total Quality Management System for transforming technology into effective management strategy. *International Journal of Management*, 30(2), 610–630.

Amasaka, K. (2014). New JIT, new management technology principle: Surpassing JIT [Special issues]. *Procedia Technology*, 16, 1–11. doi:10.1016/j.protcy.2014.10.128

Amasaka, K. (2015). New JIT, New Management Technology Principle. CRC Press, Taylor & Francis Group.

Amasaka, K. (2017a). *Toyota: Production System, Safety Analysis, and Future Directions*. NOVA Science Publishers.

Amasaka, K. (2017b). Strategic Stratified Task Team Model for realizing simultaneous QCD fulfilment: Two case studies. *The Journal of Japanese Operations Management and Strategy*, 7(1), 14–36.

Amasaka, K. (2018a). Innovation of automobile manufacturing fundamentals employing New JIT: Developing Advanced Toyota Production System, *International Journal of Research in Business. Economics and Management*, 2(1), 1–15.

Amasaka, K. (2018b). Automobile Exterior Design Model: Framework development and support case studies. *Journal of Japanese Operations Management and Strategy*, 8(1), 67–89.

Amasaka, K. (Ed.). (2019). *The fundamentals of the manufacturing industries management: New Manufacturing Theory-Operations Management Strategy 21C*. Shankei-Sha.

Amasaka, K. (2019a). Studies on New Manufacturing Theory, *Noble. International Journal of Scientific Research*, 3(1), 42–79.

Amasaka, K. (2019b). Establishment of an Automobile Optimal Product Design Model: Application to study on bolt-nut loosening Mechanism, *Noble. International Journal of Scientific Research*, 3(9), 79–102.

Amasaka, K. (2020a). Studies on New Japan Global Manufacturing Model: The innovation of manufacturing engineering. *Journal of Economics and Technology Research*, *1*(1), 42–71. doi:10.22158/jetr.v1n1p42

Amasaka, K. (2020b). Evolution of Japan Manufacturing Foundation: Dual Global Engineering Model Surpassing JIT. *International Journal of Operations and Quantitative Management*, *26*(2), 101–126. doi:10.46970/2020.26.2.3

Amasaka, K. (2021). New Japan Automobile Global Manufacturing Model: Using Advanced TDS, TPS, TMS, TIS & TJS. *Journal of Advanced Manufacturing Systems*, *6*(6), 499–523.

Amasaka, K. (2022a). New Manufacturing Theory: Surpassing JIT (2nd Edition). Lambert Academic Publishing.

Amasaka, K. (2022b). *Examining a New Automobile Global Manufacturing System*. IGI Global Publisher. doi:10.4018/978-1-7998-8746-1

Amasaka, K. (2022c). A New Automobile Global Manufacturing System: Utilizing a dual methodology, *Scientific Review. Journals in Academic Research*, *8*(4), 41–58.

Amasaka, K. (2022d). A New Automobile Product Development Design Model: Using a dual corporate engineering strategy. *Journal of Economics and Technology Research*, *3*(3), 1–21. doi:10.22158/jetr.v4n1p1

Amasaka, K. (2023a). New Lecture-Surpassing JIT: Toyota Production System-From JIT to New JIT-, Lambert Academic Publishing.

Amasaka, K. (2023b). A New Automobile Sales Marketing Model for innovating auto-dealer's sales. *Journal of Economics and Technology Research*, *4*(3), 9–32. doi:10.22158/jetr.v4n3p9

Amasaka, K., Ito, T., & Nozawa, Y. (2012). A New Development Design CAE Employment Model. *The Journal of Japanese Operations Management and Strategy*, *3*(1), 18–37.

Amasaka, K., Kurosu, S., & Morita, M. (2008). *New Manufacturing Theory: Surpassing JT-Evolution of Just in Time*. Morikita-Shuppan. (in Japanese)

Amasaka, K., & Nagaya, A. (2002). Engineering of the new sensitivity in the vehicle: Psychographics of LEXUS design profile. Development of articles over the sensitivity-The method and practice. Japan Society of Kansei Engineering, Nihon Suppan Service Press.

Amasaka, K., Nagaya, A., & Shibata, W. (1999). Studies on Design SQC with the application of Science SQC improving of business process method for automotive profile design [in Japanese]. *Japanese Journal of Sensory Evaluations*, *3*(1), 21–29.

Amasaka, K., Ogura, M., & Ishiguro, H. (2013). Constructing a Scientific Mixed Media Model for boosting automobile dealer visits: Evolution of market creation employing TMS. *International Journal of Business Research and Development*, *3*(4), 1377–1391.

Amasaka, K., & Sakai, H. (2010). Evolution of TPS fundamentals utilizing New JIT strategy – Proposal and validity of Advanced TPS at Toyota. *Journal of Advanced Manufacturing Systems*, *9*(2), 85–99. doi:10.1142/S0219686710001831

Amasaka, K., & Sakai, H. (2011). The New Japan Global Production Model "NJ-GPM": Strategic development of *Advanced TPS. The Journal of Japanese Operations Management and Strategy*, 2(1), 1–15.

Asakura, S., Kanke, R., & Tobimatsu, K. (2011). A study on Automobile Exterior Color and Interior Color Matching Model: The 20th woman example [Thesis, School of Science and Engineering, Aoyama Gakuin University].

Ebioka, E., Sakai, H., Yamaji, M., & Amasaka, K. (2007). A New Global Partnering Production Model "NGP-PM" utilizing Advanced TPS. *Journal of Business & Economics Research*, 5(9), 1–8.

Goto, T. (1999). *Forgotten Management Origin*. Seisansei-Shuppan.

Ishiguro, H., & Amasaka, K. (2012). Establishment of a Strategic Total Direct Mail Model to bring customers into auto-dealerships. *Journal of Business & Economics Research*, 10(8), 493–500. doi:10.19030/jber.v10i8.7177

Koizumi, K., Kanke, R., & Amasaka, K. (2014). Research on Automobile Exterior Color and Interior Color Matching. *International Journal of Engineering Research and Applications*, 4(8), 45–53.

Ohno, T. (1977). *Toyota Production System*. Diamond-Sha.

Okutomi, H., & Amasaka, K. (2013). Researching Customer Satisfaction and Loyalty to boost marketing effectiveness: A look at Japan's auto-dealerships. *International Journal of Management & Information Systems*, 17(4), 193–200. doi:10.19030/ijmis.v17i4.8093

Sakai, H., & Amasaka, K. (2008). Demonstrative verification study for the next generation production model: Application of the Advanced Toyota Production System. *Journal of Advanced Manufacturing Systems*, 7(2), 195–219. doi:10.1142/S0219686708001577

Taylor, D., & Brunt, D. (2001). *Manufacturing operations and supply chain management–The lean approach*, Thomson Leaning, London.

Toyoda, S., Nishio, Y., & Amasaka, K. (2015). Creating a Vehicle Proportion, Form, and Color Matching Model. *International Organization of Scientific Research*, III(3), 9–16.

Toyota Motor Corporation. (1996). *The Toyota Production System*. International Public Affairs Division Operations Management Consulting Division.

Yamaji, M., & Amasaka, K. (2009). Strategic Productivity Improvement Model for white-collar workers employing Science TQM. *The Journal of Japanese Operations Management and Strategy*, 1(1), 30–46.

Yamaji, M., Hifumi, S., Sakalsiz, M. M., & Amasaka, K. (2010). Developing a strategic advertisement method "VUCMIN" to enhance the desire of customers for visiting dealers. *Journal of Business Case Studies*, 6(3), 1–11. doi:10.19030/jbcs.v6i3.871

Chapter 1
The Demand for Advancing Management Technology in Global Production

ABSTRACT

In this chapter, the author describes the demand for advancing management technology in global production to succeed in corporate management. Recently, advanced companies in countries all over the world, including Japan, are shifting to global production. Particularly, today, the pressing management issue for Japanese manufacturers to survive in the global market is the "uniform quality worldwide and production at optimum locations." Specifically, therefore, to reconstruct world-leading, the urgent mission for Japanese manufacturers is the establishment of "advanced principles of management technology" which realizes the "simultaneous achievement of QCD (quality, cost and delivery)" for the market value creation.

MANAGEMENT TECHNOLOGY SHIFTING TO GLOBAL PRODUCTION

Advanced companies in the world, including Japan are shifting to *global production* to realize the *"uniform quality worldwide and production at optimum locations"* for survival in fierce competition (Amasaka, Ed., 2012; Amasaka, 2015, 2017a, 2022ab. 2023).

To attain successful global production, the "technical administration, production control, purchasing control, sales administration, information system, and other administrative departments" should maintain close cooperation with clerical and indirect departments while establishing strategic cooperative and creative business linkages with individual development, production and sales departments, and outside manufacturers (suppliers) (Amasaka, 2007).

Today, consumers have quick access to the latest information in the worldwide market thanks to the development of Information Technology (IT), and strategic organizational management of the production control department has become increasingly important. Simultaneous attainment of quality, cost and delivery (QCD) requirements is the most important mission for developing highly reliable new products ahead of competitors (Amasaka, 2004, 2008a,b; Amasaka, Ed., 2012).

DOI: 10.4018/978-1-6684-8301-5.ch001

This requires the urgent establishment of an innovative production control system for the next generation (called next-generation production control system). With a view to assuring that future management technology is a new leap forward for "Japanese manufacturing", the "progress of production control of plants in the manufacturing industry" made so far by the manufacturing industry is summarized in Figure 1 (Amasaka, 2004a).

Figure 1. Progress of management technology in the manufacturing industry

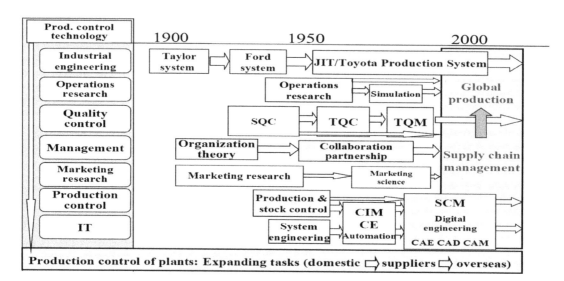

In the figure, the basis of the major production control methodologies, such as industrial engineering (IE), operations research, quality control, management of administration, marketing research, production control, and IT, are plotted along the vertical axis. Along the horizontal axis, some of the key elemental technologies, management methods, scientific methodologies, and so on are mapped out in a time series.

Since the beginning of the century, the operation of manufactures has shifted from domestic production in Japan to overseas production bases, and management technology has become increasingly complicated, as depicted in Figure 1.

For the production control department, the key to success in global production is modeling strategic supply chain management (SCM) for domestic and overseas suppliers with a systematization of its management methods. In the implementation stage, deep-plowing studies of the typical Toyota Production System (TPS) called Just in Time (JIT), Total Quality Management (TQM), partnering, and digital engineering will be needed in the future (Amasaka, 1999a,2004b, 2006, 2009).

Is Japanese Manufacturing All Right?

A straightforward look at the recent management activities of Japanese manufacturers reveals a series of quality-related recalls that were totally unexpected by the top corporations in the industry and increasing delays in technological development that may even threaten the survival of these manufacturers (Nihon Keizai Shinbun, 1999, 2000, 2001, 2002b, 2006; Nikkei Sangyo Shinbun 2000a,b).

In recent years, both developed Western nations and developing nations have advanced the study of Japanese TPS and TQM and re-acknowledged the importance of the "quality" of administrative management technology. They have also promoted the reinforcement of "quality" in manufacturing on a national level. As a result of such efforts, the "superiority in quality of Japanese products" has been gradually compromised.

One distinctive example of this is shown in Figure 2, a comparison of the "quality of automobiles" sold in the United States. Although Toyota, a leading Japanese car manufacturer, can be seen to have achieved steady improvements in the quality of its automobiles (IQS, Initial Quality Study) up to now, GM of the United States and Hyundai of Korea have also promoted quality improvements and achieved even more dramatic results (Gabor, 1990; Joiner, 1994; Nezu, 1995; JD Power, 2009; Amasaka, 2010).

Figure 2. The comparative examples of automobile quality in U.S.A
(J.D Power and Associates)

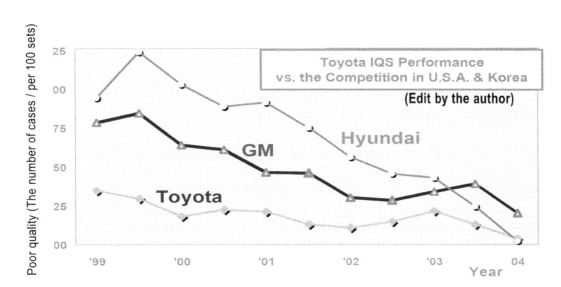

The observations above indicate that in order for Japanese manufacturers to continue to play the leading role in the world, it is urgent to reform their management technology from a fresh standpoint, rather than simply clinging to the successful experiences they have enjoyed up to now (Amasaka, 2002b, 2004b, 2014).

THE DEMAND FOR ADVANCEMENT IN MANAGEMENT TECHNOLOGY

Management Tasks for the Manufacturing Industry

The environmental changes that surround today's manufacturing industry are truly severe. The customers strictly judge the reliability of the manufacturer according to the reliability (quality and value gained from use) of their products. Therefore, for Japanese manufacturers to be successful and survive in the future marketplace, they must carry out global marketing that will focus on customers worldwide and then be able to quickly offer high quality products of the latest model that are designed to enhance the value to the customer (Amasaka, 2002b, 2004a).

They need to precisely grasp the customers' preferences and advance their "manufacturing" so that it can respond to the demands of the times or they will be eliminated from the world market. To prevail in today's manufacturing industry, which is often referred to as a worldwide quality competition, the pressing management issue is to realize the kind of global production that can achieve the so-called "worldwide uniform quality and simultaneous production".

It is not an exaggeration to say that what will ensure Japanese manufacturers' success in global marketing is the realization of "competitive manufacturing – the simultaneous achievement of QCD" ahead of their competitors in an effort to offer high value-added products (Amasaka, 2004b, 2006, 2008b, 2017b).

In the midst of the drastic changes taking place at the manufacturing site due to the use of "digital engineering", it is imperative not to fall behind in the advancement of administrative management technology that is so vital to manufacturing. The urgent mission for Japanese manufacturers is to reconstruct world-leading, an "Advanced production principles of management technology" and "administrative management technology", which will be viable even for next-generation manufacturing.

To accomplish this, it will be vital for such management-related departments as technology management, production management, sales management, and information technology which collectively comprise the core of corporate management for engineering, production and sales, as well as cooperate with their suppliers (Amasaka, 2000a,b, 2004a, 2008a).

Needs for the Reform of Japanese-Style Management Technology

The top priority issue of the industrial field today is the "new deployment of global marketing" for surviving the era of "global quality competition" (Kotler, 1999; Amasaka, 2002b, 2004a, 2008a,b).

The pressing management issue particularly for Japanese manufacturers to survive in the global market is the "uniform quality worldwide and production at optimum locations" which is the prerequisite for successful global production. To realize manufacturing that places top priority on customers with a good QCD and in a rapidly changing technical environment, it is important to develop a new production technology principle and establish new process management principles to enable global production.

Furthermore, a new quality management technology principle linked with overall activities for higher work process quality in all divisions is necessary for an enterprise to survive (Burke and Trahant, 2000; Amasaka, 2004a).

The creation of attractive products requires each of the sales, development design, and production departments to be able to carry out management that forms linkages throughout the whole organization (Amasaka, 2000b, 2004a).

From this point of view, the reform of Japanese-style management technology is desired once again. In this need for improvements, Toyota is no exception (Goto, 1999; Amasaka, 2002b, 2004a, 2006, 2008a,b).

Importance of Strategic QCD Studies With Affiliated and Non- Affiliated Suppliers

IT development has led to a market environment where customers can promptly acquire the latest information from around the world with ease. In this age, customers select products that meet their lifestyle and have a sense of value on the basis of a value standard that justifies the cost.

Thus, the concept of "Quality" has expanded from being product quality, which is oriented to business quality, to becoming corporate management quality-oriented. Customers are strict in demanding the reliability of enterprises through the utility values (quality, reliability) of their products (Amasaka, 1999b; Evans and Dean, 2003).

Advanced companies in countries all over the world, including Japan, are shifting to global production. The purpose of global production is to realize "uniform quality worldwide and production at optimum locations" in order to ensure company's survival amidst fierce competition (Amasaka, 1998, 1999a,b, 2004a,b; Doz and Hamel, 1998).

For the manufacturing industry, the key to success in global production is systematizing its management methods when modeling strategic SCM for its domestic and overseas suppliers (Yamaji and Amasaka, 2009a),

In-depth studies of the Toyota Production System called *JIT* and Lean Production, System, TQM, partnering, and digital engineering will be needed when these methods are implemented in the future (Ohno, 1977; Roos et al., 1991; Toyota Motor Corp., 1997; Amasaka, 1998, 2002b, 2004a, 2008a; Amasaka, Ed., 2012).

Above all, manufacturers endeavoring to become global companies are required to collaborate not only with affiliated companies, but also with non-affiliated companies to achieve harmonious coexistence among them based on cooperation and competition. In other words, a so-called "federation of companies" is needed (Hamel and Prahalad, 1994; Amasaka, 1999a,b, 2004a, 2008a, 2017b).

THE DEMAND FOR ADVANCING MANAGEMENT TECHNOLOGY IN GLOBAL PRODUCTION

What Are Top Management's Concerns?

For manufacturers to be successful in the future global market, they need to develop the various products that have a strong impression on consumers and supply these items in a timely fashion through effective corporate management.

Then, the author shows that it is clearly impossible to continue to lead the next generation simply by adhering to and maintaining the TPS (Toyota Production System) and TQM (Total Quality Management), which are the dual-pillars of traditional Japanese management technology. To overcome these problems, it is essential not only to advance the "Japanese Production System" (JPS) named "New Japan Production Model" (NJPM), a core technology of the production processes, but to also establish a core technology for the sales, design, and development related divisions (Amasaka, 2002b, 2004a,c, 2007a).

Given the above, the author has conducted an awareness survey of general management personnel and executives (a total of 72 people) from 12 advanced companies belonging to the Toyota Group (Amasaka, 2002b, 2007a,b).

Similarly, based on another awareness survey of other companies (Fuji Xerox and Daikin among others, with a total of 153 participants) participating in the "Study Group for Manufacturing Quality Management (a.k.a. "The Amasaka Forum")", management technology issues have been investigated from the standpoint of corporate management (Amasaka et al., 2008; Amasaka, Ed., 2012).

The overall management technology issues have been plotted in a chart as shown in Figure 3. In Figure 3, this confirms that managers responsible for development give the highest priority to "suggestion-based new merchandise and product development" as a global merchandise strategy, whereas production managers put efforts into establishing the

Figure 3. Management technology problems (Positioning of opinion)

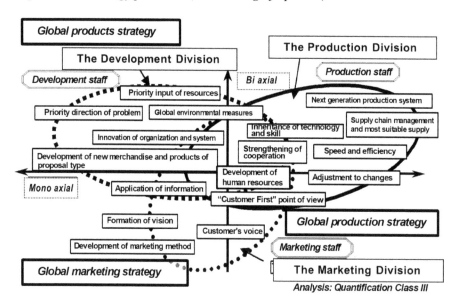

"Next-Generation Production System" in View of Global Production

Sales managers, on the other hand, prioritize the "development of new marketing methods" in order to be successful in global marketing. Moreover, the issue common to all was how to respond to globalization.

Therefore, to overcome these management issues, it will be necessary to carry out reforms of the human resources cultivation system through intelligent sharing of information, and create a new management technology for closer ties among the company's all divisions.

The above awareness surveys and analysis clarified the core technologies necessary for the next-generation management technology principle, the basis for new management technologies, and the technological elements required for linking these core technologies.

Thus, in the future, it will be important to create management technology equipped with a new concept that enables total linkage of QCD research conducted by each of the aforementioned divisions from the standpoint of strategic corporate management, and by so doing, to create an advancing management technology "New JIT, new management technology principle" named "New Manufacturing Theory: Surpassing JIT" (Amasaka et al., 2008; Amasaka, 2008b, 2019, 2022a, 2023; Amasaka, Ed., 2019).

Specifically, to offer customers high value-added products and prevail in the worldwide quality competition, it is necessary to establish the advanced production system that can intellectualize the production engineering and production management system.

This will in turn produce high performance and highly functional new products. On a concrete target, the simultaneous achievement of QCD requirements that reinforce the product appeal is required to realize this global production system (Amasaka, 2002b, 2008b, 2015, 2022b, 2023; Yamaji and Amasaka, 2009a).

To succeed in automobile global production, *"worldwide uniform quality and simultaneous launch (production at optimal locations)"* is an urgent task. Therefore, in increasing sophistication and diversification of customers' needs, the author has made the development of a form of global production that acts in concert with the overseas deployment of production bases into a pressing management issue.

Reinforcement of the Corporate Management Function for Realizing Customers' Wants

As mentioned above, the top priority issue of the industrial field today is the "new deployment of global marketing" for surviving the era of global production competition. Therefore, to continue to play the leading role in the world, the reform of Japanese-style management technology is desired once again.

Today, customers select products that suit their lifestyle and personal values. In addition, they are strongly demanding manufacturing that will enhance customer value via the products' reliability (quality and value gained from use).

For this reason, unless the manufacturer advances its manufacturing processes in such a way as to respond to these demands, while also accurately grasping the customers' preferences, it will be pushed out of the world market.

In this day and age when customers have quick and easy access to the latest information from every corner of the world due to the permeation of IT, the production management department, which is the command center of manufacturing, needs more than ever before to have a global view and deploy strategic production management as the core of corporate management (Amasaka, 2007a,b, 2015, 2022a, 2023).

The new mission of the production management department is the deployment of strategic cooperation between the on-site departments of engineering, production and sales, and general administration departments, as well as the domestic and overseas suppliers.

This is done in order to realize worldwide uniform quality and simultaneous launch (production at optimal locations) ahead of other competitors (Amasaka, 2000c; Ebioka et al., 2007).

In other words, in order to solve the various manufacturing problems both domestically and overseas, the experienced-based implicit knowledge (such as know-how, empirical rules, intelligent information, etc.) possessed by the white-collar and on-site production workers needs to be converted to "linguistically expressed knowledge" or "explicit knowledge" as shown in Figure 4 (Amasaka, 2002b, 2005; Yamaji and Amasaka, 2009b).

This will be done through the incorporation of the "Science TQM, new quality management principle" employing the "Science SQC, new quality control principle" that has been proven effective as a scientific problem solving method (Amasaka, 1998, 1999a,b, 2003, 2004a,b; Amasaka, Ed., 2000, 2012).

Moreover, this is important for all divisions to turn implicit knowledge of their business process into explicit knowledge through integrated and collaborative activities by sharing awareness.

Figure 4. Visualization of experienced-based implicit knowledge

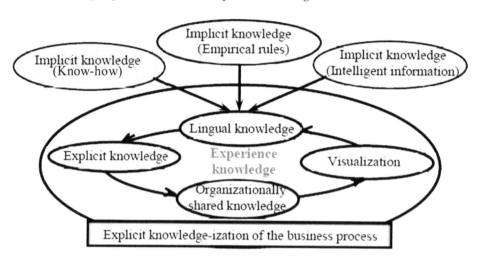

To accomplish this, a new systematic and organized "SQC promotion cycle" in the Toyota way (Refer to Appendix 2.A), so-called "Science SQC, new quality control principle" established by the author was developed under a new concept using a new methodology that applied the 4 core principles that enabled jobs to be scientifically performed as shown in Figure 5.

This conceptual diagram shows the nucleus of Toyota's and its group's TQM activities based on a new quality principle that is the secret of success for next-generation manufacturing. As determined from figure, the four core principles are incorporated into a "New SQC application system" where they are closely linked to each other.

The 1st core principle of "Scientific SQC" refers to scientific approaches at every stage of the process ranging from determination of problem to accomplishment of objectives.

Figure 5. Schematic drawing of Science SQC, new quality control principle

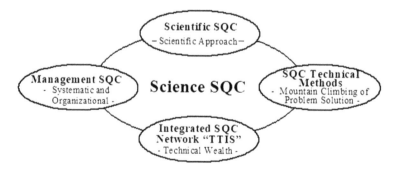

The 2nd core principle of "SQC Technical Methods", which use the seven tools (N7) for TQM, multivariate analysis (MA), design of experiment (DE), reliability analysis (RA) and others, refers to the mountain-climbing methodology for solving problems.

The 3rd core principle of SQC, "Integrated SQC Network "TTIS" (named "Toyota SQC Technical Intelligence System")", represents the networking of SQC software application by using the sub-core principles. It can turn proprietary data inheritance and development into science.

The 4th core principle of "Management SQC" is to support prompt solution of deep-rooted engineering problems.

Particularly in the practical application, the gaps between principles and rules have to be clarified scientifically as engineering problems, and general solutions have to be approached by clarifying the gaps that exist in theory, testing, calculation, and actual application. For further details on Science SQC with four core principles (Amasaka, 1998, 2000c, 2003, 2004c; Amasaka, Ed., 2012). Science SQC is vital to deploy this advancement of manufacturing in a systematic and organizational manner by deploying the personal "empirical knowledge" as "organizationally shared knowledge".

CONCLUSION

In this chapter, the author described the "Demand for advancing management technology in global production" to succeed in corporate management from the viewpoint of the management technology shifting to global production. To continue to play the leading role in the world, particularly, the author has focused on the necessity of the advanced manufacturing which realizes the "simultaneous achievement of QCD". Then, to realize the market value creation, the author has researched as follows; (1) Progress of management technology in the manufacturing industry, (2) The comparative examples of automobile quality "Toyota IQS performance vs. GM and Hyundai" in U.S.A, (3) Management tasks for the manufacturing industry, (4) Needs for the reform of Japanese-style management technology, (5) Importance of strategic QCD studies with affiliated and non-affiliated suppliers, (6) What are top management's concerns? to be successful in the future global market, and (7) Reinforcement of the corporate management function for realizing customers' wants by developing "Science TQM using Science SQC".

REFERENCES

Amasaka, K. (1988). Concept and progress of Toyota Production System (Plenary lecture). The Japan Society of Precision Engineering.

Amasaka, K. (1999a). *TQM at Toyota-Toyota's TQM Activities: to create better car (Special Lecture), A Training of Trainer's Course on Evidence- based on Participatory Quality Improvement, International Health Program (TOT Course on EPQI), Tohoku University School of Medicine (WHO Collaboration Center).* Sendai-city.

Amasaka, K. (1999b). *The TQM responsibilities for industrial management in Japan - The research of actual TQM activities for business management.* The Japanese Society for Production Management.

Amasaka, K. (Ed.). (2000). Science SQC: The reform of the quality of a business process, Japanese Standards Association.

Amasaka, K. (2000a). Basic principles of JIT – Concept and progress of Toyota Production System (Plenary lecture), *The Operations Research Society of Japan.* Strategic research group-1, Tokyo.

Amasaka, K. (2000b). *Partnering chains as the platform for Quality Management in Toyota*. Proceedings. of the 1st World Conference on Production and Operations Management, Seville, Spain.

Amasaka, K. (2000c). A demonstrative study of a new SQC concept and procedure in the manufacturing industry: Establishment of a New Technical Method for conducting Scientific SQC. *Mathematical and Computer Modelling*, *31*(10-12), 1–10. doi:10.1016/S0895-7177(00)00067-4

Amasaka, K. (2002a). Intelligence production and partnering for embodying a high-quality assurance (Plenary Lecture). The 94th Course of JSME Tokai Branch, Nagoya, Japan.

Amasaka, K. (2002b). New JIT, A New Management Technology Principle at Toyota. *International Journal of Production Economics*, *80*(2), 135–144. doi:10.1016/S0925-5273(02)00313-4

Amasaka, K. (2003). Proposal and implementation of the "Science SQC", quality control principle. *Mathematical and Computer Modelling*, *38*(11-13), 1125–1136. doi:10.1016/S0895-7177(03)90113-0

Amasaka, K. (2004a). Development of "Science TQM", a new principle of quality management: Effectiveness of Strategic Stratified Task Team at Toyota. *International Journal of Production Research*, *42*(17), 3691–3706. doi:10.1080/00207540420000203867

Amasaka, K. (2004b). *The past, present, future of production management (Keynote lecture)*. The 20th Annual Technical Conference, Nagoya Technical College, Aichi, Japan. .

Amasaka, K. (2004c). *Science SQC, new quality control principle: The quality control principle: The quality strategy of Toyota*. Springer-Verlag Tokyo. doi:10.1007/978-4-431-53969-8

Amasaka, K. (2005). Constructing a *Customer Science* application system "*CS-CIANS*"-Development of a global strategic vehicle "*Lexus*" utilizing *New JIT*. *WSEAS Transactions on Business and Economics*, *3*(2), 135–142.

Amasaka, K. (2006). A new principle, next generation management technology: Development of New JIT (Editorial), *Production Management* [in Japanese]. *Transaction of the Japan Society for Production Management*, *13*(1), 143–150.

Amasaka, K. (2007a). New Japan Production Model, an advanced production management principle: Key to strategic implementation of *New JIT*. *The International Business & Economics Research Journal*, *6*(7), 67–79.

Amasaka, K. (2007b). High linkage model "*Advanced TDS, TPS & TMS*": Strategic development of "*New JIT*" at Toyota. *International Journal of Operations and Quantitative Management*, *13*(3), 101–121.

Amasaka, K. (2008a). Science TQM, a new management principle: The quality management strategy of Toyota. *The Journal of Management and Engineering Integration*, *1*(1), 7–22.

Amasaka, K. (2008b). Strategic QCD studies with affiliated and non-affiliated suppliers utilizing New JIT. Encyclopedia of Networked and Virtual Organizations, VIII(PU-Z), 1516-1527.

Amasaka, K. (2009). New JIT, advanced management technology principle: The global management strategy of Toyota (Invitation lecture*). International Conference on Intelligent Manufacturing and Logistics System, and Symposium on Group Technology and Cellular Manufacturing*, Kitakyushu, Japan.

Amasaka, K. (2010). *Manufacturing of the 21st Century: The proposal of a "New Manufacturing Theory" surpassing JIT (Plenary Lecture), The Japan Society of Production Management.* The 23th Annual Conference, Kobe University, Kobe, Japan.

Amasaka, K. (Ed.). (2012). Science TQM, New Quality Management Principle: The quality management strategy of Toyota, Bentham Science Publishers, U.A.E, USA, The Nethrlands.

Amasaka, K. (2014). New JIT, New Management Technology Principle: Surpassing JIT. *Procedia Technology*, *16*(special issues), 1135–1145. doi:10.1016/j.protcy.2014.10.128

Amasaka, K. (2015). New JIT, new management technology principle, Taylor and Francis Group, CRC Press, Boca Raton, London, New York.

Amasaka, K. (2017a). *Toyota: Production system, safety analysis and future directions.* NOVA Science Publishers Inc.

Amasaka, K. (2017b). Strategic Stratified Task Team Model for Realizing Simultaneous QCD Fulfilment: Two Case Studies. *Journal of Japanese Operations Management and Strategy*, *7*(1), 14–35.

Amasaka, K. (2019). Studies on New Manufacturing Theory, *Noble. International Journal of Scientific Research*, *3*(1), 42–79.

Amasaka, K. (Ed.). (2019). *The fundamentals of the manufacturing industries management: New Manufacturing Theory-Operations Management Strategy 21C.* Shankei-sha.

Amasaka, K. (2022a). New Manufacturing Theory: Surpassing JIT (2nd Edition), Lambert Academic Publishing, Germany.

Amasaka, K. (2022b). *Examining a New Automobile Global Manufacturing System.* IGI Global Publisher. doi:10.4018/978-1-7998-8746-1

Amasaka, K. (2023). New Lecture-Toyota Production System: From JIT to New JIT, Lambert Academic Publishing, Germany, Printed by Printforce, United Kingdom.

Amasaka, K., Kurosu, S., & Morita, M. (2008). *New Manufacturing Theory: Surpassing JT-Evolution of Just in Time.* Morikita-Shuppan.

Amasaka, K., & Osaki, S. (1999). The promotion of New Statistical Quality Control Internal Education in Toyota Motor: A proposal of "Science SQC" for improving the principle of TQM, *The European Journal of Engineering Education, Research and Education in Reliability, Maintenance, Quality Control. Risk and Safety*, *24*(3), 259–276.

Burke, W., & Trahant, W. (2000). *Business climate shift.* Butterworth Heinemann.

Doos, D., Womack, J. P., & Jones, D. T., (1991). *The machine that changed the world–The story of lean production.* Rawson / Harper Perennial.

Doz, Y. L., & Hamel, G. (1998). *Alliance advantage.* Harvard Business School Press, Boston.

Ebioka, E., Sakai, H., Yamaji, M., & Amasaka, K. (2007). A New Global Partnering Production Model "NGP-PM" utilizing "Advanced TPS". *Journal of Business & Economics Research*, *5*(9), 1–8.

Evans, J. R., & Dean, J. W. (2003). Total quality management, organization and strategy, Thomson, South-Western, Mason, United States.

Gabor, A. (1990). *The Man Who Discovered Quality; How W. Edwards Deming Brought the Quality Revolution to America- The Stories of Ford, Xerox, and GM*. Random House, Inc.

Goto, T. (1999). *Forgotten origin of management–Management quality taught by G.H.Q, CCS management lecture*. Seisanse-Shuppan.

Hamel, G., & Prahalad, C. K. (1994). *Competing for the future*. Harvard Business School Press.

Joiner, B. L. (1994). Fourth generation management: The new business consciousness. Joiner associates, Inc., Louisville, GA.

Nezu, K. (1995). *The scenario of the U.S. manufacturing industry: Remarkable progress aimed at by CALS*. Kogyo-Chosakai Publishing.

Ohno, T. (1977). *Toyota Production System*. Diamond-sha.

PowerJD. (2009). UTL: http;//jdpower.com/

Toyota Motor Corporation. (1997). *Creation is infinite: The history of the Toyota Motor Corporation 50 year*. Toyota.

Yamaji, M., & Amasaka, K. (2009a). Strategic Productivity Improvement Model for white-collar workers employing Science TQM. *The Journal of Japanese Operations Management and Strategy, 1*(1), 30–46.

Yamaji, M., & Amasaka, K. (2009b). Proposal and validity of Intelligent Customer Information Marketing Model: Strategic development of *Advanced TMS, China - USA. Business Review (Federal Reserve Bank of Philadelphia), 8*(8), 53–62.

APPENDIX

"SQC Promotion Cycle" Activity in Toyota

The author has been engaged in *"SQC Promotion Cycle" activity* (Implementation – Practical Effort – Education – Growing Human Resources) under the banner of the *Toyota's SQC Renaissance* as shown in Figure 2.A. This was done in order to capture the true nature of making products and in the belief that the best way to develop personnel is through practical research that will raise the technological level. The aim of SQC Promotion Cycle activity that is being promoted by Toyota is to take up the challenge of solving vital technological assignments (Amasaka, 1998, 2004b).

Figure 6. A "SQC promotion cycle" activity in Toyota

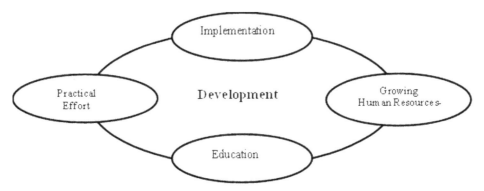

14

Chapter 2
Foundation of Automobile Production Systems for Customer Value Creation

ABSTRACT

In this chapter, the author describes the foundation of automobile production systems for customer value creation through the development of the automobile management technology in corporate strategy. Specifically, to realize this, the author has researched (i) What is the main concern of top management and the Top management and manager class?, (ii) Future management technology aiming the high reliability of company, organization, and human resources, (iii) Creation of the scientific quality management employing customer science principle, and (iv) Developing science TQM to strengthen of automobile management technology using science SQC. Concretely, as the typical example, the author illustrates Toyota's management technology which has come to represent Japanese manufacturing through the (i) Outline of Toyota's corporate management system in Japan, and (2) Implementation of a dual core fundamental "TPS and TQM".

STRENGTHENING OF THE AUTOMOBILE MANAGEMENT TECHNOLOGY IN CORPORATE SYRATEGY

The author discusses the realization of "Strengthen of the automobile management technology in corporate strategy" in the following first based on the "investigation and that knowledge of 2nd Chapter" as the "Foundation of Automobile Production System for customer value creation" through the researches of the author until now (Amasaka, 2004a, 2015, 2017, 2022a,b, 2023; Amasaka, Ed., 2007, 2012, 2019).

WHAT IS THE MAIN CONCERN OF TOP MANAGEMENT AND THE MANAGER CLASS?

A close look at recent corporate management activities can reveal various situations wherein an advanced manufacturer, which is leading the industry, is having a difficult time due to unexpected quality related problems (Amasaka, 2000, 2002; Amasaka et al., 2008)

DOI: 10.4018/978-1-6684-8301-5.ch002

Some companies have failed to see through the customers' feelings and have slowed down in product development or in their production engineering development / production management systems, and are thus facing a crisis of their own survival as a manufacturer.

On the other hand, quite a few manufacturers have been thriving and enjoying steady growth for the past few decades by actively and bravely taking a scientific approach toward (concretizing) their customers' feelings and thereby conducting a reform program in product development and the service / sales systems, thus carrying out a company-wide "total marketing" activity through partnering with their suppliers.

What is the cause of this situation wherein major differences between manufacturers have developed? When considering the main concern on the minds of top management and the manager class, it is obtaining reasonable management results by means of the "creation of a corporate environment for utilizing human resources and activating the organization" that is indispensable for realizing "quality management that gives customers top priority" (Amasaka, 2004b,c; Amasaka, Ed., 2007).

The fact that many Japanese companies recruit graduates having a similar academic ability from all over Japan makes it evident that the key to the rise and fall of a company is held by these newly employed young business persons, and whether they are engaging in creative work and have proved to be on the leading edge of corporate environment reform.

With regard to the management issues facing Japanese manufacturers, the studies conducted by the author, have pinpointed, among other things (Amasaka et al., 2008; Amasaka, 2008a,b, 2009a,b; Amasaka, Ed., 2007, 2019).

(1) The "vulnerability of management technological capability" in corporate management.

(2) More specifically therein, the lack of ability at manufacturing sites is pointed out.

(3) The lack of reliability in technological development designing is also brought up.

(4) The need for conducting reasonable marketing activity regardless of traditional methods is presented for sales / service of related departments.

(5) The need for taking a new scientific approach for creating a new market is brought to the forefront of the general planning, merchandize planning, and the product planning related departments.

(6) As the basis of the reinforcement measures for global production, "simultaneous achievement of QCD" is necessary for office administration / management departments by reforming the quality of their business process while breaking away from the conventional, reactive ways of business and performing proactive work as the core of the corporate management activity.

FUTURE MANAGEMENT TECHNOLOGY AIMING FOR THE HIGH RELIABILITY OF COMPANY, ORGANIZATION, AND HUMAN RESOURCES

In the midst of this severe worldwide quality competition for the survival of manufacturers inside and outside of Japan, the top management and manager class need to recognize anew the "way of manufacturers' quality management", particularly when considering the repeated cases of quality problems in recent years that can considerably harm customer satisfaction (CS) (Amasaka, 2004a,b,c, 2006; Amasaka et al., 2008).

Against this background, the author has focused attention on the "way of future quality management aiming for high reliability of company, organization, and human resources" which is necessary for the "creation of highly reliable products", and therefore discussed the importance of systematic and organizational cooperation among all departments involved (Amasaka, 2008ab; Amasaka, Ed., 2007, 2019).

Figure 1 is given here to present a business process which realizes the "high quality assurance manufacturing" and shows an organization chart of an onsite work department belonging to a typical manufacturer, as well as an example of its QCD activity. The figure clearly indicates the representative 13 departments ranging from engineering, to production, to sales, and their respective missions.

In each step of the cyclical business process made up of "merchandise planning- product planning - designing-design drawing -research and development-prototyping / experiment evaluation - production engineering development - production preparation (production management) - purchase and procurement–manufacturing / inspection - promotion / sales-service-marketing", it is vital to engage in called "total marketing".

That is, activities that improve the "reliability of the work performed" in close cooperation with each other in order to best perform the cyclic "missions" of: "how it is to be - what is needed - how to make it - how well it is made - how to sell it (whether the customers are pleased with it or not) - how it was all done" (Amasaka, 2004a,c, 2008a,b, 2009a,b; Amasaka et al., 2008; Amasaka, Ed., 2019).

Figure 1. A typical corporate organization chart and QCD activity

At this point, consideration shall be given to the reason of why market claims and recalls occur, and how important it is for all the departments to systematically and organizationally cooperate in order to prevent such problems.

For this reason, the author incorporates the theory of probability as the "importance of cooperative activity among departments for improving the work reliability" (Amasaka, 2008b).

Table 1 illustrates the work quality of each department converted into the degree of reliability for Cases 1 to 6. To present it simply, the theory of probability is incorporated here for interpreting the overall work quality of all the 13 departments into their degree of reliability (credibility) to show the total reliability (or total unreliability, hereafter called claim ratio).

Table 1. Job reliability of 13 departments

Case 1: 99.9 % / department
→ $0.999^{13} = 0.987$ (1.3% problems)
Case 2: 99.0 % / department
→ $0.990^{13} = 0.878$ (12.2% problems)
Case 3: 95.0 % / department
→ $0.950^{13} = 0.513$ (48.7% problems)
Case 4: 90.0% / department
→ $0.900^{13} = 0.254$ (4.6% problems)
Case 5: When 12 departments are 99.99% but 1 department is 50.0%:
→ $0.999^{12} \times 0.500 = 0.499$ (50.01% problems)
Case 6: 99.99 % / department
→ $0.9999^{13} = 99.88\%$ (0.12% problems)

On the assumption that cooperation between departments (ratio of circulating information) is indexed at 1.00 (100%), even when the reliability of Department 1 is 99.9%, the total reliability is rated at 98.7%, indicating that claims can occur in the market at the probability of 1.3%.

Similarly, in Case 2, (99.00% / department) 12.2% of recall probability can be deduced. The recall probability increases to 48.7% in Case 3 (95.00% / department) and it jumps further to 74.6% in Case 4 (90.00% / department).

In Case 5, although 12 departments are marked with a high reliability of 99.99%, 1 department performs poor quality work (50.00% / department) due to inexperienced workers or careless mistakes overlooking certain things, and the result is a devastating 50.01% claim probability, indicating that corporate reliability can be deeply undermined.

In Case 6 (99.9% / department) where the total reliability reaches 99.88%, the claim probability can finally be reduced to the order of ppm (0.12%).

The above analysis makes it easier to understand why all the departments are required to perform high quality work in order to maintain reliability. The table also makes it very clear that when the circulation of information is not sufficiently conducted between departments that the total reliability can drop even lower.

Such a situation is far too serious to categorize simply as "something like a traffic accident! (or bad luck!)" as can be seen in the recent recall-related social responsibility crisis.

CREATION OF THE "SCIENTIFIC QUALITY MANAGEMENT" EMPLOYING "CUSTOMER SCIENCE PRINCIPLE"

The mission of a manufacturer is to offer products the consumers (customers) are pleased with, as the basis for sustainable growth. Entering into a new century of product creation based on the management of global marketing, it is necessary to create the kind of products which further enhance the life stages and lifestyles of customers, as well as customer value.

To develop and offer attractive, customer-oriented products, it is vital to urgently and seriously consider "customer needs" and to establish strategic product development methods which are ahead of the times (Amasaka, 2002a, 2004a, 2005).

To directly confront the management environment today's companies are surrounded by and to implement the necessary measures to respond to it, it is indispensable to establish a "scientific approach toward customer orientation".

A reasonable business approach is needed which can be utilized for product planning and technical development through the digitalization of the hidden desires of customers, so that subjective information (about the customers) and objective information (objectified by technology) can be mutually and compatibly exchanged.

Generally speaking, though customers have both favorable and unfavorable evaluations about current products in the market, they usually do not have a clear image of what types of products they want in the future. The customers express their demands in spoken words, and therefore the product designers (planning and designing staff) need to accurately interpret such expressions and convert them into corresponding design drawings.

For this reason, the sales and service staff who are closest to the customers need to express the product image that the customers have to the planners, (research engineers / designers who think objectively in numerical terms) who engage in product development, in a scientific, common language rather than rely on an implicit, vague language (Amasaka, 2002a, 2005).

In connection with the creation of future products, it is particularly important to "offer precisely and quickly what the customers want before they realize they want it". To do this, it is vital to clearly grasp the hazy, ambiguous feelings of customers.

The product development technological method – "Customer Science principle" (CSp) shown in Figure 2 is what gives concrete shape to such customer wants (Amasaka, 2002a, 2004a, 2005, 2008b). It is intended to present a mode of (an approach to) a new business process for creating "wants" which is indispensable for manufacturing attractive products.

As depicted in the Figure 2, called objectification of subjectivity wherein the image of customers' words (implicit knowledge) is expressed in a common language (lingual knowledge) and then, by incorporating technical words (design drawings, etc.) as well as correlation techniques, it is further interpreted appropriately (into explicit knowledge).

This refers to the CSp that converts subjective information (y) and objective information ($\hat{y}$) reasonably to two-way through application of correlation technology.

When using CSp for approaching various customer-related situations, such as why the customers are satisfied or dissatisfied with a particular product, what is the underlying feeling behind a certain expression, what kind of products then need to be offered, or in what specific situation a recall case occurs, the situations can then be interpreted into a common language, and further converted into the language of technology.

Then the staff of the research & development or designing departments can digitize such situations by means of correlation techniques utilizing statistical science, simulate them in the laboratory or experiment facility, and confirm the conditions in which such situations are most likely to occur.

Figure 2. Schematic drawing of customer science principle (CSp)

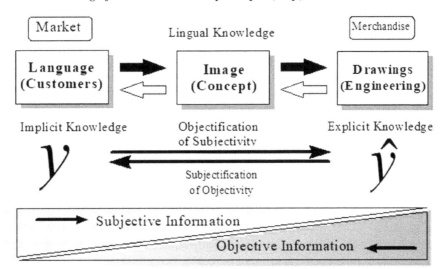

Finally, it is necessary to check whether what is represented on a drawing specifically reflects what the customers actually want and thereby confirm the accuracy of the work being performed, thus subjectifying the objectivity using correlation techniques (Amasaka et al., 1999; Amasaka, 2002a, 2003a, 2005, 2007a).

By conducting "total marketing," that is, an approach focusing on "quality management that gives customers top priority" incorporating Customer Science principle, the implicit business process, consisting of promotion / sales, product planning, designing, development designing, and production, which has been a major concern for the management class, can be clarified further.

By means of the scientific knowledge obtained from the cycle of these business processes, "accumulation of successes" or "correction of failures" can be carried out more accurately than ever, and therefore highly reliable quality management, "scientific quality management" can definitely be realized (Amasaka et al., 2005).

It is observed that well-performing manufacturers both inside and outside Japan today have maintained an attitude which prompts them to humbly repeat the process of clarifying implicit knowledge in order to grasp the customers feelings to the greatest extent possible, and then feed it back to check whether what is reflected in their product design drawings truly represents the objectified demands of customers.

Such an attitude constitutes the basis of their manufacturing activity (Amasaka, 2007a,b,c, 2008a, 2009b).

DEVELOPING SCIECE TQM TO STRENGTHEN OF "AUTOMOBILE MANAGEMENT TECHNOLOGY" USING SCIENE SQC

Then, to strengthen of "automobile management technology", the author has developed the "Scientific TQM" named "Science TQM, new quality management principle" using "Science SQC, new quality control principle" as the "Scientific methodology in CSp employment" (Amasaka, Ed., 2000, 2007, 2012; Amasaka, 2002a, 2003a, 2004a,b,c, 2005) (See to Figure 4 and 5 in Chapter 2).

"Science SQC" in Science TQM strategy utilizes to strengthen the core management technologies as a dual methodology, and to contribute systematically and organizationally to the rational activities of management tasks: "CS (Customer Satisfaction), ES (Employee Satisfaction) and SS (Social Satisfaction)" (Amasaka, 1998)/

To achieve these management tasks, it is important for all divisions to turn the implicit knowledge of their business processes into explicit knowledge through integrated and collaborative activities and sharing objective awareness. Specifically, to do the solution of various management technology issues, Science SQC consists of the "Four core principles for problem-solving of typical corporate management technologies" as shown in Figure 3 to 6 as follows; (Amasaka, 1998, 1999a, 2000b, 2003a, 2004a)

Scientific SQC (1st Core Principle)

The primary objective of SQC as applied by the manufacturing industry is to enable all engineers and managers (hereinafter referred to as businessmen) to carry on excellent QCD (Quality, Cost and Delivery) research activities through insight obtained by applying SQC to scientific and inductive approaches in addition to the conventional deductive method of tackling engineering problems.

It is important to depart from mere SQC application for statistical analyses or trial and error type analysis, and to scientifically use SQC in each stage from problem structuring to goal attainment by grasping the desirable form.

Figure 3 shows a conceptual drawing of "Scientific SQC".

Figure 3. Schematic drawing of "scientific SQC"

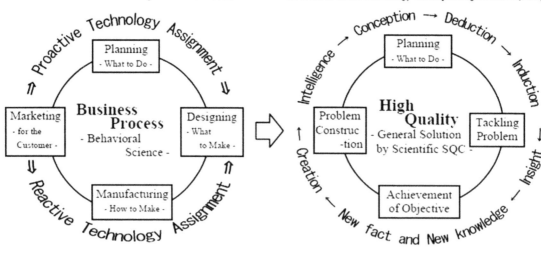

3-1 Business Process for Customer Science   3-2 Scientific SQC for Improving Technology

SQC Technical Methods (2nd Core Principle)

For solving today's engineering problems, it is possible to improve experimental and analysis designs by using N7 (new seven tools for TQC) and basic SQC methods based on investigation of accumulated technologies. Moreover, it is possible to "Mountain-climbing for Problem-Solving" by using a proactive combination of the "RA (reliability analysis), MA (multivariate analysis), DE (design of experiment) & various statistical science" as may be required using multivariate analysis amalgamated with engineering technology.

The methodology in which the SQC method is used in combination at each stage of problem-solving has spread and established as the "SQC Technical Methods" for efficiently improving the jobs of businessmen. (i) In 1st stage, N7 is used for "structuring problem and selecting topic", (ii) In 2nd stage, MA is used for "problem-solving (level-1)", (iii) In 3rd stage, DE develops for "problem-solving (level-2)" and (iv) In addition, RA is used for both stages of the "structuring problem & verification" in target achievement level".

Figure 4 shows a conceptual diagram of the established "SQC Technical Methods".

Figure 4. Schematic drawing of SQC technical methods

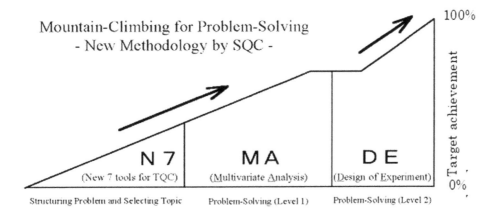

Integrated SQC Network "TTIS" (3rd Core Principle)

Various cases of successful SQC application to actual business need to be systematized in order for them to contribute to forming engineering assets and help inheritance and further development. This is a prerequisite for the development of Science SQC. This is achieved with the "Integrated SQC Network "TTIS" (Toyota SQC Technical Intelligence System) that supports engineering problem solving as shown in Figure 5 (Amasaka and Maki, 1992; Amasaka, 1995, 2000b, 2004a).

TTIS is an intelligent SQC application system consisting of 4 main systems integrated for growth by supplementing one another as follows;

1st, TSIS (Total SQC Intelligence System) can be referred to as library of SQC application examples that has been constructed as an entry consisting of 4 subsystems for facilitating accumulation of technological assets; (i) TSIS-QR (-Quick registration and Retrieval library), (ii) TSIS-RB (-Reference Book), (iii) TSIS-PM (-Practice Manual) and (iv) TSIS-ML (-Mapping Library).

2nd, TPOS (Total SQC Promotional Original Soft) is an SQC software package for a personal computer that promotes spiraling the SQC cycle developed within the company upward.1 3rd, TSML (Total SQC Manual Library) is a library of classified technical methods for scientific application of SQC techniques along the job flow. 4th, TIRS (Total Information Retrieval System) consists of the technical reports and engineering books enabling the degree of contribution of SQC to be confirmed according to the daily job results.

Figure 5. Schematic drawing of integrated SQC network "TTIS"

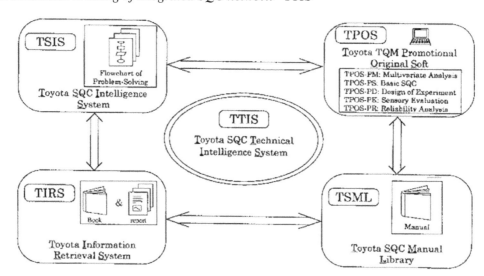

Management SQC (4th Core Principle)

The main objective of SQC in the manufacturing industry is to support quick solution of deep-rooted engineering problems. Therefore, the main objectives of Science SQC is to find a scientific solution for the gap generated between the theories (principles and fundamental rules) and reality (events).

Particularly, in the application of Science SQC, the differences from the principles and rules in an engineering problem should be scientifically analyzed to clarify six gaps that occur between theory, calculation, experiment and actual result to obtain a general solution. Filling these gaps results in an organizational problem.

For problem solving, it is necessary for the planning, design, manufacturing and marketing departments to clarify the six gaps, in other words to turn tacit knowledge on the business process to explicit knowledge for good understanding and coordination among the departments (Amasaka, 1997).

The methodology for organizationally managing the development of Science SQC is called "Management SQC". Figure 6 shows a conceptual drawing of "Management SQC" for "Science SQC" development. In Figure 6, to solve the pending issue of a technology problem in the market, it is necessary to create a universal solution (general solution) by clarifying the existing six gaps (① to ⑥ in Figure 6) in the process consisting of theory (technological design (technological design model), experiment (prototype to production), calculation (simulation), and actual result (market) as shown on the lower left of Figure 6.

To accomplish this, the clarification of the six gaps (① to ⑥) in the business processes across the divisions, described in the lower right of Figure 6 below, is of primary importance. By taking these steps, the intelligent technical information owned by the related divisions inside and outside the corporation will be fully linked, thus reforming the business processes involved in development design.

Figure 6. Schematic drawing of management SQC

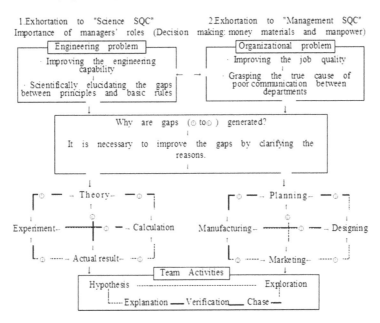

Recently, the author is developing Management SQC strategically as the "Total QA (quality assurance) High Cyclization Business Process Model" which systematically and strategically realizes high quality assurance by incorporating analyses, and he verifies that validity (Amasaka, 2004a, 2008c, 200cb; Amasaka, Ed., 2007) (See to Figure 4 in Chapter 8). Furthermore, the author discussed and studied Management SQC through demonstrative cases of "Total Task Management Team activities" at Toyota Motor Corporation (Amasaka et al. 1997; Amasaka, 2002b, 2003b) (See to Chapter 6 in detail).

FOUNDATION OF TYPICAL JAPANESE PRODUCTION SYSTEM WITH A DUAL CORE FUNDAMENTAL "TPS & TQM" IN TOYOTA

As the typical example of the "Foundation of Automobile Production System for customer value creation", the author describes Toyota's management technology which has come to represent Japanese manufacturing (Amasaka, 2015, 2017, 2022a,b, 2023). Toyota's corporate management with a dual core fundamental "Toyota Production System (TPS) and Total Quality Management (TQM)" is a successful production system that has become widespread and been developed internationally as the "JIT" (Just in Time) (Amasaka, 1988, 1989, 1999b, 2000c; Amasaka, Ed., 2007, 2012, 2019; Amasaka et al., 2008).

Outline of Toyota's Corporate Management System in Japan

What is known as the "Just in Time" (JIT) system, the Japanese production system is an automobile manufacturing system that was developed by the Toyota Motor Corporation that contributed most to the world in the latter half of the 20th century was the Japanese-style production system by the TPS and TQM (Ohno, 1977; Toyota Motor Corp., 1997).

This system is enhanced by the quality management technology principle generally referred to as "Just in Time" (JIT) and "Lean System". JIT is a production system that enables provision of what customers desire when they desire it. JIT was also introduced in a number of enterprises in the United States and Europe as a key management technology (Hayes and Wheelwright, 1984; Womack et al, 1990; Gabor, 1990; Taylor and Brunt, 2001; Womack et al., 1991, 1994;).

The basic philosophy of the Toyota's corporate management system is built on the ideas of the Toyota Group founder, Sakichi Toyoda, and his business *"mottos"*, (1) "be ahead of the times through endless creativity, inquisitiveness, and the pursuit of perfection; and (2) a product should never be sold unless it has been carefully manufactured and has been tested thoroughly and satisfactorily". The philosophy also reflects the ideas of Toyota Motor's founder, Kiichiro Toyoda, regarding improvement through inspection: (1) grasp the demands of consumers firsthand and reflect them in your product and (2) investigate the product quality and business operations and then improve them.

These are the basic concepts of Toyota's corporate management system, which aims to realize quality and productivity simultaneously by effectively applying total quality control (TQC) and total quality management (TQM) to the automobile manufacturing process. It also pursues maximum efficiency (optimal streamlining, which is called Lean Systems) while also being conscious of the principles of cost reduction and thereby improving the overall product quality (Toyota Motor Corp., 1997; Amasaka, 1988, 1999b).

Implementation of a Dual Core Fundamental "TPS and TQM"

In implementation stage, it is important to constantly respond to the customers' needs, promote flawless production activities, and conduct timely quality, cost, and delivery (QCD) research, as well as put it into practice (Amasaka, 1988, 1989).Therefore, Toyota has positioned a dual core fundamental "TPS and TQM" as the core management technologies for realizing *"reasonable manufacturing"*, and these management technologies are often likened to being the wheels of an automobile (Amasaka, 1989, 1999a,b, 2000a,b,c, 2003a; Hayashi and Amasaka, 1990).

These management technologies have been placed on the vertical and horizontal axes as shown in Figure 7. The combination of these technologies reduces large irregularities in manufacturing to the state of tiny ripples, and average values are consistently improved in the process. This strategy is an approach used by reasonable corporate management in which the so-called *leaning process* is consistently carried out. As indicated by the vertical and horizontal axes in the figure, when the "Hardware technology of the TPS" and "Software technology of TQM (TQC) are implemented, the statistical quality control (SQC) is to be effectively incorporated to scientifically promote QCD research and achieve constant upgrading of the manufacturing quality (Amasaka, 1999a, 2002a).

Figure 7. A dual core fundamental "TPS and TQM"

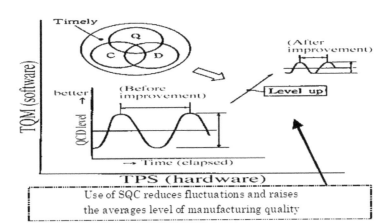

In first point, the aim of TPS deployment is to create a highly functional production system that offers better products at lower prices more quickly through the timely application of QCD activities for strengthening Value Engineering (VE) by using Value Analysis (VA) (Amasaka, 1988, 1989, 1999b).

Another point that can be understood from the figure is that TQM and SQC are the foundations of maintaining and improving the manufacturing quality, and have historically served as a basis for the advancement of TPS. In this way, the basic concept of TPS has reformed the automobile manufacturing process used at Toyota. As a result, the effectiveness of TPS has been recognized on a worldwide scale, and it is now regarded as the core concept of manufacturing (See to Chapter 4 and 5 in detail).

CONCLUSION

In this chapter, the author has discussed the "Foundation of Automobile Production Systems for customer value creation" through the "Strengthen of the automobile management technology in corporate strategy" as follows;

(i)   The main concern of the top management and manager class-it is obtaining reasonable management results by the "creation of a corporate environment for utilizing human resources and activating the organization" that is indispensable for realizing "quality management".

(ii) Future quality management aiming for high reliability of company, organization, and human resources"-What is necessary for the "creation of highly reliable products", and therefore, discussed the importance of systematic and organizational cooperation among all departments involved.

(iii) Creation of the "Scientific quality management employing Customer Science principle (CSp)-In creation of future products, it is particularly important to "offer precisely and quickly what the customers want before they realize they want it". To do this, it is vital to clearly grasp the hazy, ambiguous feelings of customers.

(iv) Developing Science TQM to strengthen of "automobile management technology" using Science SQC-Specifically, to do the solution of various management issues, Science SQC consists of the "4 core principles: Scientific SQC, SQC Technical Methods, Integrated SQC Network "TTIS", and Management SQC.

Concretely, as the typical example, the author illustrates Toyota's management technology which has come to represent Japanese manufacturing through the (1) Outline of Toyota's corporate management system in Japan and (2) Implementation of a dual core fundamental "Hardware technology of the TPS and "Software technology of TQM" using Science SQC, and develops to be effectively incorporated to scientifically promote QCD research and achieve constant upgrading of the manufacturing quality.

REFERENCES

Amasaka, K. (1988). Concept and progress of Toyota Production System (Plenary lecture). The Japan Society of Precision Engineering, The Japan Society for Technology of Plasticity, The Japan Society for Design and Drafting, and Hachinohe Regional Advance Technology Promotion Center Foundation, Hachinohe, Aomori-ken, Japan.

Amasaka, K. (1989). TQC at Toyota: Actual state of quality control activities in Japan (Special lecture). The 19th Quality Control Study Team for Europe, Union of Japanese and Engineers, Statistical Quality Control, 39, 107-112.

Amasaka, K. (1998). Application of classification and related methods to the SQC Renaissance in Toyota Motor. Hayashi, C. et al. (Eds), Data Science, Classification and Related Methods. Springer.

Amasaka, K. (1999a). A study on "Science SQC" by utilizing "Management SQC-A demonstrative study on a New SQC concept and procedure in the manufacturing industry-. *International Journal of Production Economics*, *60-61*, 591–598. doi:10.1016/S0925-5273(98)00143-1

Amasaka, K. (1999b). *The TQM responsibilities for industrial management in Japan – The research of actual TQM activities for business management.* The Japanese Society for Production Management, The 10st Annual Technical Conference, Kyushu-Sangyo, University, Fukuoka, Japan.

Amasaka, K. (Ed.). (2000). Science SQC: Revolution of Business Process Quality, Study group of the Nagoya QST Research, Japanese Standards Association, Tokyo.

Amasaka, K. (2000a). Partnering chains as the platform for Quality Management in Toyota. *Proceedings of the 1st World Conference on Production and Operations Management, Seville, Spain.*

Amasaka, K. (2000b). A demonstrative study of a new SQC concept and procedure in the manufacturing industry-Establishment of a New Technical Method for conducting scientific SQC-. *Mathematical and Computer Modelling*, *31*(10-12), 1–10. doi:10.1016/S0895-7177(00)00067-4

Amasaka, K. (2000c). Basic Principles of JIT – Concept and progress of Toyota Production System (Plenary lecture), *The Operations Research Society of Japan, Strategic research* [Tokyo.]. *Group*, 1.

Amasaka, K. (2002a). New JIT, a new management technology principle at Toyota. *International Journal of Production Economics*, *80*(2), 135–144. doi:10.1016/S0925-5273(02)00313-4

Amasaka, K. (2002b). Reliability of oil seal for Transaxle–A Science SQC approach at Toyota, Case Studies in Reliability and Maintenance. John Wiley & Sons.

Amasaka, K. (2003a). Proposal and Implementation of the Science SQC Quality Control Principle. *Mathematical and Computer Modelling*, *38*(11-13), 1125–1136. doi:10.1016/S0895-7177(03)90113-0

Amasaka, K. (2003b). *A "Dual Total Task Management Team" involving both Toyota and NOK: A cooperative team approach for reliability improvement of transaxle*. Proceedings of the Group Technology / Cellular Manufacturing World Symposium, Columbus, Ohio.

Amasaka, K. (2004a). *Science SQC, new quality control principle: The quality strategy of Toyota*. Springer-Verlag. doi:10.1007/978-4-431-53969-8

Amasaka, K. (2004b). *The past, present, future of production management (Keynote lecture)*. The Japan Society for Production Management, The 20th Annual Technical Conference, Nagoya Technical College, Aichi, Japan.

Amasaka, K. (2004c). Development of "Science TQM", A new principle of quality management: Effectiveness of Strategic Stratified Task Team at Toyota. *International Journal of Production Research*, *42*(17), 3691–3706. doi:10.1080/00207540420002038867

Amasaka, K. (2005). Constructing a *Customer Science* Application System *"CS-CIANS"* - Development of a global strategic vehicle *"Lexus"* utilizing *New JIT* -. *WSEAS Transactions on Business and Economics*, *2*(3), 135–142.

Amasaka, K. (2006). (Editorial) Development of New JIT, new management technology principle. *Production Management Transaction of the Japan Society for Production Management*, *13*(1), 143–150.

Amasaka, K. (Ed.). (2007). New Japan Model: Science TQM – Theory and practice for strategic quality management, Study group of the ideal situation the quality management of the manufacturing industry, Maruzen, Tokyo.

Amasaka, K. (2007a). The validity of *"TDS-DTM"*, a strategic methodology of merchandise - Development of *New JIT*, Key to the excellence design *"LEXUS"*-. *The International Business & Economics Research Journal*, *6*(11), 105–115.

Amasaka, K. (2007b). The Validity of *Advanced TMS*, A strategic development marketing system utilizing *New JIT* –. *The International Business & Economics Research Journal*, *6*(8), 35–42.

Amasaka, K. (2007c). High linkage model "Advanced TDS, TPS & TMS"-Strategic development of "New JIT" at Toyota. *International Journal of Operations and Quantitative Management*, *13*(3), 101–121.

Amasaka, K. (2008a). Strategic QCD studies with affiliated and non-affiliated suppliers utilizing New JIT, Encyclopedia of Networked and Virtual Organizations, III(PU-Z), 1516-1527.

Amasaka, K. (2008b). Science TQM, A New Quality Management Principle: The Quality Management Strategy of Toyota. *The Journal of Management and Engineering Integration*, *1*(1), 7–22.

Amasaka, K. (2008c). An Integrated Intelligence Development Design CAE Model utilizing New JIT. *Journal of Advanced Manufacturing Systems*, *7*(2), 221–241. doi:10.1142/S0219686708001589

Amasaka, K. (2009a). New JIT, advanced management technology principle: The global management strategy of Toyota (Invitation Lecture*). International Conference on Intelligent Manufacturing and Logistics System, and Symposium on Group Technology and Cellular Manufacturing*, Kitakyushu, Japan.

Amasaka, K. (2009b). Establishment of Strategic Quality Management - Performance Measurement Model "SQM-PPM": Key to Successful Implementation of Science TQM. *The Academic Journal of China-USA Business Review*, *8*(12), 1–11.

Amasaka, K. (2009c). An Intellectual Development Production Hyper-cycle Model: *New JIT* fundamentals and applications in Toyota. *International Journal of Collaborative Enterprise*, *1*(1), 103–127. doi:10.1504/IJCENT.2009.026459

Amasaka, K. (Ed.). (2012). Science TQM, New Quality Management Principle: The quality management strategy of Toyota, Bentham Science Publishers, U.A.E.

Amasaka, K. (2015). New JIT, new management technology principle. Taylor and Francis Group, CRC Press, Boca Raton, London, New York.

Amasaka, K. (2017). *Toyota: Production system, safety analysis and future directions*. Nova Science Publishers.

Amasaka, K. (Ed.). (2019). *The fundamentals of the manufacturing industries management: New Manufacturing Theory-Operations Management Strategy 21C*. Shankei-Sha.

Amasaka, K. (2022a). New Manufacturing Theory: Surpassing JIT (2nd Edition), Lambert Academic Publishing.

Amasaka, K. (2022b). *Examining a New Automobile Global Manufacturing System*. IGI Global Publisher. doi:10.4018/978-1-7998-8746-1

Amasaka, K. (2023). New Lecture-Toyota Production System: From JIT to New JIT. Lambert Academic Publishing.

Amasaka, K., Kurosu, S., & Morita, M. (2008). *New Manufacturing Theory: Surpassing JIT-Evolution of Just in Time*. Morikita-Shuppan.

Amasaka, K., Nagaya, A., & Shibata, W. (1999). Studies on Design SQC with the application of Science SQC - Improving of business process method for automotive profile design. *Japanese Journal of Sensory Evaluations*, *3*(1), 21–29.

Amasaka, K., Watanabe, M., & Shimakawa, K. (2005). Modeling of strategic marketing system to reflect latent customer needs and its effectiveness, *The Magazine of Research & Development for Cosmetics. Toiletries & Allied Industries*, *33*(1), 72–77.

Gabor, A. (1990). *The man who discovered quality; How W. Edwards Deming, brought the quality revolution-The stories of Ford, Xerox, and GM*. Random House.

Hayashi, N., & Amasaka, K. (1990). *Concept and progress of Toyota Production System: The physical distribution improvement in manufacture (Special lecture).* The Annual Conference of Japan Physical Distribution Management Association, Tokyo.

Hayes, R. H., & Wheelwright, S. C. (1984). *Restoring our competitive edge: Competing through manufacturing.* Wiley.

Ohno, T. (1977). *Toyota Production System.* Diamond-sha. (in Japanese)

Taylor, D., & Brunt, D. (2001). *Manufacturing operations and supply chain management–The lean approach–,* Thomson Leaning, London.

Toyota Motor Corporation. (1997). *Creation is infinite: The history of the Toyota Motor Corporation 50 year. (in Japanese).* Toyota.

Womack, J. P., & Jones, D. (1994). *From lean production to the lean enterprise.* Harvard Business Review.

Womack, J. P., Jones, D. T., & Roos, D. (1991). *The machine that changed the world – The story of Lean Production.* Rawson.

Chapter 3
Fundamentals of Toyota Production System (TPS) Using Basic JIT

ABSTRACT

In this chapter, the author describes the fundamentals of TPS using basic JIT as a dual core principle "TQM and TPS" in Toyota. Specifically, as the basic principle of manufacturing via the TPS, the author describes the simultaneous realization of quality and productivity via a lean system and application examples of fundamentals of TPS. As actual examples using basic JIT, the author focuses on the process control and process improvement using TPS fundamentals: Manufacturing methods and costs; Daily control activities at the production site; Daily improvement activities at the production site; and Innovation of the production process.

FUNDAMENTALS OF TOYOTA PRODUCTION SYSTEM (TPS)

In this section, the fundamentals of TPS that reformed the automobile manufacturing of Toyota are discussed. The basic principles and on-site implementation of manufacturing using TPS, which has been adopted as a core concept of the world's manufacturing, will be illustrated here.

This fundamental idea is the basic concept of TPS, which aims to realize quality and productivity simultaneously by effectively applying TQM to the automobile manufacturing process as a dual core principle "TQM and TPS" shown in Figure 7 (See to Chapter 1).

The TPS production also pursues maximum rationalization (optimal streamlining) to improve overall product quality while maintaining an awareness of the principles of cost reduction (Amasaka, 1988, 2000, 2004).

As concrete deployment, TPS is conducted via one-by-one (single part) production and its aim is to achieve is the simultaneous realization of quality and productivity through the timely application of QCD activities for strengthening Value Analysis (VA) based on the Value Engineering (VE).

The concept of TPS and its approach have reformed automobile production at Toyota. As a result of its effectiveness being highly praised all over the world, TPS has been established as a core concept of the world's manufacturing (Amasaka, 2009a).

DOI: 10.4018/978-1-6684-8301-5.ch003

THE BASIC PRINCIPLE OF MANUFACTURING VIA THE TPS

The high productivity and high quality of manufacturing via the TPS in Japan today are universally acknowledged (Ohno, 1978; Toyota Motor Corp., 1996; Amasaka, 1988, 2000).

Simultaneous Realization of Quality and Productivity via a Lean System

The basic principle of manufacturing via the "TPS (1)" is lean system production. As shown in Figure 1, in this system, manufacturing, is conducted via *one-by-one* (single part) *production* and its aim is to achieve is the simultaneous realization of quality and productivity (Amasaka, 1988, 2000, 2009a, 2015, 2017; Amasaka et al., 2008).

Figure 1. The basic principle of manufacturing via the TPS (1)
-The simultaneous realization of quality and productivity-

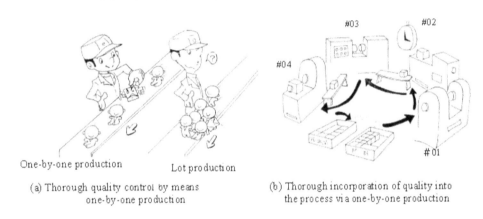

(a) Thorough quality control by means one-by-one production

(b) Thorough incorporation of quality into the process via one-by-one production

The first basic principle of manufacturing is thorough quality control by means of one-by-one production.

As Figure 1(a) illustrates, this system of one-by-one production on a manufacturing line using an assembly conveyor gives the assembly worker the ability to conduct a self-check on each piece. If a defective item comes to their assembly point from the previous process, they can then stop the conveyor and detect the defect without fail. Therefore, the assembly workers can provide 100% quality products to the downstream processes.

It is obvious that compared to lot production, this system can considerably improve the detection of from a probabilistic viewpoint. In this context, the one-by-one production is sometimes compared to a fine, quick, and pure flow of water upstream in a mountain stream, while lot production is likened to the broad, stagnant and slow water found downstream. The machining process of the Figure 1(b) is still the same, too.

The incorporation by the TPS of this one-by-one production deserves to be recognized as Lean System production.

(1) The second basic principle of manufacturing is the thorough incorporation of quality into the process via one-by-one production.

The operation of Figure 1(b) is mentioned in detail here. Figure 2 shows a machining Process. In Figure 2. this diagram illustrates an operation *One-by-one production* in a where a worker picks up a piece (work) from the parts box, conducts the machining process operation on it from processes #01-processes #04 in the order shown, and finally places the completed piece into the completed parts box.

Since the production operation is conducted according to a predetermined cycle time, the worker can consistently carry out the prepared standard operation in a rhythmical manner. Similar to the previous case of operation on an assembly conveyor, a self-check can be consistently performed and the incorporation of quality is ensured so that the stabilization of production can also be promoted.

Figure 2. The basic principle of manufacturing via the TPS
-The Simultaneous Realization of Quality and Productivities-

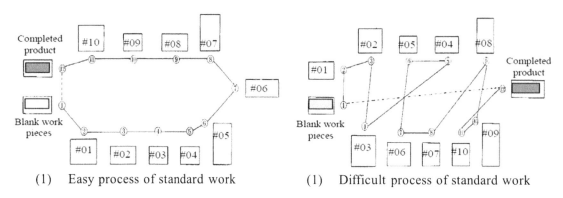

(1) Easy process of standard work   (1) Difficult process of standard work

Even in the case of a production operation that involves multiple types of work pieces, the intelligent application of production engineering and process designing can be used to support workers and allow them to conduct the standard operations in a rhythmical manner. A process like this is depicted in Figure 1.

A process like this is depicted in Figure 2, where the zigzag work flow interferes with the smooth performance of the standard operations each time a work piece change. This causes the worker to become fatigued and triggers human errors, thus resulting in the lowering of quality and productivity.

Fundamentals of TPS

In this section, the author shows the application examples of "Fundamentals of TPS" to actual strengthen manufacturing management of Toyota as follows (Amasaka, 1988, 1999a, 2000, 2015, 2017, 2022a,b, 2023; Hayashi and Amasaka, 1990; Amasaka et al., 1990, 2008; Amasaka, Ed., 2003);

(1) The basic philosophy of TPS
(2) Sale price and cost
(3) Manufacturing methods and costs via the TPS
(4) "*Heijyunka*" (Levelled production)

(5) Standardized work

(6) Standardized work chart

(7) Standardized work combination table

(8) Working sequence and operation standards

(9) The exclusion of "*Muda*" (Non-value added)

(10) True efficiency and apparent efficiency

(11) "*Takt Time*" and "*Cycle Time*"

(12) The flow of the product and the information

    (12) Examples of "*Kanban*": Improve "*Kaizen*" and manpower saving

    (13) Examples of "*Andon*"

    (14) Examples of production lead time

    (15) "*Kaizen*" (improvement) activities for VE (value engineering) using VA (value analysis) – "*Kaizen*" is eternity and infinity

    (16) "*Kaizen*" of "Isolated Jobsite": "production fluctuates up or down" and Flexible Manpower Line "*Hanare-Kojima*"

    (17) "*QC Circle Activity*" and "*Creative Suggestion System*"

    (18) Physical experiments: Demonstration of standard work and "*Kaizen*"

As the introductory remarks, the author has verified the examples of the "Basic philosophy", "*Kaizen*" of the "Shortening of reduction lead time" through the "Auto-robot painting process of rear axle unit", and "*Kaizen*" of "Isolated Jobsite" (*Hanare-Kojima*) through the "Production line of front and rear axle unit" (Amasaka and Kamio, 1985) (See Figures 7, 8, 9 in the Appendix). Furthermore, the author has verified the "VA (VE) effects of Shock absorber of front axle-unit" in the similar "*Kaizen*" approaches (Amasaka and Sugawara,1988) (See to Chapter 5).

PROCESS CONTROL AND PROCESS IMPROVEMENT USING TPS FUNDAMENTALS

At manufacturing sites, the process control and process improvement, mainly promoted by production workers under the supervision of their supervisors, and the QCD research, conducted by white-collar engineers, are equally valued in order to ensure the incorporation of design quality, which is demanded by customers (Amasaka, 1988, 1999a,b, 2000, 2002a,b, 2003, 2007, 2008, 2009b; 2015, 2017, 2022a,b, 2023; Amasaka and Sugawara, 1988; Amasaka et al., 1990; Amasaka and Sakai, 2010).

Manufacturing Methods and Costs

Consideration is given here to the method of manufacturing an automobile and the resulting manufacturing cost (initial cost) and profit based on Figure 3.

The main parts (heavy material parts) that compose an automobile are the skeletal structure, made up of the body and frame, and the parts that control the running functions, such as the engine and axles.

Figure 3. Manufacturing methods and costs

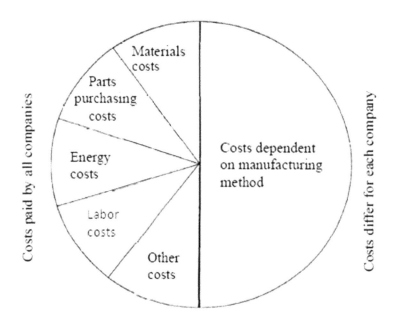

These parts are made of steel plates and cast forged steel. Automobile manufacturers procure these materials from steel manufacturers. Parts such as tires, engine ignition batteries, brake units, power steering units, and a wide variety of other parts such as windshields, lights, mirrors, seats, and so on are purchased from specialized parts manufacturers.

To produce a complete automobile, the manufacturer also purchases a variety of machining tools for processing parts, welding machines for the body and frame, and assembling robots for assembling the vehicle, and other equipment from specialized machine or equipment manufacturers.

Finally, the electric power needed to operate these machines and equipment is bought from power supply companies.

Generally speaking, the manufacturing costs listed above that are borne by the automobile manufacturer does not differ very much from those of other manufacturers. The labor cost (workers' wages) is also similar among manufacturers. Therefore, the manufacturing cost (initial cost) shown on the left half of the figure for automobiles of a similar design does not vary greatly from manufacturer to manufacturer.

On the other hand, the right half of the figure shows the manufacturing cost that results from the method of manufacturing the vehicle at the production site, and this can widely depend on how each manufacturer handles their manufacturing process control and process improvement.

The following are some poor cases of process control.

(i) Due to the inconvenient process design and production process, the worker cannot carry out their operation according to the standard operation sheet, resulting in a number of irregularities result in the operation. This in turn causes the assembly line to stop often, lowering the line operating rate.

(ii)   Due to a defective standard operation sheet, the worker's operation becomes strained, wasteful, and irregular, requiring the production line to stop: therefore, the planned production volume cannot be ensured.

(iii) Due to insufficient maintenance and upkeep of the machines and equipment, malfunctions occur, lowering the equipment operating (availability) rate.

(iv)   Although the production line is not operating during the break time, the production equipment's power is still on, resulting in wasteful power consumption.

(v)   Due to inadequate instructions from the supervisor or insufficient education and training provided to new workers regarding manual (skill required) operation or handling of machines, quality defects result from operation errors or improper machine manipulation. This causes pieces with machining defects, which entails their disposal or reworking of these pieces.

(vi)   Due to a lot of waste materials (scrap for disposal) during the stamping process or excessive cutting processing that causes the blade to be damaged, the processing line has to be stopped or defective work pieces result. This increases the cost of replacing the stamping process blades.

For these and other reasons, the manufacturing cost can exceed the expected cost, and the profit rate per vehicle that was initially calculated can also decline substantially. Cases such as these should not be neglected, and the process control at the production site needs to be thoroughly reviewed.

Process improvement and preventive measures need to be thoroughly reviewed. Process improvement and preventive measures need to be devised and implemented through ingenuity, and original suggestions should be obtained via the full participation of all the on-site staff.

If the efforts of the production site staff alone do not make progress, then the cooperation of the engineering staff from related departments should be obtained to thoroughly implement process improvements.

The ultimate objective is to reduce the manufacturing costs and improve the profit rate. The accumulation of a variety of process improvements at the production site can also conserve resources, resulting in lower material cost, parts purchasing cost, and energy consumption (items on the left side of Figure 2).

In turn, this will all lead to reduction in the number of quality defects and an improvement in the line operating rate. Furthermore, unnecessary labor cost can be eliminated by reducing excessive overtime work, and the manufacturing cost can be further improved.

In addition, continuous efforts to improve process control and process improvement will enhance the skills and motivation of the production workers and contribute greatly to the strengthening of the work environment and work culture at the production site.

Daily Control Activities at the Production Site

Daily control activities at the production site make a large difference in the level of finished quality and productivity. For this reason, the third basic principle of manufacturing is a thorough and continuous effort to employ daily control activities through cooperation between the supervisors and workers on the production site (Amasaka, 1988, 2000, 2007).

The basic elements needed for daily control activities at the production site are the maintenance and improvement of what are called the 5M-1E: Material, Machine, Man, Method, Measurement and Environment (Amasaka, 2007).

In the implementation stage of daily control activities, it is vital to organize work standards and thoroughly implement standard operations by using the check sheets, such as the standard operation sheet, important quality check sheet, worker process designation list, and new worker operation sheet.

These are designed to help ensure quality and productivity. Setting up reminder boards designed to remind workers about major quality defects in the past is also an effective preventive measure (Amasaka and Kamio, 1985).

In addition, an important quality control activity for ensuring manufacturing quality is the simultaneous confirmation of quality and incorporation of quality through utilization of control charts via a scientific process control method named "*Issei Hinshitsu Kakunin*":

"The structure that the operation of the production line is suspended in the fixed time or fixed quantity and all of the operators confirm product quality all together (Amasaka and Yamada, 1991; Amasaka et al., 2008).

Another factor that is vital to stably incorporate quality into the process is the control of conditions at the production facility. Condition controls take the following forms: (1) Condition controls for bolt tightening on the assembly line will indicate the operating pressure of air tools or the type of lubricating oil to be used. In the case of (2) Condition controls for the machining process line will be used to ensure the position and standard operation for the visual control of heat-treated process.

As an example, Figure 4 illustrates the "idea of condition control of heat-treated process in auto-drive axle-shaft (Amasaka et al., 2008) (See to Chapter 5).

Through daily control activities such as these, the reliability and maintainability of production facility can be ensured, as well as the incorporation of quality (process capability (Cp) and machine capability (Cm) and productivity by using an "Intelligence control chart") employing "Digital engineering" (Amasaka, Ed., 2003; Amasaka et al., 2008; Amasaka, 2015, 2017, 2022a,b, 2023).

Figure 4. The idea of condition control in heat-treated process

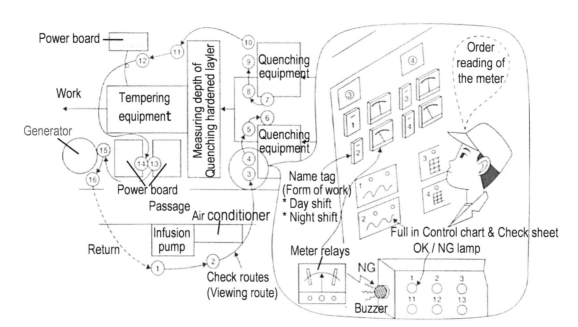

Daily Improvement Activities at the Production Site

The fourth basic principle of manufacturing is daily improvement activities at the production site. The basis for simultaneous realization of quality and productivity via the TPS is improvement of the work environment to prevent errors, improvement of control through visual confirmation, and so on.

These improvements are designed to help the workers to maintain their standard operations in a rhythmical flow. Concerted efforts are made to make small improvements that do not incur a large expense, such as safe work procedures, prioritizing quality, quickening the pace of work by one second, and so on (Amasaka and Kamio, 1985).

The main facilitators of these daily improvement activities are the workers themselves, who carry them out through small group activities, mainly in the form of QC Circles or through their own voluntary originality and ingenuity. Then, the author illustrates the "Reducing of Fatigue by Creative Suggestion System" by using "Kaizen of Hard work" employing QC Circle Activity" (Amasaka, 1999a) (See Figure 10 in the Appendix).

In the implementation stage, activities such as having an improvement team support the production site or promoting greater cooperation between the supervisors and production staff are put into practice by using Science SQC (Amasaka, 1999b, 2003).

One characteristic example of such "daily improvement activities" is a process improvement that achieves work operations that are less tiring. Similarly, the second example is a process improvement made to an inconvenient work operation.

As demonstrated by these daily improvement examples, carrying out process improvements on a daily basis gives the workers themselves the opportunity to develop more convenient work processes, make their work operations safer, and also ensure that quality is being incorporated. At the same time, redundant work processes can also be removed (Amasaka and Yamada, 1991).

Innovation of the Production Process: QCD Research Activities by White-Collar Staff

The fifth basic principle of manufacturing that is capable of quickly responding to the stringent demands of today's customers is the innovation of the production process and advancements in manufacturing that result from the QCD research activities conducted by white-collar staff (Amasaka and Kamio, 1985; Amasaka and Sugawara, 1988; Amasaka et al., 1990; Amasaka and Yamada, 1991; Amasaka, 2002a,b, 2009a, 2015, 2017, 2022a,b, 2023; Amasaka, Ed., 2003; Amasaka and Sakai, 2010).

(1)   First, the fundamental principle of TPS production is to manufacture only what can be sold, when it can be sold, in the quantity that can be sold. To accomplish this, it is essential to establish a flexible production system that will produce and transport only what is needed, when it is needed, in the quantity that is needed, as a rational production measure.

(2)   Second, as a production mechanism that will realize the conditions listed above, it is important to incorporate production leveling, shortening of the production lead time, and a pull system into the process planning and design.

(3)   Third, with a view to reasonably carrying out the above, it is vital to always use *"Kanban"* [*1], facilitate small branch conveyance to raise precision, promote the flow of the production process, determine the *Takt-Time* according to the needed volume, and then strictly adhere to these measures.

(*1) *Kanban* is a small signboard that is key control tool for Just-in-Time production. The aims of *Kanban* are (i) to check against overproduction "*Muda*", (ii) instruction for production and conveyance, and (iii) key technology for visual control of manufacturing, and the functional categories of *Kanban* are classified (iv) production instruction *Kanban* and (v) parts withdrawal *Kanban*.

(4) Fourth, it is imperative to reinforce the capabilities of the production site and to advance TPS production by actively developing new production technologies that will solve the bottleneck technological problems in production and substantially improve the quality and productivity of the production site.

Having said the above, the following is an explanation of a small part of a QCD research activity that was promoted by the white-collar engineering staff (or engineers and managers responsible for manufacturing technology, production engineering development, process planning, process designing, and production management) that contributed to innovation of the production process.

Figure 5 shows an example of a process improvement for a layout that facilitates the incorporation of quality by implementing a countermeasure for the "*outlying island*" called "*Hanare Kojima*" layout (Amasaka and Kamio, 1985) (See Figure 11 in the Appendix).

Before the improvement –"*Kaizen*", as seen in the left of Figure 6(a), the six workers at the automatic table conveyor are separated from one another by the five parts feeders that supply the parts to the assembling machines (the five automated machines, marked as A–E) located at each of the preceding processes.

This process arrangement is the *outlying island* ("*Hanare Kojima*") layout and when there is quality trouble or a small stop of the assembling machines, it is difficult for an expert worker (skilled worker) to provide instruction (assistance) to a newly assigned worker. In addition, each worker's standard operation time differs due to the variation in experience.

Therefore, an expert worker is forced to wait for work from a new worker (creating wasted time), since their operation time is shorter than that of newly assigned workers.

Moreover, in the case of a production line with this outlying island layout, the takt time cannot be set according to the fluctuations (increase and decrease) in production volume because the takt time is fixed according to the operation time of the workers.

Therefore, the takt time cannot be changed in a flexible manner according to the number of workers.

Given this background, the engineering staff (white-collar engineers) reviewed the production line where the workers' manual operations and automated machining processes are combined, so that they could reform the production process. As seen in Figure 6(b), after the improvements - "*Kaizen*" were made, the automated machining processes were consolidated.

In an effort to create a continuous processing operation from A to B to C, as well as from D to E, the set positions of the work pieces were changed and the processing jigs were altered. The parts feeders that supply the same parts were also integrated so that the number of parts feeders could be reduced from five to three.

These improvement measures rearranged the automatic table conveyor into a single line and the old work layout that had separated the workers was discarded. The above case is a characteristic example of a process improvement that realized flexible production that could react to fluctuations in production volume; it resulted in improvements in the line operating rate and stabilized the production quality. This all effectively contributes to meeting the QCD requirements. These improvement measures promoted as follows;

Figure 5. Example of a production layout that facilities built-in quality by eliminating isolated worksites

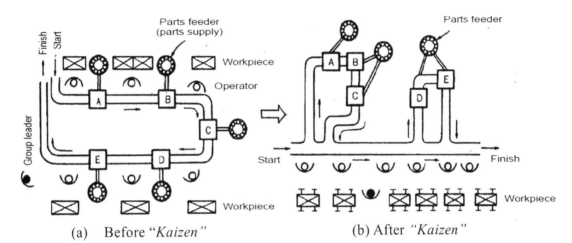

(a) Before *"Kaizen"*   (b) After *"Kaizen"*

(i)   Mutual confirmation between the workers on work progress,
(ii) Mutual assistance in case of any delay in work operation,
(iii) Observance of simultaneous confirmation of quality,
(iv) Improvement in the efficiency of material transport,
(v)   The ability to respond to the increase or decrease in the number of workers according to the changing production volume,
(vi) The ability of supervisors to respond to short stops of the automated processing line outside of their own production line, to supply parts to the parts feeders, to provide work instructions and assistance to line workers, and to create greater efficiency in the replacement work between completed work pieces and materials, and so on.

Below are a few examples of process improvements that have helped solve human errors. In recent years, it has become increasingly important to improve product quality in response to an increase in customers' sensitivity to quality;

(1)   The 1st example describes the efforts made to achieve high quality assurance for shock absorbers, a main part of the auto-drive axle-unit utilizing "Measurement of *Sensory Evaluation*" based on the "*Affective (Kansei) Engineering*" realizing Customer Science (Amasaka and Nagasawa, 2000; Amasaka, 2005).

Usually, the inspection to ensure the finished quality of assembled parts was dependent on the personal experience of the inspectors and their visual senses. Judgment of the "*Lissajous waveforms*" indicating the damping force characteristics (damping force and up-and-down strokes), which are displayed in rotation on the X-Y coordinates of an oscilloscope, is a difficult inspection process.

It often induced human error and misjudgments, even when performed by expert workers. In addition, it was time-consuming to find where on the work piece (product) the problem was or to conduct a disassembly investigation, fault diagnosis, and correction. This was a problem that could not be solved easily.

Then, the engineering staff and production site staff cooperated and developed an automatic diagnosis inspection device as shows in Figure 6. This device simultaneously conducts a quantitative evaluation of the quality into what had been a bottleneck problem [*2] in this inspection process (Refer to Amasaka and Saito (1983) and Amasaka and Kamio (1985) in detail).

(*2) Such cases as the mis-stockout, mis-attachment and part performance defect due to the existence of the wound, deformation and foreign substance" and others.
(2)   The 2nd example is of an improvement made to help solve unforced errors such as wrong or missing parts and misassembled pieces

Generally, automobile production involves manufacturing various models, and therefore, it is necessary for the vehicle or parts assembly line workers to conduct *parts assembly* while being alert to wrong or missing parts and mistaken assembly. As a process improvement measure to reduce these sorts of errors, IT was used to assist in accurately sorting out the work pieces on the assembly conveyor.

Figure 6. Example of a process improvement for the human error solution

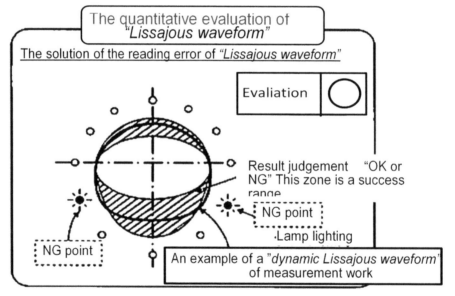

All the number inspection devices of   *Lissajous waveform*
(The mechanization of the visual check)

In this system, production instructions, electronic signboards, and radio-frequency identification tags (IC tags) are used to detect the types of work pieces. Then, it lights up the assembly parts shelf that corresponds to the intended work pieces and uses sensors and visualization equipment to match the work operation with the proper work pieces (Amasaka and Kamio, 1985).

(3)   The third example is the identification of work-piece types using this IT

Until now, the parts unit signboards were visually confirmed, and the unit numbers were written down manually, but this repeatedly resulted in human errors such as misidentification or mistaken descriptions.

After the process improvement was carried out, a bar-code signboard-reading device was developed and the unit code was automatically identified. An automatic marking device that prints out the intended code on the work-piece surface was also developed, and these devices solved the human error-based problems in this process.

These examples of process improvement are the results of QCD research activities carried out through cooperation between the white-collar and production site workers or supervisors in an effort to solve the inconvenient work operations from the standpoint of "*human behavioral science*" (Amasaka, 2008) (See to Chapter 5).

CONCLUSION

The Japan production technology principle that contributed most to the world in the latter half of the 20th century was the Japanese-style production system typified by the TPS as the Basic of JIT. TPS aimed to improve product quality while pursuing maximum efficiency through the application of TQM) into the manufacturing process, as well as applying the principle of cost reduction. As for this subsection, the author has introduced the "Fundamentals of TPS" about the whole aspect of TPS (philosophy, theory and practice) which is the core principle of the manufacturing management in Toyota. The TPS also pursues maximum rationalization (optimal streamlining) to improve overall product quality while maintaining an awareness of the principles of cost reduction through the timely application of QCD activities for strengthening value analysis (VA). As a result, the effectiveness of TPS has been recognized on a worldwide scale, and it is now regarded as the core concept of manufacturing.

REFERENCES

Amasaka, K. (1988). *Concept and progress of Toyota Production System (Plenary lecture), Co-sponsorship: The Japan Society of Precision Engineering, Hachinohe Regional Advance Technology Promotion Center Foundation and others*. Hachinohe, Aomori-ken. (in Japanese)

Amasaka, K. (1999a). *New QC Circle activities in Toyota (Special Lecture), A Training of Trainer's Course on Evidence-Based on Participatory Quality Improvement, TOT Course on EPQI, Tohoku University School of Medicine (WHO Collaboration Center)*. Sendai-city.

Amasaka, K. (1999b). A study on Science SQC by utilizing Management SQC, A demonstrative study on a new SQC concept and procedure in the manufacturing Industry. *International Journal of Production Economics, 60-61*, 591–598. doi:10.1016/S0925-5273(98)00143-1

Amasaka, K. (2000). Basic Principles of JIT–Concept and progress of Toyota Production System (Plenary lecture), *The Operations Research Society of Japan, Strategic research group-1, Tokyo*.

Amasaka, K. (2002a). New JIT, a new management technology principle at Toyota. *International Journal of Production Economics*, *80*(2), 135–144. doi:10.1016/S0925-5273(02)00313-4

Amasaka, K. (2002b). Intelligence production and partnering for embodying a high-quality assurance (Plenary Lecture), The Japan Society of Mechanical Engineers, The 94th course of JSME Tokai Branch, Nagoya, Japan, 35-42.

Amasaka, K. (2003). Proposal and implementation of the "Science SQC" Quality Control Principle. *Mathematical and Computer Modelling*, *38*(11-13), 1125–1136. doi:10.1016/S0895-7177(03)90113-0

Amasaka, K. (Ed.). (2003). *Manufacturing fundamentals: The application of Intelligence Control Charts—Digital Engineering for superior quality assurance.* Japanese Standards Association.

Amasaka, K. (2004). *The past, present, future of production management.* The Japan Society for Production Management, The 20th Annual Technical Conference, Nagoya Technical College, Aichi, Japan.

Amasaka, K. (2005). Constructing a *Customer Science* Application System *"CS-CIANS"* –Development of a Global Strategic Vehicle *"Lexus"* Utilizing *New JIT–*. *WSEAS Transactions on Business and Economics*, *2*(3), 135–142.

Amasaka, K. (2007). *Developing the high-cycled quality management system utilizing New JIT.* Proceedings of the Seventh International Conference on Reliability and Safety, Beijing, China.

Amasaka, K. (2009a). The Foundation for Advancing the Toyota Production System utilizing New JIT. *Journal of Advanced Manufacturing Systems*, *80*(1), 5–26. doi:10.1142/S0219686709001614

Amasaka, K. (2009b). An Intellectual Development Production Hyper-cycle Model-*New JIT* fundamentals and applications in Toyota-. *International Journal of Collaborative Enterprise*, *1*(1), 103–127. doi:10.1504/IJCENT.2009.026459

Amasaka, K. (2015). New JIT, New Management Technology Principle. Taylor & Francis Group.

Amasaka, K. (2017). *Toyota: Production System, Safety Analysis and Future Directions.* NOVA Science Publishers.

Amasaka, K. (2022a). *Examining a New Automobile Global Manufacturing System.* IGI Global Publishers. doi:10.4018/978-1-7998-8746-1

Amasaka, K. (2022b). New Manufacturing Theory: Surpassing JIT (2nd Edition). Lambert Academic Publishing.

Amasaka, K. (2023). *New Lecture-Surpassing JIT: Toyota Production System-From JIT to New JIT.* Lambert Academic Publishing.

Amasaka, K., & Kamio, M. (1985). Process improvement and process control: The example of axle-unit parts of the automobile, *Quality Control* [in Japanese]. *Union of Japanese and Engineers*, *36*(6), 38–47.

Amasaka, K., Kurosu, S., & Morita, M. (2008). *New Manufacturing Principle: Surpassing JIT-Evolution of Just in Time.* Morikita-Shuppan.

Amasaka, K., & Nagasawa, S. (2000). *The foundation of the Sensory Evaluation and application: For Kansei Engineering in automobile*. Japanese Standards Association.

Amasaka, K., Ohmi, H., & Murai, F. (1990). The improvement of the corrosion resistance for axle unit for vehicle: Joint QCD research activity, Coatings Technology, Japan Coating Technology Association.

Amasaka, K., & Saito, H. (1983). *Mechanization of sensory evaluation: Example of the judgment of the Lissajous waveform indicating the damping force characteristics* Proceedings of the Annual Conference of Central Japan Quality Control Association 1983, Nagoya, Japan.

Amasaka, K., & Sakai, H. (2010). Evolution of TPS fundamentals utilizing *New JIT* strategy: Proposal and validity of Advanced TPS at Toyota. *Journal of Advanced Manufacturing Systems*, *9*(2), 85–99. doi:10.1142/S0219686710001831

Amasaka, K., & Sugawara, K. (1988). QCD research activities by manufacture engineers: The improvement of the corrosion resistance of the shock absorber parts, *Quality Control* [Special issue] [in Japanese]. *Union of Japanese and Engineers*, *39*(11), 337–344.

Amasaka, K., & Yamada, K. (1991). Re-evaluation of the QC concept and methodology in auto-industry, *Quality Control* [in Japanese]. *Union of Japanese and Engineers*, *42*(4), 13–22.

Hayashi, N., & Amasaka, K. (1990). Concept and progress of Toyota Production System: The physical distribution improvement in manufacture (Special lecture). *The Annual Conference of Japan Physical Distribution Management Association, 3*.

Ohno, T. (1978). *Toyota Production System: Beyond Large-scale production*. I Japanese.

Toyota Motor Corporation. (1996). *The Toyota Production System*. International Public Affairs Division Operations Management Consulting Division.

APPENDIX

The Basic Philosophy of TPS

Figure 7. The timely QCD activities for strengthen of VA (VE) effects
(Refer to Amasaka et al. (1990) and Amasaka (2015, 2017, 2022a,b) in detail)

The **TPS** aim is to achieve is the *simultaneous realization of quality and productivity* through the **timely application of QCD activities** for strengthening of **VE (VA) effects** as follows (Amasaka, 1988, 2004, 2008a).

TPS philosophy "Motto" pursues "*maximum rationalization (optimal streamlining) to improve overall product quality*" while maintaining an awareness of the principles of cost reduction (Toyota Motor Corp., 1987, 1996).

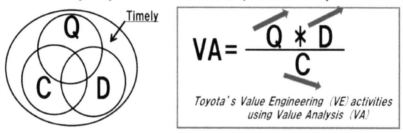

$$VA= \frac{Q*D}{C}$$

Toyota's Value Engineering (VE) activities using Value Analysis (VA)

Kaizen of the Shortening of Production Lead Time

Figure 8. Outline of the robot painting process of the rear axle unit
(Refer to Amasaka et al.(1990) and Amasaka (1988, 2000) in detail)

Ex. Outline of the painting process of the rear axle unit
This refers to the **time to produce** one product **from beginning work on raw materials to completing product** as follows[Amasaka et al., 1990].

Production Lead Time = **processing** + **non processing time (Muda)**
Ex. Painting Process Lead Time = **7 minutes** + **24 minutes** = **31 minutes**
➡ Kaizen: **Super quick drying paint reduces** non processing time !!

"Removal & installation" of unit-axle from the overhead conveyer

44

Kaizen of Flexible Manpower Line Realizing Shojinka

Figure 9. Kaizen of isolated jobsites in auto-axle unit production line
(Refer to Amasaka (1988, 2000) in detail)

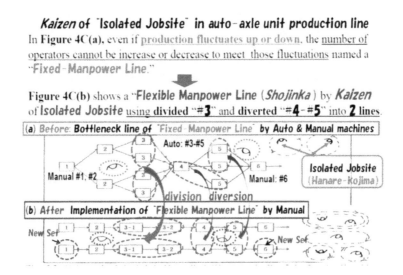

"Kaizen of Hard Work" Employing QC Circle Activity

Figure 10. Kaizen of "Hard Work–The Reduction of Fatigue to Enable Handling for a Long Time"
(Refer to Amasaka (1988) and Amasaka et al. (2008) in detail)

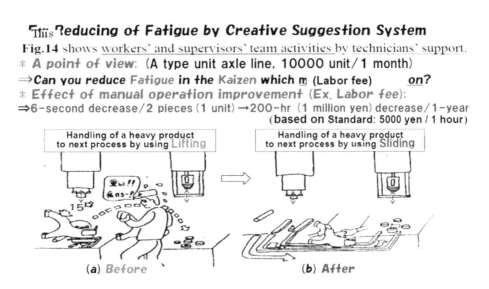

Flexible Manpower Line Using Eliminating Isolated Jobsites

Figure 11. Kaizen of eliminating isolated jobsites of automatic table conveyor
(Refer to Amasaka (1988) and Amasaka et al. (2008) in detail)

"**Flexible Manpower line**" means preparing a production line so that it can meet changing production requirements with any number of operators, without lowering productivity (Amasaka, 1988).

(a)**Before**: This refer to jobsite that are in "*Isolated Jobsites*", where the operator cannot easily work with other jobsites for flexibility during increases or decreases in production.

(b)**After**: 1) The automated machining processes were consolidated.
2) The **automatic table conveyor** re-arranged into a single line and the before work layout that had separated the workers was discarded.

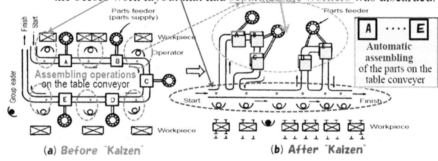

Chapter 4
Applications of TPS for Realizing QCD Studies Developing JIT Strategy

ABSTRACT

In this chapter, the author describes the foundation and effectiveness of applications of TPS for realizing QCD studies by developing JIT strategy based on the chapter, "Fundamentals of Toyota Production System (TPS) Using Basic JIT." Specifically, the author illustrates the innovation of the production in automobile rear axle unit assembly line by having cooperation between on-site white-collar engineers and supervisors and workers with affiliated and non-affiliated suppliers by the author's patens. Actually, to do the solution of bottleneck problems of manufacturing technology, the author has verified the validity of "Kaizen" through the whole of production process in welding, machining and assembling.

APPLICATIONS OF TPS FOR REALIZING QCD STUDIES

To realize manufacturing that places top priority on customers with good quality, cost, and delivery (QCD) in a rapidly changing the technical environment, it is essential to create a new manufacturing technology in order to reform super-short-term development production (Amasaka, 2008).

Furthermore, a new quality management technology linked with overall activities for higher work process quality in all divisions and SCM strategy is necessary for an enterprise to survive (Amasaka, 1988, 2000a,b, 2004a,b, 2008; Burke et al., 2000; Roger and Flynn, 2001; David and David, 2001).

Therefore, the author illustrates the foundation and effectiveness of "Applications of TPS for realizing QCD studies" based on the "Fundamentals of TPS) using basic JIT"".

This is based on effort to attain "for realizing QCD studies by developing JIT strategy" through innovative of manufacturing technology by having cooperation between "on-site white-collar engineers and supervisors & workers with affiliated and non-affiliated suppliers" as the "Collaboration" using Toyota's Partnering named "Strategic Stratified Task Team Model" (SSTTM) (Amasaka, 1986, 2000b, 2004a, 2008; Yamaji and Amasaka, 2009).

In these examples, the author has verified the effectiveness on applications of TPS through innovation of the production line of "Automobile rear axle unit: suspension and chassis parts" (Amasaka and Kamio, 1985; Amasaka, 2009, 2015, 2017, 2022a,b, 2023).

DOI: 10.4018/978-1-6684-8301-5.ch004

ACTUAL HIGH-QUALITY MANUFACTURING THROUGH INNOVATION OF JIT PRODUCTION TECHNOLOGY

To prevail in the worldwide quality competition and be successful in global production – worldwide uniform quality and simultaneous production launch (production at optimal locations), manufacturing with high quality assurance has to be realized at the manufacturing site (Amasaka, 2000a, 2009, 2015, 2017, 2022a,b, 2023; Amasaka, et al., 2008).

Cooperation Between On-Site Manufacturing Staff and White-Collar Engineers

The foundation for advancing TPS is based on efforts to attain simultaneous acheivement of QCD through inovation of manufacturing technology (Amasaka, 2008).

This in turn is accomplished by having on-site manufacturing (engineering) staff carry out innovation at the manufacturing site, as well as having white-collar engineers lead the innovation and improvement of the manufacturing technolgy (Amasaka and Kamio, 1985; Amasaka and Yamada, 1991; Amasaka, 2002, 2003, 2004a,b, 2007, 2008; Amasaka, et al., 2008).

All this is done wth the aim of conducting manufacturing with high quality assurance. Specifically, some of the features of the TPS are the improvement of laborious work operations and the creation of a safe, worker-friendly manufacturing site in which the supervisors and workers take the initiative, while the manufacturing technology staff members (white-collar engineers) also cooperate as the "Partnering" (Umezawa and Amasaka, 1999; Amasaka. 2000b).

Firstly, through such a process, the improvement of daily management and maintenance can be re-inforced, which realizes rhythmical and smooth work operations, thus enabling reliable manufacturing (ensuring the process capability and machine capability).

Secondly, TPS takes on the challenge of "simultaneous achievement of QCD" through innovation of manufacturing technology. In this case the initiative is taken by the white-collar engineers, while the supervisors and workers on the manufacturing site cooperate.

This effort has contributed to drastically improve the manufacturing technology (for the working environment, production equipment, production management, production technology, quality management technology, etc.), which has been a bottleneck obstacle of the manufacturing site.

In the implementation stage, the innovations mentioned above are to be carried out by a "Strategic Total Task Management Team" (STTMT) led by the on-site manufacturing technical staff and production engineering development staff and production engineering development staff, while "production design-ing staff, research development staff and even suppliers (parts manufacturers, production equipment, and machinery manufacturers, etc.) also participate" (Amasaka, 2000b, 2004a, 2008) (See to Chapter 6).

By means of these activities, wide-ranging business processes from the suppliers to the automobile manufacturers can be linked in a highly efficient manner, and therefore, manufacturing can be conducted while enabling JIT production (See to Chapter 4).

Below is an example of the "Realizing QCD studies by developing JIT strategy" through innovation of manufacturing technology.

In this example, innovation of the production line producing suspension and chassis-related parts for automobiles in Toyota's plant will be presented (Amasaka and Kamio, 1985).

It will demonstrate an example of solving a bottleneck manufacturing technology problem that was inhibiting JIT production (Amasaka, 1983, 1985, 1986; 1999; Amasaka, et al., 1982, 1988a,b; Amasaka and Yamada, 1991).

Innovation of the Production Line of Automobile Suspension and Chassis Parts

As a characteristic example, the author presents the "innovation of the auto-production line of the rear axle unit-related parts in Toyota's machining div. of Motomachi Plant" (in Toyota-shi, Aichi, Japan) (Amasaka, 1983, 1984a,b, 1985, 1986, 1988a,b, 1999, 2009; Amasaka and Saito, 1983; Amasaka and Kamio, 1985; Amasaka and Sugawara, 1988; Amasaka and Yamada, 1991; Amasaka et al., 1982, 1988a,b, 1990, 1993; Amasaka and Ishida, T. 1991).

The author demonstrates the typical example of solving a bottleneck manufacturing technology problem by using typical JIT production technology.

Outline of the Production Process of the Rear Axle Unit

A rear axle unit for an automobile underbody is manufactured in order of a production process that includes a welding process, machining process and assembly process based on TPS.

This process performed in order by means of JIT production called *"Tatenagashi production"* at Toyota, and is generally referred to as a lean production (production of one piece at a time) system.

After assembly is complete, the "Rear axle unit" is further combined together with various other products via a method called the *"Jyunbiki or Jyunjyo hikitori"* (sequential parts withdrawal system) and then it is taken to a vehicle assembly plant via the pull system production method.

The author indicates an example of Toyota's typical "Rear axle unit process" proceeding to assembly (before and after *kaizen*) for "Realizing QCD studies by developing JIT strategy" by white-collar engineers" by developing "Innovation production technology" as shown in Figure 1 (Amasaka and Kamio, 1985; Amasaka, 2009) (See Figure 5 in the Appendix).

Then, Figure 5 in the Appendix shows "Outlie 1: Continuous flow processing" using *"Jyunbiki"* based on the "Levelled Production" (*"Heijyunka production"*) (Refer to Amasaka (1985, 1988b)).

Figure 1 illustrates the main characteristic of the TPS is to carry out production without keeping any excess parts in stock.

Therefore, in this process, it is necessary to conduct assembly in an orderly manner according to the assembly sequence from the start of the (5) "Assembly conveyor" (Conveyor driven assembly process) to the latter process in the vehicle assembly plant.

This is done by utilizing an intra-process instruction "Shikake Kanban" (Production instruction Kanban) that clearly indicates the order of the processes in coordination with the downstream pull system processes.

In Figure 1, the main component parts of this product, namely the "Rear axle housing" and "rear axle tube", come in a wide variety of different types that are delivered from parts manufacturing suppliers, the machining and assembling process of a rear axle unit manufactured by a typical TPS (The previous process of welding is omitted here).

The main characteristic of the TPS is to carry out production without keeping any excess parts in stock.

Both parts are put together through friction welding and are subjected to machining before being supplied to the beginning of the assembly process.

Figure 1. Rear axle unit process proceeding to assembly (before and after kaizen)

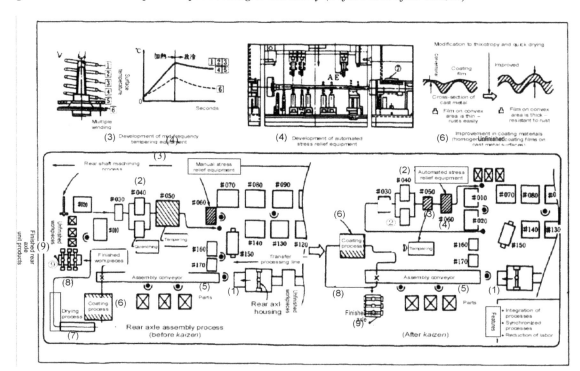

Similarly, the various types of "Rear axle shaft" delivered from a designated supplier are subjected to machining and then supplied to the beginning of the (5) "Assembly conveyor".

For both products, what is required is synchronized production called "*Doukika production*" (Synchronized production), where finished processed parts of the needed types are supplied in a timely, JIT fashion to the head of the (5) assembly conveyor, while minimizing the amount of intra-process parts and the number of finished products kept in inventory to the greatest extent possible.

As for the "Rear axle housing", after a variety of "Brackets" are welded onto it mainly by CO_2 arc welding machines (involving a mixture of manual, exclusively automated, and robot welding processes), the finishing process is applied on a one-by-one (or unit-by-unit) basis via a (1) "Transfer machine" in the downstream machining process.

The "Rear axle shaft" is also subjected to machining one at a time (2 shafts per unit) according to the production sequence table.

In the intermediate process for this part, (2) "High frequency hardening", (3) "Tempering", (4) "Stress relief", and other operations are conducted.

As can be seen above, it is not easy to carry out synchronized production of either of these machined parts as they require a long lead time and there can be a large quantity of in-process inventory.

Moreover, what make this synchronized production even more difficult are the considerable differences in the operation availability of each of the processes of machining and assembly.

Generally, the operation availability of the assembly process is relatively high, but the previous machining process has low operation availability due to bottleneck problems and other reasons.

Similarly, the operation availability of the welding process is even lower as it requires the welding "Nozzles" to be cleaned of weld spatters, the replacement of worn out welding "Tips", and so on.

As discussed, the production line for the "Rear axle units" is arranged in a one-unit-at-a-time production manner from welding, to machining, to assembly.

Therefore, *"Doukika production"* is not easy, particularly in the case where in-process inventory is not held in each process or between processes (Refer to Amasaka and Kamio (1985) and Amasaka (2009) in detail).

Before Process Improvement–Bottleneck-Like Manufacturing Technical Problems Which Influence JIT Production

In addition to the above, in each of the processes of machining, and assembly, bottleneck-like manufacturing technical problems which influences JIT production, and these have not been solved for a long time.

1. Bottleneck problems in the welding process for rear axle housing

The bottleneck problem of this process involves welding quality defects caused by forgotten welding operations, blow holes in welded areas, core deviation of the welding beads, and so on.

In addition, these defects subsequently lower the operation availability due to discontinuation of equipment operation as shown in Figure 2 (Amasaka, 1988b, 2015, 2017, 2022a,b, 2023; Amasaka, et al., 1988a).

In the CO_2 arc welding process, periodic cleaning is conducted to remove the weld spatter that sticks to the inner surface of the "Welding nozzle" used for gas-shielding the welded areas as shown in Figure 2(a).

Also, regular replacement is required to avoid causing welding core deviation, which is caused by wear on the inner surface of the welding tip that feeds to the welding wire.

As a result, as indicated in Figure 2(b), the operation availability of the arc robot or automated welding equipment can be lower than that of the machining process or assembly process in the downstream processes.

Consequently, inventory of half-processed or finished products is kept within the welding process (as indicated in Figure 1).

This bottleneck situation prevents production that is synchronized with the downstream assembly process in (5) "Assembly Conveyor", and has not been solved, so it has remained as a pending technical issue.

2. Bottleneck technology in the machining process for rear axle shaft

The bottleneck problems in this process are ensuring manufacturing quality and energy conservation, as well as the technical hindrances that prevent synchronized production for realizing *"Jyunbiki"* and "Flexible Manpower Line" (*"Shojinka"*) (Amasaka, et al., 1982; Amasaka, 1983) (Refer to Figure 6 in the Appendix).

Figure 2. CO$_2$ gas shieled arc welding

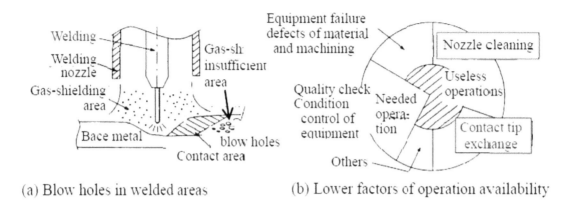

(a) Blow holes in welded areas   (b) Lower factors of operation availability

In "Reforming Rear Axle Shaft Machining Line" of the Appendix, the author shows "Outlie 2: *Kaizen* of operators management engineering" realizing "Synchronous production" to connect the "Machining of Rear axle housing, Rear axle shaft and Rear axle assembly unit line" (Refer to Amasaka and Kamio (1985) and Amasaka (2009) in detail).

As Figure 1 suggests, the upstream process of machining (#010-#030) and the downstream process conducted manually by production operators (#070-#170) are separated by the large-size, (2) "Automated high frequency quenching equipment" (#040) and (3) "Heat circulating-type tempering equipment" (#050), which consume a large amount of power and have a large amount of in-process inventory. For this reason, these processes are left as the "Isolated Jobsites" ("*Hanare Kojima*") as it were.

As a result, the one-operator work operations in the upstream process (#010-#030) require assistance from the supervisor in order to accommodate any extra cycle time in the case where the number of production units per day is increased. On the other hand, the operators are in standby mode when production volume per day is small.

In addition, the operation of (4) "Manual stress relief equipment" (Manual straigtening machine) (marked as #060) is dependent on the emprical skills and abilities of expert production operators.

The long work pieces (Rear axle shafts) supplied from the upstream process (Quenching and Tempering equipment, #040-#050) have a "stress profile of complicated spirals" after "Heat treatment".

Also, in cases where the required "Heat stress adjustment" is large, the number of adjustment operations (time needed for adjustment) becomes larger, resulting in extra cycle time.

When a production operator is overburdened and conducts excessive adjustment operations in order to avoid the above situation, cracks can be induced inside or on the surface of these long work pieces.

Moreover, if the operator has not acquired sufficient skills, this can result in human error and a supply of defective long work pieces that do not satisfy the standard stress adjustment specifications to the downstream processes (#070-#170). This will then cause a number of machining process defects.

As discussed above, many problems such as defective quality, "*Hanare Kojima*" work area arrangement, wasteful power consumption, extra cycle time, and in-process inventory have not been solved, presenting bottleneck manufacturing technical problems.

Due to this situation, half-processed or finished products are stuck within the processes as inventory and synchronized production in coordination with the downstream assembly process (5) cannot be realized, particularly in the case of the welding process.

This bottleneck situation has not yet been solved and this prevents the "realization of simultaneous achievement of QCD" by developing JIT strategy.

3. Bottleneck problems in assembly process for rear axle unit

The bottleneck problems of this process range from ensuring product quality, to energy conservation, to production response to the downstream pull system (Amasaka et al., 1982; Amasaka, 1983) (Refer to Figure 7 in the Appendix).

Specifically, the market requires that the quality of rust proofing of manufactured products be assured.

The steel surface of a rear axle housing is flat and smooth. Therefore, the rustproof coating thickness can be ensured with (6) "Automated coating robots".

However, the rustproof quality of the uneven areas on the "cast metal surface" cannot be sufficiently ensured as the coating on "convex surfaces" tends to be thinner due to the characteristics of the coating being used, as shown in (6) of Figure 1 (in the upper right corner) (Amasaka, 1986; Amasaka et al., 1990).

To avoid such uneven coating, extra coating operations are conducted to ensure sufficient thickness, and these results in a higher coating cost.

Furthermore, after the coating operation indicated by (6) in the figure, the work pieces are passed through the (7) "Electric heater-type drying machine", which requires considerable power consumption, and are then cooled down to room temperature in the cooling process area. After this, the finished products are unloaded to the (9) "Pull system cart".

For this reason, excessive inventory remains on the (8) "Overhead conveyor", and therefore, as in the cases of the upstream welding and machining processes, this process presents a hindrance to the realization of simultaneous achievement of QCD.

Concretely, in bottleneck problems in this assembly process, "Reforming rear axle unit assembly line" of the Appendix, shows the quality problem "Reliability of anti-rust treatment", (7) "Drying large power consumption and Long production lead time" between (9) "Installing of unit axle assembly and (5) Unit axle assembly conveyor" realizing for "Simultaneous achievement of QCD".

After process improvement – Simultaneous achievement of QCD by innovation of manufacturing technology

To solve the above mentioned the various bottleneck problems of this production line, it is necessary to innovate the manufacturing technology.

Given this background, the author has decided to take a systematic approach to the improvement of the fundamental production process by taking into consideration the timing of production of a forthcoming "New vehicle model" by using "Partnering" for quality management, and adaption of the "author's patents" [*1] (Amasaka and Kamio, 1985; Amasaka and Yamada, 1991; Amasaka and Sugawara, 1988; Amasaka, 2000b) (See to Chapter 5).

(*1) Examples of Japanese Patent (JP-): *1) - *6)

- 1) Arc welding "nozzle and contact tip: 1989-40910;1991-37471, -61545, -61546, -61547 -71943; 1992-18950, -42067;1993-299, -300, -4944;1993-79780;1995-25082, -34994, -87993;1996-22525, -34999
- 2) Arc preventive agent for deposit of welding spatter: 1992-53633, 60000; 1996-2515109
- 3) Coloring method of a welding bead: 1994-79780; 1996-2558493; 1998-2609889
- 4) Torque control impact wrench: 1986-22773; 1989-30267;1991-25304
- 5) Electromagnetic induction heating equipment: 1982-20974; 1983-10599,
- 6) Tempering equipment: 1989-14671

Actually, before this plan was implemented, the author has formed a "Strategic Total Task Management Team" (SSTTM) led by the technical staff on the manufacturing site, as well as the production engineering development staff based on white-color engineers (Amasaka, 2008; Yamaji and Amasaka, 2009).

The production development staff, parts manufacturers (affiliated and non-affiliated suppliers), and equipment machine manufacturers also participated, and the author attempted to realize the simultaneous achievement of QCD with revolutionary approaches such as those introduced below.

1. Solution of bottleneck problems related to welding process for rear axle housing
   (1-1) Resolving Issues of human error – Development of a weld bead coloration method

In this welding process, a lot of welding operations are conducted on large piece parts or brackets, which are important safety-related parts.

For the purpose of solving problems related to "Human error", such as forgotten welding areas or missed welding defects during the visual inspection, development efforts were made in cooperation with several suppliers who deliver weld wire and a colored bead welding method was established.

The conventional combination of black colored workpieces (without deoxidization treatment) and black weld bead was improved and changed by devising a method to add color to the weld bead (such as white, yellow, pink, etc.) and thereby this problem was solved by author's patents (Amasaka, 1988b; Amasaka et al., 1988a,b; Amasaka, 2017, 2022a,b, 2023).

(1.2) Improvement in operation availability and stabilization of quality –Applying new ceramic materials to the welding nozzle and welding tip

The conventional welding nozzle has been made of a copper base alloy (Cr-Cu alloy), and therefore welding spatter can easily get stuck on it. Similarly, the welding tip is also made of copper base alloy.

Therefore its inner surface, where the welding wire passes through, can quickly become worn out and welding spatter can get stuck to this tip area.

In either case, periodic cleaning and replacement are frequently needed and when there is considerable adhesion of welding spatter, it can cause continuous quality defects, requiring the equipment to be stopped.

Given this background, the author has collaborated separately with "Noritake" and "Toshiba Tungaloy" to improve the operation availability and assure the quality stability of the welding process.

The author focused on the high strength wear resistance and heat resistance characteristics of new ceramics technologies in an effort to solve the above problems associated with the welding process by author's patents (Amasaka et al., 1988a,b; Amasaka, 2017, 2022a, 2023).

(i)   Lengthening the welding tip wear life by improving its surface

As shown in Figure 3, the conventional "Cr-Cu alloy" welding tip (Comparison: (A) in the figure) has been improved to "Zr-Cr-Cu alloy" (Comparison: (B) in the figure).

Next, the author has developed a surface improvement technology, "New Ceramics" by the author's patents, that employed CVD (chemical vapor deposition) or PVD (physical vapor deposition) to the entire surface layer of the welding tip, as well as developed these manufacturing methods.

Furthermore, efforts were also put into improving technology by developing a variety of ceramic coating methods to be used as intermediate layers for a thick coating treatment that used plasma spraying (as indicated (C)-(G) in the figure).

Figure 3. Improvement of the wear life of welding tip by new ceramics films

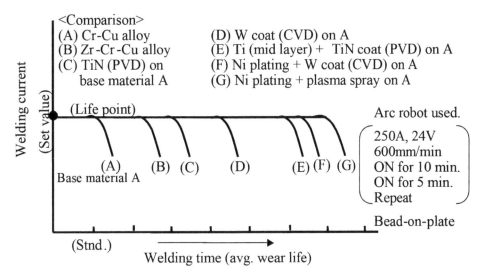

These improvement measures increased the wear life by about a dozen times and the operation availability was also improved as expected. More recently, development of conductive ceramics substrates has also been promoted.

(ii)   Development of all ceramic welding nozzle

Similarly, the above surface improvement technology was also applied to welding nozzles, and the desired results were obtained. In addition to this, the author and others developed a substrate material for all ceramic welding nozzle, as well as its manufacturing method, and thereby solved the problem of weld spatter adhesion.

This eliminated the need for periodic cleaning maintenance and lengthened the life of the welding tip. These improvements substantially enhanced the operation availability of the welding process and consequently contributed to the simultaneous achievement of QCD and synchronized production.

2.   Solution to bottleneck problems related to processing of rear axle shaft

(2.1)   Development of mid-frequency tempering equipment

In place of the aforementioned, large-scale, hot air circulating oven, panel-heater-type tempering equipment that consumes a lot of "electric power" (#050), a newly developed the (3) "Tempering" using mid-frequency ("Heat induction coil-type") was installed (Amasaka et al., 1982; Amasaka and Ishida, 1991).

As indicated in the upper section in Figure 5 (in the upper right corner), (3) "Tempering" (#05) is now conducted for one work piece at a time while concentrating on the high frequency quenching area.

This meant that power consumption and in-process inventory are drastically reduced and the lead time is also substantially shortened (Amasaka et al., 2008) (Refer to Figures 8 and 10 in the Appendix).

In "Reforming Rear Axle Shaft Machining Line" of the Appendix, the author shows the outline of "Kaizen" of "Rear axle shaft machining process" through the creation of "Tempering" (#05) and "Auto-straightening machine" (#06) (See to subsection (2.2) Development of automated stress relief equipment for the removal of the "Isolated Jobsites" ("*Hanare-Kojima*") by author's patents, and so on (Amasaka et al., 1982; Amasaka and Saito, 1983: Amasaka and Ishida, 1991) (Refer to Figures 10 and 11 of the Appendix).

Particularly, in "Medium-Frequency Tempering Equipment" of the Appendix, the author shows the two developments of a "Mid-frequency Tempering" (#05) employing "Heating Coil of Electromagnetic Induction" by the author's patents (in the upper left corner).

Specifically, in Figure 11 of the Appendix, the author shows the effect of *Kaizen* (various improvements of QCD) through the comparison of "Before" and "After" of "#05: Tempering" below. "Before" is the "products whole heating" using "Panel heater", and is the "Large hot air circulation furnace", "Large power stock" and "Long cycle time".

"After" is the "Ultra-small tempering equipment" for the "Small power and stock", "Ultra-short cycle time", and working operation of "Room temperature".

(2.2)   Development of automated stress relief equipment

In addition to the above, the aforementioned "Manual stress relief equipment" (Before, #60) that depends on the empirical skills and abilities of an expert operator was replaced with a newly developed (4) "Automated stress relief equipment" ("Auto-Straightening Machine") (After, #06) (Amasaka and Saito,1983; Amasaka, 1991) (Refer to Figure 11 in the Appendix).

In "Automatic Straightening Machine" of the Appendix, effects as follows;

(i)   Quality stability of strain correcting for realizing "Securing the process capability, Cp", (ii) Solution of "Cycle time over of strain correction time by manual work operation) as the drastic improvement "Removal of the hard-operated distortion correction),
(iii) Solution of "Cracking of strain modification" using "Cracking AE sensor".

These improvements have realized the "Risk management" through the various fault preventions "Prevention of human errors" by utilizing "statistical science: Science SQC" as the rational problem-solving approach (Toyota Motor Corp., 1993; Amasaka Ed., 2012; Amasaka, 2000b,c, 2003, 2004c, 2012) (See to Chapter 3).

Specifically, as indicated (4) "Automated stress relief equipment" (After #06) in the upper section in Figure 1 (in the center of upper), the master skills commonly acquired by expert operators are simulated in order to conduct stress relief operations in an automatic and intelligent manner.

This equipment is capable of properly conducting stress adjustment operations on work pieces with a variety of different shape profiles (lengths and diameters).

Even in the case of a sizable spiral adjustment, stress relief can be done in a cycle time that is comparable to that of skilled operators.

Also, by installing a work piece crack detector, stable quality manufacturing was ensured and uneven cycle times were successfully eliminated.

These improvements in the facility equipment made it possible for the flow of work pieces from the (3) "High frequency quenching equipment", to the "Mid-frequency tempering equipment", to the "Automated stress relief equipment (#040→#050→#060), as indicated in Figure 1, to be connected in sequence along with automatic transportation of the work pieces.

Consequently, these processes could be separated from the manual work processes. Also, by integrating the upstream processes (#010→#030) and downstream processes (#070→#170) as indicated in Figure 1, the problem of work operations conducted over "Isolated Jobsites" (*"Hanare-Kojima"*) work areas was also solved.

These process improvement measures make it possible to flexibly arrange production operators according to the increase and decrease in production volume per day, and therefore enhance productivity.

As a result, manufactured quality was improved (Amasaka et al., 1993), the lead time was considerably shortened, the problems associated with energy conservation, labor force minimization, and skill acquisition were solved, and synchronized production was realized. This substantially contributed to the simultaneous achievement of QCD (Amasaka, 1986, 2017).

3.   Solution to bottleneck problems related to processing of assembly process of rear axle unit

   (3.1)   Simultaneous achievement of QCD for corrosion resistance coating

In recent years, in an aim to improve the product value of parts related to an automobile's suspension and chassis, efforts have been made toward the simultaneously achievement of QCD in the area of "Corrosion-resistant coating technology" by drastically improving its rust proofing characteristics without raising the coating cost (Amasaka, 1986; Amasaka et al., 1988b, 1990) (See Figure 9 of the Appendix).

To implement this project [*2], the author has organized the "Strategic Total Task Management Team" (SSTTM) through collaboration with the on-site staff, production engineering staff, product development and designing staff, and coating material manufacturers (Aisin Chemical Co., Ltd., one of several affiliated coating manufacturers, and also non-affiliated manufacturer, Tokyo Paint Co., Ltd.) (See to Chapter 11).

[*2] The manufacturing technical staff took the lead in this project.

In "Reforming Rear Axle Unit Assembly Line" of the Appendix, the author patents:

(i)   "Development of painting of rust proof, quick dry and Low cost".
(ii) "Removal of drying to hold the large power consumption for realization of short "Production Lead Time".

(iii) "Removal of the "Isolated Jobsites" ("*Hanare-Kojims*") for "Flexible Manpower Line and realizing "Sequential parts withdrawal" ("*Jyunbiki*": Amasaka, 2015, 2017, 2022a,b, 2023) (See Figure 12 in the Appendix).

Specifically, Figure 12 in the Appendix aims the "*Kaizen*" of "Quick drying anti-rust coating" by "Optimization of various painting elements" using DOE design of experiment) for "Simultaneous achievement QCD (Amasaka et al., 1988).

Concretely, Figure 4 shows the development procedure (rust proof quality and coating cost) that aimed for the simultaneous achievement of QCD by improving the coating materials, coating equipment, and drying equipment.

In addition, the author also aimed to solve the technical problems in order to enhance the product value (VA is indicated by the rust proof quality/cost) of the rear axle units to which the coating materials manufactured by Tokyo Paint are applied (Improvement (1)-(8) in Figure 4) (Amasaka et al., 1990).

The coating cost in the figure indicates not the price of coating materials, but that of the coating operation, which includes the entire cost of the coating materials, energy consumed, cleaning, and maintenance cost of the coating equipment per one work piece.

Conventionally, styrene modified alkyd resin solvent coating was used, but in an effort to prevent initial rusting caused by the spread of antifreeze agents, mainly used overseas, improved phenol modified alkyd resin (Improvement (1)) was adopted, and this resulted in an increase in the coating cost due to the high material cost.

Figure 4. Improvement process of the paint

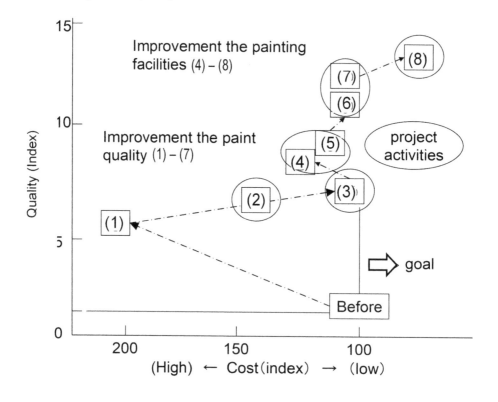

Given this background, a task team was formed that set development targets in several phases, while also anticipating the required quality improvements demanded by the market, which grow higher year after year, and systematically promoted the project through close monitoring of the process improvement.

In the early phases of improvement, mainly the coating materials were improved (Improvement (1)-(7)), and in the latter half, improvements in coating and drying equipment were promoted in parallel (Improvement (4)-(8)). More specifically, the following widely varied improvements were implemented as follows;

(i)   The aim of improving the solvent coating materials was better corrosion resistance characteristics. Air-drying alkyd resin provides drying and corrosion resistance characteristics and nonpolluting rustproof colorant and filling colorant were then properly mixed with the modified alkyd resin. These were used as the base materials and improvement of the corrosion resistance was achieved.

(ii)   Another target was to improve the quick-drying characteristics and cost, corrosion resistance, quality of the coating films, storage stability of the coating materials, and safety (prevention of spontaneous ignition by the coating mist). Improved results were obtained by using a lacquer-type phenol alkyd resin that is resistant to oxidation polymerization.

(iii)   In order to homogenize the coating films and enhance the finished exterior of the film surface, stearin acid and organic bentonite were added to the quick drying materials to improve thixotropy. This resulted in improvement of the finish quality, cost, and corrosion resistance characteristics.

(iv)   To improve the coating efficiency of the airless electrostatic coating, a polar solvent and nonpolar solvent having strong polarity and high permittivity were optimally composed in such a way as to satisfy the requirements of cost, drying characteristics, solubility, workability, and storage stability. As a result, the electric resistance of the coating materials was optimally adjusted.

(v)   The initial material cost was reduced through an optimal choice of the coating materials, filling colorants, thinners, and various compounding ingredients.

(vi)   To improve the equipment for the coating and drying processes, irregularities in the coating efficiency and films were reduced by adapting dilution-adjusted coating materials to seasonal changes, hot sprays, and so forth, so that the amount of coating materials consumed per unit was reduced by over 30%.

Moreover, by implementing collection and reuse of over spray coating and optimal compounding with new agents, the initial cost was further reduced. In addition, the adoption of robots to the coating operation realized a reduction in the running cost and improved the finish quality and coating efficiency in comparison with the conventional fixed multiunit airless electrostatic gun coating (Improvement (7)-(8)).

As a result, the final improved coating operation conducted 10 months later showed marked improvement. The rustproof characteristics were improved by 14 times (index based) and the visual quality was improved by 5 times (index based) by homogenizing the coating films. Furthermore, the drying facility was abolished because of the development of the quick-drying coating materials with room temperature drying characteristics.

These improvements reduced the in-process inventory within the process to one-third and drastically reduced the coating cost by over 30% compared to the conventional method.

Even in the case of Aisin Chemical, similar improvement approaches have been taken and resulted in the "Simultaneous achievement of QCD" (Amasaka, 1986; Amasaka et al., 1988b).

(3.2)   Improvement of the bolt tightening tools

As shown in Figure 1 (on the right), the (9) "finished rear axle unit products" (Rear Axle Unit Assembly) are mounted on pull system carts located at the final process of the (5) "Assembly conveyor" (Amasaka, 1984a,b: Amasaka and Yamada, 1991).

The work hours needed for this operation have been reduced by improving the "bolt-tightening tools" for the "brake units" and "disc wheel units" that are assembled together with the "rear axle units".

The conventional bolt-tightening operations are conducted in three steps: (i) tightening with an impact wrench via the emperical feel of the operators, (ii) additional tightening with a torque wrench, and (iii) torque checking.

These operations have been very burdensome and difficult, even for expert production operators, and tightening operations were not always uniform in quality.

Given this background, the author has developed a "high-precision tightening tool with torque control" with "tightening torque detector" by the author's patents, which reduced this operation to only a single step, thereby reducing the manual work time and also improving and stabilizing the bolt-tightening quality at the same time.

The efforts made through the task team activities employing "SSTTM" of the on-site manufacturing staff and technical engineers introduced in this discussion were undertaken to realize the simultaneous achievement of QCD and in an aim to attain high quality assurance manufacturing by innovating the manufacturing technology.

These activities have been positioned as the basis for advancement of the TPS, along with daily improvement activities at the manufacturing site (Amasaka, 2004a, 2007, 2008, 2015, 2017, 2022a,b, 2023; Amasaka et al., 2008; Amasaka, Ed. 2012).

CONCLUSION

In this chapter, the author described Through the "Realizing QCD studies by developing JIT strategy", Specifically, the "foundation and effectiveness of applications of TPS" is based on efforts to attain "Simultaneous acheivement of QCD" through inovation of production technology. Concretely, innovation of the production line producing suspension and chassis-related parts for automobiles in "Toyota's Machining Div. of Motomachi Plant" in Japan presented by having cooperation between on-site white-collar engineers, supervisors and workers with affiliated and non-affiliated suppliers. In these examples, the author has proved the effectiveness on advancing TPS through innovation of *t*he production line of automobile rear axle unit.

REFERENCES

Amasaka, K. (1983). *Mechanization for operation depended on intuition and knack – Corrective method for distortion of rear axle shaft*. The Japanese Society for Quality Control, The 26th Technical Conference, Nagoya, Aichi, Japan.

Amasaka, K. (1984a). Process control of the bolt tightening torque of automotive chassis parts (Part 1) . *Standardization and Quality Control*, *37*(9), 97–104.

Amasaka, K. (1985). Management of the quenching process of the automotive chassis parts in case measurement takes time . *Standardization and Quality Control*, *38*(3), 93–100.

Amasaka, K. (1986). *The improvement of the corrosion resistance for the front and rear axle unit of the vehicle-QCD research activities by plant production engineers.* The Japanese Society for Quality Control, The 30th Technical Conference, Nagoya, Japan.

Amasaka, K. (1988a). Electrode s of Arc welding, Surface modification technology-A dry process and its application. The Japan Society for Precision Engineering.

Amasaka, K. (1988b). *Concept and progress of Toyota Production System (Plenary lecture). Co-sponsorship: The Japan Society of Precision Engineering, and Hachinohe Regional Advance technology Promotion Center, etc.* Hachinohe.

Amasaka, K. (1991). Application of multivariate analysis in auto-production: Establishment of high-accuracy prediction formulas for the manufacture technology . *Standardization and Quality Control*, *44*(11), 91–110.

Amasaka, K. (1999). *New QC circle activities at Toyota, A training of trainer's course on evidence: Based on participatory quality improvement (Special lecture), The International Health Program (TOT Course on EPQI), Tohoku University School of Medicine (WHO Collaboration Center).* Sendai-city.

Amasaka, K. (2000a). *Basic Principles of JIT – Concept and progress of Toyota Production System (Plenary lecture).* The Operations Research Society of Japan, Strategic research group-1, Tokyo.

Amasaka, K. (2000b). *Partnering chains as the platform for quality management in Toyota.* Proceedings of the 1st World Conference on Production and Operations Management, Sevilla, Spain.

Amasaka, K. (2000c). New JIT, a new principle for management technology 21C: Proposal and demonstration of TQM-S in Toyota (Special Lecture*), Proceedings of the 1st World Conference on Production and Operations Management,* Sevilla, Spain.

Amasaka, K. (2002). *Intelligence production and partnering for realizing high quality assurance (Invited lecture).* The 94th of Japan Society of Mechanical Engineers Tokai branch, Nagoya, Aichi, Japan.

Amasaka, K. (2003). *New application of strategic quality management and SCM.* Proceedings of Group Technology/Cellular Manufacturing World Symposium, Ohio.

Amasaka, K. (2004a). Development of "Science TQM", A new principle of quality management: Effectiveness of strategic stratified task team at Toyota. *International Journal of Production Research*, *42*(17), 3691–3706. doi:10.1080/00207540420000203867

Amasaka, K. (2004b). The past, present, future of production management (Keynote lecture), *The Japan Society for Production Management, The 20th Annual Technical Conference, Nagoya Technical College, Aichi, Japan*, 1-8.

Amasaka, K. (2004c). *Science SQC, new quality control principle: The quality strategy of Toyota.* Springer-Verlag. doi:10.1007/978-4-431-53969-8

Amasaka, K. (2007). *The forefront of manufacturing technology: Evolution of Toyota Production System-The quality technological strategy by Science TQM (Special lecture).* The Small and Medium Enterprise Agency, etc., Hachinohe intelligent plaza/Hachinohe Regional Advance Technology Promotion Center Foundation, Hachinohe, Japan.

Amasaka, K. (2008). Strategic QCD studies with affiliated and non-affiliated suppliers utilizing New JIT., Encyclopedia of Networked and Virtual Organizations, III(PU-Z), 1516 -1527.

Amasaka, K. (2009). The foundation for advancing the Toyota Production System utilizing New JIT. *Journal of Advanced Manufacturing Systems, 80*(1), 5–26. doi:10.1142/S0219686709001614

Amasaka, K. (2012). *Prevention of the automobile development design, Precaution and prevention.* Japan Standard Association.

Amasaka, K. (Ed.). (2013). Science TQM, New Quality Management Principle, The quality management strategy at Toyota, Bentham Science Publishers.

Amasaka, K. (2015). *New JIT, New Management Technology Principle.* Taylor & Francis, CRC Press.

Amasaka, K. (2017). *Toyota: Production System, Safety Analysis and Future Directions.* NOVA Science Publishers, Inc.

Amasaka, K. (2022a). *Examining a New Automobile Global Manufacturing System.* IGI Global Publisher. doi:10.4018/978-1-7998-8746-1

Amasaka, K. (2022b). New Manufacturing Theory (2nd Edition). Lambert Academic Publishing.

Amasaka, K. (2023). New Lecture-Surpassing JIT: Toyota Production System. Lambert Academic Publishing.

Amasaka, K., & Ishida, T. (1991). *Tempered equipment, Aichi Invention Encouragement Prize.* Aichi Japan Institute of Invention and Innovation, Aichi-ken.

Amasaka, K., Iwata, M., & Fujii, M. (1988a). Development and the effect of the welding nozzle made from all ceramics, *Engineering Materials . Nikkan Kogyo Shinbun. Ltd., 36*(10), 60–64.

Amasaka, K., & Kamio, M. (1985). Process improvement and process control: An example of the automotive chassis parts process, *Quality Control. . Union Japanese Scientist and Engineers, 36*(6), 38–47.

Amasaka, K., Kurosu, S., & Morita, M. (2008). *New Theory of Manufacturing-Surpassing JIT: Evolution of Just-in-Time.* Morikita-Shuppan.

Amasaka, K., Mitani, Y., & Tsukamoto, H. (1993). Research into rust prevention quality assurance for plated components using SQC: Improvement of surface smoothness of rod pistons ground by centerless grinding machine, *Quality . Japanese Society for Quality Control, 23*(2), 90–98.

Amasaka, K., Ohmi, H., & Murai, H. (1990). The improvement of the corrosion resistance for axle unit of the vehicle. *Coal Technology, 25*(6), 230–240.

Amasaka, K., Okumura, Y., & Tamura, N. (1988b). Improvement of paint quality for auto chassis parts . *Standardization and Quality Control*, *41*(2), 53–62.

Amasaka, K., & Saito, H. (1983). *Mechanization of an sensory inspection - An example of viewing of a damping force – "Lissajous curve" of the front shock absorber.* Central Japan Quality Control Association, Annual Quality Control Conference 1983, Nagoya, Aichi, Japan.

Amasaka, K., & Sugawara, K. (1988). Q.C.D research activities of the engineer of the manufacture site: Improvement of paint corrosion resistance of shock absorber parts, *Quality Control . Union Japanese Scientist and Engineers*, *39*(11), 337–344.

Amasaka, K., Tsukada, H., & Ishida, M. (1982). *Development of the tempering machine using mid-frequency of the rear axle shaft.* The Energy Conservation Center, Annual Technical Conference 1982, Nagoya, Aichi, Japan.

Amasaka, K., & Yamada, K. (1991). Re-evaluation of the present QC concept and methodology in auto-industry: Development of SQC renaissance at Toyota . *Quality Control*, *42*(4), 13–22.

Burke, W., Trahant, W., & Koonce, R. (2000). *Business climate shift.* Butterworth- Heinemann.

David T., & David. B. (2001). *Manufacturing operations and supply chain management: Lean approach.* Thomson Learning, London.

Roger, G. S., & Flynn, B. B. (2001). *High performance manufacturing.* John Wiley & Sons, Inc.

Toyota Motor Corporation. (1993). *Toyota Technical Review: SQC at Toyota.* Toyota.

Umezawa, U., & Amasaka, K. (1999). "Partnering" as the platform **of** quality management in Toyota Group, *Operation Research . The Operations Research Society of Japan*, *44*(10), 24–35.

Yamaji, M., & Amasaka, K. (2009). Strategic productivity improvement model for white-collar workers employing Science TQM. *The Journal of Japanese Operations Management and Strategy*, *1*(1), 30–46.

APPENDIX: INNOVATION OF PRODUCTION TECHNOLOGY

Simultaneous Achievement of QCD by White-Collar Engineers

Figure 5. Outline of "Auto-rear axle unit" assembly process line

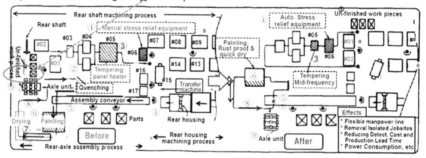

Reforming Rear Axle Shaft Machining Line

Figure 6. Outline of "Auto-rear axle unit" assembly process line

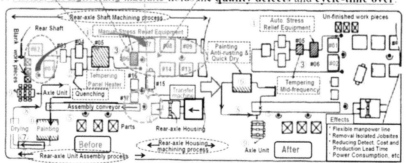

Reforming Rear Axle Unit Assembly Line

Figure 7. Outline of "Auto-rear axle unit" assembly process line

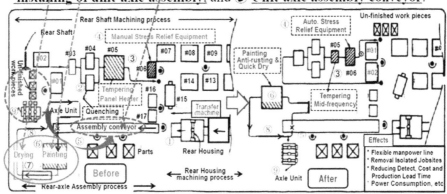

Before 2: **Bottleneck** of the **quality problem** & **long production lead time**
1. **⑥ Coating surface of Painting** holds the **reliability problem** of **anti-rust treatment**, and **painting cost** by the damage of salt and air pollution.
2. **⑦ Drying** holds the large power consumption and production lead time.
3. The existence of **Isolated Jobsites** to hold between the **⑨ Removing** and **installing of unit-axle assembly**, and **⑤ Unit-axle assembly conveyor**.

Reforming Rear Axle Shaft Machining Line

Figure 8. Outline of "Auto-rear axle unit" assembly process line

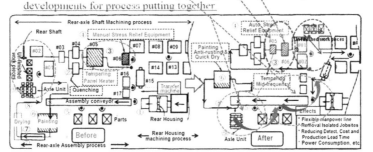

After 1: Kaizen of Tempering and Auto Stress Relief Equipment
1. ❸ The development of a **Mid-frequency Tempering (#5)** using the **Heating Coil of Electromagnetic Induction** (refer to **Appendix C**).
2. ❹ The development of an **Auto-Straightening machine** (#6) (Appendix 5B-1)
3. The **removal** of the **Isolated Jobsites** as follows; (Appendix 5B-2)
   ⇒The **blank work pieces**, "#01 and #02 machine", and #05 (**Mid-frequency tempering**) and #06 (**Auto-straightening**) are changed along with those developments for process putting together.

Reforming Rear Axle Unit Assembly Line

Figure 9. Outline of "Auto-rear axle unit" assembly process line

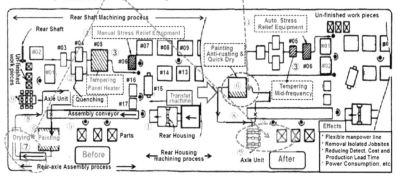

MEDIUM-FREQUENCY TEMPERING EQUIPMENT

Employing Electromagnetic Induction Coil

Figure 10. Medium-Frequency tempering equipment

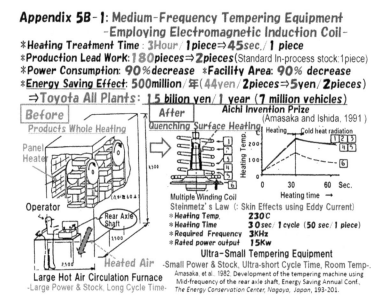

AUTOMATIC STRAIGHTENING MACHINE

Removal of the Hand-Operated Distortion Correction

Figure 11. Automatic straightening machine

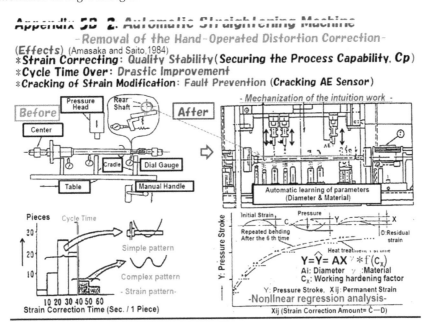

QUICK DRYING ANTI-RUST COATING

Simultaneous Achievement of QCD by White-Collar Engineers

Figure 12. Quick drying anti-coating

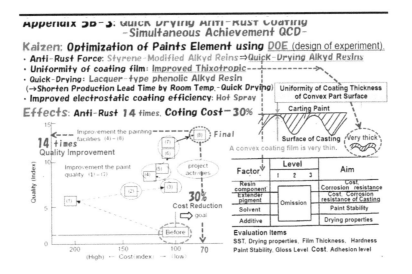

Chapter 5
Partnering of Auto Maker and Suppliers to Strengthen TPS and TQM Activity

ABSTRACT

In this chapter, the author discusses on inter-enterprise link and partnering chains that concretely carries out the platform functions of TPS and TQM activities of Toyota and Toyota group companies while depending on a framework that recognizes collaborating relation between automaker (vehicle assembler) and parts manufacturers (supplier), which is indispensable in producing good products. This study approaches the TPS and TQM activities for excellent QCD studies from a new angle of view of partnering to strengthen Japan supply system (JSS) using the strategic quality management—performance measurement model (SQM-PMM). It also touches on actual study of new development that is looked upon with expectation in near future. Specifically, this chapter studies on potential development by Toyota's Total Task Management Team (TTMT) activity among Toyota Group, which draws attention as a new development in the quality management, by quoting the typical case of the brake pad quality assurance and others.

OUTLINE OF TOYOTA MOTOR CORP. AND TOYOTA GROUP

In 1982, Toyota Motor Industry and Toyota Motor Sales merged to form Toyota Motor Corporation. Toyota Motor Corp. (Toyota) is a super world enterprise from either viewpoint. Toyota group consists of a total 14 enterprises (Toyota Motor Corp., 1987, 1999). Twelve enterprises, which branched from Toyota Automatic Loom Works, Ltd. form the nucleus of the primary group parts manufacturers (called "Kyohokai") that supply parts directly to Toyota, to which join "Hino Motors, Ltd. and Daihatsu Motor Co., Ltd.".

Each of the group companies is closely linked to Toyota in a wide and solid supplier-assembler relation "Toyota Supply System" (TSS) so-called "Japan Supply System" (JSS) as shown in Figure 1 (Amasaka, 2000, 2008, 2015, 2017a,b, 2022a,b, 2023; Amasaka et al., 2008; Amasaka, Ed., 2012). An automobile is assembled with some 20,000 parts.

DOI: 10.4018/978-1-6684-8301-5.ch005

Since it is not economical for the assembler to manufacture all the parts in-house, considerable portion of the parts are normally purchased from outside suppliers (parts manufacturers). Therefore, to years, and the situation still remains unchanged for it.

Figure 1. Japan supply system

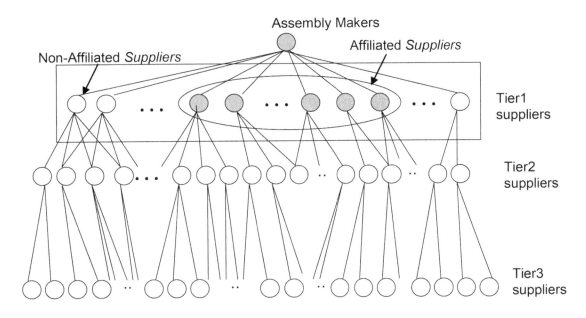

RELATIONSHIP BETWEEN VEHICLE ASSEMBLER AND SUPPLIERS

If parts purchased from the suppliers (parts manufacturers) have low dependability, vehicles assembled with them have also low dependability naturally. This is exactly the reason why "performance of a vehicle almost depends on the parts" (Amasaka, 1988, 1989, 1998, 1999a,b, 2000, 2008a,b).

In this sense, the assemblers and suppliers are the inhabitant of the same fate-sharing community. Actual supplier-assembler relation is generally quite complex and many-sided.

Relationship between Toyota and its suppliers is unique in many points compared with those of other assemblers. There is the saying that "Toyota wrings a towel even when it is dry", indicating Toyota's strict demand to suppliers for their prices and quality.

At the same time, no other assemblers are so enthusiastic as Toyota in raising strong suppliers through education and training. As early as in 1939, Toyota established its basic concept of the purchasing activities for promoting coexistence and co-prosperity.

To realize this, Toyota strengthened its suppliers by making it a rule to continue transaction forever once started. No other assemblers have their supplier groups as powerful as those of Toyota. The 14 Toyota group companies form the nucleus of the powerful supplier system.

As thus far stated, close relation between Toyota and its group members is quite cooperative in one sense while simultaneously very competitive in another. This represents the supplier-assembler relation unparalleled elsewhere.

There is no denying that the strength of Toyota that can keep on supplying popular vehicle model such as "Lexus" of high dependability originates from within Toyota's own. But it is also the fact that part of Toyota's strength comes from the strength of its supply system consisting of Toyota group and thousands of other suppliers or Toyota's skill in managing such powerful supply system.

In the following, the authors intend to zero in on "Toyota's Quality Management activities" in exploring the secret of the strength of Toyota that continues to manufacture vehicles of high dependability. This is because it is quality management activities themselves that provide important bonds to cooperation or partnering between Toyota and its group companies.

"PARTNERING" AS THE QUALITY MANAGEMENT PLATFORM AMONG TOYOTA GROUP

Utilization of "Divisional and Functional Management" by Toyota

At Toyota, functions such as product planning, product design, production engineering, purchasing, manufacturing, sale and others are called "Division" as shown in Figure 2, and control given based on this classification is called "Divisional Management".

On the other hand, quality, cost, engineering, production, marketing, human resources and administration are called "function" and management given according to this classification is called "Functional Management".

In addition, "TPS and TQM" promotion activities in Toyota's corporate management was inaugurated to establish the functional management system supported by two pillars of quality management and cost management by using statistical science named Toyota's "Science SQC" (Amasaka, 1999c, 2003, 2004a) (See to Figure 7 in Chapter 3).

The above Toyota's way of business management is based on cross – functional management (Toyota Motor Corp., 1987, 1999; Amasaka, 1999a,b).

Partnering

Partnering refers to multiple enterprises developing business by concluding equal partnership in principle while each retaining independency.

Actual form of partnering is varied from the integrated manufacture and sale, supply chain and other vertical development to the joint development of new technology and products, integrated production, joint logistics and other horizontal business tie-up and integration (Amasaka,1988, 2000, 2001, 2002, 2008a,b, 2017b; Doz and Hamel, 1998; Umezawa and Amasaka, 1999).

Toyota and its group companies are in a typical assembler-supplier relation, that is the work division relation in the vertical direction, where Toyota generally purchases and assembles parts manufactured by the parts manufacturers of the group.

Besides, there are varieties of other cases where other assembler, like Toyota Auto Body Co., purchases engines from Toyota to assembly vehicles or Toyota manufactures part of the parts in-house in a completely competitive relation to the parts manufacturer as far as the parts in question are concerned.

Basically, however, they are in the supplier-purchaser relation or in the preceding and following process relation in a vertical division of labor.

Figure 2. Divisional and functional management by Toyota
(Toyota Motor Corp., 1987)

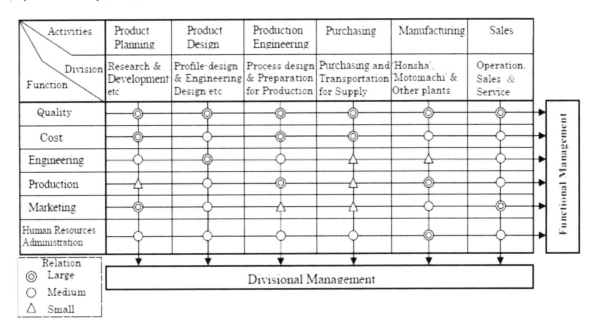

As Figure 1 is self-explanatory, Toyota is the customer = purchaser for each of the Toyota group companies. For each of them, Toyota is not only the top customer with the highest sale but it is the large shareholder of own company.

Therefore, it is quite natural to think that Toyota is special and exceptional. However, unless Toyota is supplied with quality parts at low prices, it cannot manufacture vehicles of high dependability by itself. In this sense, mutual relation can be virtually almost equal to both sides, particularly on the level of a role of persons in charge of TQM activities.

Consequently, it is correct to say that each of the Toyota group companies and Toyota are in a very partnering relationship toward the promotion of Quality Management.

Moreover, as mentioned above, respective companies in the partnering are mutually very closely related and quite cooperative. But at the same time, they can be quite competitive to each other concerning a part in question.

Such two-sidedness is the most remarkable characteristic of the partnering among Toyota group and one of the sources of Toyota's strength.

TPS and TQM Activities as "Lateral Operation of Organization"

Suppose that Figure 2 deals not with an enterprise but with a series of business process of the group as a whole, which consists of parts planning-development, manufacture, delivery to Toyota and assembly (Toyota Motor Corp., 1999).

Then the group companies replace the divisions in the figure. To manufacture vehicles of high dependability by combining all the capabilities of the group companies, a cross group enterprise-like quality control becomes essential (Amasaka, 2000, 2008b).

In other words, "functional management", which is the "partnering" of all Toyota group companies becomes indispensable for promoting Quality Management.

Understanding that Toyota style "functional management" signifies a cross-organizational operation, it is a re-engineering (Umezawa, 1994) of traditional divisional management in an enterprise or a sort of supply chain management when the object of management is the enterprises themselves that integrate operations in vertical direction.

When we proceed with the manufacturing of highly dependable vehicles by "All Toyota", it is the matter of course that quality control activities (re-engineering) will not simply remain the matter of individual enterprises but these activities will grow to the cross-group enterprise type quality control activities (a sort of supply chain management).

It is apparent that there would be a major difference in product quality depending on whether a parts manufacturer manufactures parts by thinking that Toyota, to whom he delivers the parts is his customer or this manufacture thinks that the purchaser of an assembled vehicle is the true customer for him and he makes efforts in manufacturing quality parts by working together with the assembler, Toyota (Amasaka, 1988, 2000).

Subcontractor Management and Hierarchical Supply System in Toyota Motor

Parts manufacturers of Toyota group consists of three groups; the primary supplier, who can delivery parts directly to Toyota; the secondary supplier who can deliver parts to the primary supplier; and the tertiary supplier below (Amasaka, Ed., 2007a).

As Figure 1 is self-explanatory, Toyota's supply system forms a distinctive hierarchical structure. Immediately after being awarded the "Deming Prize", Toyota Motor vigorously promoted propagation of TQC (Total Quality Control) activities and the "Kanban" system to the group parts manufacturers under a motto of "Quality Assurance by All Toyota" (Toyota Motor Corp., 1987, 1999; Amasaka, 1989, 1999a,b).

Concretely, Toyota provided frequently the members of "Kyohokai", an association of the primary suppliers, with the lecture meetings and the plant diagnoses. Toyota also dispatched the specialists to various committees of "Kyohokai" to give thorough guidance on quality control and the "Kanban" system.

In addition to this, purchasing control division of Toyota Motor held a long-term study meeting for the purchasing staff of respective primary suppliers to teach them Toyota's subcontractor control method comprehensively.

By so doing, Toyota intended to have these primary suppliers give guidance to the secondary suppliers the same way as it taught them TQC and the Kanban system. It was apparent that Toyota had still far-reaching objective to have the secondary suppliers repeat the same for the tertiary suppliers.

As the result of the organizational guidance that stretched staggering far from the 1970's to 1980's and into the 1990's, a gigantic and powerful "Toyota Supply System" of hierarchical structure was completed quality control, the Kanban system technique and know-how are beginning to proliferate autonomously everywhere on each layer.

This chapter deals Toyota group's partnering as merely seen among the top or the crown of the pyramid. It is not difficult to imagine that countless partnering does exist and function similarly to Toyota group on the secondary and tertiary supplier levels.

This is the very strength of Toyota that accomplished that surprising performance of "Lexus". Truly, "Roma was not built in one day".

ESTABLISHMENT OF STRATEGIC QUALITY MANAGEMENT – PERFORMANCE MEASUREMENT MODEL FOR JSS STRATEGY

As a key to the strategic manufacturing application of TPS and TQM, the author has established the "Strategic Quality Management – Performance Measurement Model" (SQM-PMM) in order to ensure superior corporate management achievements as the "JSS strategy" (Amasaka, 1999, 2004b,c, 2009, 2022a,b, 2023; Amasaka, Ed., 2000a, 2012; Amagai and Amasaka, 2003; Wakaizumi, 2005; Kozaki et al., 2012).

Significance of Constructing SQM-PMM

The present study is based upon the viewpoints of management at leading companies (actual quality management activities undertaken by management and the resulting achievements), rather than existing international standards or awards (Chowdhury and Zimmer, 1996; Amasaka et al., 1999; Furuya, 1999; ISO 9004, 2000; Boesen, 2000; Deming Prize Committee, 2003; Japan Quality Award Committee, 2003; EFQM, 2003).

The factors of quality management systems and practices were extracted and a model of causal relationships between these factors and QCD management achievements was constructed.

The ultimate aim was to establish an SQM-PMM that ensures effective application of strategic quality management. This "Cause and effect analysis", which directly links performance measurement to corporate management performance, is unprecedented.

Questionnaire Overview

A survey of top management at Japanese manufacturers as shown in Table 1 was conducted in order to investigate (I) TQM Activity Structures and Practices and (II) TQM Achievements (quality, CS and productivity) and future tasks (Survey targeted 898 companies and resulted in 354 valid responses for a rate of 39.4%).

Table 1. Sample composition

Number of replies		Type of industry								Total
		Machinery	Primary materials	General materials	Natural resources	Direct materials	Service	Electricity	Petrochemistry	
Number of employees	Below 100	1	1	1	0	3	0	0	0	6
	100 – 299	15	4	8	0	6	6	2	5	46
	300 – 499	9	3	6	0	6	4	0	1	29
	500 – 999	29	1	13	0	12	11	0	5	71
	1,000 – 4,999	63	6	20	1	28	5	0	6	129
	5,000 – 9,999	18	1	6	0	9	0	1	3	38
	10,000 or more	23	0	1	0	6	3	1	1	35
	Total	158	16	55	1	70	29	4	21	354

Questions in the TQM Activity Structures and Practices section (I) fell into the following categories as shown in Table 2: [1] Quality assurance and CS (8 items), [2] Product and technological development (3 items), [3] Lively corporate culture (10 items), and [4] Conventional TQM methods (7 items).

For these items, the respondents were asked to evaluate their existing activity structure and practices on a 5-point scale and state their reasons for that evaluation.

Questions in the TQM Achievements (quality, CS and productivity) and future Tasks section (II) included those in the [1] "Quality and CS (5 items)" and [2] "Productivity (4 items) categories".

The respondents evaluated their achievements and future tasks on a 3-point scale with an explanation of their reasons for that evaluation and future plans (see Table 2).

Table 2. Questionnaire content

I. TQM Activity Structure and Practice	II. TQM Achievements and Future Tasks

[1] Quality assurance, CS
    (1) Quality assurance to (8) CR structure
[2] Product and technological development
    (1) Planning analysis and results incorporation to (3) Technological development and product planning
[3] Lively corporate culture
 (i) Total participation activities
    (1) Decision making speed to (5) Intradivisional team activity
 (ii) Creative workplace
    (1) Improvement motivation to (5) ES
[4] Conventional TQM methods
    (1) Vision and strategy to (7) Top assessment and audit
	[1] Quality and CS achievement
	    (1) Customer complaint resolution cost reduction
	    (2) Customer complaint reduction
	    (3) CS improvement
	    (4) New product development
	    (5) Sales and marketing share enhancement
	[2] Productivity achievement
	    (1) Sales per head
	    (2) Hourly productivity
	    (3) Cost reduction
	    (4) Lead time reduction

Analytical survey results

(1) State of TQM implementation (Question 1, see Figure 3)

The current state of TQM activities in Japan can be summarized by looking at the following common shortcomings;

(i) expected product quality has been retained.
(ii) improvements aimed at more attractive quality are lacking.
(iii) low division-to-division coordination and organizational vitality.
(iv) structures exist for developing vision and strategies, but implementation and/or evaluation of achievements is weak.
(v) organizations only partly utilize SQC and do not accumulate techniques.
(vi) a TQM structure exists, but is not sufficiently implemented.

(2) TQM achievements and future tasks (Question 2, see Figure 4)

TQM achievements and future tasks are summarized as follows:

(i)   TQM, in principle, contributes to improvement of quality, cost and delivery; and (ii) current TQM activities do not contribute much to promote sales, improve CS, or address other important management issues.

Figure 3. TQM structure and practice

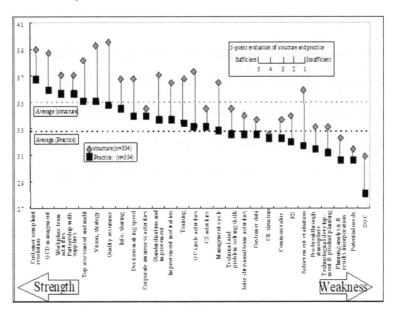

Figure 4. TQM achievements and future tasks

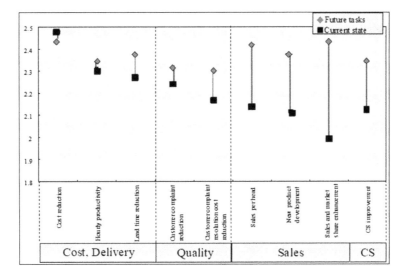

Categorical Automatic Interaction Detector (CAID) Analysis

Based on these findings, the author conducted a Categorical Automatic Interaction Detector (CAID) analysis (Murayama et al. 1982; Amasaka et al. 1998; Amasaka, 1999) in order to identify the major elements of TQM that lead a company to success. The results of this analysis are shown in Figure 5.

For example, companies that rated CS, CR (Customer Retention), and product and technological development as important contributors to sales and market share growth showed excellent achievements in terms of QCD. These results underscore the necessity of quality management performance measurements that enable top management to evaluate their company's quality management status and strategically utilize TQM activities in the future. Among the companies that implemented sufficient CS activities (N=117), 45.3% were able to increase their sales and/or market share.

Furthermore, among the companies that put a sufficient CR structure in place, performed a results analysis, and incorporated these results into their technological development and merchandise planning (N=47), about 60% were able to increase their sales and market share.

These results make it clear that companies must devote themselves to CS, CR, and results analysis, and then reflect the results in technological development and merchandise planning to improve sales and market share.

Figure 5. Causal analysis for sales and increased market share: CAID analysis

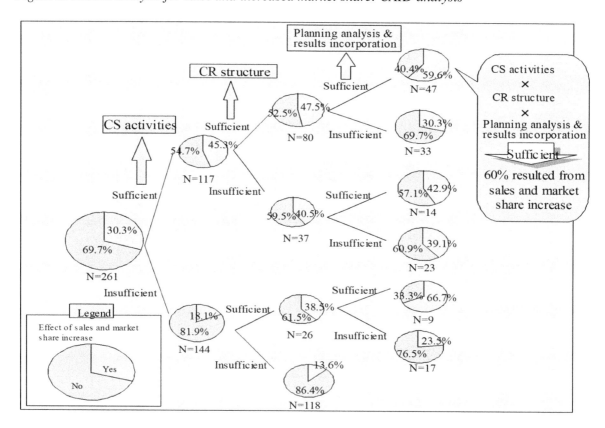

Graphical Modeling Method

In SQM-PMM formulation, further analysis was conducted using the graphical modeling method. Based on the correlations among the 28 TQM activity practice items (confirmed using a partial correlation coefficient), interrelation of the six categories—quality, CS, product and technological development, total participation, creative workshops, and TQM methods using SQC—and their contribution to corporate achievements was confirmed visually through the relation chart method shown in Figure 6.

In Figure 6, the author suggests that TQM methods, technological development, CS, quality assurance, and corporate culture are all associated with management achievement. In particular, a strong correlation was found between quality assurance and product quality as well as between technological development/CS and sales.

For example, the relationship marked with "arrow (1)" indicates that in order for their vision / strategy and top management to function effectively, companies must enhance achievement assessment, control cycles, and standardization / improvement.

The two arrows marked with "arrows (2)" indicate that a correct vision / strategy is necessary for technological development and CS to effectively contribute to sales. The "two arrows" marked "arrows (3)" signify that a creative workplace, backed by SQC, is necessary as a foundation for effective technological development.

Figure 6. Relationships among TQM practices and achievements-graphical modeling

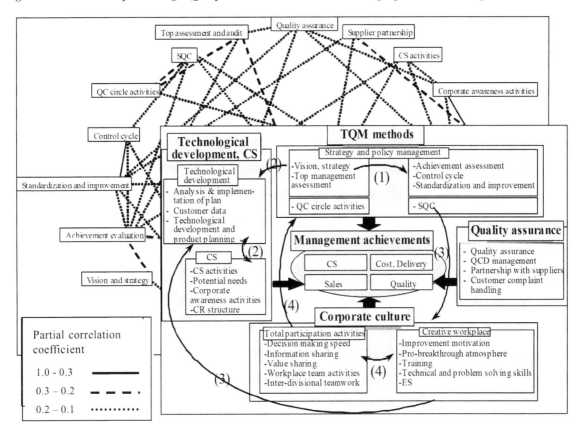

The "two arrows" marked "arrows (4)" illustrate the relationship between total participation activities and a creative workplace, as well as the importance of QC circle activities as participation opportunities for all employees.

Formulation of the SQM-PMM Through Canonical Correlation Analysis

Finally, canonical correlation analysis was conducted to clarify the correlation between the two main variables (practice items and management achievement items). The analysis focused on weighing factors of the selected first canonical variable.

As a result, SQM-PMM was successfully formulated as below: The 28 items for (I) TQM Activity Structure and Practice were sequentially determined as x_1 to x_{28}.

The first canonical variable f_1 (f_{11} to f_{16}) was divided into six categories, of which the relational expressions are indicated in (1) through to (6). Items x_1 to x_4 are quality assurance, x_5 to x_9 are CS, x_{10} to x_{12} are product and technological development, x_{13} to x_{17} are total participation activities, x_{18} to x_{22} are creative workshops, and x_{25} to x_{28} are conventional TQM methods.

(1) $f_{11}=0.418x_1+0.232x_2+0.530x_3+0.160x_4$

(2) $f_{12}=0.160x_5+0.878x_6+0.156x_7+0.494x_8+0.533x_9$

(3) $f_{13}=0.335x_{10}+0.712x_{11}+0.063x_{12}$

(4) $f_{14}=0.089x_{13}+0.477x_{14}+0.064x_{15}+0.058x_{16}+0.332x_{17}$

(5) $f_{15}=0.230x_{18}+0.316x_{19}+0.259x_{20}+0.052x_{21}+0.210x_{22}$

(6) $f_{16}=0.105x_{23}+0.254x_{24}+0.254x_{25}+0.251x_{26}+0.119x_{27}+0.190x_{28}$

Standardization of SQM-PMM

These six categories — (i) Quality Assurance (ii) TQM Methods (iii) Creative Workshops (iv) Total Participation Activities (v) Product and Technological Development, and (vi) CS—were standardized using formulas (1) to (6) and evaluated using a radar chart.

Table 3 shows the 5-point scale evaluation of category (iii), Creative Workshops. Practical application of the SQM-PMM has begun using the "Amasaka New JIT laboratory" (http://133.2.218.195/~am alab/H) website (see to Table 4 and Figure 7).

Verification Objectives and Method

Verification was conducted to see whether SQM-PMM correctly reflected the evaluators' (management of leading companies) understanding of quality management. The extent of dispersion was also verified as a function of the evaluators' positions in their organizations.

Furthermore, if evaluators made appropriate evaluations with SQM-PMM when calculating the results for (1) to (6) in 4.6, it was possible to identify the evaluation items (X_1 to X_{28} in 4.6) that should be improved to enhance management achievement items (f_{11} to f_{16} in 4.6) (Amagai and Amasaka, 2003).

Table 3. Example of SQM-PM standardization (creative workshop)

Evaluation Item		5	4	3	2	1
Creative workplace	Activities to maintain improvement motivation	Original activities implemented and are effective	Implemented generally	Implemented partially	Not implemented very much	Not implemented at all
	Activities to create pro-innovation atmosphere	Original activities implemented and are effective	Implemented generally	Implemented occasionally	Not implemented very often	Not implemented at all
	Training to enhance all employees' capabilities	Implemented regularly and is effective	Implemented generally	Implemented partially	Not implemented very much	Not implemented at all
	Practice and accumulation of special skills and problem solving capabilities	Practiced and accumulated	Practiced and accumulated generally	Practiced and accumulated to some extent	Not practiced and accumulated very much	Not practiced and accumulated at all
	Activities to enhance employee satisfaction (e.g. self realization, evaluation, treatment)	Implemented regularly and is effective	Implemented generally	Implemented partially	System exists but not implemented very much	Not implemented at all

Table 4. An example of SQM-PMM (Amagai and Amasaka, 2003)

SQM-PMM was utilized on a trial basis at five companies (A: Fuji-Xerox, B: Nihon-Hatsujyo, C: Yanma, D: Kawasaki-Seitetsu, and E: Sanden) and evaluated by eight evaluators (A_1, B_1–B_3, C_1, D_1–D_2 and E_1). Five evaluators were in positions that allowed them to grasp company-wide quality management status (top management: A_1, B_1, C_1, D_1, and E_1), while three evaluators were in positions that allowed them to observe business category–level quality management status (general managers: B_2, B_3, and D_2). Possible dispersions based on evaluator position and the validity of SQM-PM were examined.

Implementation Results and Discussion

Figure 7 is a radar chart prepared by evaluator A_1 indicating the results of the SQM-PMM trial at Company B. The chart is based on relationships (1) to (6) using the SQM-PMM.

As shown in the figure, results were balanced, exceeded the average in all aspects, and corresponded to the evaluator's prior evaluation that product development is relatively strong. Similar evaluations were also obtained from other evaluators.

In addition, evaluator positions were analyzed using the principal component analysis shown in Figure 8. This analysis was based on the SQM-PMM evaluation and performance measurement results

Figure 7. Radar chart: Sample-1

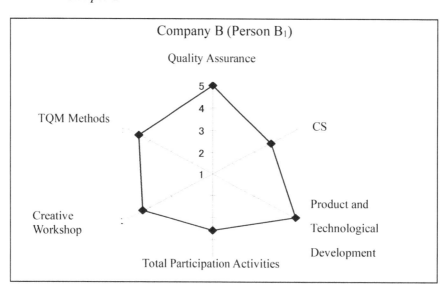

The analysis classified the evaluators into two groups—one (Group G_1) made up of those who confirmed the validity of performance measurement for company-level quality management performance (positions where it was possible to grasp company-wide quality management status), and the other (Group G_2) made up of those who wanted performance measurement designed exclusively for each business category (positions where it was possible to observe business category–level quality management status).

In Figure 8, the primary component axis (horizontal axis) represents the evaluation scores (SQM-PMM) and the secondary main component axis (vertical axis) represents the evaluation focus (company-wide perspective or business category–oriented). Judging from the positioning shown in the figure, the G_1 group (company-wide level) tended to give higher ratings while the G_2 group (business category level) tended to give lower ratings.

Moreover, business category–oriented evaluations were confirmed. While evaluator B_2, a division general manager, made his evaluation from a company-wide standpoint, even the top management evaluators B_1, D_1, and E_1 showed a business category–oriented viewpoint. Upon presenting these findings (see the positioning in Figure 8) to all evaluators, it was proven that the results correctly reflect the evaluators' implicit prior evaluation, thus demonstrating the general validity of SQM-PM.

Figure 8. Evaluator positions

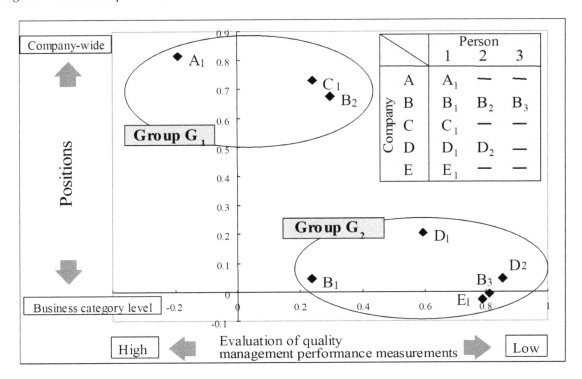

Based on the evaluation guidelines (e.g., an evaluation sheet for each of the six categories and a radar chart for the evaluation results), top management evaluated the state of their company's quality management. As a result, the strengths and weaknesses of quality management recognized in advance were expressed objectively.

Verification of SQM-PMM Validity

To verify the validity of SQM-PMM, additional investigations were conducted at companies A (Fuji-Xerox: 2 persons, 2 sections), B (Nihon-Hatsujyo: 5 persons, 2 sections), E (Sanden: 13 persons, 3 sections), F (Calsonic-Kansei: 6 persons, 3 sections) and G (Daikin-Kogyou: 17 persons, 6 sections) in the same way.

The validity of SQM-PMM was reconfirmed as shown in Figure 9, proving the effectiveness of SQM-PM as a tool to identify what TQM actions are necessary for strategic quality management.

APPLICATION EXAMPLES

Case of the Brake Pad Quality Assurance Through Partnering

In a general assembly industry like the automotive manufacturing, a through "Quality Assurance (QA)" plays an important role. It includes not only the parts and unit quality control, but also the optimization of adaptive engineering for assembly by the vehicle manufacturers and the quality of production, sale and services (Amasaka and Osaki, 1999; Amasaka, 1999c).

Figure 9. Radar chart Sample-2

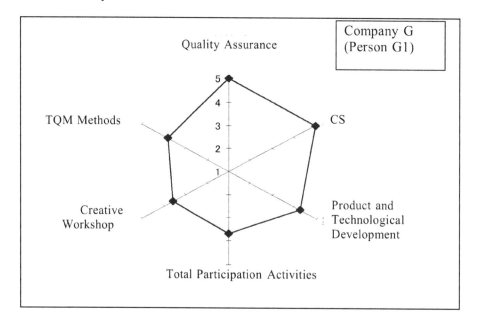

Therefore, when working on the solution of a pending engineering task, separate and independent team activities by the vehicle manufacture or parts manufacturers alone often fail to clarify the engineering area, which exists in-between or across the two teams, often turning it into implicit knowledge.

In this connection, it is important to zero-in on the end users from the view point of customer-in and to create new devices by mutually disclosing their software and hardware techniques to each other.

Here, this case intends to verify the effectiveness through the "partnering" between the vehicle assembler and the parts manufacturers on the "investigation of the mechanism of the brake pad noise and quality assurance", which has become one of the continual engineering problems of the automotive manufacturers of the world (Amasaka et al., 1977; Amasaka, 1999c, 2001, 2004, 2008a, 2015, 2017a,b, 2022a,b, 2023; Amasaka and Osaki, 1999, Amasaka and Otaki, 1999; Amasaka, Ed., 2000b, 2012).

Objectives

Disk brakes work on a principle that the pressing of pads to the rotors by the calipers generate braking force. Brake noise is generated when the pad and the rotor are in contact with each other in a delicately unstable condition.

The condition that allows noise or abnormal sound to be generated with ease and the response quality of brake are items contrary to each other.

Therefore, it is important to analyze the properties of pads and the response quality of noises and brake, and reduce dispersions in the properties of pads that affect the generation of noises and the braking effect (Miller, 1978).

The key point of this activity lies in establishing technologies that make sensitivity analysis of factors on the braking effect and noise in the aspect of design engineering, and to make these mutually contrary braking performance compatible with each other through bi-directional interaction with the manufacturing engineering.

Total Task Management Team (TTMT) Activity

To establish an organizational system that allows the vehicle design, parts design, production process design, manufacture, inspection, maintenance, sale (service, marketing quality) divisions including these of automotive manufacturer to share know how and information supports engineers' conception.

This links to the improvement of technology and improves the quality of business process. Here, this organization is realized in the form of a "Total Task Management Team" (TTMT), as shown in Figure 10.

Five teams were formed, which consisted of QA1 (Engineering design), QA2 (Production engineering), QA3 (Manufacture and Inspection), QA4 (Facility Maintenance) and QA5 (Quality and Production information). These five teams mutually cooperated in developing Total QA Network (QAT) (Amasaka et al., 1995; Amasaka and Osaki, 1999; Amasaka, 2017b).

Figure 10. Organizational outline of Total Task Management Team (TTMT)

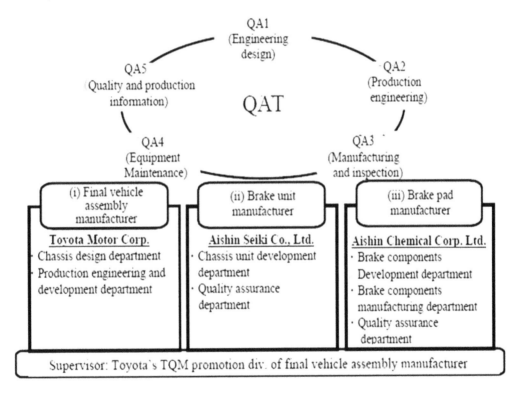

This organization has been realized by final vehicle assembly manufacturer (Toyota Motor Corporation), brake unit manufacturer (Aishin Seiki Co., Ltd. of tier-1 supplier in Figure 1), and brake pad manufacturer (Aishin Chemical Co., Ltd. of tier-2 supplier in Figure 1) described in Figure 10 (Amasaka, 2004, 2008b; Amasaka, Ed., 2012a, 2017b).

By the top policy of Toyota and Kyohokai (Toyota Motor Corp., 1987), the author (a general manager of TQM promotion div. of Toyota) constitutes the following organizations as a supervisor of TTMT activity.

TTMT consists of (i) two general managers of the chassis design department, production engineering and development department in Toyota, (ii) two general managers of the chassis unit development department and quality assurance department in Aishin Seiki, and (iii) three general managers of the brake components development department, brake components manufacturing department and quality assurance department in Aishin Chemical.

In TTMT composed of the seven general managers of three companies, each general manager assigns several section chiefs, sub-section chiefs, technical staffs, and statistics analysis staffs named "SQC special adviser" (Amasaka, 2004, 2017b) as the TTMT members. The author commissions the general manager of the quality assurance department of Aishin Seiki and Aishin Chemical as "sub-supervisor" in order to rationalize progress management of QAT.

Next, the author commissions each part chief as Generators of QA1 to QA5 as follows; QA1 is chassis design department of Toyota, QA2 is chassis unit development of Aishin Seiki, QA3 is brake components development department of Aishin Chemical, QA4 is Brake components manufacturing department, and QA5 is production engineering and development department of Toyota.

Similarly, five teams of QA1 to QA5 arrange section chief as Mentors, sub-section chief as Promoters, and technical staffs and statistics analysis staffs as the Producers or every team.

Development of Total QA Network (QAT) by TTMT Activity

For quick systematic and organizational optimization of TTMT activity, the author has developed the "Total QA Network using SQC Technical Methods" as the "Mountain-Climbing for Problem-Solving" as shown in Figure 4 (See to Figure 4 in Chapter 3).

1. Three pillars of "Total QA Network (QAT)"

As a key technology for quality assurance, three management activities- "Total technical Management (TM)" by QA1 and QA2, "Total production management (PM)" by QA3 and QA4, and "Total information management (IM)" by corporation of QA5 and two quality assurance departments- were created to promote QAT and were promoted by integrating these activities as shown in Figure 11(1).

Five teams focus on reviewing the work method and control items in process conditions by clarifying how brake performance and squeal are affected by dispersions in raw material property and process condition through sensitivity analysis below.

(1-1) Total technical management (TM)

Makes a sensitivity analysis of dispersion in both raw materials and the process condition for the brake response quality and the brake noise, and reviews the engineering method and clarifies process condition control items.

(1-2) Total product management (PM)

To realize the process conditions and control items, all divisions such as the "research and development, product development design, production engineering, manufacturing and inspection, equipment maintenance, quality assurance and others departments participate jointly" combined carry-on production activities using the "QA Network table" (Amasaka, 1995) of manufacturing. This QA Network table is based on the matrix diagram method and the process FMEA (Failure Mode and Effect Analysis), FTA (Fault Tree Analysis), and QFD (Quality Function Deployment) by utilizing statistical science.

(1-3) Total Information Management

This activity focuses on establishing the system for timely feedback of the market quality information (from dealers), next process information (from vehicle assembly manufacture) and local process information (parts manufacturer) to respective processes.

Figure 11. Development of total QA network (QAT) by TTMT activity using SQC technical methods

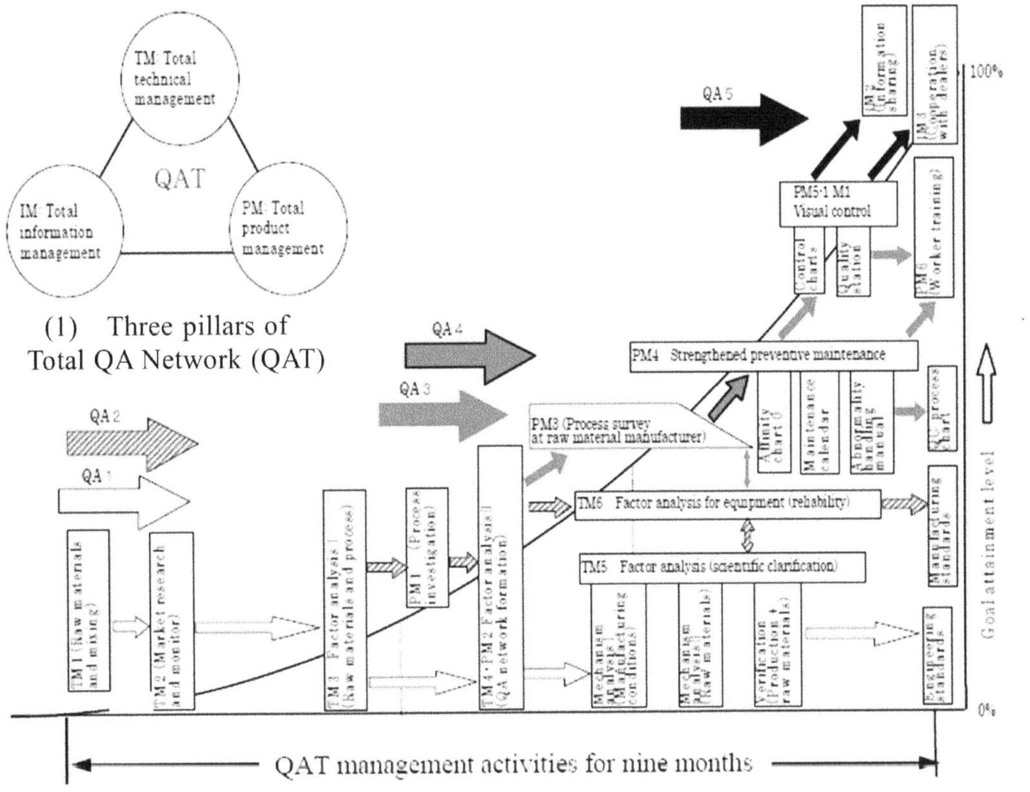

(1) Three pillars of Total QA Network (QAT)

(2) "Mountain-Climbing for Problem-Solving" approach by TTMT activity

2. "Mountain-Climbing for Problem-Solving" approach by TTMT activity

To promptly optimize the "Total QA Network" (QAT) of Figure 11(1) systematically and organically, the "SQC Technical Methods" (Amasaka, 2003), which is popularly used as the "Mountain Climbing of Problem Solution" (Figure 11(2)) is applied. Various types of arrows in the figure represent team activities of QA1 through QA5 respectively.

(2-1) Total Technical Management activity

Analyses of each raw material (TM1 in Figure 11(2)) and market research (TM2 in Figure 11(2)) were conducted, and factor analysis based on results (sensitivity analysis) was performed to screen the material in a short period.

In the factor analysis I (TM3 in Figure 11(2)) using principal component analysis, for example, it was found that the embedded material grain size and the inorganic fiber diameter are related to the abnormal sound characteristic and wear respectively as shown in Figure 12. Each "a" region where abnormal sound or wear is very conspicuous, while each "b" represents a region where its influence remains. Each "c" has been found as a region where each "b" represents a region where its influence remains. Each "c" has been found as a region where both characteristics are not contradictory.

Figure 12. Example of analyzed influences

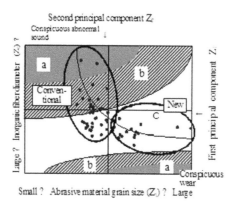

Figure 13. Causal relation between alternative characteristic of squeal and thermoforming temperature of raw material properties

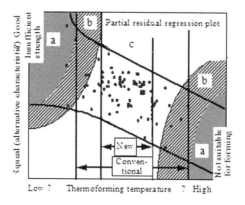

Factor analysis I has shown variation in the production process conditions (PM1 in Figure 11(2)). For the abrasive components (inorganic fibers and hard fine particles), technical analysis such as electron microscopy to observe the states of dispersion in each raw material and the pad was used in combination, to verify the factorial effect.

As a result, the mechanism of braking performance variation due to dispersion in inorganic fiber production has been clarified to enable improvement in cooperation with the material manufacturers. In the production preparation stage, product drawings were drafted using the "QA Network table" based the market quality requirement on (TM4 in Figure 11(2)) and equipment drawings were drafted using QAT based on the production quality requirement (PM2 in Figure 11(2)).

Important factors relating to raw material acceptance, production process conditions and state management were sorted out at this stage. For the sorted-out factors, the phenomena were analyzed (TM5 in Figure 11(2)) to determine the equipment condition quantitatively and the product drawing tolerances were determined by combined use on the design experiments and multivariate analysis for scientific optimization.

For example, use of the partial residual regression plot has clarified that the thermoforming temperature has a causal relation with the alternative characteristic of squeal, as shown in Figure 13. Each "a" represents a region where the strength is insufficient or a region not suitable for forming, while each "b" representing a region where the influence remains. In consideration of the strength and formability, region "c" has been determined as the control conditions of thermoforming process.

To ensure control conditions shown by region "c", factor analysis on the equipment side (TM6 in Figure 11(2)) for the molding die temperature dispersion, for example, was conducted, thus enabling unification of the pad forming temperature.

(2-2) Total product management activities

To attain the process conditions clarified by "Total technical management", the defects causes were studied by process investigation (PM1 in Figure 11(2)) and the relationships with the quality characteristics were checked by QAT (PM2 in Figure 11(2)).

Furthermore, to complete QAT, the brake unit manufacturer, brake pad manufacturer and raw material manufacturers formed a "Total Task Management Team" for mutual quality reviewing (PM3 in Figure 11(2)).

As the result, the current process capability was clarified with summarized recurrence and release prevention measures, leading to strengthened preventive maintenance (PM4: maintenance calendar in Figure 11(2)), visual control (PM5: Inline SQC in Figure 11(2)) and worker training (PM6: abnormal quality handling manual in Figure 11(2)).

(2-3) Total information management activities

A "Quality check station" (IM1 in Figure 11(2)) was formed to enable the process information as the result of "Total product management" to be seen immediately.

The route for making the market information (DAS: Dynamic Assurance System for high reliability and quality) (Sasaki, 1972) owned by the final assembly manufacturer available to parts manufacturers (IM2 in Figure 4(2)) and the route for obtaining the actual product information from dealers (IM3 in Figure 11(2)).

3.   Effect

The effect of the QCD research is remarkable through practical use of QAT as follows;

- Estimated market claim ratio reduction to 1/6 (from 2.6% to 0.4%)
- In-process fraction defective: down 60% (from 0.5% to 0.2%)
- Short convergence of initial failures: (from 9 months to 3 months)
- Cost reduction: 9.4% (156 Yen/unit)

Other Cases

The author was able to apply the TTMT activity to critical QCD studies of auto- manufacturing, including predicting and controlling the special characteristics (Toyota Motor Corp., 1993; Amasaka, Ed., 2000; Amasaka, 2004, 2015, 2017b, 2022a,b, 2023).

(i)   Preventing vehicles' rusting (Amasaka, 2004),
(ii)   Simultaneous fulfillment of QCD for improving automotive chassis (paint corrosion resistance and welding process) (Yamaji and Amasaka, 2009),
(iii) Joint task team activities between Toyota Motor Corporation and Toyota Motor Thailand (Amasaka, Ed., 2012), and others

CONCLUSION

In this chapter, the author has introduced the "Partnering of auto-maker and suppliers to strengthen TPS and TQM activity" in "Japan Supply System" for realizing "Excellent QCD studies" through the Toyota's typical "Quality management platform among Toyota Group. Specifically, the author has developed the "Total QA Network" (QAT) by Toyota's "Total Task Management Team" (TTMT) activity for realizing "investigation of the mechanism of the brake pad noise and quality assurance". As a result, the author has improved the bottleneck problems of worldwide auto-manufactures through realizing simultaneous QCD fulfilment.

REFERNCES

Amagai, M., & Amasaka, K. (2003). A study for the establishment of the quality management performance measure, *Proceedings of the 17th Asia Quality Symposium, Beijing Jiuhua Country Villa, Beijing, China,* 243-250.

Amasaka, K. (1988). Concept and progress of Toyota Production System (Plenary lecture). The Japan Society of Precision Engineering, The Japan Society for Design & Drafting and Hachinohe Regional Advance Technology Promotion Center Foundation and others Hachinohe, Aomori-ken, Japan.

Amasaka, K. (1989). TQC at Toyota: Actual state of quality control activities in Japan (Special lecture). The 19th Quality Control Study Team for Europe, Union of Japanese and Engineers.

Amasaka, K. (1999a). *TQM at Toyota-Toyota's TQM Activities: to create better car (Special Lecture). A Training of Trainer's Course on Evidence-based on Participatory Quality Improvement, International Health Program (TOT Course on EPQI), Tohoku University School of Medicine*. WHO Collaboration Center.

Amasaka, K. (1999b). *The TQM responsibilities for industrial management in Japan - The research of actual TQM activities for business management*. The Japanese Society for Production Management, The 10st Annual Technical Conference, Kyushu-Sangyo University, Fukuoka, Japan.

Amasaka, K. (1999c). A study on "Science SQC" by utilizing "Management SQC": A demonstrative Study on a New SQC concept and procedure in the manufacturing industry. *International Journal of Production Economics, 60-61*, 591–598. doi:10.1016/S0925-5273(98)00143-1

Amasaka, K. (2000). *Partnering chains as the platform for Quality Management in Toyota*. Proceedings. of the 1st World Conference on Production and Operations Management, Seville, Spain.

Amasaka, K. (Ed.). (2000a). Chapter 7: The aim of "Science SQC"– Deployment as a new TQM methodology from a management perspective. Japan Standards Association Nagoya.

Amasaka, K. (Ed.). (2000b). Science SQC: Revolution of Business Process Quality. Study group of Nagoya QST Research, Japanese Standards Association, Tokyo.

Amasaka, K. (2001). Quality management in the automobile industry and the practice of the joint activities of the auto-maker and the supplier. A change in the order system such as a part in the manufacturing industry: The condition of the supplier continuance in the automobile industry. Japan Small Business Research Institute.

Amasaka, K. (2002). Intelligence production and partnering for embodying a high-quality assurance (Plenary Lecture). The Japan Society of Mechanical Engineers, The 94th Course of JSME Tokai Branch, Nagoya, Japan.

Amasaka, K. (2003). Proposal and implementation of the *"Science SQC"* quality control principle. *Mathematical and Computer Modelling, 38*(11-13), 1125–1136. doi:10.1016/S0895-7177(03)90113-0

Amasaka, K. (2004a). *Science SQC, new quality control principle: The quality strategy of Toyota*. Springer -Verlag. doi:10.1007/978-4-431-53969-8

Amasaka, K. (2004b). *Establishment of performance measurement toward strategic quality management: Key to successful implementation of new management principle. Proceedings of the 4th International Conference on Theory and Practice in Performance Measurement*, Edinburgh International Conference Centre, UK.

Amasaka, K. (2004c). *Establishment of strategic quality management–Performance Measurement Model – Key to successful implementation of Science TQM (Part 6)*. The Japan Society for Production Management.

Amasaka, K. (Ed.). (2007a). New Japan Model–"Science TQM": Theory and practice for strategic quality management. Study group of the ideal situation the quality management of the manufacturing industry, Maruzen, Tokyo.

Amasaka, K. (2008a). Simultaneous fulfillment of QCD - Strategic collaboration with affiliated and non-affiliated suppliers. *New Theory of Manufacturing – Surpassing JIT: Evolution of Just-in-Time*. Morikita-Shuppan.

Amasaka, K. (2008b). Strategic QCD studies with affiliated and non-affiliated suppliers utilizing New JIT, Encyclopedia of Networked and Virtual Organizations, Information Science Reference, Hershey, New York, III(PU-Z), 1516-1527.

Amasaka, K. (2009). Establishment of Strategic Quality Management-Performance Measurement Model "SQM-PPM: Key to successful implementation of Science TQM,". *China-USA Business Review*, 8(12), 1–11.

Amasaka, K. (Ed.). (2012). Science TQM, new quality management principle, The quality management strategy of Toyota, Bentham Science Publishers, Sharjah, UAE, USA, THE NETERLANDS. doi:10.2 174/97816080528201120101

Amasaka, K. (2015). New JIT, new management technology principle, Taylor and Francis Group, CRC Press, Boca Raton, London, New York.

Amasaka, K. (2017a). *Toyota: Production system, safety analysis and future directions*. NOVA Science Publishers Inc.

Amasaka, K. (2017b). Strategic Stratified Task Team Model for Realizing Simultaneous QCD Fulfilment: Two Case Studies. *Journal of Japanese Operations Management and Strategy*, 7(1), 14–35.

Amasaka, K. (2022a). New Manufacturing Theory: Surpassing JIT (2nd Edition), Lambert Academic Publishing, Germany, Printed by Printforce, United Kingdom.

Amasaka, K. (2022b). *Examining a New Automobile Global Manufacturing System*. IGI Global Publisher. doi:10.4018/978-1-7998-8746-1

Amasaka, K. (2023). New Lecture-Toyota Production System: From JIT to New JIT, Lambert Academic Publishing, Germany, Printed by Printforce, United Kingdom.

Amasaka, K. (1995). The Q.A. Network activities for prevent rusting of vehicle by using SQC, *Journal of the Japanese Society for Quality Control, The 50th Technical Conference.*

Amasaka, K. (1997). The development of Total QA Network using Management SQC-Cases of Quality Assurance of Brake Pads. *Journal of the Japanese Society for Quality Control, The 55th Technical Conference*, 17-20.

Amasaka, K., Kurosu, S., & Morita, M. (2008). *New Manufacturing Theory: Surpassing JIT-Evolution of Just in Time*. Morikita-Shuppan.

Amasaka, K., & Osaki, S. (1999). The promotion of New SQC International Education in Toyota Motor - A proposal of "Science SQC" for improving the principle of TQM -, *The European Journal of Education on Maintenance Reliability. Risk Analysis and Safety*, 5(1), 55–63.

Amasaka, K., & Otaki, M. (1999). *Developing New TQM using partnering – Effectiveness of "TQM-P" employing Total Task Management Team*. The Japan Society for Product Management, The 10th Annual Conference, Kyushu Sangyo University, Fukuoka, Japan.

Boesen, T. (2000). Creating budget-less organizations with the balanced scorecard. Harvard Business School Publishing.

Chowdhury, S., & Zimmer, K. (1996). QS-9000 PIONEERS. Irwin Professional Pub., Chicago, and ASQC Quality Press, Milwaukee, Wisconsin.

Deming Prize Committee. (2003). *The bookmark of Deming Prize*. Union of Japanese Scientists and Engineers Press.

Doz, Y. L., & Hamel, G. (1998). *Alliance advantage*. Harvard Business School Press, Boston.

EFQM. (2003). *The EFQM excellence*. Zeeuw. (www.EFQM.org)

Furuya, Y. (1999). *TQM which creates Japan (2nd report): Results of the investigation of the questionnaire analysis team-The proposal of TQM which will aim at the 21 C based on the viewpoint of management*. Union of Japanese Scientists and Engineers.

ISO 9004. (2000). *Quality Management System- Guideline for performance*.

Japan Quality Award Committee. (2003). *Application form & instruction*. JQA Press.

Kozaki, T., Oura, A., & Amasaka, K. (2012). Establishment of TQM Promotion Diagnosis Model "TQM-PDM" for strategic quality management. *China-USA Business Review*, *10*(11), 811–819.

Miller, N. (1978). An analysis of disk brake squeal. *SAE Technical Paper 780332*.

Sasaki, S. (1972). Collection and analysis of reliability information in automotive industries. *The 2nd reliability and maintainability symposium*. Union of Japanese Scientist and Engineers.

Toyota Motor Corp. (1987). *Creation Unlimited, 50 Years History of Toyota Motor Corp.* Toyota.

Toyota Motor Corp. (1993). Toyota Technical Review: Special edition for SQC at Toyota. Toyota Motor Corp.

Toyota Motor Corp. (1999). *Toyota's TQM Activities-To creative better cars.* Toyota Motor Corp.

Umezawa, Y. (1994). Essential studies of re-engineering from organization theory. *Soshiki Kagaku*, *28*(1), 4–20.

Umezawa, Y., & Amasaka, K. (1999). Partnering as platform of quality management in Toyota Group, *Operations Research . Operations Research Society of Japan*, *44*(10), 560–571.

Wakaizumi, T. (2005). *Proof theory research on the Japanese quality management consultation model (A-QMDS) construction* [Master's thesis, Graduate school of Science and Engineering, Amasaka New JIT Laboratory].

Yamaji, M., & Amasaka, K. (2009). Strategic Productivity Improvement Model for white-collar workers employing Science TQM. *The Journal of Japanese Operations Management and Strategy*, *1*(1), 30–46.

Chapter 6
New JIT, New Management Technology Principle: Surpassing JIT

ABSTRACT

To be successful in the future, a global marketer must develop an excellent quality management system that can impress consumers and continuously provide excellent quality products in a timely manner through corporate management. Then, the author has established the new JIT, new management technology principle surpassing JIT for manufacturing in the 21st century. Specifically, new JIT contains a dual hardware and software system, as the next generation technical principle for the customer value creation.

THE KEY TO SUCCESS IN GLOBAL PRODUCTION

What Are the Critical Management Issues for Japanese Manufacturing?

In recent years, consumers (customers) have been selecting products that fit their lifestyles and their set of personal values. Consequently, a market environment was created in which customers strictly judge the reliability of manufacturers according to the reliability (quality and value gained from use) of their products. For this reason, it is not an exaggeration to say that manufacturers' success or failure in global marketing will depend on whether or not they are able to precisely grasp the customers' preferences and are then able to advance their manufacturing to adequately respond to the demands of the times (Amasaka, 1999, 2000a, 2003a, 2004a, 2008a).

This is being done in order to realize global production that will achieve the so-called "globally consistent levels of quality and simultaneous production worldwide (production at optimal locations)" ahead of other manufactures. Achieving this will allow the manufacturer to not be pushed out of the market (Amasaka, Ed., 2007a). In the midst of the drastic changes taking place at the manufacturing site due to use of digital engineering, the author can say that the reconstruction of world-leading, uniquely Japanese principles of management technology with administrative management technology, which

DOI: 10.4018/978-1-6684-8301-5.ch006

will be viable even for next-generation manufacturing, are urgently needed in order to keep up with this evolution in management technology (Amasaka, 1999, 2000a,b, 2008a).

This is the mission imposed on Japanese manufacturers today (Amasaka, Ed., 2007a). To accomplish this, it is imperative that management related departments, such as technical management, production management, sales management, and information technology, which make up the core of corporate management, closely cooperate with the administrative departments, such as personnel affairs, general planning, TQM promotion, and overseas business (Amasaka, 2004a, 2008a).

Furthermore, management-related departments also need to carry out strategic collaborations with on-site departments handling technical, production, and sales matters, as well as with suppliers (parts manufacturers) (Amasaka, 2004a; Amasaka, Ed., 2007a) and overseas suppliers (Yamaji and Amasaka, 2009a).

In-depth studies of the Toyota Production System called *JIT* and Lean Production, System, TQM, partnering and digital engineering will be needed when these methods are implemented in the future (Amasaka, 2002, 2004a, 2008a,b; Amasaka, Ed., 2012).

Above all, manufacturers endeavoring to become global companies are required to collaborate not only with affiliated companies, but also with non-affiliated companies to achieve harmonious coexistence among them based on cooperation and competition. In other words, a so-called "federation of companies" is needed (Amasaka, 2000b, 2004a, 2008a,b, 2009a, 2017a,b).

The Evolution of Administrative Management Technology is Now Being Demanded

In today's rapidly changing technological environment, to realize this, one of the first management technology issues that needs to be addressed in order to realize the "simultaneous achievement of QCD" and place the customer first, is the creation of a new development designing system capable of reforming the technological development business processes of the development designing (Amasaka, 2004a, 2008a,b).

Second, it is increasingly important for departments concerned with production to develop new production technologies and advance the "production management technology system" that enables global production.

Third, it is necessary for sales-related departments to concentrate their view on global marketing and to establish a "new marketing system" that breaks away from the conventional systems in an effort to strengthen ties with the customers.

It is becoming increasingly necessary, moreover, even from the standpoint of assuring "corporate reliability", for the general administration departments and management-related departments to cooperatively establish and implement a new administrative management technology which enhances the business processes of all departments involved in corporate management (Amasaka, 2004a, 2008a,b) (See to Chapter 4 and 5).

The above statements also apply, without exception, to Toyota's TPS, which has been adopted and further developed in various systems shared internationally, such as JIT or Lean Systems (Taylor and Brunt, 2001), and therefore, it is no longer Toyota's exclusive technology (Amasaka, 2009a, 2002, 2017b). In the United States as well, the importance of quality management has been increasingly recognized though studies of Japanese TQM. TQM has been actively promoted, thus encroaching on the quality superiority that Japanese products have previously enjoyed (Gabor, 1990; Joiner, 1994; Goto, 1999).

What can be deduced from these facts is that it is clearly impossible to continue to lead in the next generation of manufacturing simply by adhering to and maintaining the traditional, production-based "Japanese management technology". To overcome these problems, it is essential not only to advance TPS, a core technology of production processes, but to also establish a core technology for the service and sales, development designing, production, general administration, and management-related departments (Amasaka, 2002, 2004a; Amasaka, Ed., 2007a).

Furthermore, for the purpose of ensuring the "reform of the business process in all departments, as well as reinforcing inter-department cooperation, it is vital to reconstruct an intelligent system for sharing information. During the implementation stage of this system, it will be important to further upgrade the advantages of *Japan Supply System*, a cooperative system between the "assembly parts manufacturers and auto-manufacturers" that has long been a part of Japanese manufacturing, to a "Partnering Chain Management System on Platforms", as well as to establish the method of this system's operation (Amasaka, 2000b, 2004a, 2007a; 2008a,b, 2009b, 2009a, Amasaka, Ed. 2007a, 2012).

ESTABLISHMENT OF NEW JIT, NEW MANAGEMENT TECHNOLOGY PRINCIPLE

To reconstruct world-leading "next-generation manufacturing, the author states the developing "Japan's global manufacturing strategy 21C" as the global management technology strategy in the world (Amasaka, 2012a, 2015, 2021, 2022a,b, 2023a).

The aim of the "corporate management for realizing global manufacturing strategy" is to offer attractive products. To realize this goal, it is vital for all departments to share the same values toward work and improve the business quality of their work through internal and external cooperation in the world. This means because the existing corporate management activities are based on past successes brought about by each person's particular experience or skills, and it is not enough to encompass such diversified technologies. Furthermore, to create customer-oriented attractive products which can truly satisfy the customer's demands, a new management technology needs to be established.

Specifically, to strengthen global market creation strategy (See to Chapter 2 and 3), the author has established the "New JIT, new management technology principle-Surpassing JIT" employing "Science TQM, new quality management principle" based on the scientific TQM activity using "Science SQC, new quality control principle" (Amasaka, 2004a,b, 2008a,b: Amasaka, Ed., 2012) (See to Figure 4 in Chapter 2 and Figure 7 in Chapter 3) as shown in Figure 1 (Amasaka, 2002, 2004a, 2007a, 2008a, 2012a, 2014a,b, 2015, 2017b, 2021, 2022a,b, 2022a,b, 2023a; Amasaka Ed., 2007a, 2012).

New JIT would allow, technological development designing, production engineering / manufacturing, the advertisement / promotion and sales related departments, which are organically linked together by the management department, and office personnel / indirect department having the role of utilizing human resources in all departments and thus activating the organization, to improve the quality of their work. It is imperative that all these are linked with one another systematically and organizationally.

This principle rationalizes the high-linkage cycle for the improvement of business process of each department in a continuous circular ring employing "a dual methodology: IT (information technology) and Science SQC, new quality control principle (Amasaka, 2000a, 2003a, 2004b) (See to Figure 4 in Chapter 2).

Figure 1. New JIT, new management technology principle using total linkage of TDS, TPS, TMS, TIS, & TJS

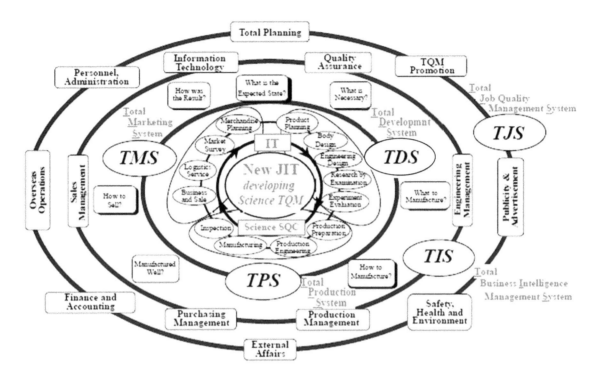

Grasping the mission of each department involved in management technology, New JIT is composed of a structured the "Total Linkage of integrated 5 core elements (systems)" as the "first hardware system" of JIT strategy as follows;

\"Total Development System" (TDS), "Total Production system" (TPS), "Total Marketing System" (TMS), "Total Business Intelligence Management System" (TIS) and "Total Job Quality Management, System" (TJS) as shown in Figure 2, so that each department is equipped with the core technology and linked with one another cooperatively. Each of TDS, TPS, TMS, TIS and TJS consists of organized "4 sub-core elements (a)-(d)" as follows;

1. The significance of TDS is to create the optimum product design based on the common knowledge by shared use of information. TDS is to systematize the development design methodologies through the (a) design based on the internal and external information with stress laid on design philosophy, (b) development-design management aimed at a reasonable design process, (c) creating general solutions based on the most advanced design technologies, and (d) clarifying the design behavior based on the design policy of a development designer (: theory-action-decision making).

2. The main objective of TPS is process management laying stress on customers and employees to realize working environment leading to skill improvement. To improve the reliability of the entire production process, TPS is composed of indispensable sub- core elements: (a) production based on information, (b) production based on management, (c) production based on engineering, and (d) production based on workshop formation.

Figure 2. Outline of new JIT using integrated five core elements

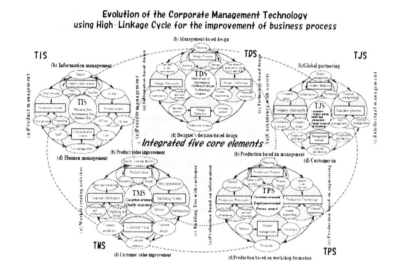

3. Similarly, TMS is to develop quality management to be relied on by customers through scientific marketing and sales not sticking to conventional concept. To realize quality assurance with an emphasis on the customer, TMS is composed of strategic sub-core elements: (a) market creating activities, (b) product value improvement, (c) building ties with customer, and (d) customer value improvement.

4. The aim of TIS has a function of new management technology system for the development design, production and sales departments in the inner circle by linkage with the indirect office department in the outer circle. TIS is composed of organized sub-core elements: (a) product management, (b) information management, (c) process management, and (d) human management based on the integrated cooperative activities.

5. Similarly, TJS has a function for improving intellectual productivity by employee training and internal/external partnering to strengthen global marketing. TJS is composed of intellectual sub-elements (a) coexistence with society, (b) global partnering, (c) intellectual management through human resource development, and (d) customer-in management activity, to grasp the importance of cooperative creation activity.

The development of New JI and validity using a dual hardware system "TDS, TPS, TMS, TIS & TJS" which has the following "Advanced TDS, TPS, TMS, TIS & TJS" are detailed to the references of Amasaka (2007b, 2011a,b, 2012a,b, 2014a,b, 2015, 2018a, 2019a, 2021, 2022a,b,c,d, 2023a,b; Sakai and Amasaka, 2008).

DEVELOPING ADVANCED TDS, TPS, TMS, TIS, AND TJS IN NEW JIT STRATEGY

To develop New JIT strategy, the author has created a high linkage of advanced management system "Advanced TDS, TPS, TMS, TIS & TJS" named "New Japan Global Manufacturing System" (NJ-AGMS) as the "second hardware system" that impresses users for providing high-value-added products using a triple software system "Science SQC, Strategic Stratified Task Team Model (SSTTM), and Patent Value Appraisal Model"(JS-PVAM) (See to Chapter 2, 11 and 12) as shown in Figure 3 (Amasaka, 2019a, 2021, 2022a,b,c,,d 2023a,b);

1. Advanced TDS, strategic development design system-In Figure 3-1, to tackle this issue, the author has created the "Advanced TDS, strategic development design model" using "organized four sub-core elements (i)-(iv)" for further updates TDS. Specifically, the author has created the (i) Intelligence Product Design Management System so as to (ii) create the High Reliable Development Design System, thereby (iii) eliminating prototypes with accurate prediction and control by means of Intelligence Numerical Simulation. Moreover, it is important to introduce the (iv) Intellectual Technology System which enables a sharing of knowledge and the latest technical information possessed by all related divisions.

Figure 3. High linkage of advanced management model "Advanced TDS, TPS, TMS, TIS & TJS" for developing New JIT strategy

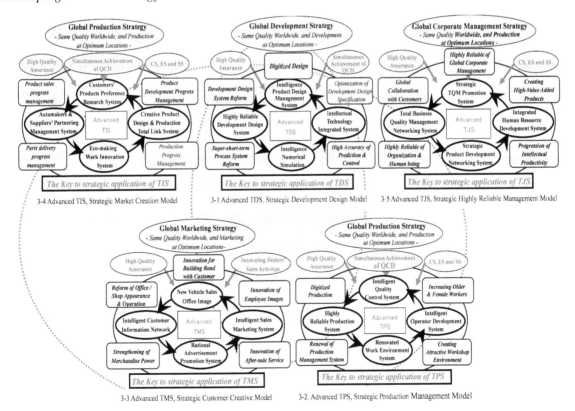

2. Advanced TPS, strategic production management system-In Figure 3-2, to solve this issue, the author has clarified the "Advanced TPS, strategic production management model" using organized "four sub-core elements (i)-(iv)" for further updates TPS.

Specifically, the author has created the (i) strengthen process capability maintenance and improvement by establishing Intelligent Quality Control System, (ii) establish the Highly Reliable Production System for high quality assurance, (iii) realize the Renovating Work Environment System in order to enhancement intelligent productivity, and (iv) develop the Bringing up Intelligent Operators through the establishing Intelligent Production Operating System. Particularly, these objectives will achieve the higher-cycled business processes through joint efforts of production technology, production preparation, manufacture.

3. Advanced TMS, strategic customer creation system-In Figure 3-3, therefore, the author has established the "Advanced TMS, Strategic Customer Creation Model" using organized "four sub-core elements (i)-(iv)" for further updates TMS. Specifically, the author has created the (i) "New Sales Office Image" is important to achieve a high cycle rate for market creation activities by "innovation for bond building with the customer" and "reform of shop appearance and operation". In the practice stage, it is more important to develop the (ii) "Intelligent Customer Information Network", (iii) "Rational Advertisement Promotion System" and (iv) "Intelligent Sales Marketing System" improving "Customer information software application know-how".

4. Advanced TIS, strategic market creation system-In Figure 3-4, to tackle this issue, the author has created the "Advanced TIS, strategic market creation model" using "organized four sub-core elements (i)-(iv)" for the strategic application of further updates TIS. Specifically, the author has created the (i) "Customers' Products Preference Research System" for progress management of product planning, (ii) "Networking system of customer-data practical use" for progress management of product development and design, (iii) "Total Link System of global development and production" for progress management of SCM and logistics, and (iv) "Operating System of intellectual manufacturing" for progress management of production and sales & marketing.

5. Advanced TJS, strategic highly reliable corporate management system-In Figure 3-5, to enhance the "cooperative creation activity, intellectual productivity and human resource development" using "organized four sub-core elements" for the strategic application of further updates TJS. Specifically, the author has created the (i) "Propulsion system of strategic human-resources design" for realizing high reliability of a company, (ii) "Technical development system and propulsion system of intellectual property" for realizing high reliability of human / organization and progression of intellectual productivity, (iii) "Intellectual operator's training system" for realizing global production and partnering, and (iv) "Deployment system of strategic product advertisement" for publicizing high-value-added product.

A DUAL SOFTWARE SYSTEM, DRIVING FORCE IN DEVELOPING NEW JIT STRATEGY

Science SQC to Strengthen Science TQM in the Customer Science Development

Supplying products that satisfy consumers (customers) is the ultimate goal of companies that desire continuous growth. Customers generally evaluate existing products as good or poor, but they do not generally have concrete images of products they will desire in the future. For new product development in the future, it is important to precisely understand the vague desires of customers.

Specifically, developing "Customer Science principle" (CSp) as shown in Chapter 3 makes it possible to concretize customer desires "Wants" (Amasaka, 2002, 2005, 2018b) (See to Figure 2 in Chapter 3 in detail). To realize this, the further expansion of the New JIT, "Science SQC, new quality control principle" by using the four core principles shown in Chapter 2 and 3" is a new principle for a next generation quality management technique for the manufacturing business, aiming at providing a "general solution" for problem solving (Amasaka, 1999, 2000a, 2002, 2003a, 2004b, 2014a,b, 2015) (See to Figure 4 in Chapter 2 and Figure 3-6 in Chapter 3 in detail).

Science SQC consists of the "four core principle" as follows;

(i) First is "Scientific SQC" as the scientific quality control approach, (ii) Second is "SQC Technical Methods" as the methodology for problem solving, (iii) Third is "Total SQC Technical Intelligence System" (TTIS) as the integrated SQC network, and (iv) Fourth is "Management SQC" as the problems existing between departments and organizations, and verbalizes the implicit understanding inherent in the business process, thus further presenting it as explicit knowledge and as a general solution for the technical problem.

Recently, as the validity of CSp employing Science SQC, the author has constructed a dual methodology "Automobile Optimal Product Design Model" (AOPDM) and "Customer Science using Customer Information Analysis and Navigation System" (CS-CIANS) for strategic product development design (Amasaka, 2005, 2019b).

Strategic Stratified Task Team Model Employing "Partnering Performance Measurement Model" to Strengthen New JIT Strategy

To develop CSp using Science SQC, as a management technology strategy to enable smooth business management for realizing high-quality assurance, the author has created the "Strategic Stratified Task Team Model" (SSTTM) employing "Partnering Performance Measurement Model" (PPMM) as shown in Figure 4 (Amasaka, 2003a, 2004a, 2005, 2008b, 2017a) (See to Chapter 11and Chapter 12 in detail).

The expected role of SSTTM and benefits it provides are not limited to cooperation among the departments inside the company. It contributes to strengthening the ties among group manufacturing companies, non-group companies, and even overseas manufacturers.

This model consists of Task-1 to - 8 teams involving the group, department, division, field, whole company, affiliated companies, non-affiliated companies, and overseasaffiliates.

Figure 4. Strategic stratified task team model (SSTTM)

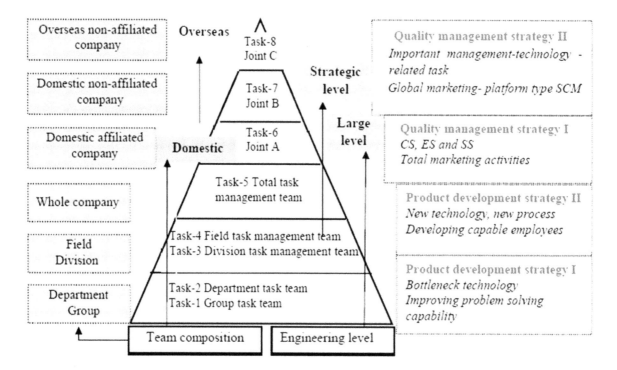

The level of problem-solving technology rises in product development strategy I and II through joint task teams of intra-company departments and divisions (Task-1 to -5, Group Task team, Department task team, Division task management team, Field task management team and Total task management team) in proportion with the improvement of stratified task level.

This technology is further expanded to quality management strategy I to II through the domestic affiliated company, domestic non-affiliated company, and overseas non-affiliated company (Task-6 to -8: Joint A to C).

Task-6 (Joint A) is aimed at establishing a collaboration with the group suppliers with whom domestic affiliated company has a capital tie-up, and Task-7 (Joint B) is aimed at a collaboration with suppliers that are not within its group. Task-8 (Joint C) is to strengthen cooperation with overseas suppliers.

As the application study of driving force of SSTTM in New JIT strategy, it is vital to reinforce Japanese-style partnering, or "Japan Supply Chain Management" between automobile manufacturers and parts suppliers (Amasaka, 2017a).

Specifically, against these backgrounds, the author has developed above PPMM (Partnering Performance Measurement Model) for strengthen "auto-assembly makers and suppliers" as the global SCM strategy (Yamaji et al., 2008).

As these results of developing a dual software system "Science SQC & SSTTM", the author has verified the effectiveness of typical studies of in New JIT strategy through Toyota's management technologies as follows;

(1) "Strategic Patent Value Appraisal Model" (SPVAM) in corporate strategy, (2) "Total Quality Assurance Networking Model" (TQA-NM) for new defect prevention, and (3) "New Global Partnering

Production Model (NJ-PPM) for overseas production (Amasaka, 2009, 2015, 2017b, 2022a,b) (See to Chapter 10 to 12 in Section 2).

Then, the author illustrates these results of New JIT strategy through the following "application examples".

APPLICATION EXAMPLES

In today rapidly changing management technology and environment, the mission of automobile manufacturers by developing New JIT is to be properly prepared for the "worldwide quality competition" so as not to be pushed out of market, and to establish new management technologies which enables them to offer "attractive highly reliable products" that are capable of enhancing customer value creation.

Developing Automobile Profile Design, Form, and Color Matching Support Methods

First, to raise the customers' delight the "*Affective Quality*" named "*Kansei Engineering*", the author has created an "Automobile Proportion, Form and Color Matching Support Methods" (APFC-MSM) using "Optimization of Profile Design (Proportion), Form & Color Matching Method" by psychographics approach in 7 Steps: (1) profile→(2) form→(3) color→(4) profile and form→(5) form and color→(6) profile and color→(7) profile, form and color based on an "Automobile Exterior Design Model" (AEDM) as shown in Figure 5 (Toyoda et al., (2015) and Amasaka (2018b) (See to Chapter xx in detail).

Actually, APFC-MSM has contributed the creation of a "prestige car LEXUS" & others as the pioneering "Auto-exterior design innovation" based on the *Kansei Engineering* using a dual methodology "CS-CIANS and Science SQC" by which all the sections collect the customer data "Wants" of the auto-dealers (Refer to Amasaka et al. (1999), Amasaka and Nagaya (2002) and Amasaka (2005, 2018a, 2022a,b,d, 2023a).

Figure 5. Automobile profile design, form and color matching support methods

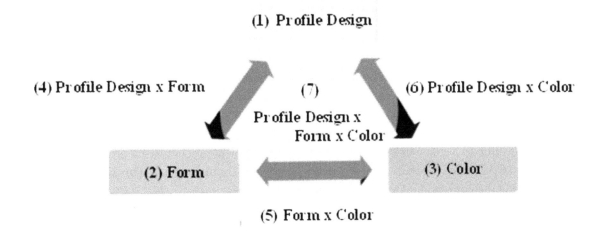

Creation of Automobile Product Design System
for the Bottleneck Technology Solution

Focusing on management technology for development and production processes, the author shows the "transitions in the automotive design and development process in Japan". For model changes in the past (development to production: approx. 4 years), after completing the designing process, problem detection and improvement were repeated mainly through the process of prototyping, testing, and evaluation as shown in Figure 6.

In some current automotive development, vehicle prototypes are not manufactured in the early stage of development due to the utilization of CAE and Simultaneous Engineering (SE), resulting in a substantially shorter development period.

Figure 6. Transitions in the automobile design and development processes in Japan

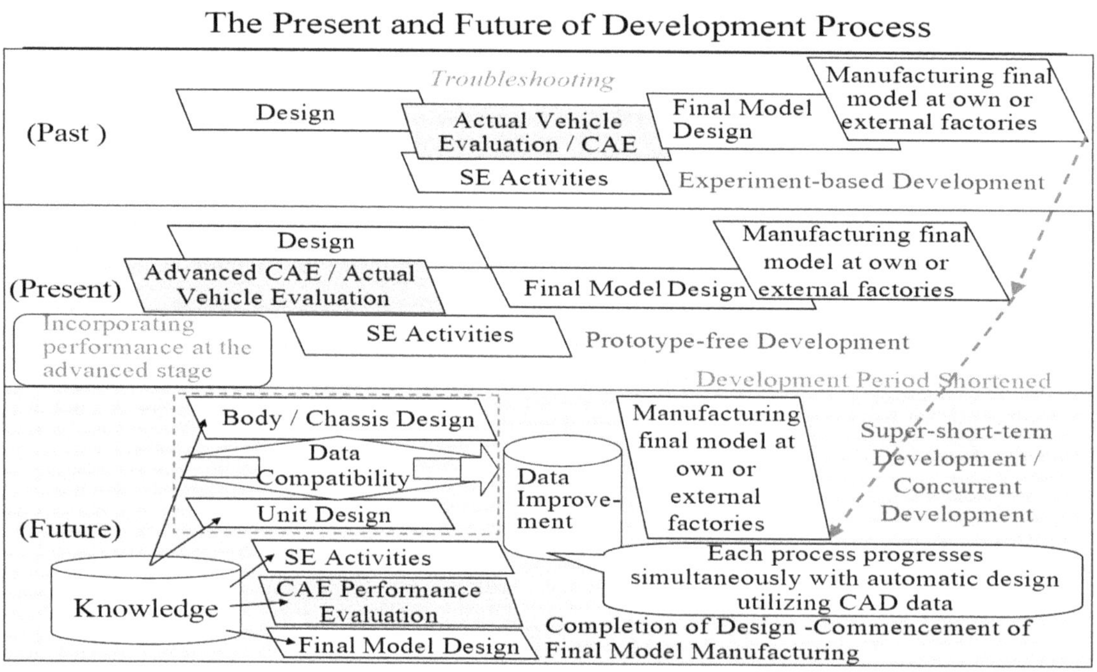

The Present and Future of Development Process

It is now possible to utilize CAE for comparative evaluation, rather than the conventional supplementary "observation" role during the testing of prototypes. This improvement means that CAE is utilized to the same extent as prototype testing.

The vehicle design and development and production process has been shortened to one year and there has been a transition to a super short-term concurrent development process based on the utilization of CAE and Solid CAD, allowing individual processes to progress simultaneously. This would be virtually impossible using the conventional repetitive testing of prototypes.

To solve the global bottleneck technologies, the author has created the "Intelligence CAE Management Approach System" as shown in Figure 7 (Refer to Amasaka (2007c,d, 2008c, 2009b), Amasaka Ed. (2007b, 2012) and Amasaka et al. (2012) in detail).

Specifically, the author has developed in 6 steps: ① Visualization→②Mechanism→③Modeling→④Navigation CG (computer graphics)→⑤Numeric value simulation→⑥Evaluation, design and improvement.

Particularly, in most suitable "Modeling", the author has created a "Highly Precise CAE Technology Model" by the optimal CAE analysis with 5 Steps as follows; (i) Problem, (ii) Modeling, (iii) Algorithm, (iv) Theory and (v) Computer (calculation technology) (Refer to Amasaka (2007c,d, 2008c, 2010, 2012b) and Amasaka, Ed. (2007b) in detail).

Actually, this system has contributed to the improvement of the automobile bottleneck technologies in the world as follows; (i) oil leak, (ii) braking effect and noise, (iii) loosening bolts & nuts, (iv) aerodynamics of body lift, (v) modeling defect of seat urethane form, (vi) vibration of door outer mirror, and others (Amasaka, Ed., 2007b;

Figure 7. Intelligence CAE management approach system

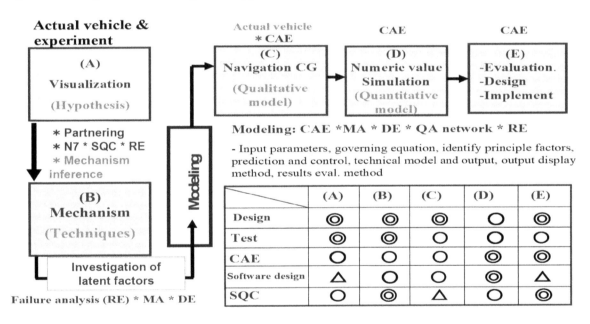

Innovation of Production Technology Employing the Connection of Whole Sections

To realize the product development of a super-short term in the global production competition, the connection of the production technological power by the whole sections of "Product development design, production engineering (by preparation in facilities) and Manufacturing (by production operators with excellent production skill) is very important.

Specifically, to proceeding of New JIT strategy, strengthening of the automobile highly reliable production becomes the most important of strategic development with digital engineering. To realize this, the author has established the "Toyota's Human Digital Pipeline System" (T-HDPS) employing a dual model "NJ-GPM and NJ-GMM" as shown in Figure 8 below (Yamaji et al., 2007; Amasaka and Sakai, 2010, 2011; Amasaka, 2015, 2020a,b, 2022b, 2023a; Amasaka et al., 2009; Amasaka, Ed., 2019) (See to Chapter 9).

Actually, in Figure 8, the author has carried out the image training of production processes in actual order of assembly automobile operations in order to the innovation of production planning. In addition, to intellectual training, this is done at the preparatory stage before beginning mass production without having a real product on hand.

T-HDPS promotes the leveling of the workload of operators in each process, and then completes the building up of production line even before launching it using "Total Linkage System" (TLS) of intellectual production information through the product design, production engineering and manufacturing (Sakai and Amasaka, 2007, 2008).

First, specifically, T-HDPS creates and supplies in advance "Standard Work Sheets" on which production operators have recorded each task in the correct order for jobs such as assembly work, by using design data for new products and facilities prepared from design through to production technology, even if there are no production prototypes.

Second, T-HDPS enables visualization training for machining processes step-by-step in the order that parts are built up, even if the actual product does not yet exist.

This system is proving to be very effective in raising the level of proficiency for processes requiring skills and capabilities at the production preparation stage.

Figure 8. Toyota's human digital pipeline system (T-HDPS)

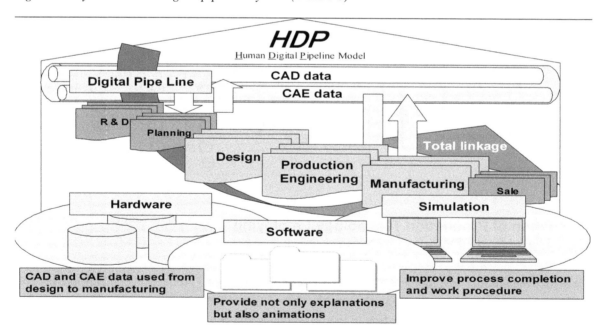

Development of Strategic Sales Marketing System

To offer a customer-oriented marketing, the author has created a "Modeling of Strategic Marketing System" (MSMS) named "Scientific Customer Creative Model" (SCCM) in Japan as shown in Figure 9 (Refer to Amasaka et al. (2005, 2007c, 2009c, 2011, 2015, 2020a,b, 2023a,b; Yamaji and Amasaka (2009b); Amasaka, Ed., 2012) in detail).

Specifically, MSMS consists of 4 core elements: "New Sales Office Image Model (NSOIM), Intelligent Sales Marketing System (ISMS), Rational Advertisement Promotion System (RAPS), and Intelligent Customer Information Network System (ICINS)".

Actually, as a way of Toyota's boosting marketing efforts, the author has created the "Customer Purchasing Behavior Model" (CPBM) and "Intelligent Customer Information

Marketing Model" (ICIMM) using "Scientific Mixed Media Model" (SMMM), and has checked that the visit rate to a dealer by Mixed Media (TV-CM, newspaper, radio, flyer, magazine, DM, poster, train-car ad., Internet, etc.) in "New car sales was 5 times to the effect of TV-CM alone (Amasaka, 2003b; Yamaji and Amasaka, 2009; Yamaji et al., 2010;Amasaka et al., 2013; Ishiguro and Amasaka, 2012a,b; Ogura et al., 2013, 2014).

Figure 9. Modeling of strategic marketing system (MSMS)

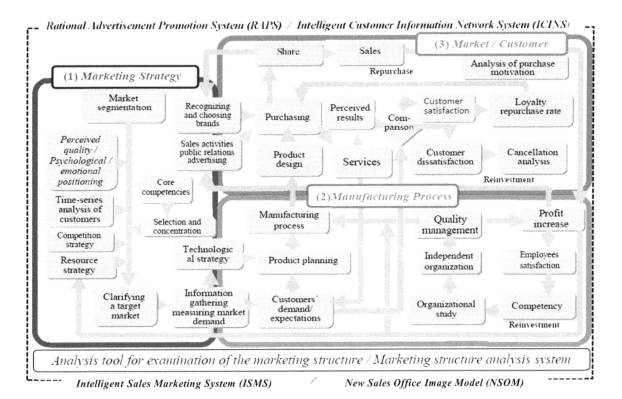

Then, the author has researched the "Customer Satisfaction (CS) and Customer Loyalty (CL) as a way of boosting marketing effectiveness, clarifying the key factors that comprise customer loyalty, and help improve the marketing strategy (Refer to Okutomi and Amasaka (2013) in detail).

As a result, the author has established the "Toyota's New Automobile Sales Marketing System" (T-NASMS) based on a dual system "VUCKMIN and TDMM" (Video Unites Customer Behavior and Maker's Designing Intentions and Total Direct Mail Model) by innovating auto-dealers' sales activities for customer creation (Amasaka, 2011; Yamaji et al., 2010; Ishiguro and Amasaka, 2012a,b). The achievements of the author's studies are currently being applied at Netz Chiba and other dealers (Amasaka, 2015, 2022a, 2023a,b).

Developing Strategic Productivity Improvement Model "NGP-PM" for Global Production Strategy

Recently, the global manufacturing companies in the world are endeavoring to grasp the information on human resources, take hold of the work, and formulate the vision in order to compete at a higher level. Therefore, to strengthen the above development of "NGP-PM" (New Japan Partnering Production Model) (See to Chapter 9), the author has established the "Strategic Productivity Improvement Model" (SPIM) so-called Toyota's "NJ-QMM" (New Japan Quality Management Model) for global production strategy as shown in Figure 10 (Yamaji and Amasaka, 2007, 2008).

In Figure 10, the following are the functions of the white-collar sections as corporate environment factors for succeeding in "global marketing for customer value creation":

1) "High quality production" based on the 2) "CS, ES and SS", and as a strategic factor to realize it, in order to realize the 3) "High productivity by evolution of business process" based on the "Development of Human resources" of "White-collar workers". By these, 4) "Global production" is realized.

Figure 10. Outline of strategic productivity improvement model (SPIM)

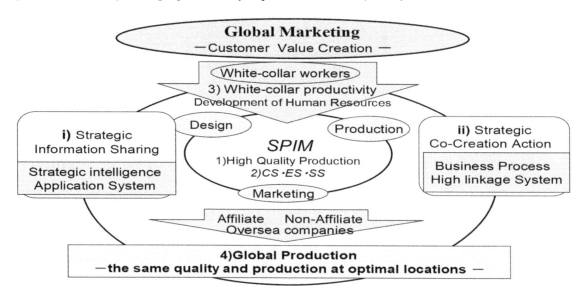

So, the "same quality and production at optimal locations" are achieved. For that purpose, the highest priority was given to the i) "Strategic information sharing" and ii) "Strategic co-creative action" so that a dual production management technology "Strategic Intelligence Application System (SIAM) and Business Process High Linkage System (BPHIS)" can effectively function.

When implementing a high cyclization of a business flow that consists of the setup of management policy, creation of a business plan, budget establishment, business deployment, optimal workforce distribution, task management, and evaluation, the relevant information needs to be shared among many departments and they need to also grasp the numerical values that show company-wide trends.

In that way, upon confirming abnormal numerical indications, the problem can be identified and solved at an early stage. The validity of SPIM is verified through application cases at the successful companies employing "NJ-GPM and NJ-GMM" in Chapter 10 below (Yamaji et al., 2009; Amasaka, 2015, 2017, 2020a,b, 2022a,b, 2023a).

CONCLUSION

In this chapter, the author has established the New JIT, new management technology principle to realize auto-manufacturing of the 21st century as the surpassing JT. Specifically, New JIT consists of the 5 hardware elements "TDS, TPS, TMS, TIS & TJS" for realizing market value creation. Concretely, moreover, the author has developed the High Linkage Model "Advanced TDS, TPS, TMS, TIS & TJS" for the advanced management strategy employing 3 software elements "Science TQM with Science SQC, SSTTM and SPVAM" developing CSp for expanding "uniform quality worldwide and production at optimum locations". The validity of New JIT has been verified based on the author's applications conducted in Toyota.

REFERENCES

Amasaka, K. (1999). A study on "Science SQC" by utilizing "Management SQC": A demonstrative study on a new SQC concept and procedure in the manufacturing industry. *International Journal of Production Economics*, *60-61*, 591–598. doi:10.1016/S0925-5273(98)00143-1

Amasaka, K. (2000a). A demonstrative study of a new SQC concept and procedure in the manufacturing industry: Establishment of a new technical method for conducting Scientific SQC. *Mathematical and Computer Modelling*, *31*(10-12), 1–10. doi:10.1016/S0895-7177(00)00067-4

Amasaka, K. (2000b). *Partnering chains as the platform for quality management in Toyota*. Proceedings of the 1st World Conference on Production and Operations Management, Sevilla, Spain.

Amasaka, K. (2002). *New JIT*, a new management technology principle at Toyota. *International Journal of Production Economics*, *80*(2), 135–144. doi:10.1016/S0925-5273(02)00313-4

Amasaka, K. (2003a). Proposal and implementation of the *Science SQC* quality control principle. *Mathematical and Computer Modelling*, *38*(11-13), 1125–1136. doi:10.1016/S0895-7177(03)90113-0

Amasaka, K. (2003b). *A demonstrative study on the effectiveness of "Television Ad." for the automotive sales.* Japan Society for Production Management, The 18th Annual Conference, Nagasaki, Japan.

Amasaka, K. (2004a). Development of *"Science TQM"*, a new principle of quality management: Effectiveness of Strategic Stratified Task Team at Toyota. *International Journal of Production Research, 42*(17), 3691–3706. doi:10.1080/0020754042000203867

Amasaka, K. (2004b). *Science SQC, new quality control principle: The quality control principle: The quality strategy of Toyota.* Springer-Verlag Tokyo. doi:10.1007/978-4-431-53969-8

Amasaka, K. (2005). Constructing a *Customer Science* application system *"CS-CIANS"*-Development of a global strategic vehicle *"Lexus"* utilizing *New JIT. WSEAS Transactions on Business and Economics, 3*(2), 135–142.

Amasaka, K. (2007a). *New Japan Production Model*, an advanced production management principle: Key to strategic implementation of *New JIT. The International Business & Economics Research Journal, 6*(7), 67–79.

Amasaka, K. (Ed.). (2007a). New Japan Model: Science TQM-Theory and practice of strategic quality management, Study group on the ideal situation on the quality management on the manufacturing, Maruzen, Tokyo.

Amasaka, K. (2007b). High linkage model *"Advanced TDS, TPS & TMS"*: Strategic development of *"New JIT"* at Toyota. *International Journal of Operations and Quantitative Management, 13*(3), 101–121.

Amasaka, K. (Ed.). (2007b). Establishment of a needed design quality assurance framework for numerical simulation in automobile production, Edited by Working Group No. 4 Studies in JSQC, Study group on simulation and SQC, Tokyo.

Amasaka, K. (2007c). The validity of *"TDS-DTM"*, A strategic methodology of merchandise-Development of *New JIT*, Key to the excellence design *"LEXUS". The International Business & Economics Research Journal, 6*(11), 105–115.

Amasaka, K. (2007d). Highly Reliable CAE Model: The key to strategic development of *New JIT. Journal of Advanced Manufacturing Systems, 6*(2), 159–176. doi:10.1142/S0219686707000930

Amasaka, K. (2007e). The validity of *Advanced TMS,* a strategic development marketing system utilizing *New JIT. The International Business & Economics Research Journal, 6*(8), 35–42.

Amasaka, K. (2008a). Science TQM, a new quality management principle: The quality management strategy of Toyota. *The Journal of Management & Engineering Integration, 1*(1), 7–22.

Amasaka, K. (2008b). Strategic QCD studies with affiliated and non-affiliated suppliers utilizing New JIT, Encyclopedia of Networked and Virtual Organizations, III(PU-Z), 1516-1527.

Amasaka, K. (2008c). An Integrated Intelligence Development Design CAE Model utilizing New JIT: Application to automotive high reliability assurance. *Journal of Advanced Manufacturing Systems, 7*(2), 221–241. doi:10.1142/S0219686708001589

Amasaka, K. (2008d). Proposal and validity of the High-Quality Assurance CAE Model for automobile development, *Journal of the Japanese Society for Quality Control, 38*(1), 38-44. doi:10.1142/S0219686709001614

Amasaka, K. (2009b). An Intellectual Development Production Hyper-cycle Model: *New JIT* fundamentals and applications in Toyota. *International Journal of Collaborative Enterprise, 1*(1), 103–127. doi:10.1504/IJCENT.2009.026459

Amasaka, K. (2009c). Effectiveness of flyer advertising employing TMS: Key to scientific automobile sales innovation at Toyota, *China-USA. Business Review (Federal Reserve Bank of Philadelphia), 8*(3), 1–12.

Amasaka, K. (2009d). Proposal and validity of Patent Value Appraisal Model "TJS-PVAM": Development of "Science TQM" in the corporate strategy. *The China-USA Business Review, 8*(7), 45–56.

Amasaka, K. (2010). Proposal and effectiveness of a High Quality Assurance CAE Analysis Model: Innovation of design and development in automotive industry, *Current Development in Theory and Applications of Computer Science. Engineering and Technology, 2*(1/2), 23–48.

Amasaka, K. (2011). Changes in marketing process management employing TMS: Establishment of Toyota Sales Marketing System, *China & USA. Business Review (Federal Reserve Bank of Philadelphia), 10*(7), 539–550.

Amasaka, K. (Ed.). (2012). Science TQM, new quality management principle: The quality strategy of Toyota, Bentham Science Publishers.

Amasaka, K. (2012a). *Strategic development of New JIT for global transforming management technology (Plenary Lecture)*. Computers and Materials, The 5th WSEAS International Conference on Sliema, Malta.

Amasaka, K. (2012b). Constructing Optimal Design Approach Model: Application on the Advanced TDS. *Journal of Communication and Computer, 9*(7), 774–786.

Amasaka, K. (2012c). *Prevention of the automobile development design, precaution and prevention.* Japanese standards Association Group, Tokyo.0000000df

Amasaka, K. (2014a). New JIT, new management technology principle. *Journal of Advanced Manufacturing Systems, 13*(3), 197–222. doi:10.1142/S0219686714500127

Amasaka, K. (2014b). New JIT, new management technology principle: Surpassing JIT [Special issues]. *Procedia Technology, 16*, 1–11. doi:10.1016/j.protcy.2014.10.128

Amasaka, K. (2015). New JIT, new management technology principle, Taylor and Francis Group, CRC Press.

Amasaka, K. (2017a). Strategic Stratified Task Team Model for realizing simultaneous QCD fulfilment: Two Case Studies. *Journal of Japanese Operations Management and Strategy, 7*(1), 14–35.

Amasaka, K. (2017b). *Toyota: Production system, safety analysis and future directions*. NOVA Science Publishers.

Amasaka, K. (2018a). Innovation of automobile manufacturing fundamentals employing New JIT: Developing Advanced Toyota Production System, *International Journal of Research in Business. Economics and Management*, 2(1), 1–15.

Amasaka, K. (2018b). Automobile Exterior Design Model: Framework development and support case studies. *Journal of Japanese Operations Management and Strategy*, 8(1), 67–89.

Amasaka, K. (Ed.). (2019). *The fundamentals of the manufacturing industries management: New Manufacturing Theory-Operations Management Strategy 21C*. Shankei-Sha.

Amasaka, K. (2019a). Studies on New Manufacturing Theory, *Noble. International Journal of Scientific Research*, 3(1), 42–79.

Amasaka, K. (2019b). Establishment of an Automobile Optimal Product Design Model: Application to study on bolt-nut loosening Mechanism, *Noble. International Journal of Scientific Research*, 3(9), 79–102.

Amasaka, K. (2020a). Studies on New Japan Global Manufacturing Model: The innovation of manufacturing engineering. *Journal of Economics and Technology Research*, 1(1), 42–71. doi:10.22158/jetr.v1n1p42

Amasaka, K. (2020b). Evolution of Japan Manufacturing Foundation: Dual Global Engineering Model Surpassing JIT. *International Journal of Operations and Quantitative Management*, 26(2), 101–126. doi:10.46970/2020.26.2.3

Amasaka, K. (2021). New Japan Automobile Global Manufacturing Model: Using Advanced TDS, TPS, TMS, TIS & TJS. *Journal of Advanced Manufacturing Systems*, 6(6), 499–523.

Amasaka, K. (2022a). New Manufacturing Theory: Surpassing JIT (2nd Edition). Lambert Academic Publishing.

Amasaka, K. (2022b). *Examining a New Automobile Global Manufacturing System*. IGI Global Publisher. doi:10.4018/978-1-7998-8746-1

Amasaka, K. (2022c). A New Automobile Global Manufacturing System: Utilizing a dual methodology, *Scientific Review. Journals in Academic Research*, 8(4), 41–58.

Amasaka, K. (2022d). A New Automobile Product Development Design Model: Using a dual corporate engineering strategy. *Journal of Economics and Technology Research*, 3(3), 1–21. doi:10.22158/jetr.v4n1p1

Amasaka, K. (2023a). New Lecture-Surpassing JIT: Toyota Production System-From JIT to New JIT. Lambert Academic Publishing.

Amasaka, K. (2023b). A New Automobile Sales Marketing Model for innovating auto-dealer's sales. *Journal of Economics and Technology Research*, 4(3), 9–32. doi:10.22158/jetr.v4n3p9

Amasaka, K., Ito, T., & Nozawa, Y. (2012). A New Development Design CAE Employment Model. *The Journal of Japanese Operations Management and Strategy*, 3(1), 18–37.

Amasaka, K., Kurosu, S., & Morita, M. (2008). *New Manufacturing Theory: Surpassing JIT-Evolution of Just in Time*. Morikita-Shuppan.

Amasaka, K., & Nagaya, A. (2002). Engineering of the new sensitivity in the vehicle: Psychographics of LEXUS design profile, Development of articles over the sensitivity-The method and practice. Japan Society of Kansei Engineering. Nihon Suppan Service Press, 55-72.

Amasaka, K., Nagaya, A., & Shibata, W. (1999). Studies on Design SQC with the application of Science SQC improving of business process method for automotive profile design [in Japanese]. *Japanese Journal of Sensory Evaluations*, *3*(1), 21–29.

Amasaka, K., Ogura, M., & Ishiguro, H. (2013). Constructing a Scientific Mixed Media Model for boosting automobile dealer visits: Evolution of market creation employing TMS. *International Journal of Business Research and Development*, *3*(4), 1377–1391.

Amasaka, K., & Sakai, H. (2010). Evolution of TPS fundamentals utilizing New JIT strategy – Proposal and validity of Advanced TPS at Toyota. *Journal of Advanced Manufacturing Systems*, *9*(2), 85–99. doi:10.1142/S0219686710001831

Amasaka, K., & Sakai, H. (2011). The New Japan Global Production Model "NJ-GPM": Strategic development of *Advanced TPS*. *The Journal of Japanese Operations Management and Strategy*, *2*(1), 1–15.

Ebioka, E., Sakai, H., Yamaji, M., & Amasaka, K. (2007). A New Global Partnering Production Model "NGP-PM" utilizing Advanced TPS. *Journal of Business & Economics Research*, *5*(9), 1–8.

Gabor, A. (1990). *The man who discovered quality; How Deming W. E., brought the quality revolution to America*. Random House, Inc.

Goto, T. (1999). *Forgotten Management Origin–Management Quality Taught by GHQ, CCS management lecture*. Seisansei-Shuppan.

Ishiguro, H., & Amasaka, K. (2012a). Proposal and effectiveness of a Highly Compelling Direct Mail Method – Establishment and Deployment of PMOS-DM. *International Journal of Management & Information Systems*, *16*(1), 1–10.

Ishiguro, H., & Amasaka, K. (2012b). Establishment of a Strategic Total Direct Mail Model to bring customers into auto-dealerships. *Journal of Business & Economics Research*, *10*(8), 493–500. doi:10.19030/jber.v10i8.7177

Joiner, B. L. (1994). *Fourth generation Management: The new business consciousness*. Joiner Associates, Inc.

Ogura, M., Hachiya, T., & Amasaka, K. (2013). A Comprehensive Mixed Media Model for boosting automobile dealer visits. *The China Business Review*, *12*(3), 195–203.

Ogura, M., Hachiya, T., Masubuchi, K., & Amasaka, K. (2014). Attention-grabbing train car advertisements. *International Journal of Engineering Research and Applications*, *4*(1), 56–64.

Okutomi, H., & Amasaka, K. (2013). Researching Customer Satisfaction and Loyalty to boost marketing effectiveness: A look at Japan's auto-dealerships. *International Journal of Management & Information Systems*, *17*(4), 193–200. doi:10.19030/ijmis.v17i4.8093

Sakai, H., & Amasaka, K. (2007). Human digital pipeline method using total linkage through design to manufacturing. *Journal of Advanced Manufacturing Systems*, *6*(2), 101–113. doi:10.1142/S0219686707000929

Sakai, H., & Amasaka, K. (2008). Demonstrative verification study for the next generation production model: Application of the Advanced Toyota Production System. *Journal of Advanced Manufacturing Systems*, *7*(2), 195–219. doi:10.1142/S0219686708001577

Taylor, D., & Brunt, D. (2001). *Manufacturing operations and supply chain management - Lean approach*, Thomson Learning, London.

Toyoda, S., Nishio, Y., & Amasaka, K. (2015). Creating a Vehicle Proportion, Form, and Color Matching Model. *International Organization of Scientific Research*, *III*(3), 9–16.

Yamaji, M., & Amasaka, K. (2007). Proposal and validity of Global Intelligence Partnering Model "GIPM-CS" for corporate strategy. *International IFIP TC 5.7 Conference on Advanced in Production Management System.* Springer.

Yamaji, M., & Amasaka, K. (2008). New Japan Quality Management Model: Implementation of New JIT for strategic management technology. *The International Business & Economics Research Journal*, *7*(3), 107–114.

Yamaji, M., & Amasaka, K. (2009a). Strategic Productivity Improvement Model for white-collar workers employing Science TQM. *The Journal of Japanese Operations Management and Strategy*, *1*(1), 30–46.

Yamaji, M., & Amasaka, K. (2009b). Proposal and validity of Intelligent Customer Information Marketing Model: Strategic development of *Advanced TMS, China - USA. Business Review (Federal Reserve Bank of Philadelphia)*, *8*(8), 53–62.

Yamaji, M., Hifumi, S., Sakalsiz, M. M., & Amasaka, K. (2010). Developing a strategic advertisement method "VUCMIN" to enhance the desire of customers for visiting dealers. *Journal of Business Case Studies*, *6*(3), 1–11. doi:10.19030/jbcs.v6i3.871

Yamaji, M., Sakai, H., & Amasaka, K. (2007). Evolution of technology and skill in production workplaces utilizing Advanced TPS. *Journal of Business & Economics Research*, *5*(6), 61–68.

Yamaji, M., Sakatoku, T., & Amasaka, K. (2008). Partnering Performance Measurement "PPM-AS" to strengthen corporate management of Japanese automobile assembly makers and suppliers. *International Journal of Electronic Business Management*, *6*(3), 139–145.

Chapter 7
New Automobile Product Development Design Model for Global Marketing

ABSTRACT

In recent years, customers have been selecting products that fit their lifestyles and their set of personal values. For this reason, manufacturers' success or failure in global marketing will depend on whether or not they are able to grasp precisely the customers' preferences and are then able to advance their manufacturing to adequately respond to the demands of the times. In this chapter, therefore, the author describes a new automobile product development design model (NA-PDDM) using a dual corporate engineering strategy for innovation of automobile product engineering fundamental employing TDS. Specifically, NA-PDDM contains both the Exterior design engineering strategy and Driving performance design engineering strategy by using customer science principle (CSP) developing advanced TDS. Concretely, the foundation of NA-PDDM consists of the automobile exterior design model with three core methods (AEDM-3CM), automobile optimal product development design model (AOP-DDM), and CSP-customer information analysis and navigation system (CSP-CIANS).

THE KEY TO SUCCESS IN AUTOMOBILE GLOBAL PRODUCTION FOR REALIZING CUSTOMER VALUE CREATION

Particularly, looking closely at the quality management issues "Unpopularity of appearance design quality and Recalls of driving performance" facing Japanese advanced auto-manufacturing industry in both of domestic and overseas, it has become clear that a new corporate management technology by focusing "Product development design strategy" is being strongly sought after (JD Power and Associates, 1998; Umezawa and Amasaka, 1999; Amasaka, 2002, 2007a,b; 2014; Nihon Keizai Shinbun, 2000, 2006, 2012).

The mission of automobile manufacturers (automakers) in this rapidly changing management technology environment is to be fully prepared for worldwide quality competition so as not to be pushed out of the market, and also to establish a new management technology model that enables them to offer highly value products of the latest design that are capable of enhancing customer value (Amasaka, 2003a, .2004a, 2005).

DOI: 10.4018/978-1-6684-8301-5.ch007

What Are the Critical Management Issues for Japanese Automobile Manufacturing?

In recent years, customers have been selecting products that fit their lifestyles and their set of personal values in marketing on the world. Furthermore, with the rapid move towards global production, it has become increasingly critical for manufacturers to drastically shorten the time it takes to move a product from "product development design to production" while ensuring quality of Customer Satisfaction (CS) (Amasaka and Nagaya, 2002; Amasaka et al., 2012).

Consequently, a market environment was created in which customers strictly judge the products' reliability (quality and value gained from use).

For this reason, to develop the marketing creation, it is not an exaggeration to say that automakers' success or failure in global marketing will depend on whether or not they are able to precisely grasp the customers' preferences and are then able to advance their production to adequately respond to the demands of the times (Amasaka, 2005, 2007a, 2011; Okutomi and Amasaka, 2013).

Particularly, the technological innovation of product development design is indispensable in order to realize strengthen global marketing that will achieve the globally consistent levels of quality and simultaneous production at optimal locations.

Necessity of Business Process Renovation in Automobile Product Development Design

To provide the attractive products with customer's orientation permanently, the establishment of "new development design technologies" to take customer's needs in advance is today's challenge and current issue. Therefore, to realize this, the renovation of the business process of automobile product development design becomes important employing TDS (Total Development System) (See to Figure 1 and 2 in Chapter 7).

Then, to strengthen the business process of product design, production and sales marketing, the author has created the "Intelligent Customer Information Marketing Model" (ICIMM) for TDS strategy as shown in the Figure 1 (Amasaka, Ed., 2007a,b, 2012).

In Figure 1, then, use this information to create "wants" as part of a market creation activity and also to establish an intellectual structure and system for development and production that is capable of offering new products.

In the implementation stage, it is important to apply the Science SQC based on the Science TQM developing "Advanced TDS, strategic development design system", via a verifiable scientific business approach, to each step "(1) Input information, (2) Information for development, and (3) Output information" of the business process of product development designing and manufacturing (Amasaka, 2004, 2007a; Amasaka, Ed., 2012) (See to Figure 4 in Chapter 2 and Figure 3 in Chapter 7).

This is done in order to effectively carry out to bring about the evolution of corporate management technology that can ensure high reliability business process which realizes the customer's "Wants" developing CSp (Customer Science principle) (Amasaka, 2002, 2005; Yamaji and Amasaka, 2009) (See to Figure 2 in Chapter 3).

To realize this, it is important to developing the "high Quality Assurance (QA), super-short-term product development design process and simultaneous achievement QCD (Amasaka, Ed., 2007a,b; Amasaka et al., 2012; Amasaka, 2017a, 2018). To carry out the above various subjects, there are two critical management issues in the Japanese product development design section to raise that position as a top-runner from now on, too.

Figure 1. Intelligent customer information marketing model (ICIMM) for TDS strategy

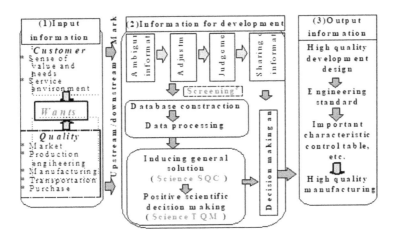

Design SQC using SQC Technical Methods realizing customers' wants

First, as customers' values become increasingly diverse, automobile exterior design is becoming one of the most critical elements influencing customer purchase behavior for automakers. Unfortunately, as people's values and subjective preferences become more varied and complex, it becomes increasingly difficult to accurately define their wants and needs.

Therefore, it is important for mapping up exterior design strategy to study on "what style of vehicles would sell in the future?" However, in many cases, automaker's vehicle designers do not have a clear idea of future vehicle styling (Amasaka et al., 1999).

Success of designing directly affects the sale of enterprises. Therefore, design business is established as a marketing strategy and its significance lies in the quality of the proposal.

True market-in should be in proposing a desirable thing before it is desired. From Figure 2, it is important for "Design SQC" to contribute to enhancing individual designer's proposing capability using "Science SQC, new quality control principle" (Amasaka, 1999a,b, 2003b, 2004b, 2018; Okazaki et al., 2000; Amasaka and Nagaya, 2002).

Conventionally, designing is generally developed directly to the profile design after analyzing the research itself (event analysis). To respond to such a trend of vehicle exterior design, it is essential to conduct a scientific approach whereby the objective preferences of customers' tastes are accurately grasped and reflected in the vehicle exterior and interior designing process for realizing "vehicle appearance quality".

Specifically, by employing both of the "affective engineering" named *"Kansei Engineering"* and statistical science named "Science SQC", the author has developed an automobile exterior and interior design tool "Intelligence Design Concept Method" (CSp-IDCM) for developing "Customer Science principle" (CSp) as shown in Figure 3 (Amasaka et al., 1999; Amasaka and Nagasawa, 2000; Amasaka, 2004, 2005, 2018).

Figure 2. Desirable relationship between "designing" and "design SQC"

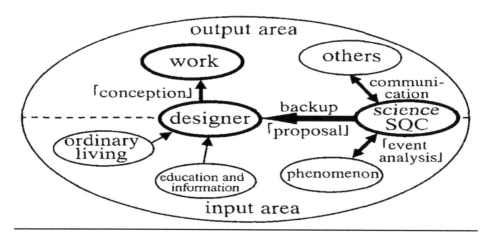

This tool is the objective of this study to find the guideline for establishing a method for scientifically supporting the designing so as to establish it expressly as a more creative activity from the state of tacit knowledge. It is considered that the very analysis process for establishing it as an activity would be the key to the successful conception making. It is necessary for us to create a particular live solution that catches the liking of the next generation.

Figure 3. Intelligence design concept method using CSp (CSp-IDCM)

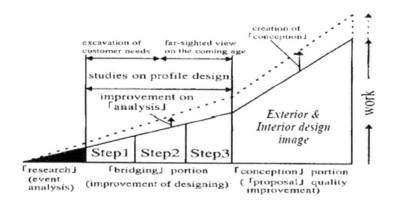

In this connection, "Science SQC" with a core method "SQC Technical Methods" is applied to the flow designing to actually enhance the quality of the designer's job named Design SQC as the scientific new methodology "Mountain-Climbing for Problem-Solving" using "Customer Science principle" (CSp) in Toyota (See to Figure 2 and 4 in Chapter 3).

In Figure 3, the author thinks that the analysis process that turns implicit knowledge into explicit knowledge constitutes the secret to the conception as the event analysis to the exterior design in three steps of researches from Step 1 to 3.

Step 1 analyzes relationship between images of vehicles desirable and those actually selected to research, and it actualizes apparent relevancy whereby a vehicle type can be specified by the desirable image.

Step 2 grasps what part of a vehicle customers observe to evaluate it. By coming down from the overall assessment, partial assessment and detailed assessment, this clarifies which design factor should better be given priority to satisfy customers.

In Step 3, the designers research the excellent exterior design by grasping the relevance of vehicle images and profile design (called proportion) data. CSp-IDCM will help improve the designing process (work) for creation of exterior and interior design image involving the matching profile design, form, and color optimization.

New Japanese product development design to realize the "simultaneous achievement of QCD"

Second, in the midst of rapid change of management technologies, a key challenge facing the automakers is important to develop the new Japanese product development design which provides the latest, highly reliable and customer-oriented products so that they can survive the worldwide quality competition.

Focusing on management technology for product development design and production processes, it is clear that there has been excessive repetition "trial-and-error" of prototyping, testing and evaluation for the preventing of "scale-up effect" in the bridging stage between product development design and mass production (Amasaka Ed., 2007a,b, 2012; Amasaka, 2007a,c, 2010a, 2017; Amasaka et al., 2012).

Above all, to realize the excellent "vehicle driving functional quality", the advanced automakers must establish the "intellectual evolution of the product development design" in order to realize the "simultaneous achievement of QCD".

Specifically, the author has developed a "Total QA High Cyclization Business Process Model" employing "Management SQC as the 4th core principle of Science SQC" as shown in Figure 4 (Amasaka, 2008) (See to Figure 2 and 6 in Chapter 2 and 3).

Figure 4. Total QA high cyclization business process model to realize the "simultaneous achievement of QCD

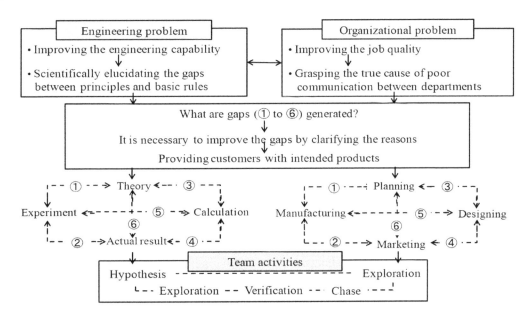

117

This model aims to shifting from the way of insisting on "business process management from experimental evaluation based on tests and prototypes" until now to the combination with "predictive evaluation" through the latest highly reliable numerical simulation (Computer Aided Engineering, or CAE". To minimize discrepancies in the results obtained from testing of actual products and CAE, it is necessary to properly formalize the expertise on the many technical analysis required for CAE analysis.

This model is created from the standpoint of verification and validation (divergence of CAE from theory and divergence of CAE from testing) in order to make possible highly reliable CAE analysis that is consistent with the market testing theory profile, which systematically and strategically realizes high quality assurance by incorporating analyses made via the core technologies of Science SQC (Amasaka, 2004b, 2007c, 2008, 2010).

Specifically, to solve the pending issue of a technology problem in the market, it is necessary to create a universal solution (general solution) by clarifying the existing six gaps (① to ⑥ in Figure 4) in the process consisting of theory (technological design model), experiment (prototype to production), calculation (simulation), and actual result (market) as shown on the lower left of Figure 4.

To accomplish this, the clarification of the six gaps (① to ⑥) in the business processes across the divisions, shown in the lower right of Figure C below, is of primary importance. By taking these steps, the intelligent technical information owned by the related divisions inside and outside the corporation will be fully linked, thus reforming the business processes involved in development design.

Concretely, to develop this model, the author has created the "Highly Reliable CAE Analysis Technology Component Model" with four components "problem-modeling-algorithm-theory- computer" as shown in Figure 5 was designed to make the shift from conventional prototype testing methods to effectively applying CAE in predictive evaluation methods.

The comprehensive issuance of this model is essential to achieving the desired shift (Amasaka, Ed., 2007b: Amasaka, 2008, 2010).

Figure 5. Highly reliable CAE analysis technology component model

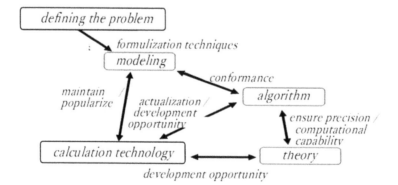

In Figure 5, the critical aspects of this model include (i) *defining the problem* (physically checking the actual item) in order to clarify the mechanism of the defect, using visualization technology to identify the dynamic behavior of the technical issue; (ii) full use of *formulization techniques* to generate logical *modeling* (statistical calculations, model application); (iii) constructing compatible *algorithms* (calculation methods); (iv) developing *theories* (establishing theories required to clarify problems) that ensure the precision of numerical calculations and sufficient computational capability; and (v) comprehensively putting the above processes in action using *computer* (selection of *calculation technology*).

Developing Customer Science principle aiming customers' demand scientific analysis

The mission of automakers is to offer products that please customers, and make this serves as a basis for sustainable growth.

In this new era, when product development design is required as a basis of global marketing, it is important to establish a "behavior science principle" for strategic product development which can dig deep into the customers' demands (needs, desires and wants), thus preempt the trend of times (Amasaka, 2002, 2003b, 2005, 2007b, 2017b, 2018).

Customers express their demands in words. Therefore, the product planners or designers engaging in product development design must properly interpret these words and draw up accurate plans accordingly.

The mentioning above CSp aims an automobile exterior design for customer value creation with a clear-cut styling concept based on research of psychographics by the viewpoint of customers' life stage and lifestyle described in Figure 2 of Chapter 3 (Amasaka, 2002, 2005, 2007b, 2018; Takimoto et al., 2010).

It is intended to indicate the desirable state of new business processes for creating wants indispensable to the development of attractive products (See to Figure 2 in Chapter 3 in detail).

ESTABLISHMENT OF A "NEW AUTOMOBILE PRODUCT DEVELOPMENT DESIGN MODEL" USING A "DUAL CORPORATE ENGINEERING STRATEGY"

To strengthen high quality assurance, super short-term products development design process and simultaneous achievement QCD for the "customer value creation", the author has established a "New Automobile Product Development Design Model" (NA-PDDM) using a "Dual Corporate Engineering Strategy" as shown in Figure 6.

NA-PDDM develops the CSp by employing "*Kansei Engineering*" and Science SQC (Amasaka & Nagasawa, 2000; Amasaka, 2004a,b, 2005).

Specifically, NA-PDDM contains a "Dual Corporate Strategy" with both of the "Exterior design engineering strategy and Driving performance design engineering strategy" for the developing "appearance quality and functional quality" that contributes to strengthening of Japanese automobile global corporate strategy.

Because of that realization, to develop the "Same quality worldwide and products development design at optimal locations", the foundation of NA-PDDM consists of the "Automobile Exterior Design Model with 3 Core Methods" (AEDM-3CM), "Automobile Optimal Product Development Design Model" (AOP-DDM) and CSp-CIANS as follows;

Automobile Exterior Design Model Employing Three Core Methods for Progressing Psychographics

To develop above the *Design SQC* using CS-IDCM for progressing *psychographics*, the author has conducted the "Advanced Exterior Design Project using Science SQC" named "ADS" (Advanced Design SQC) for raising customers worth in Toyota Motor Corp. in the following (Nunogaki et al., 1996; Amasaka et al., 1999).

Figure 6. A new automobile product development deign model using a dual corporate engineering strategy

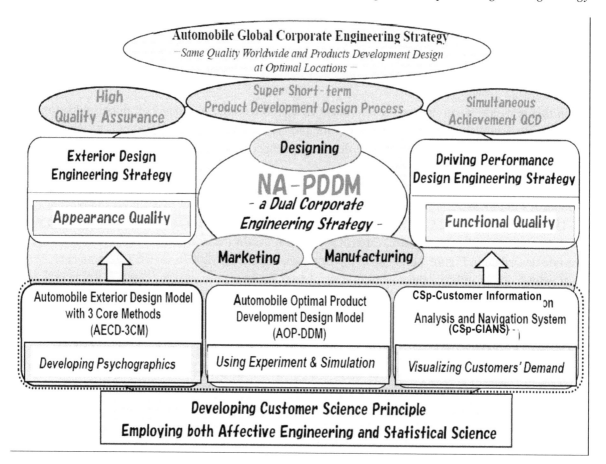

As for the development of ADS, the author as the chief examiner of TQM Promotion Div. (1992-2000). Then, "Amasaka's New JIT Laboratory" of Aoyama Gakuiin Univ. (2000- 2017) has organized the (i) Design Div. I of Vehicle Development Center I for the development of new world car "Lexus", Design Div. II of Vehicle Development Center II for other model change of various mid-size cars, and Toyota Design Laboratory Tokyo for Advanced design cars, (ii) Marketing Service Div., Dealer Marketing System Div., Auto-salon Amulux Tokyo, U.S. Office, Europe Office and others for the internal and external customers information gathering, and (iii) TQC Promotion Div. for developing Design SQC (Nunogaki, 1996; Nagaya et al., 1998; Amasaka et al., 1999).

In ADS projects, the author has developed the "Automobile Exterior Design Model with 3 Core Methods" (AEDM-3CM). This model combines 3 core methods as follows; (A) Improvement of "Business Process Methods for Automobile Profile Design" (BPM-APD), (B) Creation of "Automobile Profile Design using "Psychographics Approach Methods" (APD-PAM), and (C) "Automobile profile design, form and color matching support methods" (APFC-MSM) as shown in Figure 7 (Amasaka, 2018).

As developing CSp, the actual studies for AEDM-3CM have been applied to Toyota and others. In core technology (I), the first aim was to hold the characteristic of the profile design (proportion) such as "BMW518, Benz W123, Jaguar X16, etc." placed on the famous car of the world rationally by the concrete development of AEDM-3CM using "Design SQC and CS-IDCM".

Figure 7. Automobile exterior design model with three core methods (AEDM-3CM)

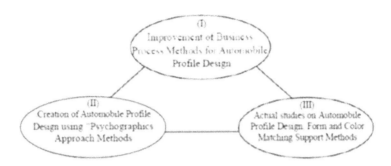

To realize the above knowledge, in core technology (II), the second aim was the realization of the profile design, which is the main elements of exterior design of Toyota's strategic prestige car "new-model, Lexus" surpassing BMW / Benz by using the "Psychographics" approach method. In core technology (III), the third aim was the realization of the profile design development of various mid-size cars by the application of Lexus exterior design development employing both of (I) and (II) (Refer to Amasaka (2018) in detail).

Specifically, then, the design work of (III) "APFC-MSM" (Automobile Proportion, Form and Color Matching Support Methods) which is the core technology of the strategic development of ADS projects at present is illustrated by developing the various psychographics approaches described in Figure 5 of Chapter 7 (Toyoda et al., 2015a; Amasaka, 2018).

In Figure 5 of Chapter 7, APFC-MSM starts with the three elements (1) profile design (proportion), (2) form, and (3) color.

Recent design work strategies make it a point to optimize business processes so that they are in line with the vehicle design concept from the product planning stage.

Next, each element must be matched: (4) profile design and form, (5) form and color, and (6) profile design and color.

Finally, (7) all three elements "profile design, form, and color" must be integrated harmoniously to address modern market demands. At present, based on the knowledge acquired by above-mentioned subsection (A) and (B), the authors are tackling the development of "AEDA-3CM" by using "APFC-MSM" currently.

These studies were carried out in "Amasaka's New JIT laboratory" by collaboration with "Toyota Motor Corp., Toyota Tokyo Design Research Laboratory, Nissan Motor Corp., Honda Motor Co., Ltd., Mazda Motor Corp., Nippon Paint Co., Ltd., Kansai Paint Co. Ltd." and others (Amasaka, 2018).

Concretely, to realize the validity of the "Design SQC and CS-IDCM", the author has expanded "AEDM-3CM employing CSP using CSP-CIANS" described in Figure 11 below, and indicates the applications (A), (B) and (C) in the next Section "Application examples" below (: subsection "Actual studies on automobile exterior design optimization") (Refer to Amasaka (2015a, 2022a,b,c, 2023) in detail).

Total Intelligence CAE Development Design Model
for Realizing High Precision and Control

The time between product design and production has been drastically shortened in recent years with the rapid spread of global production. Quality assurance, or QA, has become increasingly critical. This makes it essential that the development design process—a critical component of QA—be reformed to ensure quality (Kume, 1999; Amasaka, 2010b).

Figure 8 shows the typical product development design process currently used by many companies (Amasaka, Ed., 2007b). In Figure 8, the author shows that companies first create product development design instructions based on market research and planning.

Figure 8. CAE in the product development design process

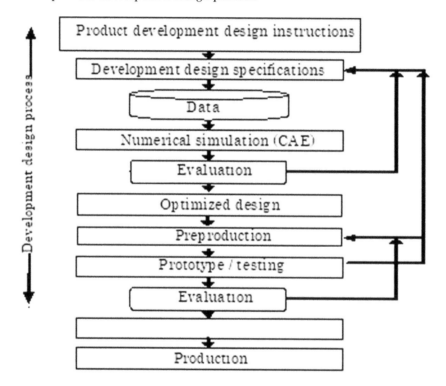

Then, they use these instructions to make development design specifications (drawings) and to promptly convert them to digital format so that they can be suitably processed and applied. The data is primarily used in numerical simulations known as computer-aided engineering, or CAE. CAE and other numerical simulations have been applied to a wide variety of business processes in recent years, including research and development, design, preproduction and testing/evaluations, production technology, production preparation, and manufacturing. These and other applications are expected to have effective results (Magoshi et al., 2003; Leo et al., 2004; Amasaka, 2010b).

In this age of global quality competition, using CAE for predictive evaluation method in design work is expected to contribute a great deal to shortening development design time and improving quality (Amasaka, 2007c, 2008, 2010b). However, generally, at the design and development stage, there is a gap (discrepancy) between prototype evaluation results and CAE analysis results (Amasaka et al., 2012). It has become evident that some manufacturers are not fully confident in CAE results.

Then, to win for the world quality competition. the author organized both of "Study Group of the Ideal Situation on the Quality Management of the Manufacturing Industry" in Union of Japanese Scientist and Engineers" (JUSE) and "Working Group No. 4 studies for establishment of a needed design quality assurance for numerical simulation at automotive industry in Japanese Society for Quality Control (JSQC).

Specifically, the author has researched the "New product development design technique employing strategic CAE application" named "New Japan Development Design Model" (NJ-DDM) (Amasaka, 2007a,b, 2015b).

As a concrete instance, to develop the above "Highly Reliable CAE Analysis Technology Component Model" using CSP, the author has developed the "Total Intelligence CAE Management Model" in order to achieve highly-accurate CAE analysis equivalent to prototype testing results as shown in Figure 9, which contributes to high quality assurance as well as QCD simultaneous achievement in automobile development design (Amasaka, 2008, 2010; Amasaka et al., 2012).

In Figure 9, many manufacturers are aware of the gap between evaluations of actual vehicles and CAE, and not fully confident in CAE results, they prefer to conduct Step (I) survey tests with actual vehicles rather than CAE evaluation. Even among leading corporations, Step (II) CAE utilization is limited to relative evaluation.

The author noticed a situation where, as shown in the figure, the application ratio of CAE to actual vehicles is about 25% for surveys and about 50% for relative evaluation revealing the dilemma that the effectiveness of CAE invested for reduction in development time has not been fully utilized.

Based on the above, in Step (III), as seen in the figure, the mechanism of the pending technical problem was clarified through visualization technology, and the technical knowledge which enables absolute evaluation through the creation of generalized models was incorporated in the CAE software. As a result, it was confirmed that the accuracy of CAE analysis had improved and the application ratio of CAE had increased to about 75%.

Based on the technical analysis derived from Steps (I) to (III), Step (IV) further incorporated a robust design which takes into consideration the influential factors and contributing ratio needed for optimal design, thus enhancing the accuracy of CAE calculation, and demonstrating a remarkable increase in the ratio of CAE application.

Then, to successfully develop the above AOP-DDMH, the author has created the "Highly-Reliable CAE Analysis Technology Component Model" required for highly precise CAE analysis software as shown in Figure 10 (Amasaka, 2007c, 2008; Amasaka, Ed., 2007b).

Moreover, from the viewpoint of achieving highly-reliable CAE analysis, the author illustrates the "Intelligence CAE software creation requirements" described in Figure 9 as follows (illustration: Know-how linkage-cycle with reciprocal action from (i) to (iv));

In linkage-cycle, Figure 10 shows the (i) Modelling visualization, (ii) Simulation (: Structural analysis, collision analysis, fluid analysis, chemical process analysis etc.), (iii) Optimization tool, (iv) statistical analyses, (v) knowledge base.

Figure 9. Total intelligence CAE management model

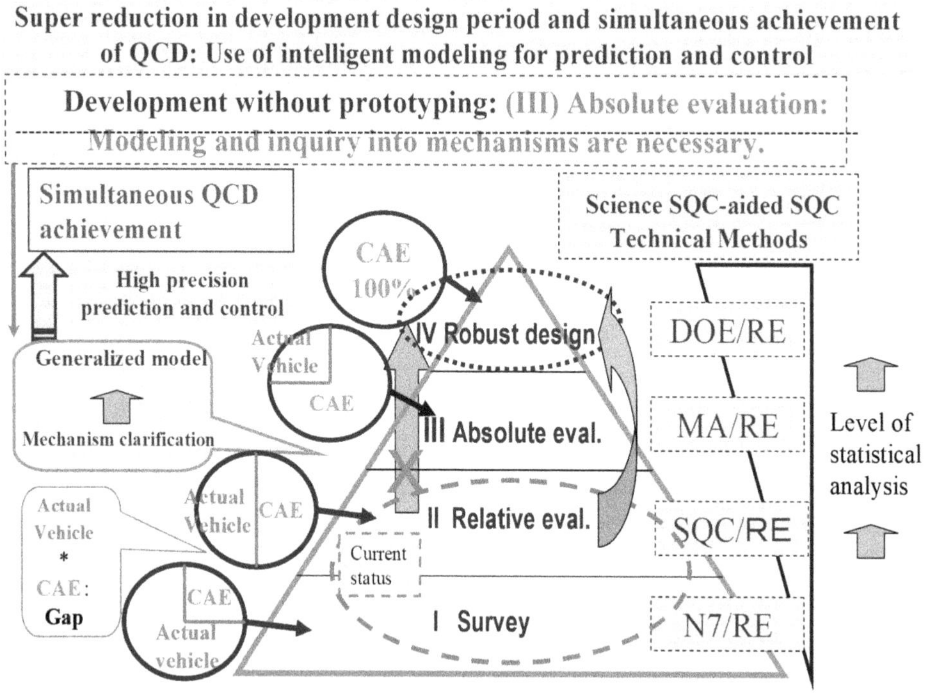

Super reduction in development design period and simultaneous achievement of QCD: Use of intelligent modeling for prediction and control

Specifically, in Figure 10, the process of CAE first starts with (1) "Problem"—Setting of problems to be solved, as well as (2) "Modeling"—Modeling of these problems as some type of mathematical formula.

In CAE, when using calculators as a means to analyze the model, such a means of analysis needs to be provided in the form of a calculation procedure, namely, (3) Algorithms—So that the software can perform calculation.

(4) "Theory"—The validity, applicable range, and performance or expected precision of such algorithms themselves can be deduced from some kind of theory.

(5) "Computer"—Needless to say, the technology related to the computer itself functioning as "hardware" to realize the algorithms, is undoubtedly a factor having a large effect on the success of CAE.

Success in AOP-DDM depends on the "collective strengths" of the elemental technologies. The formulation of such "implicit knowledge" confined to the personal know-how of the engineers is an indispensable step to be taken for sophistication of CAE as a Problem-Solving method.

Then, skilled CAE engineers are not experts in all the fields of the elemental technologies, but they understand their characteristics and interactions as the "implicit knowledge" and thus conduct selection combination to obtain favorable interactions and consequently the desired results. Therefore, it is positioned as a major theme in author's work (Amasaka, 2008, 2010; Amasaka, Ed., 2012).

Figure 10. Highly precise CAE technology component model (HR-CAE-ATCM)

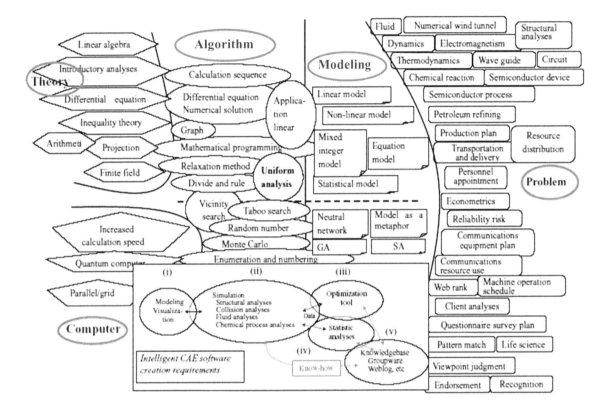

CSp-Customer Information Analysis and Navigation System for Visualizing Customer Demand

Today, growing companies both in Japan and abroad try to grasp the unprejudiced desires of their customers from the viewpoint of customer-oriented business management and to reflect these desires in future product development design.

However, the actual behavioral patterns (conception methods) of designers (new product planners and development designers) in trying to grasp latent customer desires depend heavily on the designers' empirical skills.

Accordingly, designers often worry that their current business approaches are likely to depend on job performing capabilities and on the sensitivity (intuition or knack) of individual persons, which will not improve the probability of success in the future, regardless of whether or not they have "lucky success" or "unlucky failure".

For the realization of strategic products, the collection as well as intellectual analysis of information for creating customers' demands is the core essence for success in "CSp".

Table 1 shows the levels of systematic utilization of customer information and the modes of intellectual information-sharing among the related divisions inside and outside the company that are necessary to achieve this objective.

To advance the level of execution of *"Customer Science* activities", it is necessary to evolve customer information-sharing among the "marketing, sales & service, merchandise for product planning & Product development design, production engineering, manufacturing and overseas" from off-line to on-line (Amasaka, 2005, 2015a).

Specifically, to realize the "Collecting information overseas for sheared use" described in Table 1, the author develops the "Analyzing customer information for reflection in over business employing "CSp" strategy.

Table 1. Systematic utilization levels of customer information for CSp activities

Level / Div.	Marketing, sales and service	Merchandise for product planning and Product development design	Production engineering and manufacturing	Overseas
5	Collecting information overseas for shared use			
4	Information sharing between divisions			
3	*Advanced Customer Science Level*		Unifying plan concepts based on shared information	Analyzing customer information for reflection in overseas business
2		Requesting and reporting (viewing) on the Web	Reflecting customer information analysis on sales and service	
1	Requesting and reporting (viewing) in writing		*New business process for creating "wants" indispensable to the development of attractive products*	
Operation	Intellectual Implementation of *Customer Science* by using scientific analysis approach			
Online	On-line use possible			
	Intelligent on-line use possible			

To realize this, then, for strategic product development design employing above "AEDM-3CM and AOP-DDM", it is important to explore consumer values, which are the basis for creating "demands" by employing "CSp", through the collection/analysis of customer information, and to reflect as well as exteriorize such values in product development design.

Against this background, the "CSP-Customer Information Analysis and Navigation System" (CSp-CIANS) was constructed as shown in Figure 11 (Amasaka, 2005).

As indicated therein, this system enables the networking the (1) Merchandise Div. to strengthen strategic product planning which explores customer value creation and (2) each division of the "Product Planning, Development and Design" to regularly receive customer data from (3) domestic and overseas dealers which are exposed to the front line of the customer desires through their marketing/sales/service activities.

Similarly, the collection of customer data is also possible through (4) Consulting Spaces, namely, the showrooms promoting the company's own products or public facilities for discussions and consultations from the customers. Moreover, (5) Marketing Research Companies via (6) an exclusive company WEB.

All these sections are connected through on-line networking for building (7) a Data Base (DB) via a server of the company's own information system division. Into this system utilizing statistical science approach - Science SQC (8).

Actually, the core system of SQC integration network system contains the "Total SQC Technical Intelligence System" (TTIS) including four core elements: the "Total SQC Intelligence System" (TSIS), "Total TQM Promotional Original SQC Soft" (TPOS), "Total SQC Manual Library" (TSML) and "Total Technical Information System" (TIRS) (See to Figure 5 in Chapter 3).

These are accessible for utilization from (9) Analytical case Data Base (DB). Particularly, cooperation requests for analysis can be submitted to (10) a special SQC adviser in Quality Assurance division (Refer to Amasaka (2004, 2005) in detail).

Figure 11. CSp-CIANS, networking of customer science application system

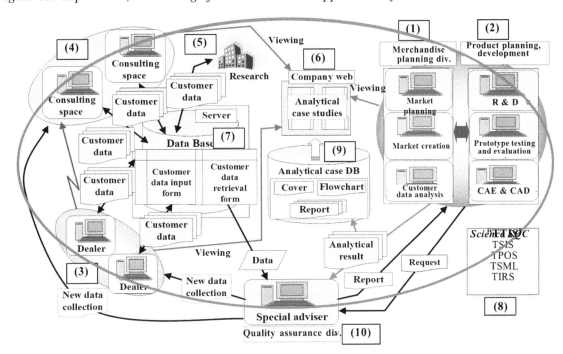

CSp-CIANS is designed in such a way that the collection of analytical results created by total linkage of the merchandise planning, product design, sales marketing and service for the successive development of analytical technology (Amasaka, Ed., 20112; Amasaka, 2015a, 2021, 2022a, 2023).

APPLICATION EXAMPLES

The aim of "NA-PDDM" above with "AEDM-3CM, AOP-DDM and CSP-CIANS" using a "Dual Corporate Strategy" in auto-makers is to be properly prepared for the "worldwide quality competition" so as not to be pushed out of market, and to establish strategic new management technologies that are capable of enhancing customer value creation in the world. Then, the author introduces the typical application examples.

Actual Studies on Automobile Exterior Design Optimization

Developing exterior design engineering strategy using APFC-MSM

In developing APFC-MSM based on AEDM-3CM, examples of actual case studies for the business process innovation that addresses optimizing the exterior design using "psychographics and *Design SQC*", corresponding to (1)–(7) in "Figure 5 in Chapter 7" by developing Figure 5 in this Chapter, are as follows (Refer to Amasaka (2015a, 2022a,b,c, 2023) in detail);

(1)   The development of prestige car "Lexus GS400 / LS430" which realized the compatibility of profile design and package design (interior space) using "Automobile Package Design Concept Support Method" (APDCSM). This research realizes a more creative product design process as an "Intelligence Profile Design Concept Method" (IPDCM) (Amasaka and Nagaya, 2002; Okabe et al., 2007).

(2)   Construction of the "Automobile Design Form Support Method" (ADFSM) in order to fully understand the visualizing relationship between form modifications as a whole (which consists of front, side and rear elements) and subjective customer's impressions using eye-tracking camera, 3D Design CAD software (CATIA V5) and *Design SQC* (Asami et al., 2010; Yazaki et al., 2013).

(3)   The author has created an "Automobile Exterior Color Development Approach Model" (AECDAM) in order to visualize the success design colors in customer demands (Muto et al., 2011; Takebuchi et al., 2012a). Specifically, this model determined the 4 factors (classy, luxurious, dignified and sporty) and 6 elements (hue, luminosity, intensity, shine, opacity and graininess) most desired by young buyers.

(4)   The author has created the "Strategic Automobile Design Support Method" (SADSM) for the profile design and form matching using "CSp-IDCM" above. Specifically, the author has visualized the preferences of younger generation through the combination of *"Design SQC* and 3D Design CAD" (Takimoto et al., 2010).

(5)   The author has constructed the "Automobile Exterior Design Approach Model" (AEDAM) for the "form and color matching", which uses biometric devices along with visualization technology, 3D Design CAD and *"Design SQC"* to establish the relationship between "form and body color" which customers observe the overall vehicle design (Takebuchi et al., 2012b; Muto et al., 2013).

(6)   The author has created the "Amasaka-Lab's Exterior Design Approach Model" (AL-EDAM) for the "profile design and color matching" using "market and preference survey, form and color analysis, creation of 3D Design CAD models and verification". This focuses on the "young women preference car—luxurious, stylish, high-end & chic" using collage boards (: girlish, elegant, casual, trendy & boyish) (Asami et al., 2011).

(7)   The author has created the "Amasaka-Lab's Profile Design (Proportion), Form, and Color Matching Model" (AL-PFCM) to analyze and catch the customers' attention using "eye camera and electro-encephalograph". Specifically, to identify ideal relationships among proportion, form, and color, the author has created the 3D-CAD model cars using "experimental design and analytic hierarchy process methods" (Toyoda et al., 2015a; Kobayashi et al., 2016).

For example, in (5) "form and color matching", Figure 12 shows a "matching of form and color optimization" using front fender panels in medium sedan car. In this study, the panels were shaped and pained like car panels to show the relevant textual expression.

Figure 12(i) shows the "varying the effect of the opacity and graininess equally to produce a total of 11 panels with the six elements of color (hue, luminosity intensity, shine, opacity, and graininess".

Figure 12. An example for matching of form and color optimization using front fender panels

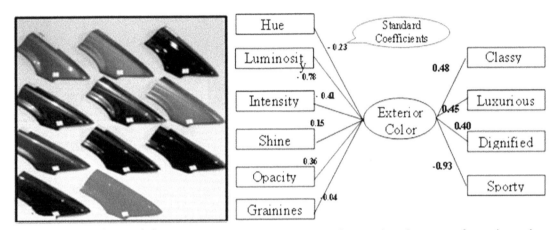

(i) Painted panels used for surveys   (ii) Path diagram from color elements of exterior color

Each panel gave the desired impression for the preferences (classy, luxurious, dignified and sporty). Aesthetic evaluation data was obtained from the results. The survey was aimed at men and women in their twenties, and the answers were obtained from a total of 94 men and women.

Figure 12(ii) shows the "Path diagram from color elements of exterior color". This path diagram of covariance structure analysis was conducted to determine the correspondence relationship between color elements and preferences. The absolute values of these standardized coefficients indicate the degree of influence while the direction of influence is indicated by positive or negative numbers.

In this path diagram used for this covariance structure analysis, the "exterior color" latent variable was chosen as the latent variable that would satisfy the four vehicle design preferences of those who value self-expression.

Respondents in the self-expression group were asked to freely evaluate the three developed colors. The respondents gave positive evaluations, saying they liked how classy the colors were and that they would like a car with colors such as these (Refer to Takebuchi et al. (2012a) in detail).

Expanding APFC-MSM for raising attractiveness to customer

Furthermore, the author has been advancing the "New deployment of APFC-MCM" in AEDA-3CM for "mid-size and small-size cars" for raising attractiveness to customer as follows;

1st is the development of "Auto-exterior and interior color matching" and "Auto-instrumentation design" (Koizumi et al., 2014; Shinogi et al., 2014).

2nd is the "Attractive exterior design concept for indifferent customers" (Kobayashi et al., 2015), "auto-instrumentation for young male customers" (Yazaki et al., 2012) and "Bicycle design model for young women's fashion" (Koizumi et al., 2013; Toyoda et al., 2015).

3rd, moreover, the author has created the "Creation of 3 typical "New profile design: (A) Advanced type, (B) Elegant type and (C) Progress type" in near future as shown in Figure 13 (Refer to Kobayashi et al. (2015, 2016) and Amasaka (2017) in detail).

Figure 13. Creation of three typical new profile design in near future

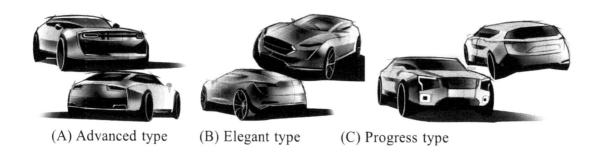

(A) Advanced type　　(B) Elegant type　　(C) Progress type

Actual Studies on Automobile Product Development Design Optimization

In general, experienced development design staff and CAE engineers understand the mechanism that is causing the bottleneck technical problem as implicit knowledge (Amasaka, 2010a). While many examples of calculation based on CAE analysis have been reported, the accuracy of estimation has not to be improved for satisfactory vehicle development (Amasaka, Ed., 2007).

Then, the author has created the "Automobile Optimal Product Development Design Model" (APFC-MSM) in an effort to help solve the bottleneck technical problem that had become a global technological issue as shown in Figure 6 of Chapter 7 (Amasaka, 2007c, 2008).

To accomplish this, as the first stage, it was important to the (A) "Visualization"- visualize the dynamic behavior of the problem" by employing "Actual vehicles and equipment and carrying out testing" using "Hypothesis". At this point the expertise of specialists from both inside and outside the company was brought together through the "Partnering" activities.

As the second stage, it was vital to deduce the (B) "Mechanism"—fault mechanism using various "Techniques". To carry out the precise fault analysis and factor analysis, "New Seven Tools (N7), Statistical Quality Control (SQC), Reliability Engineering (RE), Multivariate Analysis (MA) and Design of Experiment (DOE)" were combined and utilized to search out and identify previously unknown or overlooked latent causes.

In this way, a logical thinking process was used to carry out a logical investigation into the cause of fault mechanism for the "Modeling".

Moreover, as the third stage, all of this knowledge and information was then unified through the (C) Creation of "CAE Navigation Software" that employs "Computer Graphics" (CG) to reproduce the visualization of the actual vehicle and testing data so that it can be made consistent to a "Qualitative model".

At this stage, it was important to carry out actual vehicle and testing work so that this qualitative model could be made for the "cause and effect relationships" of the unknown mechanism. It would then become extremely important to use this model to reduce the divergence (gap) between the results from the actual vehicle testing and CAE to develop the "absolute value evaluation".

As the fourth stage, in addition, at the stage of developing the (D) "Numeric value simulation", exhaustive actual vehicle testing was carried out in order to convert the leak mechanism from implicit knowledge into precise explicit knowledge.

The information gained from these work processes would then be unified and a "highly credible numerical simulation (Quantitative model)" would be carried out to make absolute value prediction and control possible.

In the final stage, as the (E) "Evaluation, Design and Improvement", the CAE analysis results are then verified by comparing them to the actual vehicle testing results. In the case of a decentralized organization and business process (such as shown in Figure 7), it is essential that the specialists in the fields of design, testing, CAE analysis, CAE software design, and SQC carry out cooperative team activities, "partnering" (◎ Main, Sub, △ Support) at each stage of the work process (A to E).

Application to AOP-DDM for Raising Attractiveness to Customers

Then, as the application of AOP-DDM for raising attractiveness to customers, the author was able to apply the AOP-DDM to critical development design technologies for automotive production, including predicting and controlling the typical applications of the "transaxle oil seal leakage" (Amasaka, 2003b, 2008, 2010; Amasaka et al., 2012; Nozawa et al., 2013), "brake pad quality assurance" (Amasaka, 2004b, 2017a; Amasaka et al., 2012), "looseness of the bolt-nut tightening" (Amasaka, 2019) and others.

For example, Figure 14 shows a "Highly Accurate Analysis Approach Model by using Science SQC" for improvement of automobile transaxle oil seal leakage which is a bottleneck technological problem for auto-manufacturers worldwide.

Figure 14. Highly accurate analysis approach model using science SQC

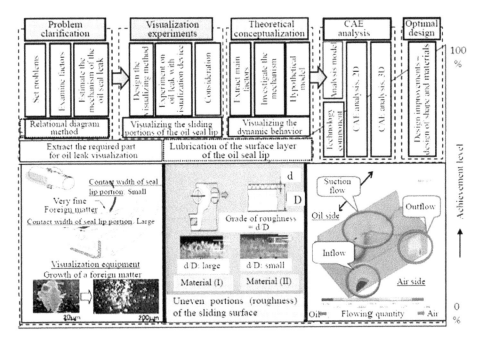

This was achieved through an analysis process involving problem clarification, visualization experiments, theoretical conceptualization, CAE analysis, and optimal design.

First, the author began by developing a device for visualizing the ascertained phenomena in order to estimate the unknown mechanisms involved in the leaks.

This made it possible to estimate the mechanism of the oil seal leaks by visualizing the dynamic behavior involved in the process whereby metal particles (foreign matter) from gear rotation wear, found around the rotating and sliding portions of the oil seal lip, become mechanically fused and accumulate.

Next, the findings obtained were used to formulate the following design countermeasures;

(i) Strengthen gear surfaces to prevent occurrence of foreign matter even after 100,000 km (improve quality of materials and heat treatments)

(ii) Formulate a design plan to scientifically ensure optimum lubrication of the surface layer of the oil seal lip (uneven portions of the sliding surface) where it rotates in contact with the drive shaft.

These general design technology elements were incorporated into the Technology Component Model for the "Oil Seal Simulator" using "Highly Precise CAE Technology Component Model" to create highly-reliable CAE analysis software capable of accurately reproducing the oil seal leak phenomena, enabling them to be identified and controlled as shown in Figure 15.

The following methods were proposed: (i) Identifying the problem: simulation of variously converging physiochemical phenomena (methodology: (1) to (3)), (ii) Modeling: building of problem-solving models (methodology: (1) to (3)), (iii) Practical algorithms: calculation methods (methodology: (1) to (2)), (iv) Rational theories: adoption of methodology (1) to (3)) and (v) Calculators: innovations enabling calculations to be made accurately within a realistic period of time (methodology: (1) to (3)).

Figure 15. Oil seal simulator using highly precise CAE technology component model

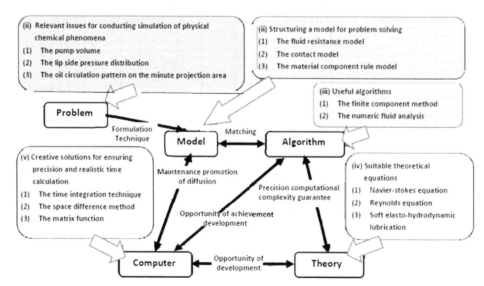

As a result, it is now possible to implement highly-reliable numerical simulation (CAE analysis, 2D and 3D), enabling the realization of the quality assurance CAE analysis model.

The CAE analysis shown in the figure is an example of numerical simulation for pump flow volume (flow of lubricant: air side [atmosphere] – oil side [gears]) around the oil seal. Oil seal leaks (market claims) have now been reduced to less than 1/20 due to the implementation of design improvements (design of shape and materials) (Refer to Amasaka et al. (2012) and Amasaka (2017a) in detail).

In each of these applications as well, discrepancy was 3%-5% versus prototype testing. Based on the achieved results, the model is now being used as an intelligent support tool for optimizing product design processes (Refer to Amasaka, 2015a, 2021, 2022a,b, 2023) in detail)

CONCLUSION

In this chapter, the author has created a NA-PDMM using a "Dual Corporate Engineering Strategy" for the innovation of automobile products engineering fundamental. Specifically, "NA-PDDM" contains both of the "Exterior design engineering strategy and Driving performance design engineering strategy" for the developing "appearance quality and functional quality" that contributes to strengthening of Japanese automobile global corporate strategy using "CSp". Concretely, to develop the "Same quality worldwide and products development design at optimal locations", the foundation of "NA-PDDM" consists of the "AEDM-3CM, AOP-DDM and CS-CIANS". The validity of "NA-PPMM" is then verified through the actual applications to automobile products development design in Toyota and others.

REFERNCES

Amasaka, K. (1999a). A demonstrative study of a new SQC concept and procedure in the manufacturing industry: Establishment of a New Technical Method for conducting Scientific SQC. *An International Journal of Mathematical & Computer Modeling, 31*(10-12), 1–10.

Amasaka, K. (1999b). A study on Science SQC by utilizing Management SQC: A demonstrative study on a new SQC concept and procedure in the manufacturing industry. *International Journal of Production Economics, 60-61*, 591–598. doi:10.1016/S0925-5273(98)00143-1

Amasaka, K. (2002). *New JIT*, A New Management Technology Principle at Toyota. *International Journal of Production Economics, 80*(2), 135–144. doi:10.1016/S0925-5273(02)00313-4

Amasaka, K. (2003a). *Development of "New JIT", Key to the excellence design "LEXUS": The validity of "TDS-DTM", a strategic methodology of merchandise*. Proceedings of the Production and Operations Management Society, Hyatt Regency, Savannah, Georgia.

Amasaka, K. (2003b). Proposal and implementation of the Science SQC Quality control Principle. *Mathematical and Computer Modelling, 38*(11-13), 1125–1136. doi:10.1016/S0895-7177(03)90113-0

Amasaka, K. (2004a). Custom*er Science: Studying consumer values*. Japan Journal of Behavior Metrics Society, The 32nd Annual Conference, Aoyama Gakuin University, Sagamihara, Kanagawa.

Amasaka, K. (2004b). *Science SQC, New Quality Control Principle—The quality control of Toyota.* Springer-Verlag Tokyo. doi:10.1007/978-4-431-53969-8

Amasaka, K. (2005). Constructing a Customer Science Application System "CS-CIANS"—Development of a Global Strategic Vehicle "Lexus" utilizing New JIT-. *WSEAS Transactions on Business and Economics, 2*(3), 135–142.

Amasaka, K. (2007a). High Linkage Model "Advanced TDS, TPS & TMS": Strategic development of "New JIT" at Toyota. *International Journal of Operations and Quantitative Management, 13*(3), 101–121.

Amasaka, K. (Ed.). (2007a). New Japan Model: Science TQM. Study Group of the Ideal Situation on the Quality Management of the Manufacturing Industry, Maruzen, Tokyo.

Amasaka, K. (2007b). The validity of "TDS-DTM", a strategic methodology of merchandise: Development of New JIT, Key to the excellence design "LEXUS". *The International Business & Economics Research Journal, 6*(11), 105–115.

Amasaka, K. (Ed.). (2007b). Establishment of a needed design quality assurance framework for numerical simulation in automobile production. Working Group No. 4 Studies in JSQC, Study group on simulation and SQC, Tokyo.

Amasaka, K. (2007c). Highly Reliable CAE Model, the key to strategic development of *Advanced TDS. Journal of Advanced Manufacturing Systems, 6*(2), 159–176. doi:10.1142/S0219686707000930

Amasaka, K. (2008). An Integrated Intelligence Development Design CAE Model utilizing New JIT: Application to automotive high reliability assurance. *Journal of Advanced Manufacturing Systems, 7*(2), 221–241. doi:10.1142/S0219686708001589

Amasaka, K. (2010a). Proposal and effectiveness of a High Quality Assurance CAE Analysis Model: Innovation of design and development in automotive industry, *Current Development in Theory and Applications of Computer Science. Engineering and Technology, 2*(1/2), 23–48.

Amasaka, K. (2010b). Chapter 4. Product Design, Quality Assurance Guidebook. Guide Book of Quality Assurance. The Japanese Society for Quality Control & Union of Japanese Scientists and Engineers Press.

Amasaka, K. (2011). Changes in Marketing Process Management Employing TMS: Establishment of Toyota Sales Marketing System, *China & USA. Business Review (Federal Reserve Bank of Philadelphia), 10*(7), 539–550.

Amasaka, K. (Ed.). (2012). Science TQM, New Quality Management Principle: The Quality Management Strategy of Toyota. Bentham Science Publisher, UAE, USA, The Netherlands.

Amasaka, K. (2014). New JIT, new management technology principle. *Journal of Advanced Manufacturing Systems, 13*(3), 197–222. doi:10.1142/S0219686714500127

Amasaka, K. (2015a). New JIT, new management technology principle. Taylor & Francis Group, CRC Press.

Amasaka, K. (2015b). Constructing a New Japanese Development Design Model "NJ-DDM": Intellectual evolution of an automobile product design. *TEM Journal Technology Education Management Informatics*, 4(4), 336–345.

Amasaka, K. (2017b). *Studies on Automobile Exterior Design Model for customer value creation utilizing Customer Science Principle*. The 7th International Symposium on Operations Management Strategy, Tokyo Metropolitan University, Tokyo.

Amasaka, K. (2017c). *Attractive automobile design development: Study on customer values (Plenary Lecture)*. Executive Lecture in Rotary Clube.

Amasaka, K. (2018). Automobile Exterior Design Model: Framework development and supporting case studies. *Journal of Japanese Operations Management and Strategy*, 8(1), 67–89.

Amasaka, K. (2019). Establishment of an Automobile Optimal Product Design Model: Application to study on bolt-nut loosening mechanism. *Noble International Journal of Scientific Research*, 3(9), 79–10.

Amasaka, K. (2022a). New Manufacturing Theory: Surpassing JIT (2nd Edition). Lambert Academic Publishing.

Amasaka, K. (2022b). *Examining a New Automobile Global Manufacturing System*. IGI Global Publishing. doi:10.4018/978-1-7998-8746-1

Amasaka, K. (2022c). A New Automobile Product Development Design Model: Using a Dual Corporate Engineering Strategy. *Journal of Economics and Technology Research*, 4(1), 1–22. doi:10.22158/jetr.v4n1p1

Amasaka, K. (2023). New Lecture-Surpassing JIT: Toyota Production System-From JIT to New JIT. Lambert Academic Publishing, Germany.

Amasaka, K., Ito, T., & Nozawa, Y. (2012). A New Development Design CAE Employment Model. *The Journal of Japanese Operations Management and Strategy*, 3(1), 18–37.

Amasaka, K., & Nagasawa, S. (2000). *Fundamentals and application of sensory evaluation: For Kansei Engineering in the vehicle*. Japanese Standards Association.

Amasaka, K., & Nagasawa, S. (2002). *Fundamentals and application of sensory evaluation: For Kansei Engineering in the vehicle*. Japanese Standards Association.

Amasaka, K., & Nagaya, A. (2002). *Engineering of the new sensitivity in the vehicle: Psychographics of LEXUS design profile, Development of articles over the sensitivity—The method and practice*. Japan Society of Kansei Engineering. Nihon Shuppan Service Press.

Amasaka, K., & Nagaya, A. (2002). *Engineering of the new sensitivity in the vehicle: Psychographics of LEXUS design profile", Development of articles over the sensitivity: The method and practice*. Edited by Japan Society of Kansei Engineering, Nihon Shuppan Service Press.

Amasaka, K., Nagaya, A., & Shibata, W. (1999). Studies on Design SQC with the application of Science SQC—Improving of business process method for automotive profile design. *Japanese Journal of Sensory Evaluations*, 3(1), 21–29.

Asami, H., Ando, T., Yamaji, M., & Amasaka, K. (2010). A study on Automobile Form Design Support Method "AFD-SM". *Journal of Business & Economics Research, 8*(11), 13–19. doi:10.19030/jber.v8i11.44

Asami, H., Owada, H., Murata, Y., Takebuchi, S., & Amasaka, K. (2011). The A-VEDAM for approaching vehicle exterior design. *Journal of Business Case Studies, 7*(5), 1–8. doi:10.19030/jbcs.v7i5.5598

Enrique, A. (2005). *Parallel, Metaheuristics: A New Class of Algorithms*. Addison Wiley.

Kobayashi, T., Yoshida, R., Amasaka, K., & Ouchi, N. (2015). A study for creating an vehicle exterior design at teaching in different customers. *Hong Kong International Conference on Engineering and Applied Science*, Hong Kong.

Kobayashi, T., Yoshida, R., Amasaka, K., & Ouchi, N. (2016). A statistical and scientific approach to deriving an attractive exterior vehicle design concept for indifferent customers. *Journals International Organization of Scientific Research, 18*(12), 74–79.

Koizumi, K., Kanke, R., & Amasaka, K. (2013). Research on automobile exterior color and interior color matching. *International Journal of Engineering Research and Applications, 4*(8), 45–53.

Koizumi, K., Kawahara, S., Kizu, Y., & Amasaka, K. (2013). A Bicycle Design Model based on young women's fashion combined with CAD and Statistical Science. *Journal of China-USA Business Review, 12*(4), 266–277.

Kume, H. (1999). Quality Management in Design Development. JUSE (Union of Japanese Scientists and Engineers), Tokyo.

Leo, J. D. V., Annos, N., & Oscarsson, J. (2004). Simulation based decision support for manufacturing system Life Cycle Management. *Journal of Advanced Manufacturing Systems, 3*(2), 115–128. doi:10.1142/S0219686704000454

Magoshi, R., Fujisawa, H., & Sugiura, T. (2003). Simulation technology applied to vehicle development. *Journal Society of Automotive Engineers of Japan, 53*(3), 95–100.

Ministry of Land. (2015). *Ministry of Land Infrastructure, Transport and Tourism*. MLIT. https://www.mlit.go.jp/jidosha/carinf/rcl/data.html/

Muto, M., Miyake, R., & Amasaka, K. (2011). Constructing an Automobile Body Color Development Approach Model. *Journal of Management Science, 2*, 175–183.

Muto, M., Takebuchi, S., & Amasaka, K. (2013). Creating a New Automotive Exterior Design Approach Model: The relationship between form and body color qualities. *Journal of Business Case Studies, 9*(5), 367–374. doi:10.19030/jbcs.v9i5.8061

Nagaya, A., Matsubara, K., & Amasaka, K. (1998). *A Study on the customer tastes of automobile profile design (Special Lecture)*. Union of Japanese Scientists and Engineers, The 28th Sensory Evaluation Sympojium, Tokyo.

Nonobe, K., & Ibaraki, T. (2001). An improved tabu search method for the weighted constraint satisfaction problem. *INFOR, 39*(2), 131–151. doi:10.1080/03155986.2001.11732431

Nozawa, Y., Takahiro Ito, T., & Amasaka, K. (2013). High precision CAE analysis of Automotive trans-axle oil seal leakage. *The China Business Review*, *12*(5), 363–374.

Nunogaki, N., Shibata, K., Nagaya, A., Ohashi, T., & Amasaka, K. (1996). *A study of customers' direction about designing vehicle's profile*. The Japanese Society for Quality Control, The 26th annual conference, Gifu, Japan.

Okabe, Y., Yamaji, M., & Amasaka, K. (2007). Research on the Automobile Package Design Concept Support Methods "CS-APDM": Customer Science approach to achieve CS for vehicle exteriors and package design. *Journal of Japan Society for Production Management*, *13*(2), 51–56.

Okazaki, R., Suzuki, M., & Amasaka, K. (2000). *Study on the sense of values by age using Design SQC*. The Japan Society for Production Management, The 11th Annual Technical Conference, Okayama University, Okayama, Japan.

Okutomi, H., & Amasaka, K. (2013). Researching Customer Satisfaction and Loyalty to boost marketing effectiveness: A Look at Japan's Auto Dealerships. *International Journal of Management & Information Systems*, *17*(4), 193–200. doi:10.19030/ijmis.v17i4.8093

Power, J. D., & Associates. (1998). *Vehicle Dependability Study*. JD Power. https://www. jdpower. com/releases/80401car.html/

Shinogi, T., Aihara, S., & Amasak, K. (2014). Constructing an Automobile Color Matching Model (ACMM). *IOSR Journal of Business and Management*, *16*(7), 7–14. doi:10.9790/487X-16730714

Takebuchi, S., Asami, H., & Amasaka, K. (2012b). An Automobile Exterior Design Approach Model linking form and color. *Journal of China-USA Business Review*, *11*(8), 1113–1123.

Takebuchi, S., Nakamura, T., Asami, H., & Amasaka, K. (2012a). The Automobile Exterior Color Design Approach Model. *Journal of Japan Industrial Management Association*, *62*(6E), 303–310.

Takimoto, H., Ando, T., Yamaji, M., & Amasaka, K. (2010). The proposal and validity of the Customer Science Dual System, *China-USA. Business Review (Federal Reserve Bank of Philadelphia)*, *9*(3), 29–38.

Takimoto, H., Ando, T., Yamaji, M., & Amasaka, K. (2010). The proposal and validity of the Customer Science Dual System, *China-USA. Business Review (Federal Reserve Bank of Philadelphia)*, *9*(3), 29–38.

Toyoda, S., Koizumi, K., & Amasaka, K. (2015b). Creating a Bicycle Design Approach Model based on fashion styles, *IOSR (International Organization of Scientific Research). Journal of Computational Engineering*, *17*(3), 1–8.

Toyoda, S., Nishio, Y., & Amasaka, K. (2015a). Creating a Vehicle Proportion, Form & Color Matching Model. *Journals International Organization of Scientific Research*, *17*(3), 9–16.

Umezawa, Y., & Amasaka, K. (1999). Partnering as a platform of Quality Management in Toyota Group. [in Japanese]. *Journal of the Operations Research Society of Japan*, *44*(10), 24–35.

Whaley, R. C., Petitet, A., & Dongarra, J. J. (2000). *Automated empirical optimization of software and the ATLAS project. Technical report, University of Tennessee*. Department of Computer Science.

Yamaji, M., & Amasaka, K. (2009). An Intelligence Design Concept Method utilizing Customer Science. *The Open Industrial and Manufacturing Engineering Journal, 2*(1), 10–15. doi:10.2174/1874152500902010021

Yazaki, K., Takimoto, H., & Amasaka, K. (2013). Designing vehicle form based on subjective customer impressions. *Journal of China-USA Business Review, 12*(7), 728–734.

Yazaki, K., Tanitsu, H., Hayashi, H., & Amasaka, K. (2012). A model for design auto- instrumentation to appeal to young male customers. *Journal of Business Case Studies, 8*(4), 417–426. doi:10.19030/jbcs.v8i4.7035

Chapter 8
New Japan Production Management Model Surpassing Conventional JIT

ABSTRACT

To advance the Japan manufacturing foundation, the author describes the new japan production management model (NJ-PMM) called advanced TPS that surpasses conventional JIT practices in order to re-construct world-leading management technologies. Specifically, the author mentions the strategic development of a dual global engineering model (DGEM) by possessing the new japan global production model (NJ-GPM) and new japan global manufacturing model (NJ-GMM) surpassing JIT. The effectiveness of DGEM was verified through the actual applications to auto-manufacturing in Toyota and suppliers.

NEEDS FOR ADVANCES IN GLOBAL PRODUCTION ENGINEERING

Recently, Japanese Production System, which is typified by the Toyota Production System (TPS) has been further developed and spread in the form of internationally shared global production systems such as Just in Time (JIT), and therefore it is no longer a proprietary technology of Japan (Ohno, 1977; Womack et al., 1990; Amasaka, 1988, 2002, 2007a).

Today, digital engineering is bringing about radical changes in the manufacturing way (Amasaka, 2007b; Amasaka, Ed., 2007; Amasaka and Sakai, 2010, 2011).

The Demand for Advances in Manufacturing Technology

The environmental changes that surround today's manufacturing industry are truly severe. It is vital for Japanese manufacturing not to fall behind in the advancement of management technologies. In order for manufacturers to succeed in the future world market, they need to continue to create products that will leave a strong impression on customers and to offer them in a timely fashion (Amasaka, 2002).

At present however, the TPS which is representative of Japanese manufacturing, has been further developed and spread in the form of internationally shared global production systems such as JIT and Lean System and therefore it is no longer a proprietary technology of Japan (Hayes and Wheelwright, 1984; Doos et al., 1991; Womack and Jones, 1994; Taylor and Brunt, 2001).

DOI: 10.4018/978-1-6684-8301-5.ch008

It is not an exaggeration to say that what will ensure Japanese manufacturers' success in global marketing is the realization of simultaneous achievement of QCD (quality, cost and delivery)–ahead of their competitors (Amasaka, 2004a, 2007a,b, 2008). The urgent mission for Japanese manufacturers is to reconstruct world-leading, uniquely Japanese principles of management technology, which will be viable even for next-generation manufacturing as the evolution system of JIT (Amasaka et al., 2008; Amasaka, 2009, 2014).

To prevail in today's competitive manufacturing industry, which is often referred to as a worldwide quality competition, the pressing management issue is to realize the kind of global production that can achieve called the "worldwide uniform quality and production at optimal locations" (Amasaka and Sakai, 2011; Amasaka, 2007a,b, 2018).

Innovation of Japan's Global Manufacturing Fundamentals

To offer customers high value-added products and prevail in the worldwide quality competition, it is necessary to establish an advanced production system that can intellectualize the production engineering and manufacturing management system. This will in turn produce high performance and highly functional new products.

The author believes that what determines the success of global production strategies is the advancement of technologies and skills that are capable of fully utilizing the above mentioned advanced production system in order to realize reliable Japan's global manufacturing fundamentals at the production sites (Amasaka, 2007a).

Then, the author posed the question, "What are some of the issues that need to be addressed in order to prevail in the 21st century?" to a total of 154 respondents chosen from the top management class (board members) and managers (of divisions and departments) of the six manufacturers that participated in the "Workshop on Quality Management of Manufacturers" (hosted by Kakuro Amasaka, Professor of Aoyama Gakuin University from May 2004 to March 2006).

The survey revealed the interests of the top management and manager classes (Amasaka, Ed., 2007). Figure 1 is an example of the summary and analysis results of these interests (or the free opinions gathered from the survey). As shown here, their interests center on technological development, human resources development, globalization, product differentiation, quality, safety, organization, and management.

Consolidating these items, the management technology issue that they commonly share is the establishment of the kind of global production system that can achieve worldwide uniform quality and simultaneous launch (production at optimal locations). Among other things, what they give top priority to is the realization of manufacturing with high quality assurance. This is achieved through high cycle-ization of the business process related to manufacturing at overseas production plants (or production sites) in "Industrialized nations in the West or developing nations, not to mention in the Japanese domestic market".

In particular, they are concerned about the current situation in which the production sites are undergoing drastic changes due to the use of digital engineering and IT (Information Technology). Moreover, it was seen as critical to realize the intellectualization of the production sites so that they do not lag behind the advancements being made in technology and skills.

Figure 1. What is the important issue which should be tackled?

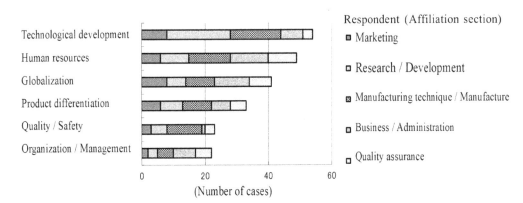

INNOVATION OF JAPAN MANUFACTURING FOUNDATION SURPASSING JIT

Creation of Total Production System Based on New Manufacturing Theory

To develop the "Advances in Global Manufacturing Engineering" above, the author has created the Total Production System which is one of core principles of the "New Manufacturing Theory" (NMT) called "New JIT, new management technology principle" using total linkage of "TDS, TPS, TMS, TIS & TJS" (Total Development System, Total Production System, Total Marketing System, Total Business Information Management System & Total Job Quality Management System) (Amasaka et al., 2008; Amasaka, Ed., 2012, 2019; Amasaka, 2014, 2019a) (Refer to Figure 1 in Chapter 7).

The aim of "Total Production System" (TPS) is to enable a focus on customers and employees as well as the reinforcement and improvement of production process control through incorporation of 4 sub-core elements "(a)-(d)" as shown in Figure 2 follows (See to Figure 2 in Chapter 7); (a) production based on information to require the reformation of the production philosophy, (b) production based on management to reform the workplace configuration, (c) production based on engineering to reinforce the production technology, and (d) production based on workshop formation to create the highly creative and active workplace.

Developing New Japan Production Management Model "NJ-PMM" Surpassing JIT

To develop NMT, furthermore, the author has developed the "New Japan Production Management Model" (NJ-PMM) named "Advanced TPS, strategic production management system" surpassing JIT in order to enable the strategic deployment of "TPS" (Total Production System) and SCM (Supply Chain Management) as shown in Figure 3 (Amasaka, 2007a,b, 2019a, 2020a,b,c, 2021, 2022, 2023) (See to Figure 3 in Chapter 7).

In Figure 3, the mission of NJ-GPMM is to contribute the worldwide uniform quality and production at optimal locations as the strategic deployment of global production and to realize Customer Satisfaction (CS), Employee Satisfaction (ES), and Social Satisfaction (SS) through high quality assurance manufacturing.

Figure 2. Schematic drawing of total production system

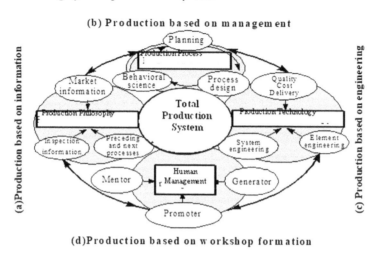

On a concrete target, this model is the systemization of a new, next-generation Japanese production management system and this involves the high-cyclization of the production process for realizing the simultaneous achievement of QCD requirements. To make this model into a reality it will be necessary to adapt it to handle digitalized production and reform it to realize a renewal production management system.

Moreover, other prerequisites for realizing this include the need to create an attractive working environment that can accommodate the increasing number of older and female workers at the production sites and to cultivate intelligent production operators.

One of the technical elements necessary for fulfilling these requirements is the reinforcement of maintenance and improvement of process capabilities by establishing an intelligent quality control system. Second, a highly reliable production system needs to be established for high quality assurance. Third, reform is needed for the creation of a next generation working environment system that enhances intelligent productivity. Fourth, intelligent production operators need to be cultivated that are capable of handling the advanced production system and an intelligent production operating development system needs to also be established.

DUAL GLOBAL ENGINEERING MODEL TO ADVANCE NJ-PMM

Then, to advance NJ-PMM, the author has created a "Dual Global Engineering Model" (DGEM) processing "New Japan Global Production Model" (NJ-GPM) and "New Japan Global Manufacturing Model" (NJ-GMM) through the all sections "production plan, technology & preparation, manufacturing, inspection and SCM" (Amasaka, 2015, 2017a, 2020b,c,2022a,b,c, 2023).

Figure 3 Schematic drawing of new Japan production management model (NJ-PMM)

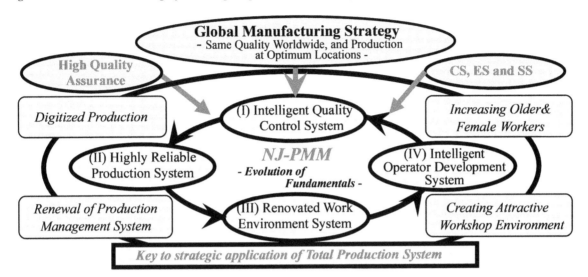

New Japan Global Production Model "NJ-GPM"

Global production must be deployed in order to establish the kind of manufacturing that is required to gain the trust of customers around the world by achieving a high level of quality assurance and efficiency and shortening lead times to reinforce the simultaneous achievement of QCD requirements.

The vital key to achieving this is the introduction of a production system that incorporates production machinery automated with robots, skilled and experienced workers (operators) to operate the machinery and, further, production information to organically combine them. Thus, having recognized the need for a new production system suitable for global production, the author has created the "New Global Production Model" (NJ-GPM) to realize the strategic deployment of the "NJ-PMM" above as shown in Figure 4 (Amasaka and Sakai, 2011).

In Figure 4, the purpose of this model is to eradicate ambiguities at each stage of the production process as the production planning and preparation through production itself and process management, and between the processes in order to achieve a highly reliable production system for global production which will improve the reliability of manufacturing through the clarification and complete coordination of these processes.

More specifically, the model is intended to (i) employ numeric simulation (Computer Aided Engineering, CAE) and computer graphics (CG) right from the production planning stage to resolve technical issues before they occur, (ii) reinforce production operators' high-tech machine operating skills and manufacturing capabilities, and (iii) visualize the above using IT in order to reform production information systems to create a global network of production sites around the world.

Figure 4. New Japan global production model (NJ-PPM)

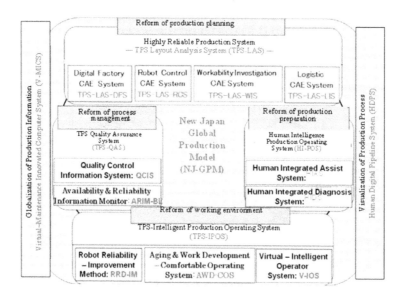

The six core management technologies that constitute this model and their characteristics are described below (Amasaka, 2017a, 2020b,c, 2023).

1. Reform of production planning: "TPS Layout Analysis System" (TPS-LAS) is a production optimization intended to realize a highly reliable production system by optimizing the layout of both of the production site as a whole and each production process with regard to production lines (logistics and transportation), robots (positioning), and production operators (allocation and workability) through the use of numeric simulation (Sakai and Amasaka, 2006a).

2. Reform of production preparation: "Human Intelligence Production Operating System" (HI-POS) is an intelligent operator development system intended to enable the establishment of a new people-oriented production system whereby training is conducted to ensure that operators develop the required skills to a uniform level, and diagnosis is then carried out to ensure that the right people are assigned to the right jobs (Sakai and Amasaka, 2006b). HI-POS is made up of two sub-systems: "Human Integrated Assist System (HIAS) and Human Intelligence Diagnosis System" (HIDS).

3. Reform of the working environment: "TPS-Intelligent Production Operating System" (TPS-IPOS) is intended to lead to a fundamental reform of the work involved in production operations by raising the technical skills level of production operators and further improving the reliability of their skills for operating advanced production equipment within an optimized working environment. TPS-IPOS is made up of three sub-systems: "Virtual –Intelligent Operator System" (V-IOS), "Aging & Work Development–Comfortable Operating System" (AWD-COS), and "Robot Reliability Design-Improvement Method" (RRD-IM) (Amasaka et al., 2000: Sakai and Amasaka, 2003, 2007a; Amasaka, 2007c).

4. Reform of process management: "TPS Quality Assurance System" (TPS-QAS) is an integrated quality control system intended to ensure that quality is built into production processes through scientific process management that employs statistical science to secure process capability (Cp) and machine capability (Cm) (Amasaka and Sakai, 1998, 2009). TPS-QAS is made up of two sub-

systems: "Quality Control Information System" (QCIS) and "Availability & Reliability Information Monitor System" (ARIM).

5. Visualization of production processes: "Human Digital Pipeline System" (HDPS) ensures that top priority is given to customers through manufacturing with a high level of quality assurance (Sakai and Amasaka, 2007b). This involves the visualization of intelligent production information throughout product design, production planning and preparation, and production processes, thereby facilitating the complete coordination of these processes.

This system enables the high-cyclization of business processes within manufacturing.

6. Globalization of production information: "Virtual-Maintenance Innovated Computer System" (V-MICS) is a global network system for the systemization of production management technology necessary to realize a highly reliable production system, which is required to achieve worldwide uniform quality and production at optimal locations (Sakai and Amasaka, 2005).

New Japan Global Manufacturing Model "NJ-GMM"

To be successful of global manufacturing in the near future, the Japanese manufacturing industry must develop an excellent manufacturing management technology employing the advances in manufacturing engineering that can continuously provide high value products in a timely manner surpassing "*Kaizen*" (improvement) of manufacture site symbolized by TPS based on the three actuals of the actual place, actual part and actual situation (Amasaka, 1988, 2009) (See to Chapter 4 and 5).

Moreover, a key of global manufacturing is the systematic deployment of SCM on a global scale that encompasses cooperative manufacturing operations with overseas suppliers employing a newly global partnering model (Amasaka, 2004a, 2008; Ebioka et al., 2007). To realize the world uniform quality-simultaneous model launches, the author has created the "New Japan Manufacturing Management Model" (NJ-GMM) by using six core models based on the structure of manufacturing engineering as shown in Figure 5 (Amasaka, 2020a,b,c, 2022a,b,c, 2023).

In "NJ-GMM", the main characteristics of each core model contribute to the advancement of manu-facturing management through actual QCD research by using statistical science, as follows (Amasaka, 2008, 2017a,b, 2020b, 2023);

1. "Intellectual Working Value Improvement Management Model" (IWV-IMM) as the basic principle of working value intends that the design, production, quality assurance, marketing, human resource development, and administration cooperate and must reform the technologies and skills of the workers for the revolution of operating technology and skill in workplaces (Yamaji et al., 2007a; Tsunoi et al., 2010). In particular, boosting morale, reduction of fatigue, development of physical strength, development of tools & devices, improvement of thermal environment, and prevention of illness & injury are required for evolution.

2. "Partnering Performance Measurement Model" (PPMM) for assembly makers and suppliers serves as a formulation model using radar chart for visualization for evaluating the actual status of Japanese partnering between automobile assembly makers and parts suppliers, which has been somewhat implicitly carried out in the past (Yamaji et al., 2008). In connection with the PPMM development,

the author creates PPMM-A for assembly makers, PPM-S for suppliers, and PPM-AS for assembly makers and suppliers, as a comprehensive dual performance measurement.

3. "Strategic Stratified Task Team Model" (SSTTM) for the driving force of problem-solving contributes to the strengthening of the ties among group manufacturing companies, non-group companies, and even overseas manufacturers (suppliers) (Amasaka, 2004a,b, 2008, 2017b). To realize this, the level of problem-solving technology rises in product development strategy I and II through joint task teams of intra-company departments and divisions (Task-1 to -6, Group task team, Department task team, Division task management team, Field task management team, Total task management team) called "Internal Partnering". This technology is further expanded to quality management strategy I to II through the domestic affiliated company, domestic non-affiliated company, and overseas non-affiliated company (Task-6 to -8, Joint A to C) called "External Partnering".

4. "Intelligence High-cycle System of Assembly Maker Production Process" (IHS-AMPP) for preventing defects and achieving quality is effective management of the advanced production process (Amasaka et al., 2008; Amasaka and Sakai, 2010; Amasaka, 2018). To improve the intelligent productivity of production operators and to consolidate the information about highly cultivated skills and operating skills, IHS-AMPP contains the advanced facilities into commonly shared fore core systems as follows; (I) "Highly Reliable Production System" (HRPS), (II) "Intelligent Quality Control System" (IQCS), (III) "Renovated Work Environment System" (RWES), and (IV) "Intelligent Operators Development System" (IODS).

5. "Strategic Quality Management using Performance Measurement Model" (SQM-PMM) is the criterion of strategic development of NJ-GMM. At a stage of creation of SQM-PMM, specifically, and a survey of top management at Japanese manufacturers was conducted in order to investigate the management achievements as a key to strategic global manufacturing (Survey targeted 898 companies) (Amasaka et al., 1999; Amasaka, 2009; Kozaki et al., 2012). First, the author conducted the graphical modeling method based on the correlations among the 28 practice items of six categories – TQM methods, technological development, CS, quality assurance, corporate culture, and total participation activities. Second, the author performed the formulation of SQM-PMM through canonical correlation analysis. Moreover, the author verified the validity of SQM-PMM standardization.

6. "Working Value Evaluation Model" (WVEM) for strengthening of manufacturing technology makes the basic of "Creative workplace" as moving power of SQM-PMM, and evaluates the awareness of the working value of workers by statistically analyzing data collected through actual condition survey on Toyota, Nissan and other three companies (Uchida et al., 2012). This model systematically covers 20 key factors based on the fatigue reduction, disease prevention, comfort, organizations, and intelligence ability. Then, the author received 25 sets of answers from workers in the manufacturing industry, and analyzed their weightings using covariance structure analysis. Also, standardization of evaluation formulas made it possible to compare the strength of each working value evaluation axis in order to deployment and verify the validity of WVEM.

Figure 5. New japan global manufacturing model (NJ-GMM)

APPLICATION EXAMPLES

In this section, the author illustrates some research examples of Toyota's pioneering technology as applications of the "NJ-PMM" based on the "NMT". "NJ-PMM" is contributing to the advancement of automobile management technology at Toyota and suppliers by developing "DGEM" with "NJ-GPM and NJ-GMM" above, and is proving to be effective both in Japan and overseas (Amasaka and Sakai, 2011; Amasaka, 2015, 2017a, 2018, 2019a, 2020a,b,c, 2021, 022a,b,c, 2023).

Innovation of Automobile Production Engineering Based on the "NJ-GPM"

Production planning employing TPS-LAS, HDPS and HI-POS

In TPS-LAS application based on production engineering, the author has developed a simulation of main body conveyance using its four constitute sub-systems ((i)-(iv)) to illustrate a highly liable production process that has contributed to the reform of production planning as shown in Figure 6 (Sakai and Amasaka, 2006a).

Firstly, specifically, a hypothetical production line is set up within a "digital factory" on a computer.

(i) TPS-LAS-DFS is then used to reproduce the flow of people and parts within the production site. This enables any interference between production machinery and production cycle times to be checked in advance using simulations. One type of advance simulation uses (ii) TPS-LAS-RCS for the optimum placement of welding robots for the main body to ensure that no interference occurs.

Next, advance verification is performed using (iii) TPS-LAS-WIS to ensure that the predetermined work (standardized work) is carried out within the predetermined cycle time with no waste (*muda*) or overburdening (*muri*). Then, (iv) TPS-LAS-LIS is used to establish optimized conveyance routes between processes and determine optimum buffer allocations.

Figure 6. TPS-LAS, developing the simulation of main body conveyance

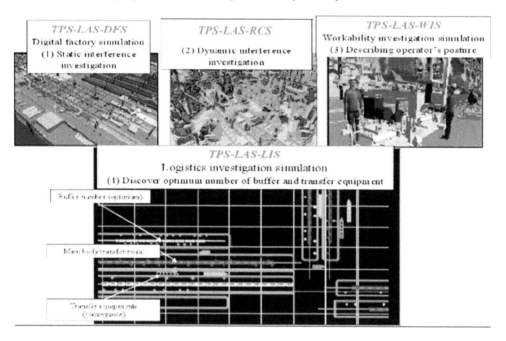

In addition, in HDPS application using production engineering, the author has carried out the image training of production processes in actual order of automobile assembly operations in order to the innovation of production planning as shown in Figure 7 (Sakai and Amasaka, 2007b).

To intellectual training, this is done at the preparatory stage before beginning mass production without having a real product on hand. HDPS promotes the leveling of the workload of operators in each process, and then completes the building up of production line even before launching it using Total Linkage System (TLS) of intellectual production information through the product design, production engineering and manufacturing.

Firstly, specifically, HDPS creates and supplies in advance "*Standard Work Sheets*" on which production operators have recorded each task in the correct order for jobs such as assembly work, by using design data for new products and facilities prepared from design through to production technology, even if there are no production prototypes.

Next, HDP enables visualization training for machining processes step-by-step in the order that parts are built up, even if the actual product does not yet exist. This system is proving to be very effective in raising the level of proficiency for processes requiring skills and capabilities at the production preparation stage.

Figure 7. HDPS, raising the level of skills and capabilities

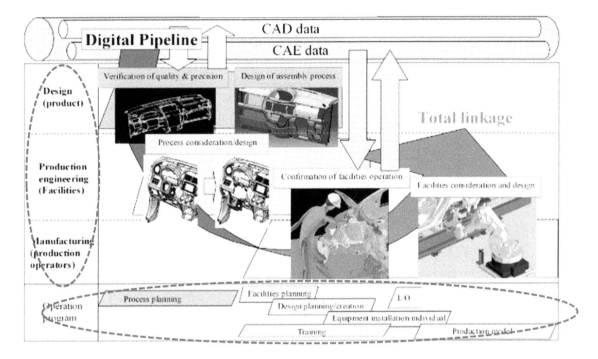

Furthermore, in HI-POS application employing production engineering, the author has created the intelligent operator development system intended to establish a new people - oriented production system as shown in Figure 8 (Sakai and Amasaka, 2003, 2005, 2006b, 2007a,b, 2008, 2015).

Firstly, the author has developed three constituent sub-systems–(a) HIAS for the evolution in production and manufacturing, (b) HIDS for the visualization of integrated production process based on the above (c) HDPS for the sharing of intelligent production information.

Next, to strengthen the integration of HI-POS, this system is integrated with two support systems–the following (d) V-IOS for the improvement in the advanced skill, and (e) V-MICS for the improvement in the operation technology of advance equipment.

The combined applications of TPS-LAS, HDPS and HI-POS are currently being contributed as a part of global production, and proving to be effective both in Japan and overseas.

Production preparation employing V-MICS, IPOS and TPS-QAS

Furthermore, the author has developed the production preparation for the strengthening of production process employing V-MICS, I-POS and TPS-QAS.

First, V-MICS takes a server and client system configuration employing DB (data base) and CG in order to browse information whenever necessary from the client computers at each maintenance station via the network, and can also input any special items as necessary as shown in Figure 9 (Sakai and Amasaka, 2005).

In V-MICS application using HDP, TPS-LAS, HI-POS, V-IOS and RRD-IM, the author has developed combining both the (i) "Production Process Optimization Simulation" (PPOS) and (ii) "Work Operation Combination Simulation" (WOCS) for the high productivity (Sakai and Amasaka, 2007a,b).

Figure 8. HI-POS, strengthening of the intelligent operator development system

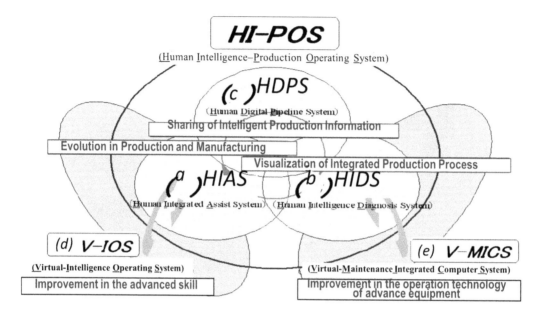

Figure 9. V-MICS, systematizing managerial technique of production facilities

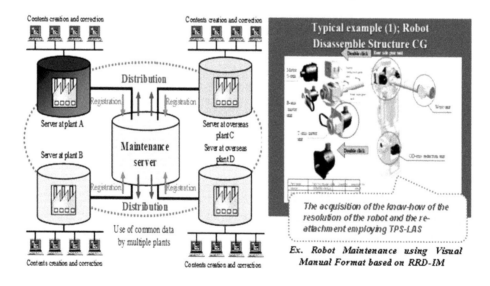

(i) PPOS shows the verification of the operators interfering with one another when determining which combinations of vehicle models cause the overlapping of operators in the time series in order to evaluation and problem finding as shown in Figure 10.

This is an aerial view of the process in which a vehicle is carried on the conveyor at a fixed speed, and the walking routes of the operators performing assigned work operations are displayed in real time in straight lines.

When multiple operators performing different operations in one process are overlapped (that is when operators' interference occurs), it is displayed instantly, and the information on which combination of models is also output immediately. Verification will be conducted through simulation as to whether operators can complete their entire operation of one process in the set "Takt-Time" and "Heijunka" (Amasaka, 1988, 2009).

Based on the simulation, verification results are displayed showing which process, on which vehicle model, caused the operator to stop the conveyor.

Figure 10. PPOS example of verification of operators' interfering with one another

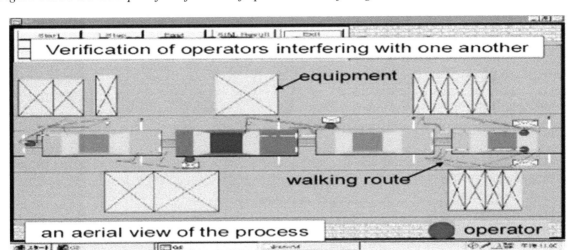

(ii) WOCS can be carried out in the real *Takt-Time* based on *Heijunka* as shown in Figure 11. In Figure 11(left), by sorting out the working time and walking time from the accumulated time results, the uneven distribution of the net-working time and the non-working time, such as time spent walking, are stratified to serve as a guideline for reviewing the process layout.

Then, Figure 11(right) shows the re-calculation of the work accumulation after reshuffling the basic work operations between processes. Such changes can be easily made on the accumulative simulation screen by dragging and dropping with the mouse, immediately confirming the work points of previous and subsequent operations while automatically adjusting the walking time involved.

Second, to employ I-POS strategically, the author has developed the (i) V-IOS, (ii) RRD-IM and (iii) TTD-IM employing HDPS, HI-POS and TPS-LAS. (i) V-IOS and (ii) RRD-IM using HIDS and *Visual Manual* (VM) is intended to improve the skills of new (inexperienced) production operators both in Japan and overseas.

Figure 11. WCOS example of accumulated work operations

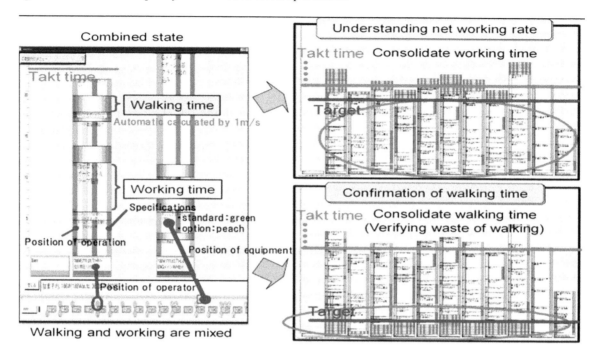

In Figure 12. at special training centers with simulations of actual assembly lines, the author shows the outline of (i) V-IOS: (a) training processes for assembly work, and (b) work training systems for assembly work are employed in the training of operators.

Then, once a certain level of skills has been mastered, operators progress to actual assembly lines where they are promptly and methodically developed as highly skilled and experienced technicians using (c) *standard work sheets* extracted (Sakai and Amasaka, 2003).

(iii)   AWD-COS constitutes a fundamental reform of work and labor named "Toyota's Epoch-Making Evolution in the Environment Model" as shown in Figure 13 (Amasaka et al., 2000; Amasaka, 2004a, 2007c).

In Figure 13, the author has initiated a company-wide project called "Aging & Work Development 6 Programs Project" (AWD6P/J) in order to combat the effects of aging as follows;

Project I is arousing motivation in workers

Project II is reviewing working styles to reduce fatigue

Project III is creating physical strength under the self-help efforts

Project IV is improving heavy work with user-friendly tools and equipment

Project V is creating thermal environments suited to the characteristics of assembly work, Project VI is reinforcing illness and injury prevention

For example, Figure 14 illustrates a result of Project II activity as the "Study work of changing the rest pattern to reduce fatigue".

In Figure 14(a), fatigue during operation gradually increases with time and decreases after each break shown in Normal Rest Pattern.

Figure 12. V-IOS example for the skill strengthening of operators using virtual operation

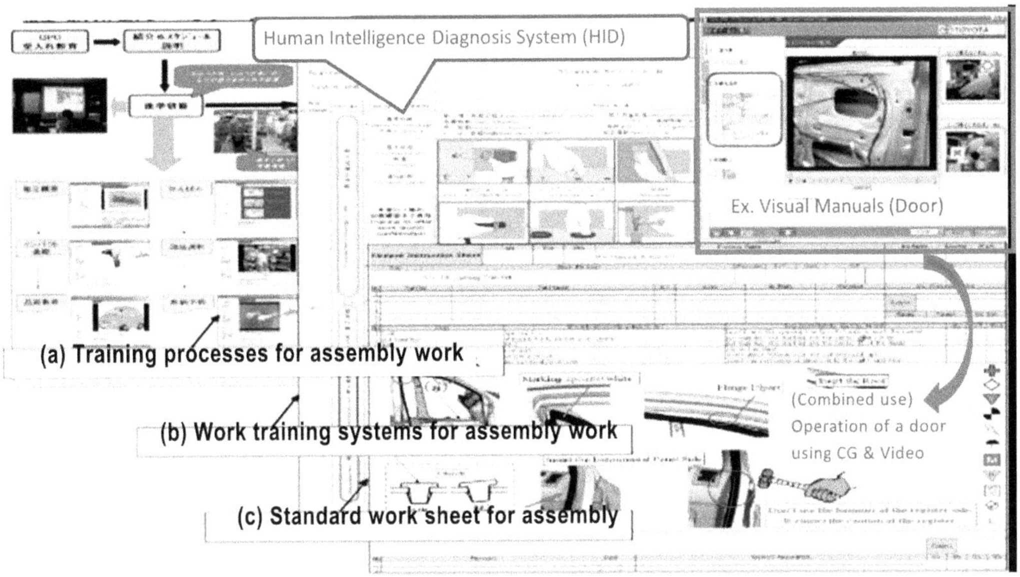

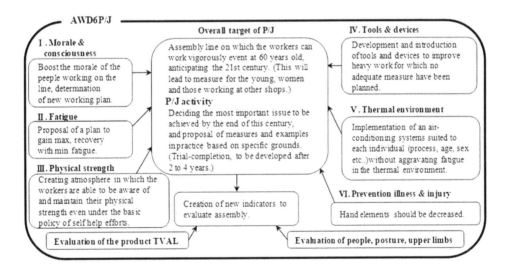

Figure 13. AWD-COS example for the aging and work development six programs project

In Figure 14(b), as the way to minimize fatigue, both of the trial "New 2 Rest Patterns: A & B" were tested in order to analyze the difference with the fatigue level of "Normal Rest Pattern".

In Figure 14(c), a line stop was cut by half by the decrease in quality defects and human error (from these experimental effects, Toyota adopts the trial "New Rest Pattern A in Japan and overseas at present)

Third, the author has established the TPS-QAS with two sub-systems–(i) QCIS and (ii) ARIM (Amasaka and Sakai, 1996, 1998, 2009) as follows:

Figure 14. Project II example -study work of changing the rest pattern to reduce fatigue

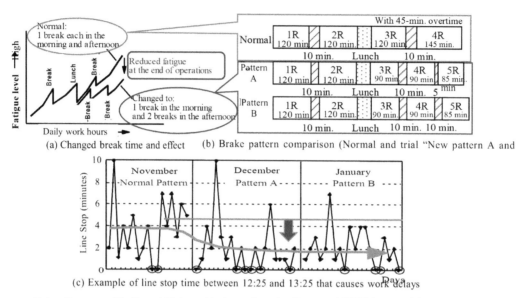

(a) Changed break time and effect (b) Brake pattern comparison (Normal and trial "New pattern A and

(c) Example of line stop time between 12:25 and 13:25 that causes work delays

Notes: Two assembly lines at Motomachi plant, where the Crown and IPSUM car models, were selected as experimental lines for 2-month trial.
       Fatigue level was calculated by both subjective and objective indicators (e.g., physiological).

(i)   QCIS enables the development of manufacturing with superior quality and productivity by integrating the high-precision quality control systems as shown in Figure 15.

In Figure 15(a), this illustrates a hardware system of trans-axle assembly line holding the (a) main-system, (b) sub-system, and connects (c) internal and external networks. In Figure 15(b), this shows a software system using intelligence quality control charts, and analyzes the diagnosis of process management abnormalities holding the (1) Scroll-function, (2) Display of grouped and raw data, (3) Hierarchical factorial analysis, (4) Kaizen, improvement of history database, (5) Abnormal diagnosis and (6) Data link with other application software.

Figure 15. An example of QCIS application using quality control charts

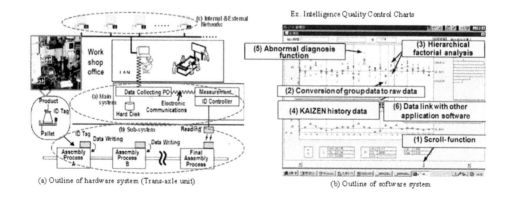

(a) Outline of hardware system (Trans-axle unit) (b) Outline of software system

(ii) ARIM enables the strengthen reliability analysis of facility operation and maintenance of information management as shown in Figure 16.

In Figure 16(a), this illustrates a hardware system for gathering the operating efficiency and failures using *Andon system*, and clusters of machinery on each production line in Japan and overseas. In Figure 16(b), for developing the preventive maintenance, this software system performs the *Weibull analysis*, failure diagnosis and failure life predicting employing Table 1 based on the reliability engineering.

Figure 16. An example of ARIM application using Weible analysis

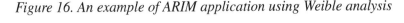

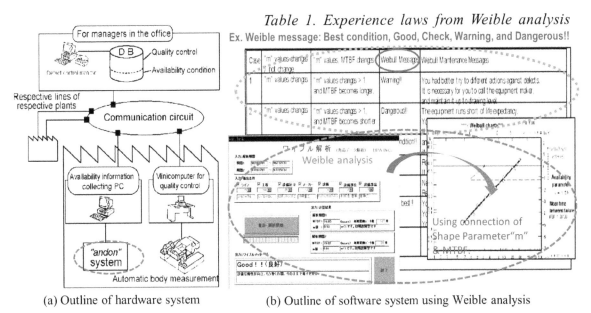

(a) Outline of hardware system   (b) Outline of software system using Weible analysis

Innovation of Automobile Manufacturing Engineering Based on the "NJ-GMM"

Improving skills of experienced and inexperienced machine workers

To develop IWV-IMM and WVEM based on the "NJ-GMM", the author has clarified the factors involved in improving the skills of experienced and inexperienced machine workers engaged in typical lathe work with 6 processing stages of setup, rough cutting, semi finishing, making adjustments, finishing and completion by using S45C workpiece (Refer to Yanagisawa et al. (2013) in detail).

Specifically, using electroencephalography (EEG) and statistical science, the physical and mechanical characteristics of a workpiece are those tensile strength, elongation, hardness, thermal conductivity, density, Young's modulus, shear elasticity modulus and yield strength, and the relative influence of these characteristics are clarified in accordance with the cutting conditions.

Through these investigations, the author clarified the important criteria for making decisions based on the chip color and chip shape.

For example, using "Covariance Structure Analysis", Figure 17 illustrates an example of clarifying factors for the "Chip color during finishing" by experienced machine workers, and clarifies the influence these elements receive from each factor.

This research clarifies points where inexperienced workers need guidance such as cutting tool selection, which is beneficial for both the person giving guidance and the person receiving guidance.

Moreover, the author has investigated how intuition and knowhow make a difference between the skills of experienced and inexperienced workers based on the validity of other researches (Refer to Kawahara et al. (2016) in detail).

These studies have contributed to technical training of inexperienced workers.

Figure 17. Chip color during finishing using covariance structure analysis

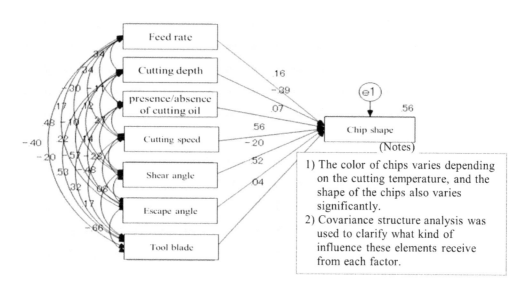

Accelerated Technical Training System for the highly skilled operators

To realize the shortened training of highly skilled operators in auto-assembly line employing "IWV-IMM and WVEM", the author has developed an "Accelerated Technical Training System" (ATTS) by development of "HI-POS" with "HIAS and HIDS" above (Sakai and Amasaka, 2006b, 2007b, 2008; Amasaka et al., 2008).

Specifically, "ATTS" contains the four sub-system of vehicle assembly process – (i) "Assessment system" of aptitude / inaptitude using aptitude test with eight categories of fundamental skills (: tightening, screw grommet, attachment, connector, hose, plug hole, tube and fitting), (ii) "Optimization system" of training steps with flowchart of training program, (iii) "Skill training system" for newly employed production operators, and (iv) "Shortened training system" for new overseas production operators as shown in Figure 18.

In Figure 18, the author illustrates an application case of "Toyota's shortened training for the new overseas production operators" with four step programs based on the courses during 5 courses during five days from "Day 1" to "Day 5"– Classroom lecture, Skill training, Dynamic training with simulating movement and Actual line training.

The deployment of "ATTS" for training newly employed production operators at domestic and overseas manufacturing plants reduced the conventional training period by more than half, from two weeks to five days, leading the full-scale production to a good start. In this case, the launch of an overseas production plant, the target operating rate was achieved in four months after the start-up.

Figure 18. Toyota's shortened training for the new overseas production operators in ATTS

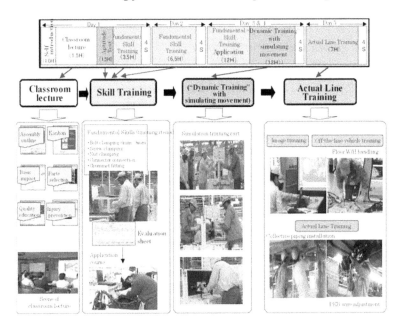

Global Intelligence Partnering Model for the global manufacturing strategy

To develop "PPMM" above, the authors has established the "Global Intelligence Partnering Model" (GIPM) for automobile global manufacturing strategy which improves the intellectual productivity of the affairs and management sections (Yamaji and Amasaka, 2007b).

Figure 19 shows the functions of the affairs and management sections as the corporate environment factors for succeeding in global marketing, customer-first, 1) CS, ES and SS, for 2) high quality product and as a strategic factor to realize it, in order to 3) product reliability (high quality assurance) and corporation reliability (excellent company), and success in 4) intellectual productivity and human resource development. Moreover, 5) global production is realized by these. So, the same quality and production at optimal locations are achieved.

For that purpose, the highest priority was given to the "i) Intellectual information sharing" and "ii) Strategic Co-Creative Action" so that the a) "Strategic intelligence application system", and b) "Business process high linkage system" can effectively function.

To actualize GIPM, the author has developed the strategic application examples for the epoch-making improvement of QCD employing "Toyota Supply System" (TSS) (Toyota, 1993; Amasaka, 2004b, 2008, 2009, 2018, 2019b) as follows (See to Chapter 4, 6, 7 and 8);

(a) Creation of "Automobile Exterior Design Model" (AEDM) for raising customers' worth of world's prestige car "Lexus", (b) Improvement of "Paint corrosion resistance and protect plating parts from corrosion" for raising auto-unit-axle reliability, (c) Development of "Mid-frequency tempering equipment using the heating coil of electromagnetic induction", (d) "Automated straightening equipment" for the realizing high productivity of auto-unit-axle shaft, and (e) "New ceramic materials to the welding nozzle and welding tip" for the improving high operational availability of auto-unit-axle housing".

Figure 19. GIPM for the global manufacturing strategy

Strategic Cooperative Task Team Model for raising management technology

To organically promotion of above "SSTTM" based on the JSS (Japan Supply System), the author has structured the "Strategic Cooperative Task Team Model" (SCTTM) between the auto-maker and affiliated/non-affiliated suppliers as shown in Figure 20 (Amasaka, 2004a, 2008, 2017b) (See to Chapter 6, 7 and 10).

In Figure 20, to purchase the necessary parts, it will be important for the manufacturer (maker) to mutually cooperate with (a) Supplier I, in-house parts maker (own company)), (b) Supplier II, (affiliated maker (capital participation)), (c) Supplier III, non-affiliated maker, and (d) Supplier IV, overseas maker (capital participation).

Specifically, in the actual deployment of "SCTTM" strategy, it is important to strategically organize the stratified task teams for raising management technology as follows: (i) Product strategy, (ii) Engineering strategy, (iii) Quality strategy, (iv) QCD effect, (vi) Human resource strategy and (v) Value of task teams.

Concretely in "SCTTM" for the "evolution of automobile manufacturing, the important job for the manufacturer's general administrator is to select jointly from his own company and suppliers described in Figure 20:

(1)   Generators gifted with a special capacity for creating ideas,
(2)   Mentors having the ability to give guidance and advice,
(3)   Producers with the capability to achieve and execute,
(4)   Promoters capable of implementing things as an organization.

Then, as the key to successful "SCTTM", the team leader (administrator) should select the members who have at least one of the capabilities for above (1) to (4), commission authority and responsibilities to the members, and has himself / herself concentrate on risk management.

Figure 20. SCTTM for the evolution of Japanese automobile manufacturing

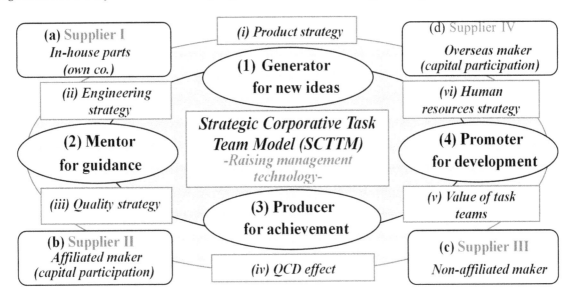

To develop this, the author presents typical case studies as the improving of the bottleneck problems of worldwide auto-makers employing "Toyota's Total Task Management Model" (T-TTMM) (Amasaka, 2004a). Specifically, to employes "T-TTMM" strategy, the author develops both of the "PPMM" (Partnering Performance Measurement Model) and "SQM-PPM" (Strategic Quality Management——Performance Measurement Model)" for realizing simultaneous QCD fulfilment (Yamaji et al., 2008; Amasaka, Ed., 2012; Amasaka, 2017b, 2022b,c, 2023).

For example, the typical cases of Suppliers I, II and III activities, the author has developed the establishment of "Brake pad quality assurance" using "Total QA Network" (QAT) activity by Toyota, Aishin Seiki, Aishin Chemical and Akebono Brake Industry.

The main objective of "QAT" activity is to establish a technology for attaining satisfactory "braking performance" while minimizing squeal through the "result of sensitivity analysis on the causes of squeal and braking performance".

The result of "QAT" is remarkable effects as follows; (i) Estimated market claim ratio reduction to 1/6 (from 2.6% to 0.4%), (ii) In-process fraction defective: down 60% (from 0.5% to 0.2%), (iii) Short convergence of initial failures: (from 9 months to 3 months), and (iv) Cost reduction: 9.4% (156 Yen/unit).

Moreover, in other cases of Suppliers I-IV, the author was able to apply the SCTTM activity as follows; (i) "TQM and SQC training for growing human resources" of Toyota group in Japan and overseas, (ii) "Specific engine fuel consumption improvement", (iii) "High reliability assurance of the transaxle oil seal leakage", (iv) "Rust preventive quality assurance of rod piston plating parts", (v) "Various anti-rusting methods to various section of the vehicle body", (vi) "Improvement of the total curvature spring back in stamped parts with large curvature", (vii) "Improvement of the "reliability of body assembly line equipment", and (viii) "Bolt-nut loosening solution by developing Automobile Optimal Product Design Model" (AOPDM) (Toyota, 1993; Amasaka, 2004b, 2019a,b; Amasaka et al., 2012).

New Global Partnering Production Model for expanding overseas manufacturing

To develop "IHS-AMPP" using "IWV-IMM and WVEM" above, the author has established a "New Global Partnering Production Model" (NGP-PM) for Japan's expanding overseas manufacturing strategy as shown in Figure 21 (Ebioka et al., 2007; Amasaka and Sakai, 2010, 2011).

The mission of "NGP-PM" is the simultaneous achievement of QCD in order to realize high quality assurance. The essential strategic policies include the following items:

Figure 21. NGP-PM for Japan's expanding overseas manufacturing strategy

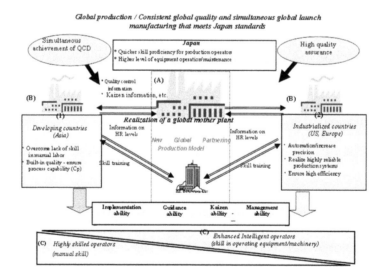

First of all, (A) establishment of a foundation for global production, "realization of global mother plants-advancement of Japanese production sites";

Second, (B) achieving the "independence of local production sites" through the incorporation of the unique characteristics (production systems, facilities, and materials) of both developing countries (Asia) and industrialized countries (US, Europe);

Third, (C) necessity of developing intelligence operators to promote knowledge sharing among the production operators in Japan and overseas as well as for the promotion of higher skills and enhanced intelligence.

To actualize NGP-PM, it is essential to create a spiraling increase in the four core elements by increasing their comprehensiveness and high cycle-ization.

Specifically, in realizing global mother plants, if Japanese and overseas manufacturing sites are to share knowledge from their respective viewpoints, the core elements must be advanced. To achieve this, a necessary measure is to design separate approaches suited to developing and industrialized countries.

Concretely, in developing countries (1), the most important issue is increasing the autonomy of local manufacturing sites.

At these sites, "training for highly skilled operators" that is suited to the manual- labor based manufacturing sites is the key to excellent QCD studies.

Similarly, in industrialized countries (2), where manufacturing sites are based on automatization and increasingly high-precision equipment, "training of intelligence operators" resulting in "realizing highly reliable production control systems and ensuring high efficiency" is the key to excellent QCD studies.

Moreover, production operators trained at "global mother plants" (3) can cooperate with operators at overseas production bases, and in order to generate synergistic results, can work to "localize global mother plants" in a way that is suited to the overseas production bases.

As the typical examples of NGP-PM deployment, the author describes the typical case studies of new integrated local production by partnering Toyota and overseas as follows;

(i) "New Turkish Production System" (NTPS), an integration and evolution of "both of the Japanese and Turkish Production System", (ii) New Malaysia Production Model (NMPM), a new integrated production system of Japan and Malaysia, (iii) New Vietnam Production Model (NVPM), developing hybrid production of Japan and Vietnam, and (iv) Developing Advanced TPS at Toyota Manufacturing USA (Amasaka, 2007c, 2016; Yeap et al., 2010; Shan, et al., 2011: Miyashita and Amasaka, 2014).

Deployment and validity of these studies are detailed to Amasaka (2015, 2017a, 2022,b,c, 2023).

Total Quality Assurance Networking Model for Toyota's new defect prevention

To survive globalization and worldwide quality competition, Japanese manufacturing must work in order to shorten development times, ensure high quality and lower costs for the shift response to market changes (Amasaka, 2007a). Therefore, to develop the above IHS-AMPP (Intelligence High-cycle System of Assembly Maker Production Process), the author has created the "Total Quality Assurance Networking Model" (TQA-NM) as shown in Figure 22 (Amasaka, 2004b; Kojima and Amasaka, 2011).

In Figure 22, TQA-NM focuses on the subject of (A) Clear QA standards that make use of quantitative values (process capability, Cp), (B) Systematic use of know-how and experience resolving part issue, (C) Development of staff (workers) and managers familiar with the site, (D) QA tools for staff (workers) and managers and (E) Process visualization enabled" which turned their attention especially to the (F) "stronger partnerships between assembly manufacturers and suppliers"

Then, this model is done through the "Prevention of defect occurrence and by supporting the simultaneous achievement of QCD" that come from "Strategically deploying High-level quality assurance process" described in Figure 22 as follows;

(I) One of the technological components in achieving these goals is strengthening QA networks by "development of QA tools",

(II) Second is the establishment of "Clear quantitative QA standards" that are not affected by worker experience,

Figure 22. TQA-NM for Toyota's new defect prevention technique

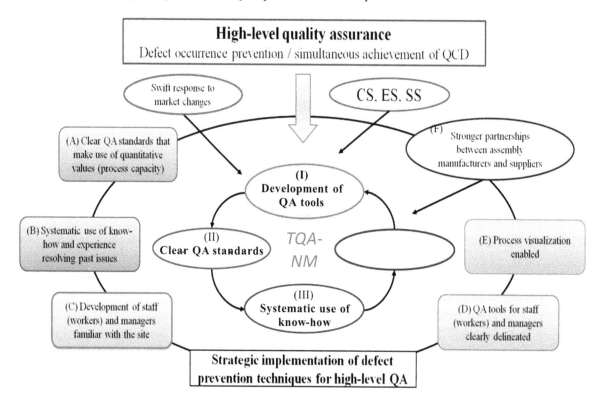

(III) Third is the development of "Systematic use of know-how" that can make use of worker expertise and information on past defects in an organized manner,

(IV) Fourth is the development of workers and frontline staff as the "High quality HRD" (Human Resource Development) who are familiar with the site.

To realize TQA-NM, specifically, the "Total QA Networking Chart" shown in Figure 23 is a defect occurrence prevention technique featuring a combination of QA tools. It also uses the "Quality Function Deployment" (QFD) or "Failure Mode and Effect Analysis" (FMEA), and "matrix diagrams" to deploy partnerships.

In Figure 23, the "vertical axis" identifies joint processes by suppliers and assembly manufacturers, from the arrival of goods to shipment, and the "horizontal axis" lists individual defects.

The degree of QA is indicated where the two axes intersect, which allows a quantitative value indicating the importance of process management tasks to be assigned to each.

Each process is comprehensively evaluated on the basis of what the defect occurrence prevention tasks are for that process and how highly they are ranked in terms of QA level using QA matrix.

The task of this chart is ensured based on the "Ranking Occurrence and Outflow prevention", and is established a "High-level QA system".

As the pioneering examples of TQA-NM deployment of "Toyota and Toyota's SCM strategy", the author presents the typical case studies "(i) –(vi)" as follows;

(i)   Establishment of the "optimum casting conditions for compact DOHC cylinder head",
(ii) Development of ferrite heat resistant cast steel,
(iii) Measures to be taken for difficulties in the "manufacture of aluminum die-casted head cover",
(iv) "Rust prevention quality assurance of plated components",
(v)   Improving the "brake pads quality of trans-axle unit,"
(vi) CAE analysis for "oil leakage mechanism of transaxle oil seal" etc. (Toyota, 1993; Amasaka, 2004b, Amasaka Ed.,2012; Amasaka, et al., 2012).

Deployment and validity of these studies are detailed to Amasaka (2015, 2017a, 2020b,c, 2022b,c, 2023).

Figure 23. An example of the total QA networking chart

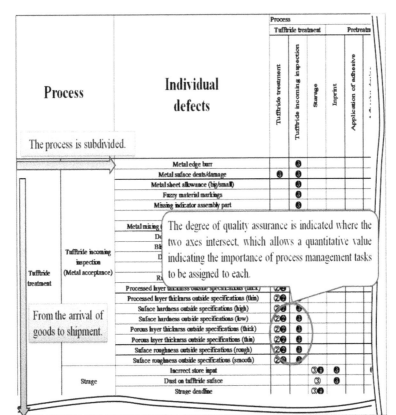

CONCLUSION

In this study, to re-construct world-leading management technologies, the author has described the "innovation of Japan manufacturing foundation NJ-PPM (New Japan Production Management Model) called "Advanced TPS, strategic production management system" for developing "New JIT strategy" that surpasses conventional JIT practices. On a concrete target, the author focused on the strategic development of DGEM (Dual Global Engineering Model) possessing the NJ-GPM (New Japan Global Production Model) and NJ-GMM (New Japan Global Manufacturing Model) surpassing JIT. As these results, the effectiveness of DGEM was verified through the actual applications to automobile manufacturing in Toyota and suppliers.

REFERENCES

Amasaka, K. (1988). *Concept and progress of Toyota Production System (Plenary lecture), Co-sponsorship: Japan Society of Precision Engineering and others*. Hachinohe, Aomori-ken.

Amasaka, K. (2002). New JIT, a new management technology principle at Toyota. *International Journal of Production Economics, 80*(2), 135–144. doi:10.1016/S0925-5273(02)00313-4

Amasaka, K. (2004a). Development of Science TQM, a new principle of quality management: Effectiveness of strategic stratified task team at Toyota. *International Journal of Production Research, 42*(7), 3691–3706.

Amasaka, K. (2004b). *Science SQC, new quality control principle: The quality strategy of Toyota*. Springer-Verlag Tokyo. doi:10.1007/978-4-431-53969-8

Amasaka, K. (Ed.). (2007). New Japan Model "Science TQM"– Theory and practice for strategic quality management. Study group of ideal situation on quality management at manufacturing industry, Tokyo: Maruzen.

Amasaka, K. (2007a). *New Japan Production Model*, an advanced production management principle: Key to strategic implementation of *New JIT. The International Business & Economics Research Journal, 6*(7), 67–79.

Amasaka, K. (2007b). High Linkage Model *"Advanced TDS, TPS & TMS"*: Strategic development of *"New JIT"* at Toyota. *International Journal of Operations and Quantitative Management, 13*(3), 101–121.

Amasaka, K. (2007c). Applying New JIT–Toyota's global production strategy: Epoch-making innovation in the work environment. *Robotics and Computer-integrated Manufacturing, 23*(3), 285–293. doi:10.1016/j.rcim.2006.02.001

Amasaka, K. (2008). Strategic QCD studies with affiliated and non-affiliated suppliers utilizing New JIT, Encyclopedia of Networked and Virtual Organizations, 3(PU-Z), 1516-1527.

Amasaka, K. (2009). The foundation for advancing Toyota Production System utilizing New JIT. *Journal of Advanced Manufacturing Systems, 80*(1), 5–26. doi:10.1142/S0219686709001614

Amasaka, K. (Ed.). (2012). Science TQM, new quality management principle: The quality management strategy of Toyota. Bentham Science Publishers, Sharjah, UAE, USA, THE Netherlands.

Amasaka, K. (2014). New JIT, new management technology principle. *Journal of Advanced Manufacturing Systems*, *13*(3), 197–222. doi:10.1142/S0219686714500127

Amasaka, K. (2015). New JIT, New Management Technology Principle. Boca Raton, USA: Taylor and Francis Group, CRC Press.

Amasaka, K. (2016). *Innovation of automobile manufacturing fundamentals employing New JIT: Developing advance Toyota Production System at Toyota Manufacturing USA*. Proceedings of the 5th Conference on Production and Operations Management, Habana International Conference Center, Cuba.

Amasaka, K. (2017a). *Toyota: Production system, safety analysis and future directions*. NOVA Science Publishers.

Amasaka, K. (2017b). Strategic Stratified Task Team Model for realizing simultaneous QCD fulfilment: Two case studies. *The Journal of Japanese Operations Management and Strategy*, *7*(1), 14–35.

Amasaka, K. (2018). Innovation of automobile manufacturing fundamentals employing New JIT: Developing Advanced Toyota Production System, *International Journal of Research in Business. Economics and Management*, *2*(1), 1–15.

Amasaka, K. (Ed.). (2019). *Fundamentals of manufacturing industry management: New Manufacturing Theory – Operations management strategy 21C*. Sankei-sha.

Amasaka, K. (2019a). Studies on New Manufacturing Theory, *Noble. International Journal of Scientific Research*, *3*(1), 1–20.

Amasaka, K. (2019b). Establishment of an Automobile Optimal Product Design Model: Application to study on bolt-nut loosening mechanism. *Noble International Journal of Scientific Research*, *3*(9), 79–102.

Amasaka, K. (2020a). Studies on New Japan Global Manufacturing Model: The innovation of manufacturing engineering. *Journal of Economics and Technology Research*, *1*(1), 42–71. doi:10.22158/jetr.v1n1p42

Amasaka, K. (2020b). Evolution of Japan Manufacturing Foundation: Dual Global Engineering Model Surpassing JIT. *International Journal of Operations and Quantitative Management*, *26*(2), 101–126. doi:10.46970/2020.26.2.3

Amasaka, K. (2020c). *New Manufacturing Theory: Surpassing JIT*. Lambert Academic Publishing.

Amasaka, K. (2021). New Japan Automobile Global Manufacturing Model: Using Advanced TDS, TPS, TMS, TIS & TJS. *Journal of Business Management and Economic Research*, *6*(6), 499–523.

Amasaka, K. (2022a). A New Automobile Global Manufacturing System: Utilizing a Dual Methodology, *Scientific Review, Journals in Academic Research Publishing. Group*, *8*(3), 51–58.

Amasaka, K. (2022b). *Examining a New Automobile Global Manufacturing System*. IGI Global Publisher. doi:10.4018/978-1-7998-8746-1

Amasaka, K. (2022c). New Manufacturing Theory: Surpassing JIT (2nd Edition). Lambert Academic Publishing.

Amasaka, K. (2023). *New Lecture-Surpassing JIT: Toyota Production System -From JIT to New JIT.* Lambert Academic Publishing.

Amasaka, K. et al. (1999). The preliminary survey in 68QCS: Questionnaire result of TQM which creates Japan (Special Lecture), *Union of Japanese Scientists and Engineers, Hakone, Kanagawa-ken, Japan.*

Amasaka K. et al. (2000). *AWD6P/J Report of First Term Activity 1996-1999: Creation of 21C production line in which people over 60's can work vigorously.* Toyota Motor Corporation.

Amasaka, K., Ito, T., & Nozawa, Y. (2012). A New Development Design CAE Employment Model. *The Journal of Japanese Operations Management and Strategy, 3*(1), 18–37.

Amasaka, K., Kurosu, S., & Morita, M. (2008). *New Manufacturing Theory: Surpassing JIT – The evolution of Just-In-Time.* Morikita-Shuppan.

Amasaka, K., & Sakai, H. (1996). Improving the reliability of body assembly line equipment. *International Journal of Reliability Quality and Safety Engineering, 3*(1), 11–24. doi:10.1142/S021853939600003X

Amasaka, K., & Sakai, H. (1998). Availability and Reliability Information Administration System "ARIM-BL" by methodology of Inline-Online SQC. *International Journal of Reliability Quality and Safety Engineering, 5*(1), 55–63. doi:10.1142/S0218539398000078

Amasaka, K., & Sakai, H. (2009). TPS-QAS, new production quality management model: Key to New JIT–Toyota's global production strategy. *International Journal of Manufacturing Technology and Management, 18*(4), 409–426. doi:10.1504/IJMTM.2009.027774

Amasaka, K., & Sakai, H. (2010). Evolution of TPS fundamentals utilizing New JIT strategy–Proposal and validity of Advanced TPS at Toyota. *Journal of Advanced Manufacturing Systems, 9*(2), 85–99. doi:10.1142/S0219686710001831

Amasaka, K., & Sakai, H. (2011). The New Japan Global Production Model "NJ-GPM": Strategic development of *Advanced TPS. The Journal of Japanese Operations Management and Strategy, 2*(1), 1–15.

Doos, D., Womack, J. P., & Jones, D. T. (1991). *The Machine that Changed the World – The Story of Lean Production.* Rawson / Harper Perennial.

Ebioka, K., Sakai, H., Yamaji, M., & Amasaka, K. (2007). A New Global Partnering Production Model "NGP-PM" utilizing Advanced TPS. *Journal of Business & Economics Research, 5*(9), 1–8.

Hayes, R. H., & Wheelwright, S. C. (1984). *Restoring Our Competitive Edge: Competing through Manufacturing.* Wiley.

Kawahara, F., Kogane, Y., Amasaka, K., & Ouchi, N. (2016). A skill map based on an analysis of experienced workers' intuition and knowhow. *IOSR Journal of Business and Management, 18*(12), 80–85.

Kojima, T., & Amasaka, K. (2011). The Total Quality Assurance Networking Model for preventing defects: Building an effective quality assurance system using a Total QA Network. *International Journal of Management & Information Systems., 15*(3), 1–10. doi:10.19030/ijmis.v15i3.4637

Kozaki, T., Oura, A., & Amasaka, K. (2012). Establishment of TQM Promotion Diagnosis Model "TQM-PDM" for strategic quality management. *China-USA Business Review*, *10*(11), 811–819.

Miyashita, S., & Amasaka, K. (2014). Proposal of a New Vietnam Production Model (NVPM), a new integrated production system of Japan and Vietnam. *IOSR Journal of Business and Management*, *16*(12), 18–25. doi:10.9790/487X-161211825

Ohno, T. (1977). *Toyota Production System*. Diamond-sha.

Sakai, H., & Amasaka, K. (2003). *Construction of "V-IOS" for promoting intelligence operator: Development and effectiveness for visual manual format.* The Japan Society for Production Management, The 18th Annual Conference, Nagasaki Institute of Applied Science, Nagasaki, Japan.

Sakai, H., & Amasaka, K. (2005). V-MICS, Advanced TPS for strategic production administration: Innovative maintenance combining DB and CG. *Journal of Advanced Manufacturing Systems*, *4*(6), 5–20. doi:10.1142/S0219686705000540

Sakai, H., & Amasaka, K. (2006a). TPS-LAS Model using process layout CAE system at Toyota, Advanced TPS: Key to global production strategy *New JIT*. *Journal of Advanced Manufacturing Systems*, *5*(2), 1–14. doi:10.1142/S0219686706000790

Sakai, H., & Amasaka, K. (2006b). Strategic HI-POS, intelligence production operating system: Applying Advanced TPS to Toyota's global production strategy. *WSEAS Transactions on Advances in Engineering Education*, *3*(3), 223–230.

Sakai, H., & Amasaka, K. (2007a). The robot reliability design and improvement method and Advanced Toyota Production System. *The Industrial Robot*, *34*(4), 310–316. doi:10.1108/01439910710749636

Sakai, H., & Amasaka, K. (2007b). Human Digital Pipeline Method using total linkage through design to manufacturing. *Journal of Advanced Manufacturing Systems*, *6*(2), 101–113. doi:10.1142/S0219686707000929

Sakai, H., & Amasaka, K. (2008). Human-Integrated Assist Systems for intelligence operators, Encyclopedia of Networked and Virtual Iorganizations, 2(G-Pr), 678-687.

Shan, H., Yeap, Y. S., & Amasaka, K. (2011). *Proposal of a New Malaysia Production Model "NMPM": A new integrated production system of Japan and Malaysia*. International Conference on Business Management 2011, Miyazaki Sangyo-Keiei University, Miyazaki, Japan.

Taylor, D., & Brunt, D. (2001). *Manufacturing Operations and Supply Chain Management - Lean Approach*, Boston, USA: Thomson Learning.

Toyota Motor Corporation. (1993). Toyota Technical Review [hosted by Amasaka, K.]. *SQC at Toyota*, *43*(Special issue), 1–172.

Tsunoi, M., Yamaji, M., & Amasaka, K. (2010). A study of building an Intellectual Working Value Improvement Model, IWV-IM. *The International Business & Economics Research Journal*, *9*(11), 79–84. doi:10.19030/iber.v9i11.33

Uchida, K., Tsunoi, M., & Amasaka, K. (2012). Creating Working Value Evaluation Model, WVEM. *International Journal of Management & Information Systems*, *16*(4), 299–306.

Womack, J. P., Jones, D., & Roos, D. (1990). *The machine that change the world – The story of Lean Production.* Rawson/Harper Perennial.

Womack, J. P., & Jones, D. T. (1994). From Lean Production to the Lean Enterprise. *Harvard Business Review*, (March-April), 93–103.

Yamaji, M., & Amasaka, K. (2007b). Proposal and validity of Global Intelligence Partnering Model for Corporate Strategy, GIPM-CS. *International IFIP-TC 5.7 Conference on Advanced in Production Management System.* Springer.

Yamaji, M., Sakai, H., & Amasaka, K. (2007a). Evolution of technology and skill in production workplaces utilizing Advanced TPS. *Journal of Business & Economics Research*, *5*(6), 61–68.

Yamaji, M., Sakatoku, T., & Amasaka, K. (2008). Partnering Performance Measurement "PPM-AS" to strengthen corporate management of Japanese automobile assembly makers and suppliers. *International Journal of Electronic Business Management*, *6*(3), 139–145.

Yanagisawa, K., Yamazaki, M., Yoshioka, K., & Amasaka, K. (2013). Comparison of experienced and inexperienced machine workers. *International Journal of Operations and Quantitative Management*, *19*(4), 259–274.

Yeap, Y. S., Murat, M. S., & Amasaka, K. (2010). Proposal of New Turkish Production System, NTPS: Integration and evolution of Japanese and Turkish Production System. *Journal of Business Case Study*, *6*(6), 69–76.

Chapter 9
New Automobile Sales Marketing Model for Innovating Auto Dealer's Sales

ABSTRACT

In this chapter, the author describes the new automobile sales marketing model (NA-SMM) using four core elements based on a dual corporate marketing strategy for innovation of auto-dealers' sales. To realize this, the author develops both the customer science principle (CSp) and science SQC, new quality control principle. Specifically, foundation of NA-SMM consists of the scientific customer creative model (SCCM), networking of customer science principle application system (NCSp-AM), video unites customer behavior and maker's designing intentions (VUCMIN) and scientific mixed media model (SMMM). The validity of NA-SMM is then verified through each model of actual applications to customer creation in Toyota.

CHANGE IN AUTO-SALES MARKETING

In today's rapidly changing Japanese corporate management in "auto-sales marketing environment", one of the first management technology issues that place the "customer first, is the "creation of new auto-sales marketing system" that breaks away from the conventional systems in an effort to strengthen ties with the customers (Amasaka, et al.,1998, 2005; Amasaka, 2001, 2005, 2007a,b).

This new understanding should form the basis of business in "auto-sales marketing innovation" for "customer value creation" (Amasaka, 2011). Therefore, a "auto-sales marketing management model" in automotive industry needs to be established so that *Business / Sales / Service* divisions, which are also in the closest position to customers, can organizationally learn customers' tastes and desires by means of the continued application of objective data and scientific methodology (Amasaka, 2003a, 2004, 2005).

However, at present, the organizational system in sales marketing has not yet been fully established in these divisions; in some cases, even the importance of this system has not been commonly realized (Amasaka, 2005).

The above research suggests that it is necessary to promote scientific activities involving sales and sales-related departments. In particular, the author believes that a better quantitative understanding of the effect of publicity and advertising in "auto-sales marketing" will enable more effective sales activities (Amasaka, 2001, 2009; Amasaka, Ed. 2007).

DOI: 10.4018/978-1-6684-8301-5.ch009

Considering recent changes in the "auto-sales marketing environment", it is now necessary to implement innovation of business and sales activities to accurately grasp the characteristics and changes of customer preferences independently of convention.

IMPORTANCE OF AUTO-SALES MARKET CREATION DEVELOPING A NEW AUTO-SALES MARKEING MODEL

Need for a Marketing Strategy Which Considers Market Trends

After the collapse of the bubble economy, the competitive environment in the market has drastically changed. Since then, companies that have implemented strategic marketing quickly and aggressively have been the only ones enjoying continued growth (Okada et al., 2001).

Therefore, a marketing management model needs to be established so that business / sales / service divisions, which are carrying out development and design for appealing products projects, and which are also in the closest position to customers, can organizationally learn customers' tastes and desires by means of the continued application of objective data and scientific methodology (James and Mona, 2004; Amasaka, 2005).

However, at present, the organizational system has not yet been fully established in these divisions; in some cases, even the importance of this system has not been commonly realized (Gary and Arvind, 2003; Amasaka, 2005).

That is, close observation of recent changes in the field of marketing by the author and other researchers has led us to the conclusion that it is necessary to place more emphasis on communicating with customers in order to gain an adequate understanding of their changing characteristics, unbiased by established concepts (Niiya and Matsuoka, 2001; Okada et al., 2001; Amasaka, 2001, 2007a,b).

Importance of an Auto-Sales Marketing Management Model Innovating Auto-Dealers' Sales Activities

Therefore, to realize the above-mentioned subjects, considering recent changes in the marketing environment, it is now necessary to implement innovation of business and sales activities to accurately grasp the characteristics and changes of customer preferences independently of convention. Then, an "auto-sales marketing management model" needs to be established so that business, sales, and service divisions, which are developing and designing appealing products and are also closest to customers, can organizationally learn "customer tastes and desires" (Amasaka et al., 2005; Amasaka, 2008).

Specifically, pursuing improvements in product quality by the continued application of objective data and scientific methodology is increasingly important (Amasaka et al., 1998; James and Mona, 2004: Amasaka, 2003a,b, 2004, 2005).

At present, the organizational system and rational methodology that allows them to analyze data on each customer using a scientific analysis approach has not yet been fully established in these divisions; in some cases, the importance of this system has not even been widely recognized (Ikeo, Ed., 2006; Amasaka, 2007c, 2011; Amasaka, Ed., 2012).

Significance of Strategic Marketing Development Model Based on Total Marketing System

Recently, in light of recent changes in the auto-marketing environment "customer orientation and quality assurance", the author believes it is now necessary to develop innovative business and sales activities that adequately take into account the changing characteristics of customers who are seeking to break free from convention (Amasaka, 2011; Amasaka et al., 2005).

Then, to strengthen the "customer orientation and quality assurance" in marketing of automobile, the author has adapted the "Total Marketing System" (TMS) with 4 sub-core elements "(a)-(d)" in "New JIT" which is composed of integrated 5 core systems as shown in Figure 1 as follows (Amasaka, 2002, 2007b, 2008, 2009, 2010a, 2014, 2015, 2022a,b, 2023a,b);

(a) Market creating activities through collection and utilization of customer information.
(b) Improvement of product value based on the understanding that products are supposed to retain their value in product design.
(c) Establishment of marketing system from the viewpoint of building ties (bonds) with customers.
(d) Realization of the customer focus using customer information network for the CS (customer satisfaction), CD (customer delight) and CR (customer retention) elements needed for the corporate attitude (behavior norm) to enhance customer focus.

Figure 1. Total marketing system (TMS)

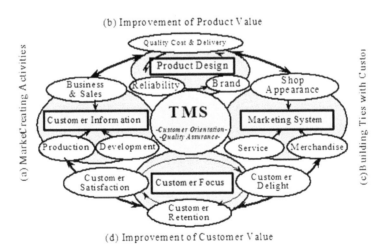

Particularly, therefore, to realize market creation with an emphasis on the customer by developing CL (Customer Royalty) to boost marketing effectiveness, the author has developed the "Strategic Marketing Development Model" (SMDM) named "Advanced TMS, strategic customer creative system" based on TMS as shown in Figure 2 (Amasaka, 2007a,b,c; Yamaji and Amasaka, 2009; Amasaka Ed. 2012; Okutomi and Amasaka, 2013) (See to Figure 9 in Chapter 7).

Figure 2. Strategic marketing development model (SMDM)

SMDM is to promote market creation and to realize quality management through scientific marketing and sales, not by sticking to conventional concepts by using organized "4 sub-core elements (i)-(iv)" as described in Figure 3.

Aim of SMDM is strengthen of "high quality assurance and innovating dealers' sales activity" in "global marketing strategy: "Same quality worldwide, development design and production at optimum locations".

Specifically, to develop the SMDM strategy, the author has created a "Modeling of Strategic Marketing System" (MSMS) named "Scientific Customer Creative Model" (SCCM). MSMS consists of 4 core elements: (i) "New Sales Office Image Model (NSOIM)" is important to achieve a high cycle rate for market creation activities by "innovation for building bonds with customer" and "reform of shop appearance & operation".

In the practice stage, it is more important to develop the (ii) "Intelligent Customer Information Network Model (ICINM)" by "strengthening of merchandise power", (iii) "Rational Advertisement Promotion System (RAPS)" and (iv) "Intelligent Sales Marketing System (ISMS)" by "innovation of sales and after-service" and "innovation of employee images". By these elements, SMDM innovates for bonding with the customer and reforms office-shop appearance and operation (Amasaka, 2015, 2022a,b, 2023a,b).

ESTABLISHMENT OF A NEW AUTOMOBILE SALES MARKETING MODEL FOR INNOVATING AUTO-DEALERS' SALES

To develop SMDM in TMS strategy, the author has established a "New Automobile Sales Marketing Model" (NA-SMM) for innovating auto-dealers' sales using a "dual corporate marketing strategy" as shown in Figure 3.

To realize this, the author develops both of the "Customer Science principle" (CSp) and "Science SQC, new quality control principle" in order to strengthen CS, CL & CR activities" for evolution of marketing process management (Amasaka, 2002, 2003a,b, 2004, 2005, 2010a, 2011, 2014) (See to Figure 2 in Chapter 3 and Figure 5 in Chapter 2).

"CSp and Science SQC" aim to convert "customer's opinion (implicit knowledge)" scientifically in "engineering language (explicit knowledge)" by partnering in marketing sections described in Figure 3. Specifically, to strengthen "customer value creation", the foundation of NA-SMM consists of the 4 core elements (Amasaka, 2005, Amasaka et al., 2005; Amasaka, Ed. 2007, 2012; Yamaji et al, 2010):

1. Scientific Customer Creative Model (SCCM) for innovating auto-office/shop appearance and operation.
2. Networking of Customer Science principle Application System (NCSp-AS) for realizing customer's "Wants".

Figure 3. New automobile sales marketing model for innovating auto-dealers' sales

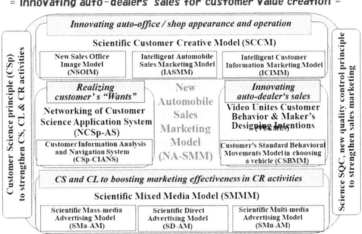

3. Video Unites Customer Behavior & Maker's Designing Intentions (VUCMIN) for innovating auto-dealers' sales".
4. Scientific Mixed Media Model (SMMM) to boosting marketing effectiveness in CS, CL and CR". Concretely, the author develops each of 4 sub-core elements "(1)-(4)":
    (1) New Sales Office Image Model (NSOIM), Intelligent Automobile Sales Marketing Model (IASMM) and Intelligent Customer Information Marketing Model (ICIMM) in SCCM.
    (2) Customer Information Analysis and Navigation System (CSp-CIANS) called Toyota's Intellectual Customer Data Collection/Analysis Integrated Model in NCSp-AS.
    (3) Customer's Standard Behavioral Movements Model in choosing a vehicle (CSBMM) in VUCMIN.
    (4) Scientific Mass-media Advertising Model (SMa-AM), Scientific Direct Advertising Model (SD-AM) and Scientific Multi-media Advertising Model (SMu-AM) in SMMM.

Scientific Customer Creative Model to Strengthen Auto-Products Plan and Marketing Strategy

The author presents a *"Scientific Customer Creative Model"* (SCCM) which takes the form of strategic marketing using above sub-core elements "NSOIM, IASMM and ICIMM" (See to Figure 9 in Chapter 7).

In SCCM, the entire structure consists of 3 domains as the *Open marketing activities* that can be performed through steady linkage with all other divisions in a company-wide framework; (1) *Marketing Strategy*, (2) *Manufacturing Process* and (3) *Market and Customers*. In each domain, the key marketing items are linked by paths to show how they are associated (Amasaka et al., 2005; Shimakawa et al., 2006).

First of all, in the (1) *Marketing Strategy* domain, the key point is how the market segment and the target market are determined. In general, the target market is determined based on the company's core competencies, competition strategy, and resource strategy over the medium and long-term basis. By introducing a scientific analysis approach that uses IT, it clarifies a potential target market from the changing market or the customer structure analysis.

Secondly, in the (2) *Manufacturing Process* domain, the key point is to collect/analyze customers' demands and expectations precisely. At this time, it is important to consider what value the customers want. When implementing information collection / analysis, customer value is described in numerical form from many different viewpoints, and a new product which is aimed at enhancing customer value is implemented through the flow of planning→ development→ production.

Thirdly, in the (3) *Market/Customer* domain, the key point is to learn the structure of the customer's motivation to buy products, CS and CL. Then, it is necessary to extract the elements for CR from this data and utilize it for specific kaizen activities such as reflecting it in future products.

It is important to develop an "analysis tool for close examination of the marketing structure" and a "marketing structure analysis system" that will support marketing activities in these three domains stated above from a strategic marketing viewpoint below.

(I)   New Sales Office Image Model (NSOIM)

In 1st sub-core element, to evolute the auto-business and sales, NSOIM develops the innovation of office/shop appearance and operation: (a) Innovation for building ties with customer, (b) Innovation of business negotiation, (c) Innovation of after-sales service, and (d) Innovation of employee images as shown in Figure 4 (Amasaka et al., 1998; Amasaka, 2011).

(II)   Intelligent Automobile Sales Marketing Model (IASMM)

In 2nd sub-core element, to increase the rate of dealer visits and vehicle purchases by current loyal users, IASMM develops based on customer type by classifying high-probability customers (HPCs), medium-probability customers (MPCs) and low-probability customers (LPCs) into those who visit the shop and those who must be visited by the author's staffs, taking characteristics at new car purchase into account as shown in Figure 5.

Furthermore, IASMM adapts the various advertisements and telephone calls for customers, and can also be made use of when visiting customers, and to help acquire new customers at the time they visit a dealer. Now, IASMM is being developed as the "Toyota Sales Marketing System" (TSMS) (Amasaka, 2001, 2003b, 2011).

Figure 4. New sales office image model (NSOIM)

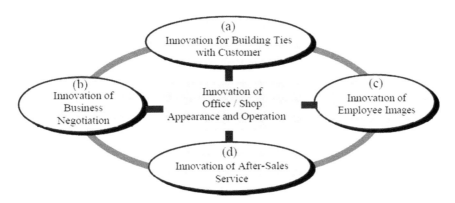

Figure 5. Intelligent automobile sales marketing model (IASMM)

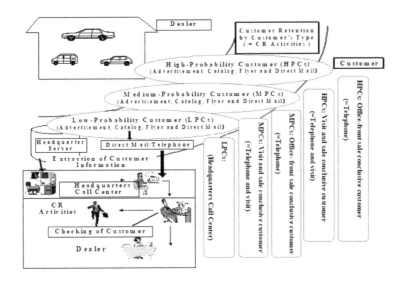

(III) Intelligent Customer Information Marketing Model (ICIMM)

In 3rd core element, to strengthen auto-marketing, ICIMM (Intelligent Customer Information Marketing Model) for TMS strategy develops to create customer's "Wants" as part of the market creation activity and also to establish a structure for development and production that is capable of offering new products as shown in Figure 6 (Amasaka, Ed., 2007; Amasaka, 2008, 2010b).

This model is done in order to effectively carry out by each step of the (1) Input information, (2) Information for marketing and (3) Output information for evolution of the business process of "auto-shop, products development design and manufacturing.

Figure 6. Intelligent customer information marketing model (ICIMM) for TMS strategy

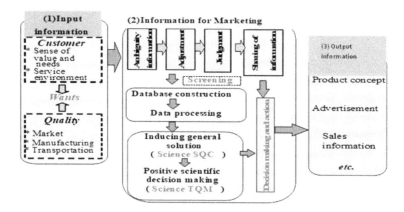

Networking of Customer Science Application System for Creating Customers' Demands

To strengthen SCCM development, the author introduces a "Customer Science Application System" (NCSp-AS) in order to evolve customer information-sharing through the collection / analysis of customer information.

Specifically, from Figure 7 in Chapter 8, the author develops a typical sub-core element "Customer Information Analysis and Navigation System! (CSp-CIANS). This system enables the networking of the (1) Merchandise planning, (2) Product planning, Development, and design, (3) domestic and overseas dealers, (4) Consulting spaces, (5) Marketing research company via (6) Exclusive company WEB, (7) Data base D/B, (8) Science SQC, (9) Analytical case D/B, and (10) Quality assurance (Amasaka, 2005, 2018).

Video Unites Customer Behavior and Maker's Designing Intentions for Investigating Customers' Preferences to Exhibition Vehicles

To realize SCCM development, the author illustrates a "Video Unites Customer Behavior & Maker's Designing Intentions (VUCMIN)" that focuses on the standard behavioral movements of customers who visit dealers as shown in Figure 7 (Yamaji et al, 2010) (See to Chapter 18 in detail).

Specifically, the author develops a typical sub-core element "Customer's Standard Behavioral Movements Model" (CSBMM) in choosing a vehicle" by developing "Customer behavior analysis using motion pictures: Customer Motion Picture – Flyer Design Method (CMP-FDM) (Koyama et al., 2010) (See to Figure 3 in Chapter 18).

As a concrete instance, the author conducted the following survey in order to investigate customer behaviors. For example, Figure 7(a) shows the enlargement of and the target car (vehicle) model is 1, gender 2, age 3, standing positions 4, and vehicle part focused on is 5. Among those items, standing positions are categorized as in Figure 7(b).

As a result, front is 1, front fender (driver seat) 2, rear fender (driver seat side) 3, trunk is 4, rear fender (passenger seat side) 5, front fender (passenger seat) 6, handle 7, shift lever 8, near passenger seat is 9. In total, all customer behaviors (standing positions, getting in and out, operation, walking time, etc.) are categorized into 85 distinct types of behaviors (Yamaji et al., 2010; Amasaka et al., 2013).

Figure 7. An example of video unites customer behavior and maker's designing intentions

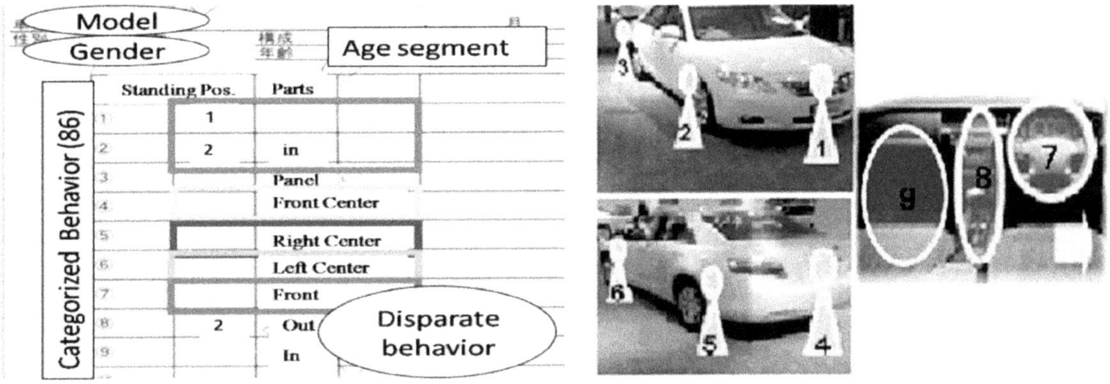

(a)   Survey sample (The enlargement of Sample 1) (b) Sample 1 of customer standing positions

Scientific Mixed Media Model to Realize the Automobile Market Creation

Recently, to strengthen TMS strategy, the author develops a "Scientific Mixed Media Model (SMMM)" named "New Mixed Media Model" which takes the form of auto-marketing creation using 3 sub-core elements "SMa-AM, SD-AM & SMu-AM" that each of the "mass-media ads. and direct ads. and multi-media ads." were optimized rationally due to "CSp and Science SQC" as shown in Figure 8 (Yamaji et al., 2010; Koyama et al., 2010; Ishiguro and Amasaka, 2010, 2012; Amasaka et al., 2013; Ogura et al., 2013a,b).

Specifically, Figure 8(a) illustrates a graphical representation of customer motives for visiting auto-dealers using a typical mixed media with the various advertising. Area A: Mass-media advertising (television (TV), radio broadcasting, flyers, public transportation (train cars), newspapers, magazines, etc.), Area B; Direct advertising (catalogs, direct mail (DM), handbills (directly handed (DH) to customers), telephone calls, etc.) and Area C: Multi-media advertising (internet, CD-ROMs, etc.).

Figure 8(b) shows the verification results from application of "New Mixed Media" (NMM) by SMMM for raising the percentage of people affected, and use as the new strategic advertisement in nine media elements (TV, radio, newspapers, internet, train cars, flyers, magazines, DM and DH) designed.

A field survey on vehicle advertising was conducted to identify 3 sub-core elements of each media type to visualize the relationship between those elements and the media as well as the causal relationships between each media type and (i) vehicle attention, (ii) vehicle interest, and (iii) desire to visit dealers.

A total of 318 valid responses (197 male and 121 female, generally uniform age balance) were collected. The investigation period was the five months leading up to the release of the "new Q model" by an advanced manufacturer: Toyota.

In Figure 8(b), the author shows the result of a follow-up survey using SMMM, where 16 people (percentage of people affected: 11.8%) actually visited the dealer, while 8 people signed a sales contract.

Comparative verification was done by looking at the results of the usual experience of mixed media" when the dealer in the figure announced the old model Q car (: 4 years ago in a survey of similar size). In this case, the percentage of people affected was just 1.1%, thus validating the effectiveness of SMMM (Amasaka, 2015, 2022a,b, 2023a,b).

Figure 8. An effectiveness of scientific mixed media model (SMMM)

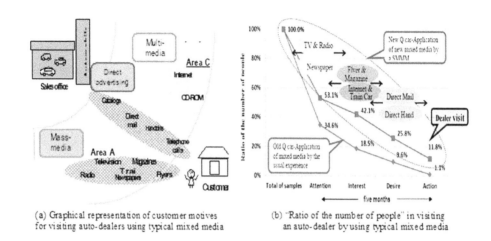

(a) Graphical representation of customer motives for visiting auto-dealers using typical mixed media

(b) "Ratio of the number of people" in visiting an auto-dealer by using typical mixed media

ACTUAL APPLICATIONS

In this section, the author describes the effectiveness of NA-SMM for innovation of the mode of business and sales operation through the actual cases in Toyota.

Establishment of the Toyota Sales Marketing Model (TSMM)

As typical example, the author has established the "Toyota Sales Marketing Model" (TSMM) to improve the repeat customer ratio for Toyota vehicles as the scientific CS, CL and CR activities below (Amasaka, Ed., 2012; Amasaka, 2015, 2022a,b, 2023a,b).

(1)   Trial for increasing sales through CR based on customer type

The achievements of the present study are currently being applied at Netz Chiba and other Toyota dealers using CSp and Science SQC.

Then, the author solves problems by using scientific approaches such as "Mountain climbing for problem-solving" (development of specific models for customers of high replacement probability) following the steps from (i) to (ix) shown in Figure 9.

In steps (i) and (ii) described in Figure 10, consideration was given to the development of specific models for "high replacement probability customers" as shown in Figure 10, an "Application type" association diagram (Step (i)-(iii)).

Then, in step (iii), a scenario of implementation plans for about a year was established (Amasaka et al., 1998; Amasaka, 2001, 2007a, 2009, 2010a,b).

In steps (iv) through (vi), the graphical "Categorical Automatic Interaction Detector" (CAID) analytical method was implemented (Murayama, 1998; Amasaka, 2001).

In CAID, the author has been developed as the new multivariate analysis, and was necessary for the qualitative and categorical data analysis required for questionnaire design, implementation and analysis.

Figure 9. Mountain climbing for problem-solving using SQC technical methods

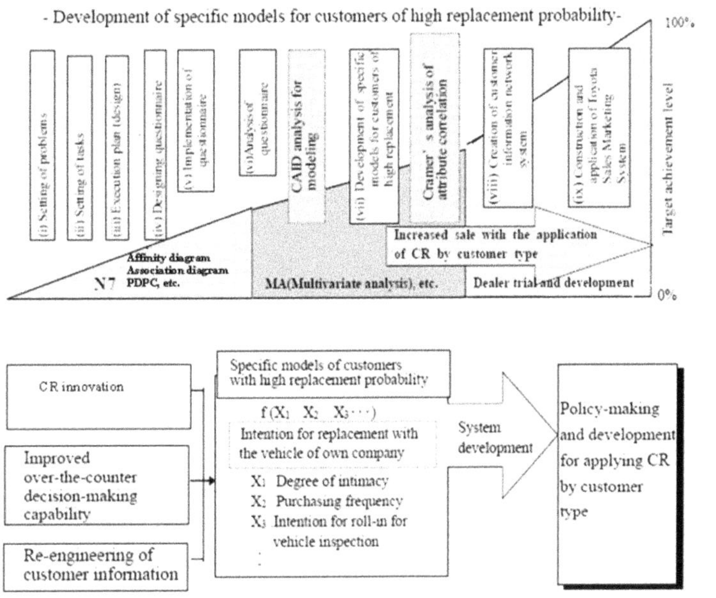

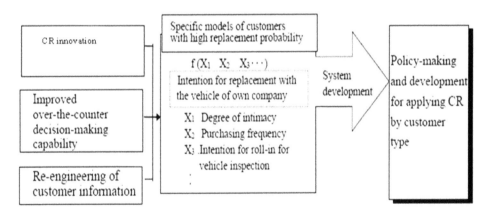

Figure 10. Application type association diagram (Step (i)-(iii))

In the next step (vii), the sales method capable of deploying CR based on customer type was obtained using the Cramer's analysis of attribute correlation as the base for developing the customer information network system in step (viii).

In the final step (ix), the scenario at Toyota "Netz Chiba" was examined as the basis for deployment to all Toyota dealers in Japan for the establishment of T-ASMM.

(2) Development of specific models for high replacement probability customers

    (2-1) Objective and explanatory variable for the planned questionnaire form

        1. Objective variable: Intention to replace with Toyota vehicle: (Yes, No)

2. Explanatory variable: Records of roll-in (oil change, inspection and maintenance, fault repair, accident repair, vehicle inspection), number of new cars purchased, referral for new car purchase, voluntary insurance contract, degree of intimacy, degree of Toyota card usage, sex, age, etc. (categories 3 to 6).

(2-2) 4 measures to achieve design and implementation of effective questionnaire

Prior telephone notification was given to 4,000 customers who bought new Toyota vehicles within the five years on the planned questionnaire.

To improve the recovery rate of the questionnaire, a simply-designed questionnaire was adopted with one section of questions laid out on one page as shown in Table 10A (Case-1) (See to Appendix 10A).

(2-3) Questionnaire analysis with CAID & Cramer's analysis of attribute correlation

After analysis of the questionnaire data, the results of analysis of causal relations were indicated graphically to accurately show the proposed measures and decision-making process that led to increased sales through the application of CR based on customer type.

Then, CAID analysis and Cramer's analysis of attribute correlation was applied to enable collation using empirical rules to form "Analysis I and Analysis II" as follows;

1. "Analysis I" involves arranging customers having a high-probability of replacement with Toyota vehicles into a model using intelligent CAID analysis.

Factors affecting replacement by high-probability customers are rearranged in the same manner as the variable designation method of multi regression analysis.

This is conducted repeatedly based on empirical techniques of the staff and managers of business/sales divisions (so as to match their experience).

Then, the characteristics and changes in the customers' orientation are ascertained on the basis of actual contact with customers.

Customers are stratified into customer types (customers of high, medium and low repeat business probabilities) from the customer CR point of view.

2. "Analysis II" involves conducting factorial analysis using "creation of ties with customers" as the key point to map out our business and sales policies (Cramer's analysis of attribute correlation, etc.).

In practice, correlation among influential factors is extracted by the intelligent CAID, including the degree of intimacy and roll-in for vehicle inspection and all other question items using the "Cramer's analysis of attribute correlation".

For example, factors which improve the degree of intimacy with customers are identified from the sales activity and after-service activity viewpoints, aimed at deployment for sales policies.

(2-4) Analytical result with CAID (Analysis I; from Step vi to Step vii)

Figure 11 shows the legends of the analytical results. Analysis a in the figure indicates that 62% of 1,610 users who answered the questionnaire intend to replace their vehicles with a Toyota vehicle, while 38% do not.

Next, analysis b is the division by the primary influential factor of the "degree of intimacy". The upper setting of having "intimacy" (customers having good acquaintance with sales staff) indicates that 75% intend to buy Toyota for replacement, and the lower setting of having "no intimacy" (customers not having good acquaintance with the sales staff) indicates that 48% intend to buy Toyota.

Figure 11. CAID[1] *analysis (step (vi))*
[1] Multiple Cross-section Analysis (Categorical Automatic Interaction Detector)

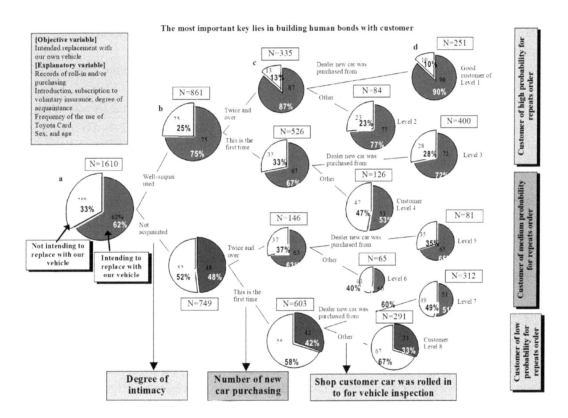

The difference between them is as much as 27%. Thereafter, c indicates the analytical result for users who bought Toyota for the first time and those for two times and over (no significant difference among 2nd to 5th time purchasers). Similarly, analysis d stratifies the users by the "intention for roll-in for vehicle inspection service".

From Figure 11, it is known that 90% of customers of level 1 (regular customers) indicated in the top position of d intend to buy a Toyota vehicle for replacement.

Figure 11 combines customer types of whom 70% intend to buy Toyota (b, c and d) on customer levels 1 through 3 (regular customers), and classifies them as customers of high probability. Likewise, customer types of whom 50% intend to buy Toyota on the customer levels 4 through 7 are classified as customers of medium probability.

Customers on level 8 are classified as customers of low probability since they fail to hold a majority.

The author does not discuss other influential factors (such as **e**: introduction, **f**: sex, and so forth) where difference is noted between two dealers (: Toyota "Netz Chiba and Netz Ehime").

(2-5) Analytical result with Cramer's analysis of attribute correlation (Analysis II: Step vii - Step viii)

Practical and detailed analysis is conducted from the sales policy standpoint aimed at increasing the frequency of contact with customers.

Here, the correlation between the degree of intimacy extracted in step (vi) and all other questions is explained using a factor and result diagram based on the Cramer's analysis of attribute correlation as shown in Figure 12.

Area 1 in the figure contains factors a through i influential to the degree of intimacy and area 2, factors j through w affecting the roll-in destination for vehicle inspection. "Index" in the figure represents the customer information numerically.

For example, the 0.14 of "sincere action against failure or accident" is an index when the Cramer's factor correlation coefficient is assumed to be 100.

It is technically possible to correlate all factors in area 1 with the 6 key data shown in the figure. Based on the information obtained as a result of these analyses, practical policies can be established for promoting sales and after-service activities capable of improving the degree of intimacy with customers who can be handled by a dealer.

(3)   Construction and application of Toyota Sales Marketing System (TSMS) (Step ix)

The information can also be used for simulation for sales expansion, which is the basis of innovation for creating strong contact between the dealer and its customers.

As a result, in step (ix), the author has constructed the "Toyota Sales Marketing System (TSMS)" as shown in Figure 5 above.

For practical application, the questionnaire in step (vi) is reanalyzed at trial stages of the system in steps (vii) and (viii) of Figure 9 above in order to ensure replacement by Toyota vehicles by adding the following strategies.

The CR activities based on customer type are adopted by classifying high- and middle-probability customers into those who visit the shop and those who must be visited by our staff, taking characteristics at new car purchase into account.

1.   A system is established so that the shop manager directly receives "Medium-Probability Customers (MPCs)" upon their visit to the shop without fail in order to promote visits to the shop by "High-Probability Customers (HPCs)". Thus, the frequency of contact with customers is increased. Further, sales and service activities focus on telephone calls for customers who visit the shop, and telephone calls and home visits for those who require visits by our staff, as shown in the figure.
2.   As for "Low-probability customers (LPCs)" who have less contact with the sales staff, a telephone call center is established within the dealer as shown in the figure to accumulate know-how related to the effective use of customer information software.

Figure 12. Factor analysis of Cramer's analysis of attribute correlation step (vii))

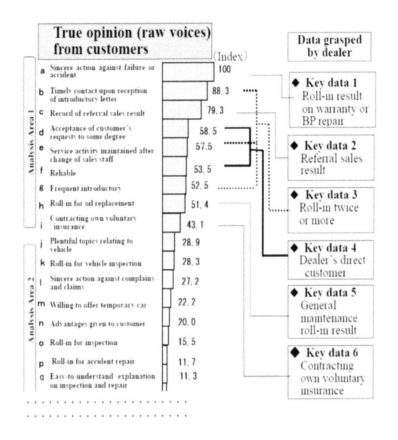

The two-step approach is adopted as the practical sales policy where telephone calls are used to follow up on the effect of publications, advertisements, catalogs, fliers and direct mail.

As expected, excellent results have been reported at "Netz Chiba" and other Toyota dealers who introduced this system by applying the TSMS constructed as above.

Recently, in parallel to this study, the author studied so called "Database Marketing" where the effects of publications, advertisements, catalogs, fliers and direct mail are quantitatively analyzed to enable effective support for this system (Amasaka, 2001, 2007a,b,c,d; Kojima et al., 2010; Ishiguro et al., 2010, 2012).

Then, the application of TSMS has recently contributed to an increase in the sales share of Toyota vehicles in Japan (40% in1998 to 46% to 2008) (Nikkei Institute, 1999).

Application to Scientific Automobile Sales Innovation at Toyota

Dynamics of the effect of publicity and advertising

The kind of information dealers need to know in order to continuously increase the number of customers attracted following the launch of a new model is (i) how many months the effect of the new model introduction lasts, (ii) the best timing (day of the week) and volumes for the most effective flyer advertising, (iii) the ratio of the "Purchase Intention Customers" (PIC) who show an interest in purchasing a new car, (iv) the percentage of "PIC" who actually purchase new cars, and (v) variation among dealers. It is preferable to know the actual causes and effects involved (Amasaka, 2009, 2015, 2022a,b, 2023; Amasaka, Ed., 2012).

Analytical methods have not been established for analyzing the dynamic effect of publicity and advertising due to the large scale and level of difficulty of market experiments for obtaining quantitative information in time series.

Therefore, there are few reports on this subject relating to automobile marketing (Amasaka et al., 1998).

For this reason, the author presents in the next section factorial analysis conducted using Science SQC, in the hope that it will assist readers in strengthening their information-based marketing, and promoting the TSMS (Amasaka, 2007a,b, 2010a,b, 2015, 2022a,b, 2023; Amasaka et al., 2013; Ogura et al., 2013).

Effectiveness of advertising in automobile sales

This subsection demonstrates the effectiveness of the "flyer advertising effect", "day of the week effect", "new car effect".

These effects are acknowledged among those on the front lines of automobile sales as having a high level of investment efficiency and resultant customer traffic, achieved through Science SQC, the called "Marketing SQC" (Amasaka et al., 1998; Amasaka, 2001, 2003a,b, 2004; Amasaka, Ed. 2012).

(1)   Publicity and advertising media and sales activities

Figure 13 shows a graphical representation of the causal relationship between sales and order taking activities (dealer visit → PIC → order placement), and publicity and advertising media (mixed media).

The diagram indicates that the dealer's publicity and advertising activities (causal system: X1 to Xn) attract customers' interest in purchasing, encourage them to visit the dealer (result system: Y0), and ultimately lead to the purchase of a vehicle (PIC: Y1→order placement: Y2).

Figure 13. Relationship between sales activities and publicity and advertising

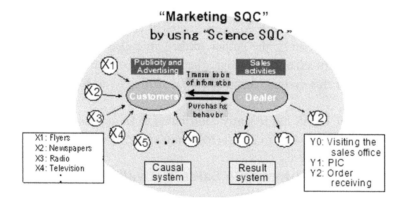

(2) Importance of mixed media

Next, the authors will comparatively evaluate the importance of the factorial effect of publicity and advertising on the basis of the subjective criteria (evaluation point: median) of representative sales staff from Toyota group dealers A and B.

"Comparison of publicity and mixed media" indicates the extent of the effect gained from each of the respective media used to promote customer visits to dealers as shown in Table 1.

This table suggests that the effect of mixed media delivered together with newspapers is expected to be significant, which agrees with the empirical knowledge (sales people's experience) fostered through years of sales activities.

(3) Survey on the effects of mixed media using factorial analysis

To verify the points made in "Importance of innovating auto-dealers' sales activities) above, analysis has been carried out on what customers say prompted them to consider buying a car at Toyota group's representative dealers A and B.

Figure 14 shows that mixed media is a significant factor in influencing a customer's decision to purchase a new car.

Table 1. Comparison of publicity and mixed media

	Flyer ad	News paper	Television	Radio	···
Information capacity	O	O△	△	×	···
Information quality	O	O△	△	×	
Frequency of acquiring information	O△	O△	△	O△	
Ease of acquiring information	O	O	△	△	
Residual effect	O	O	△	×	
Total point	174	144	60	30	···

Rating

⊘ : 5 points
O : 4 points
O△ : 3 points
△ : 2 points
× : 1 point
× × : 0 point

(4) An example of the "Flyer Advertising" effect
    (4-1) Content of the field survey

A field survey was conducted as follows by designating six sales offices of Toyota group's dealer B, which is located in an urban area.

Figure 14. Survey on what prompts customers to consider buying a car

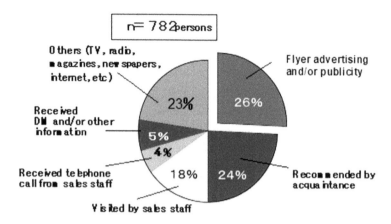

1) A visitor questionnaire was used to determine customers' motives for visiting the dealer's sales office, and to create a customer profile (age, desired vehicle type, timing of purchase, budget, etc.)
2) Analysis was carried out based on the results of the study in time series of distribution of the flyers (quantity, etc.) over the nine-month period starting before and ending after the latest introduction of a new car, and covering the order taking activities at the respective dealers (number of visitors → number of PIC → number of cars sold).
(4-2) Science SQC approach

The core techniques of Science SQC using SQC Technical Methods are used to objectify (modeling: explicit knowledge) the subjective evaluation (empirical knowledge: implicit knowledge) of the "flyer advertising" effect.

Here, the Science SQC approach shown in Figure 15 is applied to conduct a factorial analysis on the effect of flyer advertising in the sequence of Step 1 (survey) and Step 2 (analysis). The author intends to report on Step 3 (demonstration) at a later time.

(4-3) Analysis of customer motives for visiting dealers

Figure 16 represents graphically an analysis of customer motives for visiting dealers. This analysis was carried out on the basis of Quantification Class III in order to objectify subjective evaluations.

The categorical quantities of publicity and advertising media such as flyers, newspapers and magazines shown in the diagram, lead us to the conclusion that those customers prompted by flyers to visit dealers are likely to at some time need to replace their old car, and will be on the lookout for a bargain.

With other media, comparatively younger male customers may, for example, visit a dealer upon seeing some new RVs featured in a magazine. These results corroborate the information on publicity and advertising media described in Figure 15 and Table 1, and the results analysis of their respective effectiveness.

This enables us to acquire useful knowledge for developing future business and sales activities.

Figure 15. Factorial analysis of the effect of flyer advertising

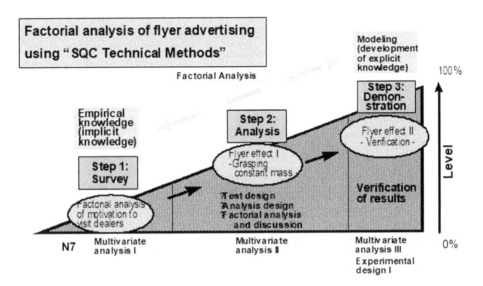

Figure 16. Analysis of motives for visiting dealers
(Quantification Class III scatter diagram showing category quantities)

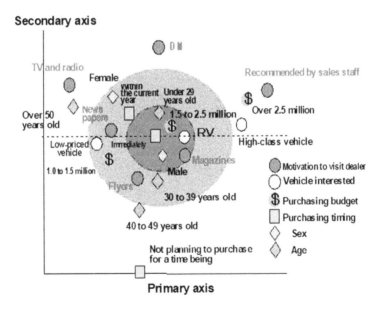

(4-4) Analysis of sales activities in time series

Figure 17 shows an overview of the number of visitors to the shop (Y0), the number of PIC (Y1) and actual orders received (Y2) in the two months following the start of a new car sale within the total nine-month period of the field survey.

The chart gives an outline of the distribution effect () of flyer advertising. Moreover, the author can assume from the multivariate association diagram in Figure 18, which covers the total nine-month survey period, that the number of visitors (Y0) is the principal factor in increasing the number of purchases (PIC (Y1) to order placement (Y2)).

Now let us look at the relationship between the number of visitors, number of HPC customers and the number of units sold, as shown in Figure 19.

This diagram shows that (i) as the number of visitors (Y0) increases, the number of PIC (Y1) and the number of units sold (Y2) also increase on a regression slope of the respective exponents (regression coefficient) a 1 and a 2.

Figure 17. An example of the records of sales activities

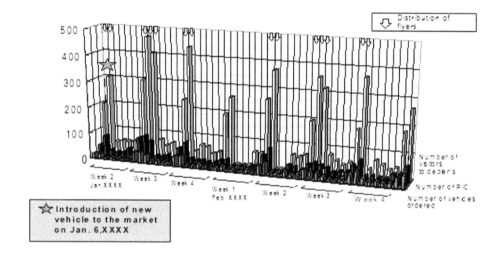

Figure 18. Relationship between the number of dealer visitors, PIC and vehicle orders placed

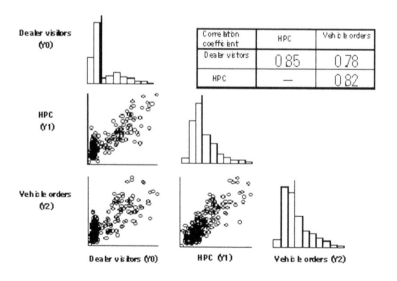

Also, the diagram confirms that (ii) the ratio of the number of units sold is about one-half of the number of PIC.

Figure 19. Scatter diagram showing the number of dealer visitors, PIC, and vehicles orders placed

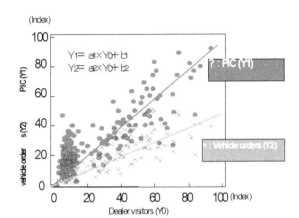

Furthermore, the author can assume from the information in the diagram that (iii) due to the considerable dispersion in the data from the regression line, there is also a considerable dispersion in the ratios of the number of PIC and units sold to the number of customers visiting the respective sales offices.

In this connection, a survey and analysis were carried out to determine whether such dispersion factors in the data are dependent on the characteristics of each sales office.

Six representative sales offices were selected from dealer B and analysis was carried out on differences in flyer distribution, the number of visitors to each of the sales offices, the number of PIC, and the number of orders received for each office.

The results of this analysis are summarized in Figure 20. This figure indicates that there is considerable variation in the exponents of orders received among the sales offices.

For example, in comparing the exponents of orders received (η= number of units ordered/number of PIC) between the main sales office located in an urban area and the "Maizuru sales office" located in an outlying area, the figure shows that the exponent "η" for the Head Office is 0.48, while that for "Maizuru sales office" is 0.71.

Although it goes beyond the scope of this discussion, the author is able to obtain new activity guidelines for formulating new sales strategies by gaining an understanding of the data dispersion among the sales offices on the basis of such analysis.

(4-5) Flyer effect, Day of the week effect, and New car effect

Due to time constraints in the object period for analysis, the author used quantification class I for the current factorial analysis of the survey data in time series.

Figure 21 shows the "flyer effect", "day of the week effect", and "new car effect", which influence the number of visitors (Y0) to the sales office as follows;

Figure 20. Map showing location of Toyota dealer B sales offices

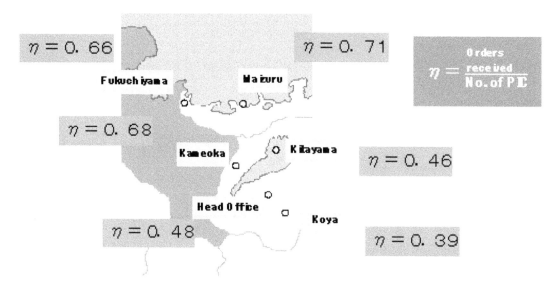

Figure 21. Quantification class I graph of the category quantities

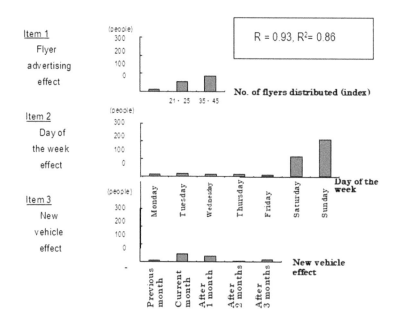

(i)   Item 1 shows that the greater the number of flyers distributed, the more customers visit the dealer.

(ii) Item 2 shows that the effect is greater on Saturdays and Sundays, while the number of visitors is more or less constant on weekdays.

(iii) Item 3 shows that the "new car effect" continued for two months in the current analysis, including the first month the new car went on sale.

The effect contracted in the period from three to 5 months after the new car went on sale.

(4-6) Considerations

As a result of the survey and analysis conducted thus far, the author has been able to quantitatively clarify the acknowledged opinions of sales staff (empirical knowledge) on the "flyer effect", "day of the week effect" and "new car effect".

Based on this information the author believes that, in order to efficiently enhance the effectiveness of flyer ad., it is necessary to carry out demonstrative research (Figure 17: Step 3 Demonstration) to determine the saturation point of flyer ad. (limit the number of flyers distributed), optimal timing (best days of the week for distribution), the elements of effective flyer design (layout and impact), and the location and scale of dealers.

Similar Cases

With its effectiveness verified, the author has developed the similar cases as follows;

(1) "Change in automobile marketing for improving business and sales using Science SQC approach for focusing on CS as a way of boosting marketing effectiveness"

At a stage of execution, the key factors comprising CS and CL among core customers at the 6 target dealerships, each of whom represent major automakers in Japan: 4 Japanese (Toyota, Nissan, Honda and Mitsubishi) and 2 foreign (Mercedes-Benz and Volkswagen) were identified in order to determine the level of impact each of carries.

Specifically, the author has analyzed sales information from core customers to identify the four key factors of CS and CL among them using "covariance structure analysis".

These results point to a customer mindset whereby those looking to replace an existing vehicle have a desire to go back to the dealership where they purchased their old car because they have a lasting impression of the courteous manner with which a salesperson treated them before (Okutomi and Amasaka, 2013).

(2) "Development of Automobile Exterior Design Model (AEDM) to strengthen the auto- products plan in auto-market value creation"

As customer values become increasingly diverse, the auto-exterior design is becoming one of the most critical elements influencing customer purchase behavior for automakers.

To address this issue, the author has developed the "Intelligence Design Concept Method (CS-IDCM)" using above "CSp-CIANS and Science SQC", and has verified the validity of AEDM through the development of the typical New model cars (: *"Lexus"* and others) (Nikkei Business. 1999; Motor Fun, 1997, 2000; Amasaka, 2005, 2007d, 2018) (See to Chapter 8).

CONCLUSION

In this study, to strengthen "customer value creation", the author has created a NA-SMM (New Automobile Sales Marketing Model) using 4 core elements: "SCCM, NCSp-AS, VUCMIN and SMMM" for innovating "auto-office / shop appearance and operation". To realize NA-SMM strategy, moreover, the author has developed the SMDM based on the TMS adapting a Dual Methodology "CSp and Science SQC". in order to boosting marketing effectiveness in CS, CL and CR activities". Concretely, the author has developed each of 4 sub-core elements as the formation of NA-SMM: (1) "NSOIM, ICIMM and IASMM" in SCCM, (2) "CSp-CIANS in NCSp-AS, (3) "CSBMM" in VUCMIN, and (4) "SMa-AM, SD-AM and SMu-AM" in SMMM. The validity of NA-SMM is then verified through the actual applications to automobile customer creation in Toyota.

REFERENCES

Amasaka, K. (2001). Proposal of Marketing SQC to revolutionize dealers' sales activities. *Proceedings of the 16th Int. Conf. on Production Research, Prague, Czech Public.* IEEE.

Amasaka, K. (2002). *New JIT*, a new management technology principle at Toyota. *International Journal of Production Economics, 80*(2), 135–144. doi:10.1016/S0925-5273(02)00313-4

Amasaka, K. (2003a). Proposal and Implementation of the Science SQC Quality Control Principle. *Mathematical and Computer Modelling, 38*(11-13), 1125–1136. doi:10.1016/S0895-7177(03)90113-0

Amasaka, K. (2003b). A demonstrative study on the effectiveness of "Television Ad." for the automotive sales, *Japan Society for Production Management, The 18th Annual Conference., Nagasaki, Japan,* 113-116.

Amasaka, K. (2004). *Science SQC, new quality control principle: The quality strategy of Toyota.* Springer-Verlag. doi:10.1007/978-4-431-53969-8

Amasaka, K. (2005). Constructing a *Customer Science* Application System *"CS-CIANS"*– Development of a global strategic vehicle *"Lexus"* utilizing *New JIT–. WSEAS Transactions on Business and Economics, 2*(3), 135–142.

Amasaka, K. (Ed.). (2007). New Japan Model: Science TQM. Study Group of the Ideal Situation on the Quality Management of the Manufacturing Industry, Maruzen, Tokyo.

Amasaka, K. (2007a). Proposal of Marketing SQC to revolutionize dealers' sales activities-A demonstrative study on Customer Science by utilizing Science SQC, *Proceedings of the 16th International Conference Production Research, Praha, Czech.*

Amasaka, K. (2007b). The validity of *Advanced TMS,* A strategic development marketing system utilizing *New JIT –. The International Business & Economics Research Journal, 6*(8), 35–42.

Amasaka, K. (2007c). High Linkage Model *"Advanced TDS, TPS & TMS"*: Strategic development of *New JIT* at Toyota. *International Journal of Operations and Quantitative Management, 13*(3), 101–121.

Amasaka, K. (2007d). The validity of *"TDS-DTM"*, a strategic methodology of merchandise: Development of *New JIT*, Key to the excellence design *"LEXUS"*. *The International Business & Economics Research Journal*, *6*(11), 105–115.

Amasaka, K. (2008). Science TQM, a new quality management principle: The quality management strategy of Toyota. *The Journal of Management & Engineering Integration*, *1*(1), 7–22.

Amasaka, K. (2009). The effectiveness of flyer advertising employing TMS: Key to scientific automobile sales innovation at Toyota, *China-USA. Business Review (Federal Reserve Bank of Philadelphia)*, *8*(3), 1–12.

Amasaka, K. (2010a). Chapter 4. Product Design, Quality Assurance Guidebook. Guide Book of Quality Assurance. The Japanese Society for Quality Control, Union of Japanese Scientists and Engineers Press.

Amasaka, K. (2010b). Proposal and effectiveness of a High-Quality Assurance CAE Analysis Model, *Current Development in Theory and Applications of Computer Science. Engineering and Technology*, *2*(1/2), 23–48.

Amasaka, K. (2011). Changes in marketing process management employing TMS: Establishment of Toyota Sales Marketing System, *China & USA. Business Review (Federal Reserve Bank of Philadelphia)*, *10*(7), 539–550.

Amasaka, K. (Ed.). (2012). Science TQM, new quality management principle: The quality management strategy of Toyota, Bentham Science Publishers, UAE, USA, the Netherlands.

Amasaka, K. (2014). New JIT, new management technology principle. *Journal of Advanced Manufacturing Systems*, *13*(3), 197–222. doi:10.1142/S0219686714500127

Amasaka, K. (2015). New JIT, new management technology principle. Taylor and Francis Group, CRC Press.

Amasaka, K. (2018). Automobile Exterior Design Model: Framework development and supporting case studies. *The Journal of Japanese Operations Management and Strategy*, *8*(1), 67–89.

Amasaka, K. (2022a). *Examining a New Automobile Global Manufacturing System*. IGI Global Publisher. doi:10.4018/978-1-7998-8746-1

Amasaka, K. (2022b). New Manufacturing Theory: Surpassing JIT (2nd Edition). Lambert Academic Publishing.

Amasaka, K. (2023a). New Lecture-Toyota Production System: From JIT to New JIT. Lambert Academic Publishing, Republic of Moldova Europe.

Amasaka, K. (2023b). A New Automobile Sales Marketing Model for innovating auto-dealer's sales. *Journal of Economics and Technology Research*, *4*(3), 9–32. doi:10.22158/jetr.v4n3p9

Amasaka, K., Kishimoto, M., Murayama, T., & Ando, Y. (1998). The development of Marketing SQC for Dealers' Sales Operation System. *Journal of the Japanese Society for Quality Control, The 58th Technical Conference*, 76-79.

Amasaka, K., Ogura, M., & Ishiguro, H. (2013). Constructing a Scientific Mixed Media Model for boosting automobile dealer visits: Evolution of market creation employing TMS. *International Journal of Business Research and Development*, *3*(4), 1377–1391.

Amasaka, K., Watanabe, M., & Shimakawa, K. (2005). Modeling of strategic marketing system to reflect latent customer needs and its effectiveness, *Fragrance Journal, The Magazine of Research & Development for Cosmetics. Toiletries & Allied Industries*, *33*(1), 72–77.

Nikkei Business. (1999). Renovation of Shop, Product and Selling Method-targeting Young Customer by Nets. Nikkei Business.

Motor Fan. (1997). *All of new-model "Aristo", a new model prompt report, 213*, 24-30. Motor Fan.

Motor Fan. (2000). *All of new-model "Celsior", a new model prompt report, 268*, 23-24. Motor Fan.

Gary, L. L., & Arvind, R. (2003). *Marketing Engineering: Computer-assisted marketing analysis and planning*. Pearson Education, Inc.

Ikeo, K. E.-C. (2006). *Feature-marketing innovation*. Japan Marketing Journal.

Ishiguro, H., & Amasaka, K. (2012). Proposal and effectiveness of a Highly Compelling Direct Mail Method: Establishment and deployment of PMOS-DM. *International Journal of Management & Information System*, *16*(1), 1–10.

Ishiguro, H., & Amasaka, K. (2012). Establishment of a Strategic Total Direct Mail Model to Bring Customers into Auto Dealerships. *Journal of Business & Economics Research*, *10*(8), 493–500. doi:10.19030/jber.v10i8.7177

Ishiguro, H., Kojima, T., & Matsuo, I. (2010). A Highly Compelling Direct Mail Method "PMOS-DM": Strategic applying of statistics and mathematical programming, *The 4th Spring Meeting of Japan Statistical Society, Poster Session (Student Poster Award)*.

James, A. F., & Mona, J. F. (2004). *Service Management*. McGraw-Hill Companies Inc.

Kojima, T., Kimura, T., Yamaji, M., & Amasaka, K. (2010). Proposal and development of the Direct Mail Method "PMCI-DM" for effectively attracting customers. *International Journal of Management & Information Systems*, *14*(5), 15–22. doi:10.19030/ijmis.v14i5.9

Koyama, H., Okajima, R., Todokoro, T., Yamaji, M., & Amasaka, K. (2010). Customer behavior analysis using motion pictures: Research on attractive flyer design method. *China-USA Business Review*, *9*(10), 58–66.

Murayama, Y. (1982). Analyzing CAID marketing review. Japan Research Center.

Niiya, Y., & Matsuoka, F. (Eds.). (2001). *Foundation Lecture on the New Advertising Business*. Senden-Kaigi.

Ogura, M., Hachiya, T., & Amasaka, K. (2013a). A Comprehensive Mixed Media Model for boosting automobile dealer visits. *The China Business Review*, *12*(3), 195–203.

Ogura, M., Hachiya, T., & Amasaka, K. (2013b). Attention-grabbing train car advertisements. *The China Business Review, 12*(3), 195–203.

Okada, A., Kijima, M., & Moriguchi, T. (Eds.). (2001). *The mathematical model of marketing.* Asakura-Shoten. (in Japanese)

Okutomi, H., & Amasaka, K. (2013). Researching Customer Satisfaction and Loyalty to boost marketing effectiveness: A look at Japan's auto dealerships. *International Journal of Management & Information Systems., 17*(4), 193–200. doi:10.19030/ijmis.v17i4.8093

Shimakawa, K., Katayama, K., Oshima, K., & Amasaka, K. (2006). *Proposal of Strategic marketing model for customer value maximization.* The Japan Society for Production Management, The 23th Annual Technical Conference, Osaka, Japan.

Yamaji, M., & Amasaka, K. (2009). Proposal and validity of Intelligent Customer Information Marketing Model: Strategic development of *Advanced TMS, The Academic Journal of China - USA. Business Review (Federal Reserve Bank of Philadelphia), 8*(8), 53–62.

Yamaji, M., Hifumi, S., Sakalsis, M. M., & Amasaka, K. (2010). Developing A Strategic Advertisement Method "VUCMIN" to Enhance the Desire of Customers for Visiting Dealers. *Journal of Business Case Studies, 6*(3), 1–11. doi:10.19030/jbcs.v6i3.871

APPENDIX

Four Measures to Achieve Design and Implementation of Effective Questionnaire

The subjects were required to answer methodically within about 15 minutes as shown in Table 10A. Furthermore, introduction of a Questionnaire Information Box helped raise the recovery rate to over 40% (normally 20%) and the valid reply ratio to 98% (normally 70%) (Amasaka et al., 1998; Amasaka, 2001, 2007, 2009, 2010).

Table 2. Typical example of questionnaire form

——— We like to inquire on the after-sales services of your dealer. ———
Q1: Please let us know the present state of the maintenance, fault, repair and other your vehicle-related after-sales service
(please give us an answer for each item from ① to ④).

	① Dealer used	② Reason for using	③ Method of payment	④ Method of roll-in
1. Oil change (please name a principal shop / station you use)	1. Dealer you purchased a new vehicle from 2. Other dealer 3. Vehicle maintenance service shop 4. Car shop 5. Gas station 6. Other ( )	1. Because of acquaintance 2. High level of technique 3. Low price 4. Privilege offered 5. Nearness of location 6. Other ( )	1. Cash 2. TOYOTA CARD 3. Other cards 4. Other ( )	1. Drive to the shop 2. Have the vehicle picked up 3. Other ( )
2. Inspection and maintenance service of your vehicle (Please name a principal shop you use)	1. Dealer you purchased a new vehicle from 2. Other dealer 3. Vehicle maintenance service shop 4. Car shop 5. Gas station 6. Other ( )	1. Because of acquaintance 2. High level of technique 3. Low price 4. Because of guidance 5. Nearness of location 6. Other ( )	1. Cash 2. TOYOTA CARD 3. Other cards 4. Other ( )	1. Drive to the shop 2. Have the vehicle picked up 3. Other ( )
3. Repair service of fault of your vehicle (please name a principal shop you use)	1. Dealer you purchased a new vehicle from 2. Other dealer 3. Vehicle maintenance service shop 4. Car shop 5. Gas station 6. Other ( )	1. Because of acquaintance 2. High level of technique 3. Low price 4. Good attending attitude 5. Nearness of location 6. Other ( )	1. Cash 2. TOYOTA CARD 3. Other cards 4. Guarantee 5. Other ( )	1. Drive to the shop 2. Have the vehicle picked up 3. Other ( )
4. Repair service of your vehicle after an accident (please name a principal shop you use)	1. Dealer you purchased a new vehicle from 2. Other dealer 3. Vehicle maintenance service shop 4. Car shop 5. Other ( )	1. Because of acquaintance 2. High level of technique 3. Low price 4. Good attending attitude 5. Nearness of location 6. Other ( )	1. Cash 2. TOYOTA CARD 3. Other cards 4. Guarantee 5. Other ( )	1. Drive to the shop 2. Have the vehicle picked up 3. Other ( )
5. Next vehicle inspection (what type of service shop would you select?)	1. Dealer you purchased a new vehicle from 2. Other dealer 3. Vehicle maintenance service shop 4. Car shop 5. Other ( )	1. Because of acquaintance 2. High level of technique 3. Low price 4. Because of guidance 5. Nearness of location 6. Other ( )	1. Cash 2. TOYOTA CARD 3. Other cards 4. Other ( )	1. Drive to the shop 2. Have the vehicle picked up 3. Other ( )
6. Purchase of a new vehicle (what type of shop would you select to purchase from?)	1. A shop you purchased your present car from 2. A shop other than one you purchased your present car from 3. Other ( )	1. Because of acquaintance 2. Sales staff is enthusiastic 3. Good attending attitude 4. Using the shop for the check-up and maintenance service of your vehicle 5. Good purchasing condition 6. Other ( )	1. Cash 2. Installment 3. Lease 4. Other ( )	1. Drive to the shop 2. Have a sales staff come to see me 3. Other ( )

Chapter 10
Strategic Stratified Task Team Model for Excellent QCD Studies

ABSTRACT

In this chapter, the author introduces the application studies of driving force in new JIT strategy that contributes to the evolution of Japanese automobile production. The author believes that the key to successful global production is the excellent QCD studies by total task management team (TTMT) activities between the assembly-maker and suppliers. Specifically, the author has established the strategic stratified task team model (SSTTM) using partnering performance measurement model (PPMM) for global SCM strategy. To actualize this, the author introduces typical case studies of how this model improved the bottleneck problems of auto-manufactures in the world for realizing simultaneous QCD fulfilment in Toyota and suppliers.

NEEDS FOR THE REFORM OF JAPANESE-STYLE MANAGEMENT TECHNOLOGY

Looking at the recent automobile recall problems, the author see rapidly increasing manufacturing quality issues with their roots in technological product design and product (Joiner, 1994; Nihon Keizai Shinbun, 2000, 2006; Amasaka, 2000a, 2001, 2002, 2008a,b,c; Amasaka, Ed., 2007).

The author cannot be content with simply resolving individual technical issues. Rather, it is necessary to evolve core technologies that result in the overhaul of every business process from development and production to SCM, and establish and systematically apply a new management technology model that intelligently links them together (Amasaka, 2004a,b).

The top priority issue of the industrial field today is the "new deployment of global marketing" for surviving the era of "global quality competition" (Kotler, 1999; Amasaka, 2004a). The pressing management issue particularly for Japanese manufacturers to survive in the global market is the "uniform quality worldwide and production at optimum locations" which is the prerequisite for successful global production.

DOI: 10.4018/978-1-6684-8301-5.ch010

To realize manufacturing that places top priority on customers with a good QCD and in a rapidly changing technical environment, it is essential to create a core principle capable of changing the technical development work processes of development and design divisions (Amasaka, 2004b, 2005a).

Furthermore, a new quality management technology principle linked with overall activities for higher work process quality in all divisions is necessary for an enterprise to survive (Burke and Trahant, 2000; Amasaka, 2004b).

The creation of attractive products requires each of the sales, engineering/design, and production departments to be able to carry out management that forms linkages throughout the whole organization (Seuring et al., 2003; Amasaka, 2004b). From this point of view, the reform of Japanese-style management technology is desired once again. In this need for improvements, Toyota is no exception (Goto, 1999; Amasaka, 2004a).

Similarly, it is important to develop a new production technology principle and establish new process management principles to enable global production. Furthermore, new marketing activities independent of past experience are required for sales and service divisions to achieve firmer relationships with customers.

In addition, a new quality management technology principle linked with overall activities for higher work process quality in all divisions is necessary for an enterprise to survive (Amasaka, 2007a).

STRATEGIC STRATEFIED TASK TEAM MODEL (SSTTM) FOR EXCELLENT QCD STUDIES

Japan Supply System-Partnering Chains as the Platform-Type

Toyota group consists of a total 14 enterprises (Toyota, 1987; Amasaka, 1988, 2000b; Amasaka, Ed., 2007a,b, 2019). Twelve enterprises, which branched from "Toyota Automatic Loom Works, Ltd." form the nucleus of the primary group parts manufacturers (called "Kyohokai") that supply parts directly to Toyota. In addition, Hino Motors, Ltd. and Daihatsu Motor Co., Ltd. jointed the group.

Each of the group companies is closely linked to Toyota in a wide and solid supplier-assembler relation called "Toyota Supply System". The author generalizes it as "Japan Supply System" (JSS) as shown in Figure 1 of Chapter 6 (Amasaka, 2000b, 2008a).

An automobile is assembled with some 20,000 parts. Since it is not economical for the assembler to manufacture all the parts in-house, considerable portion of the parts are normally purchased from outside suppliers (parts manufacturers). This situation still remains unchanged.

If parts purchased from the supplier (parts manufacturer) have low dependability, vehicles assembled with them have also low dependability naturally. This is exactly the reason why "performance of a vehicle almost depends on the parts" (Amasaka, 2000b).

In this sense, the assembly makers (automobile manufactures) and parts manufacturers (suppliers) are the inhabitant of the same fate-sharing community. Actual supplier-assembler relation is generally quite complex and many-sided.

Relationship between Toyota and its suppliers is unique in many points compared with those of other assemblers. There is the saying that "Toyota wrings a towel even when it is dry." indicating Toyota's strict demand to suppliers for their prices and quality.

At the same time, no other assemblers are so enthusiastic as Toyota in raising strong suppliers through education and training. As early as in 1939, Toyota established its basic concept of the purchasing activities for promoting coexistence and co-prosperity.

To realize this, Toyota strengthened its suppliers by making it a rule to continue transaction forever once started. No other assemblers have their supplier groups as powerful as those of Toyota. The "14 Toyota group companies" form the nucleus of the powerful supplier system. As thus far stated, close relation between Toyota and its group members is quite cooperative in one sense while simultaneously very competitive in another. This represents the supplier- assembler relation unparalleled elsewhere.

There is no denying that the strength of Toyota that can keep on supplying popular vehicle model such as "Lexus" of high dependability originates from within Toyota's own.

But one part of Toyota's strength comes from the strength of its supply system consisting of Toyota group and thousands of other suppliers, and the other comes from Toyota's skill in managing such powerful supply system.

In the following, the author intends to zero in on Toyota's quality management activities in exploring the secret of the strength of Toyota that continues to manufacture vehicles of high dependability. This is because it is quality management activities themselves that provide important bonds to cooperation or partnering between Toyota and its group companies (Amasaka, 2000b, 2001).

Strategic Stratified Task Teams Model, the Driving Force in Developing Japan Supply System

In a general assembly industry like the automotive manufacturing, a thorough quality assurance (QA) plays an important role. It includes not only the parts and unit quality control, but also the optimization of adaptive engineering for assembly by the automobile manufacturers and the quality of production, sales and services.

Therefore, when working on the solution of a pending engineering task, separate and independent team activities by the automobile manufacture or parts manufacturers alone often fail to clarify the engineering area, which exists in-between or across the two teams (both companies).

In this connection, it is important to zero-in on the end users from the view point of customer-in and to challenge for realizing "Simultaneous QCD fulfilment" by mutually disclosing their software and hardware techniques to each other (Amasaka, 2000b).

As a management technology strategy to enable smooth business management for realizing high-quality assurance, the author has created a "Strategic Stratified Task Team Model" (SSTTM) as shown in Figure 4 of Chapter 7 (Amasaka, 2004b, 2014a,b; Amasaka, Ed., 2012).

The expected role of the SSTTM and the benefits it provides are not limited to cooperation among the departments inside the company. It also contributes to strengthening the ties among group manu-facturing companies, non-group companies, and even overseas manufacturers. Two measures must be taken in order to realize this proposal.

The first is to eliminate the work methods that rely too heavily on the techniques and experiences of individuals. The second is to revolutionize the business process through the SSTTM that places emphasis on cooperation among the departments, and that with suppliers. SSTTM formed from partnering link-ages are developed systematically and organizationally to promote the strategic development of New JIT strategy.

This model consists of Task 1 to Task 8 teams involving the group, department, division, field, whole company, affiliated companies, non-affiliated companies, and overseas affiliates as described in Figure 4 of Chapter 7 (Amasaka, 2001, 2003a, 2004a,b).

Strategic Cooperative Creation Team Activity to Strengthen Employing SSTTM

The formation of "Strategic Cooperative Creation Team" (SCCT) activity indicated by Figure 1 is required in order to strengthen employment of SSTTM created in Figure 4 of Chapter 7, and solve issues of management technology (Amasaka, 2008a,b, 2015a,b; Yamaji and Amasaka, 2009).

Figure 1. Formation of strategic cooperative creation team (SCCT)

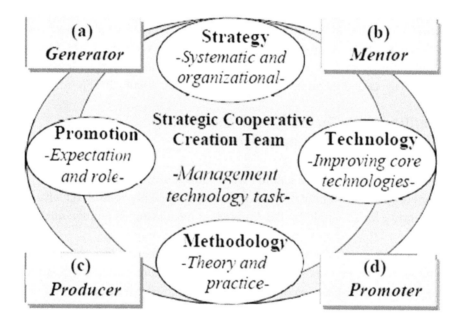

To empower the SCCT, task team members should collectively have the capabilities in (1) "Strategy" for systematic and organizational activities, (2) "Technology" to improve core technologies, (3) "Methodology" to practically identify the gap between theories and the actual, and (4) "Promotion" to fulfill the expectations and roles of the team.

If the task team tackles a strategic issue requiring high technologies, the members have to be ingenious as (a) "Generators," and at the same time they have to be able to perform strategic analysis as (b) "Mentors".

In addition, to infuse effective drive force in the team activities, creativity as (c) "Producers" and leadership to orchestrate all members' ideas as (d) "Promoters" toward target achievement are important.

As the key to the successful team activities, the team leader (Administrator) should select the members who have at least one of the capabilities for (a) to (d), commission authority and responsibilities to the members, and have himself/herself concentrate on risk management.

For this reason, as the leader, a person who has an experience of clearing business obstacles should be appointed, so that the leader is capable of leading the team overcoming difficulties.

REALIZATION OF SIMULTANEOUS QCD FULFILMENT

In recent years, leading manufacturers in Japan have been deploying a new production strategy called "globally consistent levels of quality and simultaneous global launch (production at optimal locations)" in order to get ahead in the "worldwide quality competition", and "high quality assurance in manufacturing - simultaneous achievement of QCD" (Amasaka and Sakai, 2010, 2011).

This is the key to successful global production, and has become a prerequisite for developing New JIT. It has been observed that, despite the fact that overseas plants have the relevant production systems, facilities, and materials equivalent to those that have made Japan the world leader in manufacturing, the building up of quality - assuring of process capability (Cp) has not reached a sufficient level due to the lack of skills of the production operators at the manufacturing sites.

Under such a circumstance, there are many studies abroad for globalization (Lagrosen, 2004; Burke et al., 2005; Hoogervorst et al., 2005). To realize the key to global production, the authors create the "simultaneous QCD fulfillment" for developing New JIT as shown in GIPM (Global Intelligence Partnering Model) (See to Figure 19 in Chapter 9) (Yamaji and Amasaka, 2009; Amasaka, 2009a).

From Figure 19 in Chapter 9, the function of the (i) Quality Assurance (QA) and (ii) TQM promotion as corporate environment factors for succeeding in "global production" are (1) Customer Satisfaction (CS), ES (Employee Satisfaction), and SS (Social Satisfaction), (2) High Quality Assurance, (3) simultaneously achievement of QCD, (4) success in global partnering and (5) evolution of quality management and intellectual productivity.

More specifically, the (i) QA division needs to promote manufacturing of high reliable manufacturing, and cooperative activity across the organization is indispensable to achieve that. In the (ii) TQM promotion division, it is important to cultivate human resources that have even higher skills, knowledge, and creativity. Therefore, the value of intelligent human resources must be promoted in an effort to improve the productivity of white-collar workers.

To realize the above, global partnering which enables a strategic cooperation among divisions, such as designing, production, marketing and administration as well as the entire company, affiliated companies, non-affiliated companies, and overseas corporations, must be achieved. Improvement of intellectual productivity is simultaneous achievement of QCD utilizing this model.

PARTNERING PERFORMANCE MEASUREMENT MODEL TO STRENGTHEN SSTTM IN GLOBAL CORPORATE STRATEGY

To strengthen SSTTM activity in global corporate strategy, the author has created the "Partnering Performance Measurement Model" (PPMM) for assembly makers and suppliers (Sakatoku, 2006; Yamaji et al., 2008; Amasaka, 2007e, 2009, 2011, 2015a, 2019, 2022a,b, 2023; Amasaka, Ed., 2012).

The aim of PPMM is to evaluate the "employment and justification of JSS (Japan Supply System) for developing global SCM and approach as the application study of driving force in New JIT strategy below (See to Chapter 6 and 7).

According to a survey conducted by the author, assembly makers in general assessed that they are "performing SCM with their suppliers well overall".

On the other hand, suppliers' answers indicate that they are not necessarily conducting SCM the way they would like to with some of their assembly makers, revealing a difference in evaluation, or so-called "disparity (deviation or gap) in evaluator awareness".

These SCM evaluations are often based on the evaluators' own implicit empirical knowledge, and (as far as the author know) evaluation scales are not usually shared (formulated as a model) between them.

The author therefore attempts to "formulate evaluation causes and effects" that account for the difference in implicit evaluations on both sides - the so-called "gap in awareness". Then, the author attempted to carry out "diagnosis through visualization methods" by utilizing "Science SQC, new quality control principle" (Amasaka, 1999, 2004c, 2009) (See to Chapter 2 and 3).

Under such a circumstance, there are many studies abroad for globalization (Gabor, 1990; Lagrosen, 2004; Manzoni and Islam, 2007) and SCM (Joiner, 1994; Taylor and Brunt, 2001; Pires and Cardoza, 2007).

According to the authors' investigation, a quality management survey (JUSE, 2006) is being jointly conducted by Nihon Keizai Shinbun and the Union of Japanese Scientists and Engineers. It targets 528 corporations, mainly manufacturers, and does not cover the topic of partnering with suppliers.

The OEM-Supplier Working Relations Index (WRI) (Milo Media, 2006) defines the relationship between Automobile Manufacturers and their Suppliers using index numbers, but specific evaluation items are not indicated.

In connection with the PPMM creation, the author develops (1) PPMM-A, Partnering Performance Measurement Model for assembly makers, and (2) PPMM-S, Partnering Performance Measurement for suppliers. The author also integrates the two and develops (3) PPMM-AS, Partnering Performance Measurement Model for assembly makers and suppliers, as a comprehensive Dual Performance Measurement to be shared by both.

Preparation of evaluation sheets for PPMM-A, PPMM-S, and PPMM-AS

The author extracted "evaluations of cause-and-effect elements" necessary for the preparation of the three types of evaluation sheets (PPMM-A, PPMM-S and PPMM-AS), based on the response of staff members engaged in practical affairs at SCM-related departments (both assembly makers and suppliers). More specifically, evaluators from related departments on both sides were given a survey. The questions (Evaluation sheet) are shown in Table 1. The survey comprised of a total of 19 evaluation factors (Factor X_1 to X_{19}, with evaluation contents) extracted from an affinity diagram.

The diagram was made based on meetings with assembly makers and suppliers.

(1)　Evaluators of assembly makers

Toyota Motor Corporation (Sales div. 1, Production Preparation div. 2, TQM Promotion div. 1), Hino Motors Ltd. (Procurement div. 1, TQM Promotion div. 1), Fuji Heavy Industries Ltd. (TQM Promotion div. 1, Procurement div. 2), Honda Motors Co., Ltd. (Quality Assurance div. 1), Nissan Motor Co., Ltd. (Procurement div.1), GM (Procurement div. 1), Ford (TQM Promotion div. 1).

(2)   Evaluators of suppliers

Jatco (Quality Assurance div. 2), JFE (Quality Assurance div. 2, Procurement div. 1), NHK Spring Co., Ltd. (Quality Control div. 1, SQC Promotion div. 1).

Table 1. Evaluation sheet

Factor	Contents	Factor	Contents
X_1	Corporate strategy	X_{11}	Price setting considering labor
X_2	Bias in selection of Suppliers	X_{12}	Parts inspection standards
X_3	Selection of affiliated Supplier	X_{13}	Parts inspection items
X_4	Weakening of group affiliation	X_{14}	Response to recalls
X_5	Weight of Suppliers' opinion	X_{15}	Response to contractual failure
X_6	Spirit of cooperation	X_{16}	Ability to improve
X_7	Appropriate evaluation	X_{17}	Technical capabilities
X_8	Considerate relation	X_{18}	Total satisfaction rate
X_9	Demand on discounting	X_{19}	Gained technology and know-how
X_{10}	Methods of price setting		

Formulation model of PPMM-A, PPMM-S and PPMM-AS

Using this evaluation sheet, the author attempted to derivate a formulation model that incorporated multivariable statistical analysis.

First, "cause and effect relationships" were extracted by conducting "Cluster Analysis (Ward method and Euclidean distance squared) and Principal Component Analysis (PCA)".

Next, using categorical canonical correlation analysis, formulation models for PPMM-A, PPMM-S and PPMM-AS were derived, while coefficients of each group's factors were calculated and a radar chart was designed for each group. One hundred (100) was set as the highest score.

(1)   Derivation of PPMM

PPM-A for Assembly Makers was derived, with the results of cluster analysis and principal component analysis shown in Figure 2 and Figure 3. Based on these results, evaluation factors could be categorized into five elements: "Supplier follow-up", "Degree of quality inspection", "Corporate capability", "Supplier decisions", and "Price setting".

Consideration was then given to PPMM-A, PPMM-S, and PPMM-AS so that formulation models could be derived. With that in mind, categorical canonical correlation analysis was conducted, employing optimum scaling methods in an effort to grasp the relationship between the five evaluation elements and seventeen evaluation factors.

Figure 2. Grouping by cluster analysis

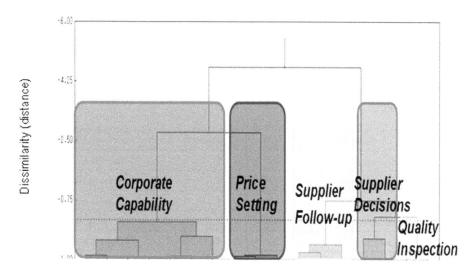

Figure 3. Grouping by principal component analysis (PCA)

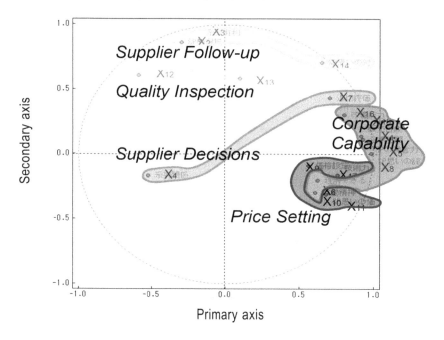

A calculation was conducted for PPMM-A using evaluation element axes (a) through (e). The weighted coefficients shown here were calculated based on component loading values for each of the five evaluation elements, resulting in a full score of 100 points.

(a) Supplier follow-up = $5.24X_4 + 8.86X_7$
(b) Degree of quality inspection = $2.28X_{12} + 12.01X_{13}$
(c) Corporate capability = $1.91X_1 + 2.44X_{16} + 1.85X_6 + 2.46X_{17} + 1.89X_5 + 1.81X_8 + 1.91X_{15}$

(d)   Supplier decisions $= 5.39X_2 + 7.23X_3 + 1.67X_{14}$
(e)   Price setting $= 6.80X_9 + 5.84X_{10} + 1.64X_{11}$

Using the above equations, evaluation points for each of the five evaluation elements were calculated and visualized in a radar chart, as seen in Figure 4.

Figure 4. PPMM-A radar chart

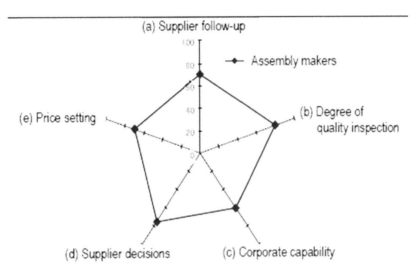

(2)   Derivation of PPMM-S

Derivation was conducted in a similar way for Suppliers' PPMM-S. As a result, the relationship between the 5 evaluation elements:

(f)   Assembly maker follow-up, (g) Degree of quality inspection, (h) Corporate capability, (i) Spirit of cooperation, and (j) Price setting and the seventeen evaluation factors was grasped. Calculation was conducted for PPMM-S using 5 evaluation element axes (f) through (j) below.

(f)   Assembly maker follow-up $= 5.83X_1 + 5.59X_4 + 2.75X_{14} + 0.12X_{15}$
(g)   Degree of quality inspection $= 1.54X_{12} + 12.75X_{13}$
(h)   Corporate capability $= 5.23X_2 + 2.27X_3 + 1.34X_7 + 0.63X_{16} + 4.81X_{17}$
(i)   Spirit of cooperation $= 6.46X_5 + 3.26X_6 + 4.56X_8$
(j)   Price setting $= 4.17X_9 + 5.41X_{10} + 4.70X_{11}$

Using the equations, evaluation points for each of the five evaluation elements were similarly calculated and visualized by a radar chart, as shown in Figure 5.

Figure 5. PPMM-S radar chart

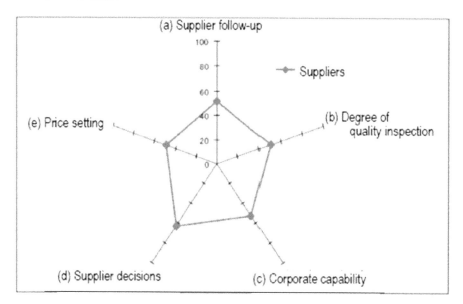

(3)   Derivation of PPMM-AS

The PPM-AS was derived in order to visualize the gap in the evaluations of both sides. This was done by closely comparing the evaluation measurements of PPMM-A and PPMM-S. Corporations using PPMM-AS can see the gap between their evaluations and those of their Suppliers at a glance. This allows conventional subjective evaluation based on each side's empirical rules to be converted into objective analysis, while also clarifying specific problems using the component evaluation factors.

One of the features of the PPMM-AS formulation model is the overlapping of the five evaluation element axes of PPMM-A and PPMM-S. This is due to the fact that both PPMM-A and PPMM-S thus far consist of five evaluation elements and seventeen evaluation factors.

The three axes of "Degree of quality inspection, Corporate capability and Price setting" are common to both, allowing these elements to be overlapped for direct comparison. In addition, the two new axes, Assembly manufacturer follow-up and Supplier follow-up are integrated to create a new axis of "Spirit of mutual support". Similarly, Spirit of cooperation and Supplier decisions are integrated to establish a new axis of Relationship with business partners.

As a result, these evaluation elements, made up of "(k) Spirit of mutual support, (i) Degree of quality inspection, (m) Corporate capability, (n) Relationship with business partners and (o) Price setting" are visualized in a radar chart as shown in Figure 6.

The figure presents example evaluations of assembly makers and suppliers.

Verification of PPMM-AS

The author has applied the created PPMM-AS to three Japanese assembly makers (Toyota Motor Corporation and Nissan Motor Co., Ltd.), one overseas assembly manufacturer (General Motors), and two Suppliers (Jatco and NHK Springs Co., Ltd.).

Figure 6. PPMM-AS radar chart

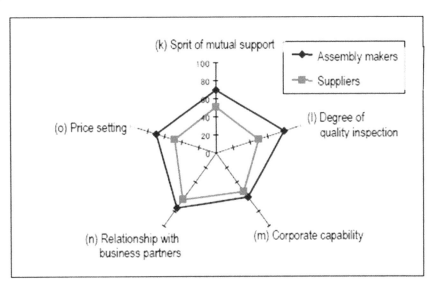

Its effectiveness was thereby verified. As pointed out in this section "PPMM for assembly makers and suppliers", it can be generally confirmed that assembly makers' evaluations are higher than those from suppliers. The suppliers' evaluations are more severe than the Assembly Manufacturer evaluations assume.

The following sections provide examples of PPMM-AS at various corporations.

(1)   Toyota Motor Corporation

The survey results used for verification are shown in Table 2(1) (Toyota Motor Corp.), Table 2(2) (NHK Springs Co. Ltd. (hereafter, NHK)) and Figure 7 (Toyota and NHK). The evaluation example of Toyota Motor Corp. and NHK. is further described in Figure 8.

As the figure indicates, evaluations on both sides are generally high. There are no large differences between elements, but a slight gap can be observed in "Spirit of mutual support" and "Price setting". The following subsections provide an interpretation of the obtained evaluation results. NHK's low evaluation in "Price Setting" and "Spirit of Mutual Support" stems from thorough cost price reduction carried out by Toyota, based on the company's Customer First principle.

However, there is almost no gap is in the Relationship with Business Partners category. NHK evaluates Toyota's attitude towards problem-solving and QCD improvement highly, and these activities are carried out with mutual support. Therefore, NHK names Toyota as the corporation they would most like to do business with.

(2)   Nissan Motor Co., Ltd.

The evaluation example of Nissan and Jatco is shown in Figure 9. One outstanding feature is a generally lower evaluation than in the case of Toyota above. Gaps can be observed in both of the "Price Setting and Corporate Capability". These results can be interpreted as showing partnering activity that is not necessarily based on mutual trust, thus generating gaps in Price Setting and Corporate Capability.

Table 2. Verification examples of PPMM-AS in assembly makers and suppliers

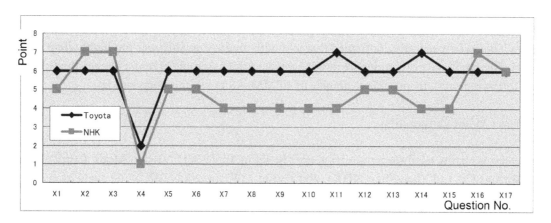

(1) Ex.1 Evaluation Sheet of Toyota (2) Ex.2 Evaluation Sheet of NHK

Figure 7. Questionnaire result of Toyota and NHK

Figure 8. Radar chart of Toyota and NHK

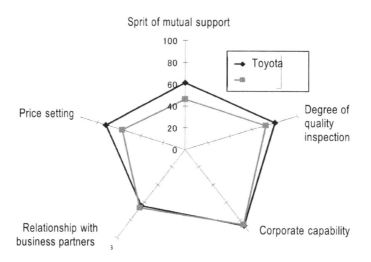

Figure 9. Radar chart of Nissan and Jatco

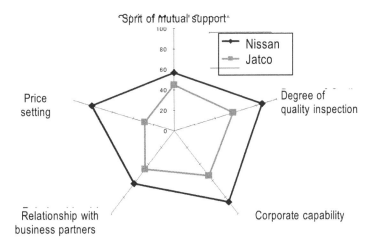

(3)   General Motors, Fuji Heavy Industries and Honda Motors

Figure 10 shows the examples of GM and Jatco. Characteristic of these results is a large gap in the relationship between Price Setting and Relationship with Business Partners. This is because Jatco cannot fully respond to GM's severe, one-sided demands regarding QCD. This leads to a gap in awareness regarding good business partnerships based on mutual trust.

Similarly, evaluation was implemented in Fuji Heavy Industries Ltd. and Honda Motors Co., Ltd., and the expected results were obtained with regard to its effectiveness.

Figure 10. Radar chart of GM and Jatco

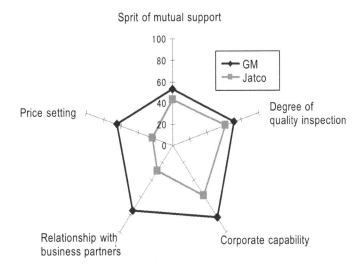

APPLICATION EXAMPLES

High Reliability Assurance of the Transaxle Oil Seal Leakage

The oil seal for the drive system works to seal lubricant inside the transaxle unit. The cause-and-effect relationship between the oil seal design parameters and sealing performance is not necessarily fully clarified. As a result, oil leakage from the oil seal is not completely eliminated, presenting a continual engineering problem (Lopez et al., 1997).

So far, oil seal quality improvement has been made as follows; a development designer having empirical engineering capability recovers the leaking oil seal parts from the market, analyzes the cause of leakage with proper technology, and incorporates countermeasures into the design.

Many of the recent leaking parts, however, exhibit no apparent problem and the cause of the leakage is often undetectable. This makes it difficult to map out permanent measures to eliminate the leakage (Amasaka, 2003a, 2008c,d, 2010, 2012a, 2017; Amasaka et al., 2012; Amasaka, Ed., 2012; Nozawa et al., 2013).

Dual Total Task Management Team Utilizing SSMMT and SCCT

Organization of a "Dual Total Task Management Team" (D-TTMT)

This case study is a joint task team activity between manufacturer and non-affiliated supplier as stated above Task-7 (See to SSTTM in Figure 4 of Chapter 7) for problem-solving "simultaneous QCD fulfilment" (See to GIPM in Figure 19 of Chapter 9).

Effective solution of technical problems requires also the formation of SCCT as state above Figure 1 and understanding of the essence of problems by the teams as a whole. To ensure high reliability of product design and quality assurance, the author uses SCCT named "Dual Total Task Management Team" (D-TTMT).

This organization has been realized in the form of D-TTMT consisting of the final vehicle assembly manufacture (Toyota Motor Corporation) with transaxle unit, and oil seal manufacture (NOK Corporation) of tier-1.

Deployment of D-TTMT consists of SCCT named "DOS-Q" (Drive-train Oil Seal - Quality Assurance Team: "T DOS-Q5" of Toyota and "N DOS-Q8" of NOK) is as shown in Figure 11.

"T DOS-Q5" consists of Q1 (Unit manufacturer quality team), Q2 (Mechanism study team), Q3 (Design improvement team), Q5 (Shaft manufacturer quality improvement team) and Q5 (install-ability improvement team).

Similarly, "N DOS-Q8" consists of Q1 (Production engineering team), Q2 (Mechanism study team), Q3 (Design improvement team), Q4 (Quality assurance team), Q5 (Receiving/Shipping team), Q6 (Affiliated company), Q7 (Production team) and Q8 (Production design team).

In DOS-Q, the author also commissions the thirteen general managers of two companies, each general manager assigns several section chiefs, sub-section chiefs, technical staffs, and statistics analysis staffs named "SQC adviser" as D-TTMT members as state above "Organizational outline of Total Task Management Team (TTMT) (See to Figure 10 in Chapter 6).

Figure 11. Dual total task management team using T DOS-Q5 and N DOS-Q8

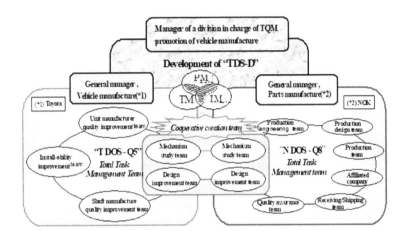

Development of D-TTMT using "T DOS-Q5" and "N DOS-Q8"

In "T DOS-Q5" and "N DOS-Q8", the author also develops SCCT utilizing three management activities utilizing TM, PM and IM named QAT in enforcement of D-TTMT as described in Figure 11. Thereby, D-TTMT was created to transform the implicit knowledge (relating to product and processes in both organization) into explicit knowledge, and to create new technology of interest to both organizations of "T DOS-Q5" and "N DOS-Q8". By the top policy of Toyota and NOK, the author constitutes the following organizations as a supervisor of Task-7 as follows;

In SCCT, five teams of "T DOS-Q5" constitute Q1 (Chassis design department), Q2 (Drive train engineering department), Q3 (Product evaluation and Engineering department) for TM, Q4 (Production engineering and development department and Machine engineering department) for PM and Q5 (Quality assurance department and TQM promotion department) for IM (Fukuchi et al., 1998).

Eight teams of "N DOS-Q8" are also the same organization and formation of Q1-Q8 (Amasaka and Otaki, 1999). Specifically, "T DOS-Q5" constituting teams comprise Q1 and Q2 in charge of investigation into the cause of the oil leakage and Q3-Q5, which handled manufacturing problems relating to drive shafts, vehicles and transaxles.

Similarly, "N DOS-Q8" formed teams Q1 through Q8. Q1 and Q2 at Toyota interacted closely with their counterparts at NOK to improve the reliability of the oil seal as a single unit and, likewise, Q3-Q8 handled the manufacturing problems for quality assurance.

Furthermore, in order to strengthen the cooperation activities of A and B, a writer forms "Cross-functional team" of "Q2 and Q3" of Toyota, and "Q2 and Q3" of NOK, solves the mechanism of an oil seal leak, and strives for an improvement of quality of design.

In addition, the author forms the cross-functional team between Q2 and Q3 of both companies, and, thereby, strengthens improvement of transaxle design by the clarification of oil leakage mechanism. Accordingly, the thirteen teams shared their individual knowledge (relating to empirical techniques and other technical information) to apply them to solving the problems under consideration. Each team had a general manager and the joint team was led by the general manager of Toyota's TQM promotion division for the vehicle reliability assurance.

The methodology of "TDS-D" (Total Design System for Drive-train Development) involving TM, PM and IM was used by utilizing three management activities for QAT as shown in Figure 11(1) of Chapter 6. Moreover, to realize optimization of "DOS-Q" business process, problem-solving is formulated using "QAT using SQC Technical Methods" by using the same approach[*1] as shown in Figure 11(2) of Chapter 6 (*1: Refer to Amasaka (2001) in detail).

Development of QAT using D-TTDT

Fault analysis and Factor analysis

In the conventional cases of oil seal unitary sample collection process, it was observed from time to time that there was no information on the mating part or the collected sample was attached with foreign matters which hampered the determination of the cause of leakage. To prevent this, recovery process was improved along with the acquisition of the background or the history of the market recovery.

Thus, Weibull analysis for the fault analysis could be made in Figure 12(1) with credibility based on the comparison result with the actual parts by recovering non-defectives as well as defectives or by recovering whole of the trans-axle unit in order to reproduce oil leakage.

As the result, the author has obtained a new knowledge that the failure type of "oil leakage" is a mixed model of three failures of the "Decreasing failure rate", "Constant failure rate" and "Increasing failure rate" which was not found in the rule of thumb. Moreover, the author could understand the correct particle size distribution and the composition of foreign matters attached to the oil leaking parts by recovering them in a special recovery case. It was also newly discovered that two oil seals of apparently similar degree of wear produce difference in pumping quantity depending on the mileage.

For the factor analysis to research the oil leakage phenomenon and mechanism of the failure type, not only defective parts (oil leaking parts) but non-defectives (non-oil leaking parts) were used. As the result, the difference of two parts was detected through the discriminating analysis as shown in Figure 12(2).

Figure 12. Fault analysis and factor analysis of the transaxle oil seal leakage

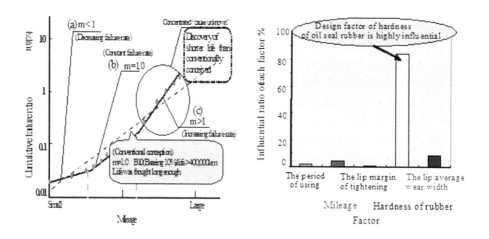

Relation between the condition of sliding surface & sealing performance

An oil seal is a machine part built into machinery that plays an important role in preventing oil from leaking and foreign matters from entering the machinery. An oil seal on an automobile's transaxle prevents the oil lubricant within the drive system from leaking from the drive shaft as shown in Figure 13.

Its basic components include a sealing lip (rubber lip grip), which consists of both a metal case providing rigidity to the outside surface and synthetic rubber (round metal casing), and a garter spring (round spring). The sealing edge has direct contact with the drive shaft to prevent leakage of oil. Why doesn't the fluid on the oil side leak out to the air side?

Figure 13. Outline of automotive transaxle oil seal

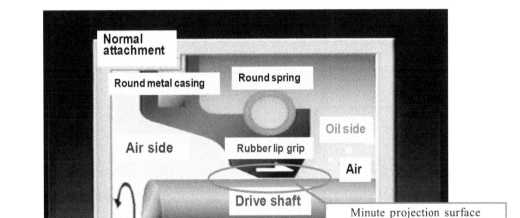

In the stage of problem clarification, the author has investigated the relation between the condition of the sliding surface and the sealing performance of the oil seal (Amasaka and Otaki, 1999; NOK Corp., 2000; Amasaka, 2005b, 2007b, 2010, 2012a,b). To grasp the features of the sliding and surface condition, the author takes a quantification approach as shown in Figure 14.

First of all, the author will consider two typical characteristic values (A_R and X_G). In Figure 14(1), the author shows roughness in the form of contour lines on a map, and the shaded area indicates the area which is actually in contact with the drive shaft.

As the characteristic value representing the degree of roughness, the author defines A_R which is the ratio of the actual (real) contact area to the apparent contact area. A_R is a characteristic value indicating roughness on the seal sliding side level, and shows the proportion of the visible touch area that is the real contact area shown in equation (1).

The smaller A_R values signify minute roughness on the sliding surface, and conversely, larger A_R values show evenly flat contact. Next, the author defines the characteristic value X_G, which indicates whether the real contact area (dA_i) is closer to the oil side or the air side as described in Figure 14(2).

X_G is the distance that the true contact area center of gravity is biased from the sliding side center to either the oil side ($1 > X_i > 0$) or the air side (atmospheric side) ($-1 < X_j < 0$); it is the characteristic value that shows the extent true contact area distribution bias axially shown in equation (2). X_G can be interpreted as the maximum pressure position of the contact pressure distribution.

If the X_G value is positive, it means that the actual contact area is biased toward the oil side. On the other hand, if the X_G value.

Figure 14. The condition of the sliding surface and the sealing performance of the oil seal

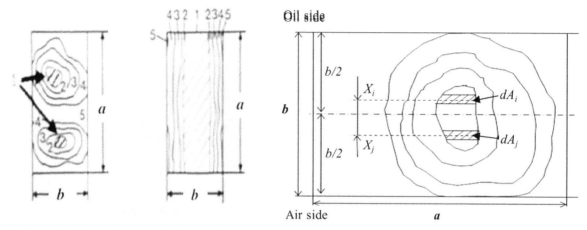

(*a*: Length of the minute projection surface, *b*: Contact width of the sliding surface)

(1) Ratio of the real contact area A_R (2) Bias of the true contact area X_G

Figure 15 shows the characteristic seal values and the outcome of the experiments for the four types of seals made of different materials and shapes.

In addition, leakage stopping can be classified according to both characteristic values by looking at the relationship between these two characteristic values and the observed state of the leakage stopping.

Sealing is achieved only in Seal type 1 described in Figure 15, and sliding surfaces show the characteristic values $A_R < 0.05$ (5%) and $X_G > 0$. The domain representing complete seal is defined as having an A_R value under 0.05 and a positive X_G value.

Then, the author observes the existence of an oil leak experimentally using the two characteristic values, A_R and X_G as shown in Figure 16. In the case of seal type 1, A_R shows a small value, indicating that the sliding surface is minutely uneven. And the X_G value is positive, meaning that the actual contact area is closer to the oil side.

Under such a sliding surface condition, there will be no oil leak (sealing). In the case of seal type 3, although the A_R value is small with an uneven sliding surface, the X_G value is negative, suggesting that the actual contact area is inclined toward the air side.

Under such a sliding surface condition, oil can easily leak through. It follows that the sliding surface must be slightly uneven and the actual contact area should be biased toward the oil side in order to realize absolute sealing.

Figure 15. Two seal characteristic values, A_R and X_G, and outcome of the experiments

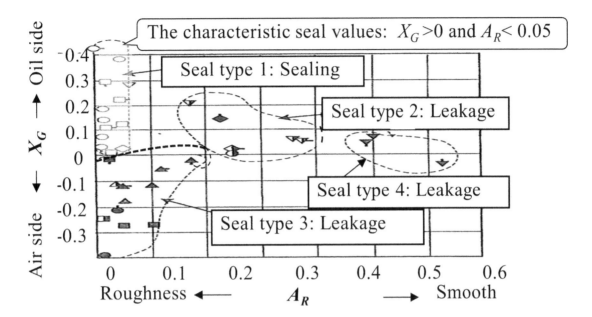

It is evident that the unique sliding side structure to ensure sealing has minute surface roughness (existence of minute projections) and true contact area is biased towards the oil side (Sato et al., 1999: Kameike et al., 2000).

Figure 16. Relation between the characteristic values, A_R and X_G and the seal condition

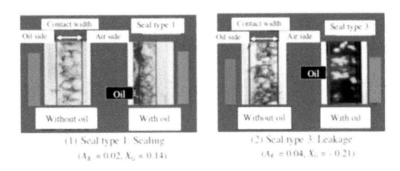

Clarification of oil leakage mechanism by visualization

The author has arranged the information obtained from the above-mentioned activities, which have boiled down to the following new facts:

(1) When compared the recovery parts with the forced wear parts, sealing performance differs because of the difference in the lip surface condition,

(2) Sealing performance drops if the axis has greater surface roughness. On the basis of such findings and quantitative analysis data, the team generated another affinity and association diagrams concerning oil leakage and estimated the oil leakage mechanism to discover the presence of "unknown mechanism" as shown in Figure 17(1).

Accordingly, a device was developed to visualize the dynamic behavior of the oil seal lip to turn this "unknown mechanism" into explicit knowledge as shown in Figure 17(2) and Figure 18.

Figure 17. Clarification of oil leakage mechanism by visualization

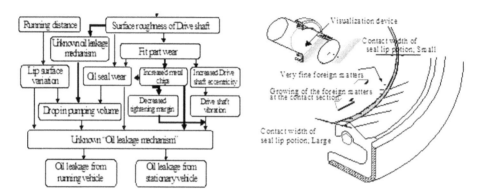

As the result of observation of the recovered parts from the market with this visualization device, a process was observed by which very fine foreign matters which were conventionally thought not to affect the oil leakage grow at the contact section (: Test-1 and Test-2 in Figure 17(2) and Figure 18(i)). As

Figure 18. Oil leakage mechanism (Tests two and three)

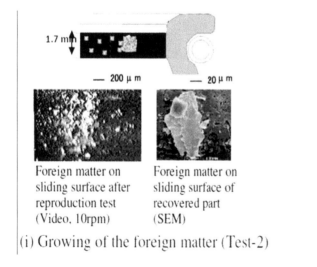

(i) Growing of the foreign matter (Test-2)

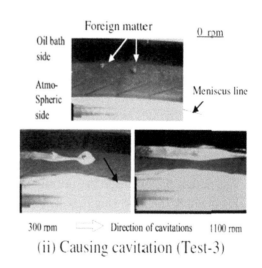

(ii) Causing cavitation (Test-3)

a result of the component analysis, it is confirmed that the fine foreign matter is the powder produced during engagement of gears inside the transaxle gear box.

These fine changes to the microscopic pressure distribution eventually degrading the sealing performance. Also, the presence of a mechanism was confirmed from a separate observation result that those foreign matters that had bitten into the lip sliding surface caused the aeration (; cavitation) to be generated to the oil flow on the lip sliding surface to deteriorate the sealing performance (Test-3 in Figure 18(ii)).

As far as the author knows, such knowledge was not given consideration conventionally and only discovered through the current team activities.

Applying Optimal CAE Design Approach Model

The author addressed the technological problem of oil seal leakage in automotive drive trains as a way to construct an "Optimal CAE Design Approach Model" for quality assurance.

The model is used to explain "cavitation caused by the metal particles (foreign matter)" generated through transaxle wear which is a pressing issue in the auto-industry.

Oil Seal Simulator using Highly Reliable CAE Analysis Technology Component Model

The author used the knowledge obtained from the visualization experiment to logically outline the faulty as the above "Clarification of oil leakage mechanism by visualization".

This was done in order to capture the problem employing the "Oil Seal Simulator using Highly Reliable CAE Analysis Technology Component Model" as shown in Figure 19 (Amasaka, 2007c).

Using this process, the author was able to arrive at a hypothesis for why the cavitation was occurring; namely, factors like low pump volume and seal damage had compromised the tightness of the seal and lead to oil leaks.

As the figure indicates, the designs are optimized by integrating several aspects of the calculation process, including problem (root cause) identification, conceptualizing the problem logically, and calculation methods (precision of calculators).

Once the root causes of the problem are identified, it is critical that there is no discrepancy between the mechanism described and the results of prototype evaluations.

The visualization experiment revealed that cavitation was occurring due to a weakening of the oil seal in areas (surfaces) that were in contact with the rotating drive shaft. This weakening was causing oil seal leaks.

The "Rayleigh Plesset Model" for controlling steam and condensation was used as a CAE analysis model that could explain the problem. The finite element method and non-stationary analyses were used as convenient algorithms.

The "Reynolds-averaged Navier-Stokes equation", "Bernoulli's principle", and "Lubrication theory" were appropriate theoretical formulas. Accuracy was ensured, and the time integration method was used to perform calculations in a realistic timeframe.

Each of the above elements was used to construct the "Oil Seal Simulator" (Amasaka et al., 2012; Amasaka, Ed., 2012; Nozawa et al., 2013).

CAE analysis examples

A cavitation is generated at the following steps; Oil collides with a foreign substance - the flow velocity rise near a foreign substance - the fall of pressure - decreased pressure is carried out to below saturated vapor pressure - emasculation of oil - generating of a cavitation as shown in Figure 20.

Figure 19. Oil seal simulator using highly reliable CAE analysis technology component model

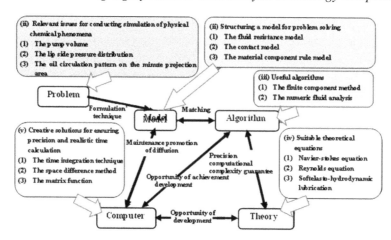

This analysis results at a rotation speed of 1100 rpm.

(i)   The fluid speed analysis was then conducted in order to look more closely at the mechanism causing cavitation. The analysis revealed that rapid changes in fluid speed were occuring in the vicinity of foreign particles, and that fluid speed drops immediately before the oil collides with foreign matter. This led to the conclusion that the presence of foreign particle was having an effect on oil flow.

Figure 20. Cavitation analysis around foreign matter

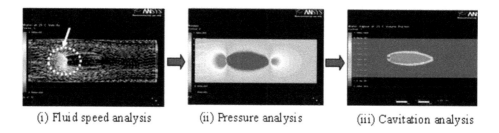

(i) Fluid speed analysis   (ii) Pressure analysis   (iii) Cavitation analysis

(ii)   Comparing cavitation and the fluid speed analysis results against the results of the pressure analysis reveals that in areas of reduced pressure, oil was disappearing inside the cavities being formed— meaning that drops in pressure were likely being caused by these concave areas.

(iii) Cavitation analysis confirmed the cavitation occurring around foreign matter, thus replicating the results of the visualization experiment.

At the same time, the finding that cavitation becomes more significant as the rotation speed of the drive shaft increases was similarly replicated.

Verification and consideration

The above CAE analysis allowed us to clarify the faulty mechanism causing cavitation; namely, the presence of metal foreign particles was affecting the strength of the oil flow, causing drops in pressure in areas with faster oil flow and creating cavities. In addition, a similar analysis of changes in the shape and size of the foreign particles revealed that these changes were also causing changes in cavitation. These CAE analysis results indicate a close link between particle size/shape and cavitation.

Preproduction and testing/evaluation of prototypes add a significant amount of time and cost to the development process. However, precise CAE allowed manufacturers to eliminate preproduction (as well as prototype testing/evaluation) and still predict the mechanism causing cavitation and oil leaks. Though gaps such as minute surface variations caused by foreign particles and the shape of the oil film model exist, the CAE analysis allowed the authors to recreate the changes in flow speed and pressure around the foreign metal particles that were causing cavitation - changes which typically cannot be identified.

The deviation between the CAE analysis results and the results of the prototype testing were less than 5%, attesting the usefulness of precise CAE analysis in certain cases.

Design changes and process control for improving reliability

These results led to two measures in order to improve design quality (shape and materials): (1) strengthen gear surfaces to prevent occurrence of foreign matter even after the B10 life (L10 Bearing to MTBF) to over 400,000 km (improve quality of materials and heat treatments) and (2) formulate a design plan to scientifically ensure optimum lubrication of the surface layer of oil seal lip where it rotates in contact with drive shaft.

The result of these countermeasures was a reduction in oil seal leaks (market complaints) to less than 1/20th their original incidence as shown in Figure 21. The author believes that this research result will contribute to period shortening of the development design and simultaneous fulfillment of QCD from now on (Amasaka, 2014a,b, 2017).

Figure 21. Reduction in market complaint rate

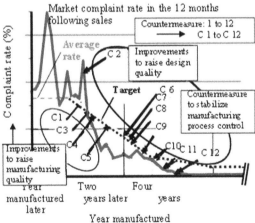

Other Cases

The author was able to apply the SSTTM to critical QCD studies of automobile manufacturing developing New JIT strategy, including predicting and controlling the special characteristics employing SCCT as follows;

(i) Preventing vehicles' rusting (Amasaka et al., 1995: Amasaka, 2004c), (ii) automotive braking performance (Amasaka, 2003b), (iii) Intelligence production operating system (Sakai and Amasaka, 2006), (iv) Epoch-making Innovation of the work environment (Amasaka, 2007d), (v) Partnering performance measurement of Japanese automobile assembly makers and suppliers (Yamaji et al., 2008), (vi) Simultaneous fulfillment of QCD for improving automotive chassis (paint corrosion resistance and welding process) (Yamaji and Amasaka, 2009), (vii) Automobile Body Color Development Approach Model (Muto et al., 2011), (viii) Total Quality Assurance Networking Model for preventing defects (Kojima and Amasaka, 2011), (ix) Prevention of bolt looseness (Hashimoto et al., 2014), (x) New Global Partnering Production Model (Ebioka et al., 2007), (xi) New Turkish Production System, New Malaysia Production Model and New Vietnam Production Model (Siang et al., 2010; Miyashita and Amasaka, 2014), (xii) Joint task team activities between Toyota Motor Corp. and Toyota Motor Thailand (Amasaka, Ed., 2012), and (xiii) Developing Advanced Toyota Production System at Toyota manufacturing USA (Amasaka, 2016).

Furthermore, the result of this research has spread in many other manufacturing industries. In the "Study group of the ideal situation the quality management of the manufacturing industry" (Union Japanese Scientists and Engineers), and "Establishment of a needed design quality assurance framework for numerical simulation" (Japanese Society for Quality Control) which the author led, "Bridgestone Corp., Daikin Industries LTD., Fuji Xerox Co. Ltd., Futaba Corp., Hino Motors LTD., NEC Corp., NHK Spring Co., NTT DATA Mathematical Systems Inc., Sanden Holdings Corp., Toyota Motor Corp., Mizuho Research Institute Ltd.", etc. indicate the case studies which can hold the validity of SSTTM, and are indicating the predetermined result (Amasaka, Ed., 2007a,b; Amasaka, 2015a, 2017, 2022a,b, 2023; Amasaka, Ed., 2019).

CONLUSION

In this chapter, the author has investigated the significance and effectiveness of SCCT activities for realizing simultaneous QCD fulfilment, necessary for the driving force in the development of New JIT. In concrete terms, the author has described the development of SSTTM using the "Partnering Performance Measurement Model" (PPMM) for Global SCM strategy, Then, the author has verified within advanced automobile manufacturer Toyota and its affiliate/non-affiliate suppliers by typical case studies. Furthermore, the author has applied the SSTTM to similar technical problems, and has verified its effectiveness. Through the positivism research taken up by these studies, to solution of a peculiar technical problem, the systematic business approach and problem-solving utilizing SSTTM has dynamism and pliability, and the author believes that all have expressed the technical general solution.

REFERENCES

Amasaka, K. (1988). *Concept and progress of Toyota Production System (plenary lecture), Co-sponsorship: The Japan Society of Precision Engineering, Hachinohe Regional Advance Technology Promotion Center Foundation, etc.* Hachinohe.

Amasaka, K. (1999). A study on "Science SQC" by utilizing "Management SQC": A demonstrative study on a New SQC concept and procedure in the manufacturing industry. *International Journal of Production Economics, 60-61*, 591–598. doi:10.1016/S0925-5273(98)00143-1

Amasaka, K. (2000a). *New JIT, a new principle for management technology 21^C: Proposal and demonstration of "TQM-S" in Toyota (special lecture).* The 1st World Conference on Production and Operations Management, Sevilla, Spain.

Amasaka, K. (2000b). Partnering chains as the platform for quality management in Toyota, *Proceedings of the 1st World Conference on Production and Operations Management, Sevilla, Spain.*

Amasaka, K. (2001). Quality management in the automobile industry and the practice of the joint activities of the auto-maker and the supplier. Japan Small Business Research.

Amasaka, K. (2002). *New JIT*, a new management technology principle at Toyota. *International Journal of Production Economics, 80*(2), 135–144. doi:10.1016/S0925-5273(02)00313-4

Amasaka, K. (2003a). *New application of strategic quality management and SCM-A Dual Total Task Management Team involving both Toyota and NOK.* Proceedings of the Group Technology / Cellular Manufacturing World Symposium, Columbus, Ohio.

Amasaka, K. (2003b). Proposal and Implementation of the *"Science SQC"* Quality Control Principle. *Mathematical and Computer Modelling, 38*(11-13), 1125–1136. doi:10.1016/S0895-7177(03)90113-0

Amasaka, K. (2004a). *Applying New JIT—A Management Technology Strategy Model at Toyota-Strategic QCD studies with affiliated and non-affiliated suppliers.* Proceedings of the 2nd World Conference on Production and Operations Management Society, Cancun, Mexico.

Amasaka, K. (2004b). Development of "Science TQM", a new principle of quality management-Effectiveness of Strategic Stratified Task Team at Toyota-. *International Journal of Production Research, 42*(17), 3691–3706. doi:10.1080/0020754042000203867

Amasaka, K. (2004c). *Science SQC, new quality control principle, The quality strategy of Toyota.* Springer-Verlag Tokyo. doi:10.1007/978-4-431-53969-8

Amasaka, K. (2005a). *New Japan Production Model*, an innovative production management principle: Strategic implementation of *New JIT.* Proceedings of the 16th Annual Conference of the Production and Operations Management Society, Michigan, Chicago IL.

Amasaka, K. (2005b), *Interim Report of WG4's studies in JSQC research on simulation and SQC(1) - A study of the high quality assurance CAE model for car development design, Transdisciplinary Federation of Science and Technology. 1st Technical Conference*, Nagano, Japan.

Amasaka, K. (2007a). High Linkage Model "*Advanced TDS, TPS & TMS*: Strategic development of *New JIT* at Toyota. *International Journal of Operations and Quantitative Management, 13*(3), 101–121.

Amasaka, K. (Ed.). (2007a). New Japan Model – "Science TQM": Theory and practice for strategic quality management, Study group of the ideal situation the quality management of the manufacturing industry, Maruzen, Tokyo.

Amasaka, K. (2007b). *Final report of WG4's studies in JSQC research activity of simulation and SQC (Part-1)-Proposal and validity of the High Quality Assurance CAE Model for automobile development design.* Transdisciplinary Science and Technology Initiative, The 2nd Annual Technical Conference, Kyoto University, Kyoto, Japan.

Amasaka, K. (Ed.). (2007b). Establishment of a needed design quality assurance framework for numerical simulation in automobile production, Working Group No. 4 studies "Study group on simulation and SQC", in Japanese Society for Quality Control.

Amasaka, K. (2007c). Highly Reliable CAE Model, the key to strategic development of *Advanced TDS. Journal of Advanced Manufacturing Systems, 6*(2), 159–176. doi:10.1142/S0219686707000930

Amasaka, K. (2007d). *Applying New JIT-* Toyota's global production strategy: Epoch-making innovation of the work environment. *Robotics and Computer-integrated Manufacturing, 23*(3), 285–293. doi:10.1016/j.rcim.2006.02.001

Amasaka, K. (2007e). *New Japan Production Model*, an advanced production management principle: Key to strategic implementation of *New JIT. The International Business & Economics Research Journal, 6*(7), 67–79.

Amasaka, K. (2008a). *Simultaneous fulfillment of QCD-Strategic collaboration with affiliated and non-affiliated suppliers.* New theory of manufacturing-Surpassing JIT: Evolution of Just-in-Time, Morikita-Shuppan, Tokyo.

Amasaka, K. (2008b). Strategic QCD studies with affiliated and non-affiliated suppliers utilizing New JIT. Encyclopedia of Networked and Virtual Organizations, Information Science Reference, Hershey, New York.

Amasaka, K. (2008c). Science TQM, a new quality management principle: The quality management *of Management & Engineering Integration, 1*(1), 7-22.

Amasaka, K. (2008d). An Integrated Intelligence Development Design CAE Model utilizing New JIT, Application to automotive high reliability assurance. *Journal of Advanced Manufacturing Systems, 7*(2), 221–241. doi:10.1142/S0219686708001589

Amasaka, K. (2009). New JIT, advanced management technology principle: The global management strategy of Toyota (special lecture). *International Conference on Intelligent Manufacturing & Logistics Systems and Symposium on Group Technology and Cellular Manufacturing*, Kitakyushu, Japan.

Amasaka, K. (2010). Proposal and effectiveness of a High-Quality Assurance CAE Analysis Model: Innovation of design and development in automotive industry, *Current Development in Theory and Applications of Computer Science. Engineering and Technology, 2*(1/2), 23–48.

Amasaka, K. (2011). *New JIT, new management technology principle (Plenary Lecture)*. VIII Siberian Conference Quality Management-2011, Krasnoyarsk, Russia, Siberian Federal University.

Amasaka, K. (Ed.). (2012). Science TQM, new quality management principle, The quality management strategy of Toyota, Bentham Science Publishers, UAE, USA, The Netherlands. doi:10.2174/97816080 528201120101

Amasaka, K. (2012a). Constructing Optimal Design Approach Model: Application on the Advanced TDS. *Journal of Communication and Computer*, 9(7), 774–786.

Amasaka, K. (2012b). A New Development Design CAE Employment Model. *The Journal of Japanese Operations Management and Strategy*, 3(1), 18–57.

Amasaka, K. (2014a). New JIT, new management technology principle. *Journal of Advanced Manufacturing Systems*, 13(3), 197–222. doi:10.1142/S0219686714500127

Amasaka, K. (2014b). New JIT, new management technology principle: Surpassing JIT. *Procedia Technology*, 16(Special Issues), 1135–1145. doi:10.1016/j.protcy.2014.10.128

Amasaka, K. (2015a). New JIT, new management technology principle. Taylor & Francis Group, CRC Press.

Amasaka, K. (2015b). Global manufacturing strategy of New JIT: Surpassing JIT (Keynote Lecture). *International Conference on Information Science and Management Engineering, Phuket, Thailand*.

Amasaka, K. (2016). *Innovation of automobile manufacturing fundamentals employing New JIT: Developing Advance Toyota Production System at Toyota Manufacturing USA*. Proceedings of the 5th Conference on Production and Operations Management, Havana, Cuba.

Amasaka, K. (2017). Strategic Stratified Task Team Model for realizing simultaneous QCD fulfilment: Two Case Studies. *Journal of Japanese Operations Management and Strategy*, (1), 14–35.

Amasaka, K. (2019). Study on New Manufacturing Theory, *Nobel. International Journal of Scientific Research*, 3(1), 1–20.

Amasaka, K. (Ed.). (2019). *The fundamentals of the manufacturing management: New Manufacturing Theory-Operations Management Strategy 21C*. Sankei-Sha.

Amasaka, K. (2022a). New Manufacturing Theory: Surpassing JIT (2nd Edition). Lambert Academic Publishing.

Amasaka, K. (2022b). *Examining a New Automobile Global Manufacturing System*. IGI Global Publisher. doi:10.4018/978-1-7998-8746-1

Amasaka, K. (2023). New Lecture—Surpassing JIT: Toyota Production System—From JIT to New JIT. Lambert Academic Publishing, Germany, Printed by Printforce, United Kingdom.

Amasaka, K., Igarashi, M., Yamamura, N., Fukuda, S., & Nomasa, H. (1995). The QA Network activity for prevention rusting pf vehicle by using SQC. *Journal of the Japanese Society for Quality Control*, 35-38.

Amasaka, K., Ito, T., & Nozawa, Y. (2012). A New Development Design CAE Employment Model. *The Journal of Japanese Operations Management and Strategy*, *3*(1), 18–37.

Amasaka, K., & Otaki, M. (1999). *Developing New TQM using partnering – Effectiveness of "TQM-P" employing Total Task Management Team.* The Japan Society for Product Management, The 10th Annual Conference, Kyushu Sangyo University, Fukuoka, Japan.

Amasaka, K., & Sakai, H. (2010). Evolution of TPS fundamentals utilizing New JIT strategy-Proposal and validity of Advanced TPS at Toyota. *Journal of Advanced Manufacturing Systems*, *9*(2), 85–99. doi:10.1142/S0219686710001831

Amasaka, K., & Sakai, H. (2011). The New Japan Global Production Model "NJ-GPM: Strategic development of Advanced TPS. *The Journal of Japanese Operations Management and Strategy*, *2*(1), 1–15.

Burke, R. J., Graham, J., & Smith, F. (2005). Effects of reengineering on the employee satisfaction: Customer satisfaction relationship. *The TQM Magazine*, *17*(4), 358–363. doi:10.1108/09544780510603198

Burke, W., & Trahant, W. (2000). *Business Climate Shift.* Oxford Butterworth –Heinemann.

Corporation, N. O. K. (2000). *The history of NOK's Oil Seal-Oil Seal Mechanism.* Promotion Video.

Ebioka, K., Sakai, H., Yamaji, M., & Amasaka, K. (2007). A New Global Partnering Production Model "NGP-PM" utilizing Advanced TPS. *Journal of Business & Economics Research*, *5*(9), 1–8.

Fukuchi, H., Arai, Y., Ono, M., Suzuki, S. T., & Amasaka, K. (1998). *A proposal TDS-D by utilizing Science SQC: An improving design quality of drive-train components.* The Japanese Society for Quality Control, The 60th Technical Conference, Nagoya, Japan.

Gabor, A. (1990). *The man who discovered quality; How Deming W. E., brought the quality revolution to America.* Random House, Inc.

Goto, T. (1999). *Forgotten origin of management – Management quality taught by G.H.Q, CCS management lecture.* Productivity Publications.

Hashimoto, K., Onodera, T., & Amasaka, K. (2014). Developing a Highly Reliable CAE Analysis Model of the mechanisms that cause bolt loosening in automobiles. *American Journal of Engineering Research*, *3*(10), 178–187.

Hoogervorst, J. A. P., Koopman, P. L., & van der Flier, H. (2005). Total Quality Management: The need for an employee-centered, Coherent approach. *The TQM Magazine*, *17*(1), 92–106. doi:10.1108/09544780510573084

Joiner, B. L. (1994). Fourth Generation Management: The New Business Consciousness. Joiner Associates, Inc., McGraw-Hill, New York.

JUSE. (2006). *Quality management survey: JUSE Home page.* JUSE. https://www.juse.or.jp/

Kameike, M., Ono, S., & Nakamura, K. (2000). The Helical Seal: Sealing Concept and Rib Design. *Sealing Technology International*, *77*(77), 7–11. doi:10.1016/S1350-4789(00)88559-1

Kojima, T., & Amasaka, K. (2011). The Total Quality Assurance Networking Model for preventing defects: Building an effective quality assurance system using a Total QA Network. *International Journal of Management & Information Systems.*, *15*(3), 1–10. doi:10.19030/ijmis.v15i3.4637

Kotler, F. (1999). Kotler marketing. The Free Press, a Division of Simon & Schuster Inc., New York.

Lagrosen, S. (2004). Quality management in global firms. *The TQM Magazine*, *16*(6), 396–402. doi:10.1108/09544780410563310

Ljungström, M. (2005). A model for starting up and implementing Continuous improvements and work development in practice. *The TQM Magazine*, *17*(5), 385–405. doi:10.1108/09544780510615915

Lopez, A. M., Nakamura, K., & Seki, K. (1997). *A study on the sealing characteristics of lip seals with helical ribs*. Proceedings of the15th International Conference of British Hydromechanics Research Group Ltd., Fluid Sealing.

Manzoni, A., & Islam, S. (2007). Measuring collaboration effectiveness in globalized supply networks: A data envelopment analysis application. *International Journal of Logistics Economics and Globalization*, *1*(1), 77–91. doi:10.1504/IJLEG.2007.014496

Media, M. (2006). *WRI: MRO today Home page*. Progressive Distributer. http://www.Progressivedistributor. com/mro_today_body.htm

Miyashita, S., & Amasaka, K. (2014). Proposal of a New Vietnam Production Model, NVPM: A new integrated production system of Japan and Vietnam. *IOSR Journal of Business and Management*, *16*(12), 18–25. doi:10.9790/487X-161211825

Muto, M., Miyake, R., & Amasaka, K. (2011). Constructing an Automobile Body Color Development Approach Model. *Journal of Management Science*, *2*, 175–183.

Nozawa, Y., Ito, T., & Amasala, K. (2013). High precision CAE analysis of automotive transaxle oil seal leakage. *China-USA Business Review*, *12*(5), 363–374.

Pires, S., & Cardoza, G. (2007). A study of new supply chain management practices in the Brazilian and Spanish auto-industries. *International Journal of Automotive Technology and Management*, *7*(1), 72–87. doi:10.1504/IJATM.2007.013384

Sakai, H., & Amasaka, K. (2006). Strategic *HI-POS*, Intelligence Production Operating System: Applying *Advanced TPS* to Toyota's global production strategy. *WSEAS Transactions on Advances in Engineering Education*, *3*(3), 223–230.

Sakatoku, T. (2006). *Partnering performance measurement for assembly maker and suppliers* [A Thesis for the master's degree, Graduate School of Science and Engineering, Aoyama Gakuin University].

Sato, Y., Toda, A., Ono, S., & Nakamura, K. (1999). A study of the sealing mechanism of radial lip seal with helical robs–measurement of the lubricant fluid behavior under sealing contact. *SAE Technical Paper Serie*s.

Seuring, S., Muller, M., Goldbach, M., & Schneidewind, U. (Eds.). (2003). *Strategy and Organization in Supply Chains*. Physica-Verlag Heidelberg.

Siang, Y. Y., Sakalsiz, M. M., & Amasaka, K. (2010). Proposal of New Turkish Production System (NTPS): Integration and evolution of Japanese and Turkish production system. *Journal of Business Case Study*, *6*(6), 69–76. doi:10.19030/jbcs.v6i6.260

Taylor, D., & Brunt, D. (2001). *Manufacturing operations and supply chain management- Lean Approach* (1st ed.). Thomson Learning.

Toyota Motor Corporation. (1987). *Creation is infinite: The history of Toyota-50 years*. Toyota Motor Corporation.

Yamaji, M., & Amasaka, K. (2009). Strategic Productivity Improvement Model for white-collar workers employing Science TQM, *JOMS*. *The Journal of Japanese Operations Management and Strategy*, *1*(1), 30–46.

Yamaji, M., Sakatoku, T., & Amasaka, K. (2008). Partnering Performance Measurement "PPM-AS" to strengthen corporate management of Japanese automobile assembly makers and suppliers. *International Journal of Electronic Business Management*, *6*(3), 139–145.

Chapter 11
Strategic Patent Value Appraisal Model for Corporate Management Strategy

ABSTRACT

In this chapter, to strengthen the New JIT strategy more, the author has created a strategic patent value appraisal model (SPVAM) that contributes to corporate management employing strategic enhancing corporate reliability model (SECRM) based on the developing TJS (total job quality management system). Specifically, to strengthen management technology in corporate innovation, improvement of patent value signifies engineers' value creation at work (invention by white-collar workers). This model consists of several elements each for inventive technique and patent right and has verified the validity of the model at Toyota and other leading corporations. Furthermore, standardization has been carried out in order to spread the effectiveness of SPVAM, and Amasaka new JIT laboratory - patent performance model (A-PPM) has been created and its effectiveness will be investigated through applications at Toyota and others.

OUTLINE

This study provides new insight that contributes to the quality patent creation process for innovating intellectual property function and enhancing engineering strategy.

Looking to attain a high-performance business model for corporate management technology in Japanese companies, the author establishes and verifies the validity of the "New JIT, new management technology principle" using the 5 core elements "TDS, TPS, TMS, TIS & TJS" (Amasaka, 2002a, 2004a, 2014, 2015, 2021, 2022a,b, 2023; Amasaka Ed., 2012) (See to Chapter 7).

Among these, the intellectual property divisions are the think tank of corporate strategy and play a key role in acquiring patent rights. Therefore, the author enhances the intellectual productivity of the company through collaborative activities employing "Strategic Stratified Task Team Model" (SSTTM) (Amasaka, 2017) (See to Chapter 11).

In this study, therefore, the author has created the "Strategic Patent Value Appraisal Model" (SPVAM) in order to strengthen the development of "New JIT" strategy more (Amasaka, 2002b, 2003a, 2007a,b, 2009, 2015, 2022a,b, 2023; Tsunoi et al., 2009; Anabuki et al., 2011, Amasaka, Ed., 2012).

DOI: 10.4018/978-1-6684-8301-5.ch011

It is generally stated that a "good patent" refers to an invention and/or a right that is useful for the entity that owns it for retaining their main business, while affecting the business of others.

What is available in this field, however, is mere analysis of simple metrical statistics that are available from patent information. In other words, no qualitative analysis is available of the engineers' level of consciousness about the depth of "what is a good patent?" for them, the patent inventors.

Specifically, the author undertook the task of formulating measures to define the concept of SP-VAM, and defined it by grasping individual engineers' recognition of the present status of their patents as linguistic information and understanding it quantitatively using "Science SQC, new quality control principle" (Amasaka, 2003b, 2004b) (See to Chapter 2 and 3).

The improvement of patent value signifies engineers' value creation at "work" *(invention by white-collar workers)*. Then, the author has developed the S-PVAM that consists of several elements, each for a strategic patent.

In recent years, the author has continued to apply the S-PVAM successfully and proved the validity of the proposal with advanced car manufacture Toyota and other leading corporations.

Moreover, standardization has been carried out in order to spread the effectiveness of SPVAM, and "A-PPM" (Amasaka New JIT Laboratory - Patent Performance Model) has been created and its effectiveness will be investigated through trial application at major enterprises; (Kaneta and Kuniyoshi, 2003; Amasaka, 2003a, 2009; Anabuki et al., 2011; Amasaka, Ed., 2012).

CORPORATE STRATEGY THROUGH NEW JIT IMPROVING PATENT VALUE

New JIT, the Key to Innovating Management Technology

The author has developed the new management technology principle "New JIT" with the aim of innovating management technology (Amasaka, 2002a, 2007b, 2014, 2015, 2021, 2022a,b, 2023) (See to Chapter 7).

This principle is underpinned by 5 core principles: "TMS" (Total Marketing System), "TDS" (Total Development System), "TPS" (Total Production System), "TIS" (Total Intelligence Information System) and "TJS" (Total Job Quality Management System).

These principles are organically linked by the "Science SQC, new quality control principle" in order to realize "Customer Science" for quality management that positions customers as the 1st priority (Amasaka, 2003b, 2004b, 2005) (See to Chapter 2 and 3).

Then, the author applied the New JIT to the quality management activity of leading corporations, Toyota and others, and verified the validity of the New JIT through observation of engineers enhanced intellectual productivity.

The most important current issue is the establishment and execution of "strategic patent acquisition measures" in order not to fall behind various companies.

"TJS", the Core of Corporate Strategy Improving Patent Value

Then, the administration/indirect divisions located on the outer circle of "New JIT" as shown in Figure.1 of Chapter 7 are the think tank of the corporate strategy for improvement of patent value as the developing TJS.

TJS, located in the outer circle, strategically integrates the TMS, TDS, TPS & TIS cores, and plays the role of enhancing intellectual productivity through the cooperative and creative activities of all divisions. As shown in Figure 1, TJS consists of 4 core elements (See to Figure 2 in Chapter 7).

Figure 1. TJS, total job quality management system

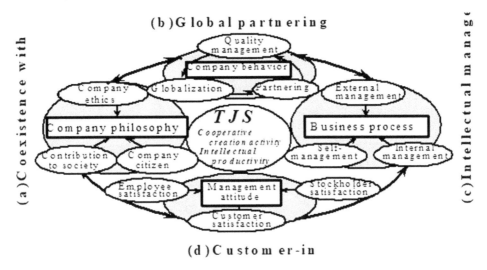

Specifically, firstly, those divisions the "transform external information" (domestic and international) into valuable management and technological resources. Therefore, they are expected to perform strategic internal management.

In collaboration with purchasing control, engineering control, production control, and information systems, these administrative divisions should promote a commercialization strategy that relates to sales, and production of the thirteen divisions positioned on the inner circle.

At the implementation stage, the company must ensure high reliability of corporate actions through (c) intellectual management through business process quality improvement and (b) global partnering.

Secondly, these administrative divisions must carry out collaborative activities with the engineering divisions for external management by developing the management technology information of the individual engineering divisions into a corporate strategy.

At the implementation stage, (a) coexistence with society and (d) customer-in management attitude is essential.

Particularly, therefore, to strengthen "Global corporate management strategy" the author has developed the "Strategic Enhancing Corporate Reliability Model" (SECRM) named "Advanced TJS, strategic highly reliable corporate management model" as shown in Figure 3 (Amasaka, 2007a, 2009, 2021; Yamaji and Amasaka, 2009; Amasaka Ed. 2012).

This model is based on "TJS" strategy with an emphasis on the customer by developing "High Quality assurance" to boost "Publicizing high-value-added product" (See to Figure 3-5 in Chapter 7).

Concretely, in next section, to realize "Evolution of corporate management technology", the author considers the "Patent acquisition in corporate management strategy" by the "Product intellectual productivity" developing both of "Product development design and Production Engineering".

Figure 2. Strategic enhancing corporate reliability model (SECRM)

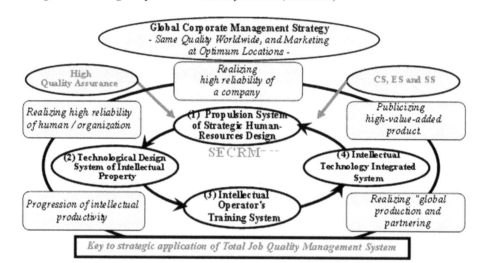

SIGNIFICANCE OF PATENT ACQUISITION IN CORPORATE MANAGEMENT STRATEGY

Current Situation of Patent Application in Japan

Patent application by companies accounts for a large percentage of total applications and is growing annually (Japan Patent Office, 2005).

Japan had the highest share of global applications in 2005, with approximately 500,000 patent applications filed, which is about 10% of the total number of global applications.

However, only 30% of the patent technology was utilized and the remaining 70% are so-called "sleeping patents". In fact, over 40% of patent applications are "defensive patents" which does not assume any practical use of the patent.

The present state, in which approximately 150,000 patents are registered annually, signifies an approximately 1 trillion-yen deficit between patent maintenance costs and royalty income from patent utilization. On the other hand, although total patent application in the United States is only half that of Japan (second largest share in the world), its trade balance in the past ten years has resulted in a 17.5 trillion-yen surplus.

Compared with the 4.1 trillion-yen deficit of the Japanese technical trade balance, the US surplus implies remarkably high patent quality (Amasaka, 2003a).

Strengthen Patent Acquisition for Corporate Innovation

The observation above confirms that the recent propagation of "Corporate innovation through competitive patents" is a matter of course (Nikkei, 2002).

The current task for a company's technological strategy is to promote development of epoch-making innovative technology that will lead to quality patents and the acquisition of competitive, unique patents. A patent must be innovative and original.

Specifically, to strengthen management technology innovation, improvement of "patent value" signifies engineers' value creation at "work" (invention by white-collar workers). Therefore, engineers based on the white-collar workers should therefore develop strategic technology with epoch-making ideas (originality), which naturally necessitate the enhancement of the engineer's creativity.

An engineer's pursuit of new technology toward the creation of "patent technology" encourages individual growth and corporate innovation and development (Amasaka, 2003a). Then, as engineer's intellectual productivity increases, "invention and acquisition of patent right" will serve as a core of corporate management technology strategy.

NEED FOR THE "STRATEGIC PATENT VALUE APPRAISAL MODEL (SPVAM)

Past Study of Qualitative Patent Value Appraisal

Generally, a "good, strategic patent" is an invention and right that advances the main business of the company, influences other companies and benefits the company itself (Amasaka et al. 1996).

To the author's knowledge, no study has been presented for qualitative patent value appraisal (Kusama, 1992; Taketomi et al., 1997; Umezawa, 1999; Kevin and David, 2000; Ueno, 2003; Japan Patent Office, 2005).

In many cases, evaluation of the patent application only involves statistical analysis, mainly classification and stratification of the technological development contents and characteristics of the company or inventor. There was one case of intellectual property value (cash flow) appraisal, but it only covered qualitative theory.

According to interviews (Japan Patent Attorneys Association, "Patent collateral value appraisal", 7 major enterprises: Toyota, Fuji-Xerox, JFE Steel, Daikin Kogyo, NEC, Sanden and NHK Spring) and research (Japan Patent Office, "Patent evaluation index", 2005) by Kanerta (2005) and Amasaka (2007a) of "Amasaka New JIT Laboratory", most appraisal cases rely on intuitive and subjective evaluations and the results are likely to deviate. From this background, establishment of a consistent "patent value appraisal model" is strongly desired.

The Challenge for Intellectual Property Divisions: Qualitative Evaluation of Patent Value

As already explained, the intellectual property divisions positioned on the outer circle play a key role in acquiring patent rights. The author researched and analyzed the example of a leading company, Toyota (97 engineers and managers from 7 divisions: 1 development, 1 design, 2 production engineering, 1 machinery, 1 research and 1 control) by using a questionnaire as shown in Table 1 and multivariate analysis (Amasaka, 2003a, 2003a).

This was done to discover the "expectations and roles of the intellectual property divisions", which are held by the managers and engineers of administrative divisions that create business model patents, and the engineering divisions that create technological patents for new technology, materials, production engineering and manufacturing methods.

To clarify the question of "What is a good patent?" and how engineers and

managers carry on their patent activities (what do they place priority on?), the author conducted a questionnaire to study their consciousness.

Then, the author mainly divided questions into the "content of invented technology" and the "content of patent right". The questionnaire is based on a multiple-choice reply form (marking method, 6 points: very important – 1 point: not needed at all) of eleven questions and one free opinion each.

Each question consisted of a hearing conducted with the staff in charge of the intellectual property, the engineers, and the managers. As done previously, the author surveyed Toyota engineers who had acquired patents in the past in order to find out "what makes a strategic, good patent at a company" (Amasaka, 2003a).

From the analysis of collected language data (cluster analysis and factor analysis), the author clarified the necessary factor structures. In general, these consist of (1) Group (A): Realist (profit-minded), (2) Group (B): Advance (pioneer-minded) and (3) Group (C): Future (innovative-minded) as shown in Figure 3.

Moreover, depending on the divisions that the engineers belong to (R&D, production engineering, manufacturing) and also their job history (practicality, competitiveness, advancement), engineers' value (awareness) of patents differs.

This new insight served as a stepping-stone to establish the "patent value appraisal model". To achieve a breakthrough on these issues, New JIT will be strategically incorporated using Science SQC applications (Amasaka, 2002a, 2014).

To be specific, the Intellectual Property Division and the engineers and managers of other indirect divisions will cooperate to (i) establish rational "patent value appraisal methods" utilizing a rule of thumb that can be shared by both divisions.

As a result, the expected effect would be the establishment of a strategic patent acquisition process model to reduce the number of sleeping patents and the mistaken rejection of inventions filed by engineers and managers.

Establishment of "Strategic Patent Value Appraisal Model" (SPVAM)

Based on the finding of the above analysis, the author has establishment and verified the "Strategic Patent Value Appraisal Model" (S-PVAM) which assists creation of strategic patents as a core of the corporate strategy.

To implement the method, the author performed research and analysis by the same methods stated in the "Challenge for intellectual property divisions: "Qualitative evaluation of patent value" above, targeting 248 engineers and managers of 16 divisions including research, development, design, production technology, production, and intellectual property, at 7 companies from a group of leading enterprises (Toyota, Fuji-Xerox, JFE Steel, Daikin Kogyo, NEC, Sanden and NHK spring) (Amasaka, 2003a).

Grasping the relationship between good invention and good patent right

Table 1. Questionnaire on "What is a good patent?"

Table 1 Questionnaire on "What is a good patent?"

(Question A): What is a good patent for you? What do you routinely place priority and/or importance on to the patents?

* Please circle a number out of 1 to 6 in the following questions. (Questions may be similar to each other, but please answer them all. For any question, contact _____ .)

If you have any other items than stated below you routinely feel the need for some action or measure, please state it in the space given to ⑫ and ㉔.

[1] The following are questions concerning the "Content of Inventions."

	very important	important	not important not unnecessary	not necessary / needed	not needed at all
① Practicality (for the present and future)	6	5	4	3 2	1
② Cost (inexpensive, makes good profit)	6	5	4	3 2	1
③ Overthrowing established technologies (unexpectedness)	6	5	4	3 2	1
④ Merchandizability (sellable, marketable with high added value)	6	5	4	3 2	1
⑤ Can make good PR on our engineering capabilities	6	5	4	3 2	1
⑥ Performance (product performance, productivity)	6	5	4	3 2	1
⑦ Epoch-making (new, and much valuable)	6	5	4	3 2	1
⑧ Advancedness (ahead of competitors)	6	5	4	3 2	1
⑨ Idea and conception (not necessarily having to implement)	6	5	4	3 2	1
⑩ Uniqueness (differentiate others, pioneer)	6	5	4	3 2	1
⑪ General-purpose, developable and applicable	6	5	4	3 2	1
⑫ State your free opinion ( )	6	5	4	3 2	1

[2] The following are questions concerning the "Right."

	very important	important	not important not unnecessary	not necessary / needed	not needed at all
⑬ Conceptual range of patent application and simple expression (unlimiting and can be widely interpreted)	6	5	4	3 2	1
⑭ Large system (combination of subsystems)	6	5	4	3 2	1
⑮ Rich with concrete cases and examples (many cases of embodiment)	6	5	4	3 2	1
⑯ Elementary technology (system component)	6	5	4	3 2	1
⑰ Coverage of patent right with multiple inventions (including improvements and peripheral points)	6	5	4	3 2	1
⑱ Possible prompt embodiment (early license fee receivable)	6	5	4	3 2	1
⑲ Toyota's embodiment technologies (confined to Toyota products)	6	5	4	3 2	1
⑳ Basic idea (nucleus of multiples, principle and rule)	6	5	4	3 2	1
㉑ Other companies are eager to acquire (others unable to bypass)	6	5	4	3 2	1
㉒ Internationally applicable (possible development overseas)	6	5	4	3 2	1
㉓ Embarrassing to others (effective in suppressing their development will)	6	5	4	3 2	1
㉔ State your free opinion ( )	6	5	4	3 2	1

Figure 3. Grouping to common recognition of strategically good patent (cluster analysis and factor analysis)

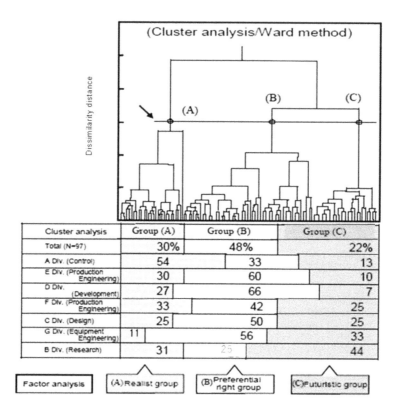

A patent is comprised of inventive technique (X) and patent right (Y). To identify the relationship between X and Y that is necessary for "strategic patents", the author extracted and summarized the necessary elements (explanatory variables: x_1 to x_{22}) from the language data obtained in the above survey and created a 6-point scale survey (6: extremely important, 5: very important, 4: important, 3: medium, 2: slightly unimportant, 1: unimportant) described in Table 1 above.

Aiming to understand the relationship between inventive technique (x_1: "practicality" to x_{11}: "expansive") and patent right (x_{12}: "scope of right is conceptual and simple" to x_{22}: prevent other companies), the author conducted a survey for Toyota engineers and managers (total 97 people from 7 divisions) and obtained the canonical correlation analysis results shown in Figure 4.

From the main component relation scatter diagram of the figure, "inventive technique" is classified into "Profit-minded: $X_①$ (prioritize profit with actual products at hand: x_1, x_2, x_4, x_5, x_6)", "Pioneer-minded: $X_②$ (prioritize advanced technology with next-generation products: x_3, x_7, x_{10}, x_{11})", and "Innovative-minded: $X_③$ (prioritize technical innovation with high-potential products: x_8, x_9).

Corresponding to these groups, "patent right" is similarly classified into "effective-minded: $Y_①$ (effective right that protects company development: x_{14}, x_{17}, x_{18})", "competitive- minded: $Y_②$ (practical right that attracts competitors: x_{13}, x_{15}, x_{20})", and "monopoly-minded: $Y_③$ (exclusive right that is globally applicable: $x_{12}, x_{16}, x_{21}, x_{22}$)".

These findings were consistent with the experiences and techniques of the inventors, managers, patent evaluators, and patent attorneys.

Figure 4. Correlation between the contents of invented technology and the patent right (Canonical correlation analysis: Scatter diagram of correspondence between components)

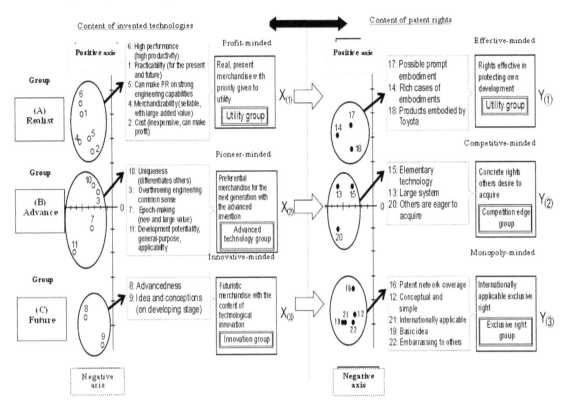

Strategic Patent Value Appraisal Model (SPVAM)

The appraisal model obtained from the canonical correlation analysis enabled qualitative patent value evaluation through an overall evaluation of inventive technique and patent right and also typifying the two elements into three clusters.

(1) Overall evaluation of inventive technique: $X=X_{①}+X_{②}+X_{③}$
   (Standardization: Perfect score is 6)
   "Profit-minded": $X_{①}=0.25x_1+0.22x_2+0.24x_4+0.20x_5 0.19x_6$
   "Pioneer-minded": $X_{②}=0.30x_3+0.32x_7+0.18x_{10}+0.20x_{11}$
   "Innovative-minded": $X_{③}=0.52x_8+0.48x_9$

(2) Overall evaluation of patent right: $Y=Y_{①}+Y_{②}+Y_{③}$
   (Standardization: Perfect score is 6)
   "Effective-minded": $Y_{①}=0.28x_{14}+0.35x_{17}+0.37x_{18}$
   "Competitive-minded": $Y_{②}=0.40x_{13}+0.38x_{15}+0.22x_{20}$
   "Monopoly-minded": $Y_{③}=0.20x_{12}+0.19x_{16}+0.18x_{19}+0.23x_{21}+0.20x_{22}$

Therefore, the author considered diffusion of the patent value appraisal model gained through the Toyota study and made revisions to the appraisal model in an approach similar to the above in cooperation with the 3 major enterprises (240 staff members in 16 divisions of Toyota, Xerox, and Daikin).

The example described in Figure 5 is the evaluation by inventors at Xerox, JFE Steel, and Toyota (Evaluators: a, b, and c) of the (1) "inventive techniques" and (2) "patent rights", expressed by radar chart. All of the inventions investigated possessed features of the contents of inventive techniques and patent rights, and received sufficient favorable comments from the inventors of the 3 enterprises.

Figure 5. Example of the qualitative evaluation map for visualized "inventive technique / patent right" using "S-PVAM"

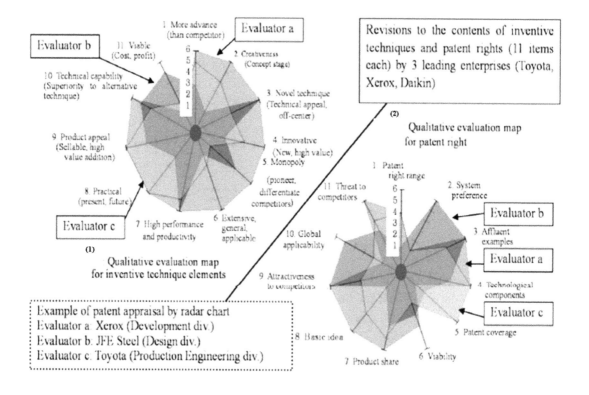

Verification of "SPVAM" validity

Therefore, the author will validate the effectiveness of PVAM with Toyota. Patent value evaluation can be strategically visualized from the appraisal model using PVAM as shown in Figure 6 (Amasaka, 2003a, 2007a).

Similarly, Figure 7 indicates an example of patent right evaluation. When the inventor applied the patent, the qualitative value of the patent had only a clearly above-average value for monopoly and competitive.

To raise the strategic value of the patent, the inventor revised the patent content with the support of the patent evaluator (intellectual property division) and patent attorney. Consequently, the patent's strategic value significantly increased.

Figure 6. Strategic patent value appraisal model (SPVAM) (Ex. 1 visualization of inventive technique)

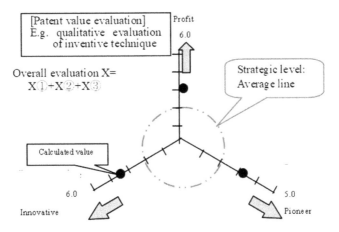

Figure 7. Improved strategic value of patent right using SPVAM (Ex. 2 visualization of patent right)

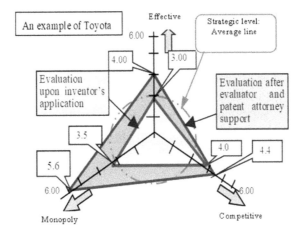

APPLICATION

This section contains a study carried out on the effectiveness of creating software for standardized "A-PPM" (Amasaka's Laboratory Patent Performance Model) at leading enterprises, as a means of effectively spreading the SPVAM created by the author (Amasaka, 2003b; Ishigaki and Niihara, 2004; Kaneta and Kuniyoshi, 2005).

Outline of the A-PPM

S-PVAM was further expanded and applied in an industry-academy collaboration study group "Quality Management in the Manufacturing Industry" led by the author, and a group of 12 leading enterprises (Toyota, Fuji-Xerox, JFE Steel, Daikin Kogyo, NEC, Sanden, NHK Spring, YANMAR, Yokogawa-Electric, NTT-Docomo, Nissan and others).

At this time, further revision was made in consideration of versatility and convenience (Amasaka, 2007a).

A proposal was made, as shown in Figure 8. Figure 8(a) outlines the "standardized as A-PPM software". Its structure and features are clearly expressed. Reflecting the effectiveness stated previously, the contents of inventive techniques and patent rights are summarized and visualized in the combination of outer and inner circles.

For example, Figure 8(b) the "General evaluation of the A-PPM effect of technique" is specified in 3 representative items: (i) "Profit" is subcategorized as "Practical, Performance and Cost" and, in a similar manner, (ii) "Pioneer" is subcategorized into "Superior to competitor, Technique is advance and alterable", and (iii) "Innovative" is subcategorized into "Creative, Expansive, and innovation of Technique" and represented in a radar chart accordingly.

Similarly, the general evaluation of (3) "Effect of Right Acquisition" is also expressed in 3 representative items: (iv) "Effective" is subcategorized into "Feasible, Product appeal, and contribution to innovation", (v) "Competitive" is subcategorized into "Attractive to competitor, Element technology, and Complete system", and (vi) "Monopoly" is subcategorized into "International, Patent coverage and Principle discovery".

The 6 representative categories (i) to (vi) are appropriately placed under 3 well-balanced elements, which enhance the versatility and convenience of the A-PPM evaluation model.

As described in Figure 8, (4) the 3 groups; namely (A) "Realist", (B) "Advanced", and (C) "Future", evaluated the contents of inventive techniques and patent rights, which are then totaled and their features (age aspect) visualized in a scale as shown in the figure.

"Realist" is the total of (i) Profit and (iv) Effective scores; "Advance" is the total of (ii) "Pioneer" and (v) "Competitive scores"; and "Future" is the total of (iii) Innovative and (vi) Monopoly scores.

Furthermore, these evaluations by the 3 groups were then totaled as (5) "Evaluation criteria", which is then placed on a 5-level scale (A: Excellent, B: Good, C: Fair, D: Poor and E: Bad). Rank C is the industrial average for patent application.

The A-PPM established in this manner automatically tabulates evaluation scores and expresses them in a radar chart, offering convenience to users.

As a result, it is now possible to perform an appraisal of patent value for each invention in approximately 20 minutes.

Validation of "A-PPM" Effectiveness

The author applied the established A-PPM and investigated its effectiveness at the previously mentioned 12 leading enterprises. Figure 9 shows the example of Fuji-Xerox.

It shows in the figure that (a) evaluation upon inventor's application was later (b) supported by a patent attorney in the intellectual property division, thereby strategically enhancing the contents of inventive techniques and patent rights.

Similarly, the authors are able to illustrate this effectiveness at other leading enterprises, confirming the results obtained.

Figure 8. Outline of A-PPM, Amasaka's laboratory patent performance model

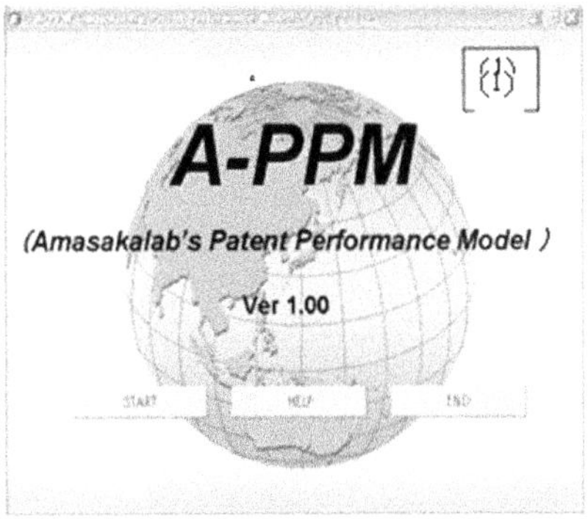

(a) "A-PPM" software

Data source: http://www.ise.aoyama.ac.jp/newjit/

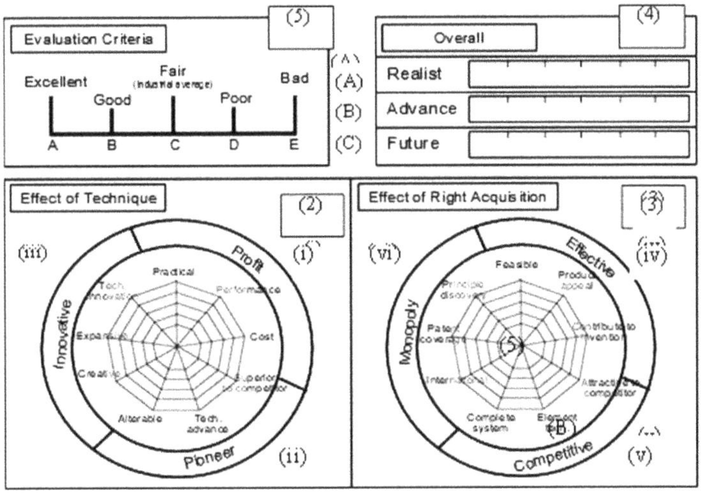

(b) General evaluation of the "A-PPM" effect of technique

CONCLUSION

In this chapter, to strengthen New JIT strategy, the author created the SPVAM (Strategic Patent Value Appraisal Model) that contributes to corporate strategy employing SECRM (Strategic Enhancing Corporate Reliability Model based on the developing TJS. *Specifically, to strengthen management technology in corporate innovation, improvement of "patent value" signifies engineers' value creation at "work" (invention by white-collar workers).* This model consists of several elements each for "inventive technique" and "patent right". In recent years, the author has continued to apply the SPVAM successfully and proved the validity of the proposal at leading enterprises such as Toyota (Amasaka, 2007a, 2009, 2015). Furthermore, to spread the effectiveness of SPVAM, the author standardized A-PPM (Amasaka-lab's Patent Performance Model) as software and validation of its effectiveness was made at 12 leading enterprises. This study provided a new insight that contributes to the quality patent creation process for innovating the intellectual property function and enhancing engineering strategy (Amasaka, Ed. 2012; Amasaka, 2022a,b, 2023).

Figure 9. Example of improved strategic value of "inventive technique" and "patent right" using "A-PPM" by Fuji-Xerox

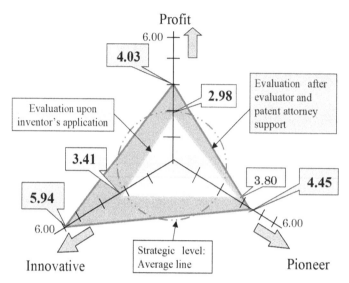

(a) Case of visualization of inventive technique

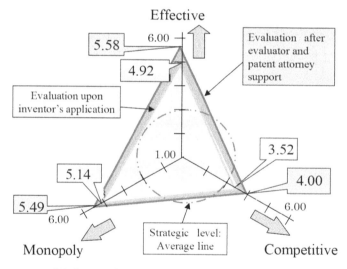

(b) Case of visualization of patent right

REFERENCES

Akimoto, S., & Shimoda, S. (2000). Modeling of the conceptual composition of invention and right: Research on objectification of the patent value by opinion poll of engineers [Thesis, School of Science and Engineering, Amasaka New JIT Laboratory].

Amasaka, K. (2002a). *New JIT*, a new management technology principle at Toyota. *International Journal of Production Economics, 80*(2), 135–144. doi:10.1016/S0925-5273(02)00313-4

Amasaka, K. (2002b). *Quality of engineer's work and significance of patent application.* The 2nd patent seminar, Aoyama Gakuin University, Research Institute, Shibuya, Tokyo.

Amasaka, K. (2003a). *The validity of "TJS-PPM", Patent Value Appraisal Method in the corporate strategy: Development of "Science TQM", a new principle for quality management (Part 3).* The Japan Society for Production Research, The 17th Annual Conference, Gakushuin University, Tokyo, 189-192.

Amasaka, K. (2003b). Proposal and Implementation of the *Science SQC*, quality control principle. *Mathematical and Computer Modelling, 38*(11-13), 1125–1136. doi:10.1016/S0895-7177(03)90113-0

Amasaka, K. (2004a). Development of "Science TQM", A New Principle of Quality Management: Effectiveness of Strategic Stratified Task Team at Toyota-. *International Journal of Production Research, 42*(17), 3691–3706. doi:10.1080/00207540420002003867

Amasaka, K. (2004b). *Science SQC, new quality control principle: The quality control principle: The quality strategy of Toyota.* Springer-Verlag Tokyo. doi:10.1007/978-4-431-53969-8

Amasaka, K. (2005). Constructing a *Customer Science* Application System *"CS-CIANS"* –Development of a global strategic vehicle *"Lexus"* utilizing *New JIT*–. *WSEAS Transactions on Business and Economics, 2*(3), 135–142.

Amasaka, K. (2007a). Proposal and validity of Patent Value Appraisal Model "TJS-PVAM": Development of *"Science TQM"* in the corporate strategy. *Proceedings of the International Conference Risk, Quality and Reliability.* Technical University of Ostrava.

Amasaka, K. (2007b). High Linkage Model *"Advanced TDS, TPS & TMS":* Strategic Development of *"New JIT"* at Toyota. *International Journal of Operations and Quantitative Management, 13*(3), 101–121.

Amasaka, K. (2009). Proposal and Validity of Patent Value Appraisal Model "TJS-PVAM"-Development of "Science TQM" in the Corporate Strategy, *China-USA. Business Review (Federal Reserve Bank of Philadelphia), 8*(7), 45–56.

Amasaka, K. (Ed.). (2012). Science TQM, new quality management principle: The quality strategy of Toyota, Bentham Science Publishers, UAE, USA, The Netherlands.

Amasaka, K. (2014). New JIT, New Management Technology Principle: Surpassing JIT. *Procedia Technology, 16*(special issues), 1135–1145. doi:10.1016/j.protcy.2014.10.128

Amasaka, K. (2015). *Strategic employment of the Patent Appraisal Method, New JIT, new management technology principle, Taylor & Francis Group.* CRC Press.

Amasaka, K. (2017). Strategic Stratified Task Team Model for realizing simultaneous QCD fulfilment: Two case studies. *Journal of Japanese Operations Management and Strategy*, *7*(1), 14–35.

Amasaka, K. (2021). New Japan Automobile Global Manufacturing Model: Using Advanced TDS, TPS, TMS, TIS & TJS. *Journal of Business Management and Economic Research*, *6*(6), 499–523.

Amasaka, K. (2022a). *Examining a New Automobile Global Manufacturing System*. IGI Global Publisher. doi:10.4018/978-1-7998-8746-1

Amasaka, K. (2022b). New Manufacturing Theory: Surpassing JIT (2nd Edition). Lambert Academic Publishers, Germany.

Amasaka, K. (2023). New Lecture-Surpassing JIT:Toyota Production System-From JIT to New JIT, Lambert Academic Publishers.

Amasaka, K., Nitta, S., & Kondo, K. (1996). An investigation of engineers' recognition and feelings about good patents by new SQC method. *Journal of the Japanese Society for Quality Control*, 17-24.

Anabuki, K., Kaneta, H., & Amasaka, K. (2011). Proposal and validity of Patent Evaluation Method "A-PPM" for corporate strategy. *International Journal of Management & Information Systems*, *15*(3), 129–137. doi:10.19030/ijmis.v15i3.4649

Hirota, M., & Miyamoto, T. (2000). A proposal of the patent measuring model "A-PAT" (Amasakalab's patent) as the corporate strategy [*Thesis, School of Science and Engineering, Aoyama Gakuin University*].

Ishigaki, K., & Niihara, K. (2001). A study for objective evaluation of patent value – Proposal and validity of A-PPM, School of Science and Engineering [Thesis, Aoyama Gakuin University, Amasaka.]

Japan Patent Office. (2005). *Patent Evaluation Index*. JPO. http:// www.jpo.go.jp/

Kaneta, H. (2005). *The research on construction of the patent evaluation model* [Thesis, Graduate School of Science and Engineering, Aoyama Gakuin University].

Kaneta, Y., & Kuniyoshi, M. (2003). A Study on Establishment of the Qualitative Valuation Modeling of Patent Value [Bachelor thesis, School of Science and Engineering, Aoyama Gakuin University].

Kevin, G. R., & David, K. (2001, July). *Discovering New Value in Intellectual Property*, Diamond. *Harvard Business Review*, 98–113.

Kusama, M. (1992). Analysis of Engineering Trend Utilizing Patent Information, *The 29th Information Science Technology Conference*, (pp. 177-182). Research Gate.

Nikkei. (2002). *Corporate Innovation through Competitive Patent*. Nikkei.

Taketomi, T., Horiguchi, Y., & Hirabayashi, T. (1997). Evaluation of Intellectual Property Value and Recovery of Invested Resources. *R & D Management*, *2*, 32–43.

Tsunoi, M., Anabuki, K., Yamaji, M., & Amasaka, K. (2009). *A study of Patent Evaluation Method "A-PPM" for Corporate Strategy*. The 11th Annual Conference of Japan Society of Kansei Engineering, Shibaura Institute of Technology, Tokyo..

Ueno, H. (2003). Formation of a strategic patent group, and a patent portfolio strategy. *Japan Marketing Journal, 80*, 25–35.

Umezawa, K. (1999). Miscellaneous Impressions of Intellectual Property Rights. *Intellectual Property Management, 49*(3), 353–36.

Yamaji, M., & Amasaka, K. (2009). Strategic Productivity Improvement Model for White-Collar Workers Employing Science TQM. *The Journal of Japanese Operations Management and Strategy, (JOMS-Special Issue), 1*(1), 30-43.

Chapter 12
Developing Automobile Exterior Design Model for Customer Value Creation

ABSTRACT

In this chapter, to develop new JIT strategy, the author develops a scientific approach to identifying customers' tastes employing automobile exterior design model (AEDM) for customer value creation. To realize this, the author uses the customer science principle (CSp) aiming to achieve the intelligence design concept method (CSp-IDCM). AEDM improves the design business process so that implicit knowledge on customer is turned into explicit knowledge. To strengthen automobile exterior design, the author has developed the Automobile Exterior Design Model with 3 Core Methods (AEDM-3CM) as follows; (A) Improvement of design business process methods for automobile profile design, (B) Creation of automobile profile design using psychographics approach methods, and (C) Actual studies on automobile profile design, form and color matching support methods (APFC-MSM) employing automobile profile design, form and color optimal matching model (APFC-OMM). The validity of AEDM was verified through case studies of the actual application examples.

DEVELOPING CUSTOMER SCIENCE PRINCIPLE

Issue in Automobile Marketing, Product Planning, and Designing

Today, an increasing number of companies both in Japan and abroad try to grasp the unprejudiced desires of their customers from the viewpoint of customer-oriented business management and to reflect these desires in future product development (Amasaka, 1995; Evans and Lindsay, 2004).

However, the actual behavioral patterns of designers in trying to grasp latent customer desires depend heavily on their empirical skills (Amasaka, 1999a,b, 2003a, 2004a).

Designers of vehicle designing department often proceed with product development using implicit business processes (Amasaka et al., 1999; Amasaka, Ed., 2007).

Accordingly, their performance is measured by sales results, and their efforts to improve business processes for future jobs may be insufficient as implicit prescriptions.

DOI: 10.4018/978-1-6684-8301-5.ch012

Furthermore, designers often worry that their current business approaches are likely to depend on job performing capabilities and, on the sensitivity (or intuition or knack) of individual persons, which does not improve the probability of success in future (Mori, 1991; Shinohara et al., 1996; Moriya and Sugiura, 1999*).

It is, therefore, important to establish a scientific approach that improves powers of product conception or a new model that assists the conception of strategic product development and tests its validity (Amasaka et al., 1999; Yamaji and Amasaka, 2009).

Customer Science Principle Aiming Customer Scientific Analysis

In this new era when product development is required as a basis of global marketing, it is important to establish a "behavior science principle" for strategic product development, which can dig deep into the customers' needs and thus preempt the trend of times (Amasaka, 2002, 2003a, 2005).

Customers express their desires in words. Therefore, the product planners or designers engaging in product development must properly interpret these words and draw up accurate plans accordingly. The "Customer Science principle" (CSp) aims for customer value creation in an automobile exterior design with a clear-cut styling concept, which is based on psychographics by the viewpoint of customers' life stage and lifestyle as described in Figure 2 of Chapter 3 (Amasaka, 2002, 2005, 2007; Takimoto et al., 2010).

In Figure 2 of Chapter 3, the image of customer's words (implicit knowledge) is translated first into common language (lingual knowledge) and then into engineering language (design drawings as explicit knowledge) by means of appropriate correlation. In other words, objectification of subjective information is realized in product development.

It is also important to transform objective drawing into subjective information through correlation to check where engineering successfully reflects customer requirements. The CSp converts subjective information, y, to objective information, , and vice versa with a two-way application of correlation technology.

The author surmises that the implementation of CSp improves the precision of "idea- product creation" in the business process "Designing" of product plan, development design divisions and thus further ensures the accumulation of successful cases as well as the failure cases more than ever before.

Taking an approach using CSp can convert a variety of issues - why the customers are pleased with this particular product, why they complain about it, what is the underlying factor in this expression from the customers, what type of products must be offered next time, and in what situations defective products are manufactured - into a common language as well as the technical terms relevant to the manufacturing field.

The designing or development staffs can convert such verbal expressions into numerical terms using correlation techniques in their laboratory or experiment room in order to confirm whether the customers' demands are satisfied.

Moreover, to double-check whether the objectification is properly conducted, it is vital for them to subjectify the results of objectification using correlation techniques.

It has been observed that thriving manufacturers, today in Japan and in the world, endeavor to convert implicit knowledge into explicit knowledge in an effort to grasp the customers feelings, and to feed-back what has been drawn up to consider whether the original objective has been realized.

That is subjectifying the objectification is done: Such a humble attitude seems to be the manufacturers' essential growth base.

Automobile Exterior Design Tool "CS-IDCM" Utilizing SQC Technical Methods

Guidelines and approaches to the study

In the 1980s, the small-size cars made in Japan had the merit of excellent run performance for reasonable price, and were exported successfully to Western countries. Thereafter, the development of the first noted car (i.e., stylish flag ship car as a world car), made in Japan, and accepted in Western countries, became pressing need.

In general, the automobile exterior design of new model car depended on the empirical rules and knowledge of experienced designers. Designers collected the surface reputation of the visitors in the car shops, and then attempted the conception and analysis in exterior design based on the implicit language information.

The creation of new methodology with the precise design idea process was a pressing need (Nunogaki et al., 1996). Therefore, the author generalized this conversion of "implicit knowledge and know how that is dependent on individual expertise" (Nagaya et al., 1998).

Although the automobile exterior design is a creative artistic activity, there must be some part to which a scientific approach can be applied as long as the authors deal with manufacturing. An objective grasp of customer tastes and sense of values for practical application to the exterior design process is important in predicting the vehicle styles that are liked well in future.

The author has conducted and reported several studies to reach the following design development methodology (Suzuki et al., 2000).

In this section, the author describes the "Intelligence Exterior Design Concept Method by developing CSp" (CSp-IDCM) utilizing SQC Technical Methods with an example of the development of "Lexus exterior design" for raising customers' worth of world's prestige car (Amasaka et al., 1999; Amasaka, 2015, 2018) (See to Figure 3 in Chapter 8).

Conception and analysis in automobile exterior design

A concept of "*Kansei Engineering*" has been in recognition for a long time (Shinohara et al., 1996: Amasaka and Nagasawa, 2000). Examples of study are introduced where it is possible to apply a statistical method to the development business of exterior design (Mori, 1991, 1993; Amasaka et al., 1999; Amasaka and Nagaya, 2002; Amasaka, 2004a).

When it comes to the application to actual business, however, there are few concrete examples of analysis worthy of introduction as an exterior design development story of new vehicles model.

This is attributable to the fact that the exterior design business (hereinafter referred to as designing) often ends up paying more importance to the end result, and that proposing good exterior designs in the design planning or development process is based on the conception, which is not closely related to the analysis.

It is apparent that the higher the analysis develops, the more important the conception becomes. But the key point to the matter is how to yield a new conception, and the process of developing conception becomes critical.

In the advanced information society where identical conditions and/or data are shared, it hardly occurs to have remarkable difference in the environment. Under the circumstances, the method for proposing a leading conception cannot be the same as before.

One may be prone to use human power for what a machine can handle, and may commit an error of misunderstanding that a conception has been gained.

Positioning of CS-IDCM

It is the objective of this study to find a guideline for establishing a method for scientifically supporting the designing so as to establish it expressly as a more creative activity from the state of tacit knowledge. It is considered that the analysis process for establishing it as an activity would be the key to the successful conception making. It is necessary for us to create a particular live solution that catches the liking of the next generation.

In this connection, "Science SQC" with a core method "SQC Technical Methods" is applied to the flow designing to enhance the quality of the designer's job called "Design SQC" actually in Toyota (Amasaka et al., 1999; Amasaka, 1996, 2003b, 2004b, 2007; Amasaka and Nagasawa, 2002) (See to Figure 2 in Chapter 8).

In developing Design SQC, SQC Technical Methods is applied as the scientific new methodology "Mountain-Climbing for Problem-Solving", which is an effective solution to the various business tasks, by analyzing the bridge portion between analysis (research) and conception (creation of contour images) (Amasaka, 2003b, 2004a,b, 2007, 2015; Amasaka and Nagasawa, 2002; Amasaka, Ed., 2007).

Designing is applied as the automobile concrete conception tool "CSp-IDCM" described in Figure 3 of Chapter 8 (Amasaka et al., 1999; Amasaka, 2017, 2018).

In CSp-IDCM, the analysis process that turns implicit knowledge into explicit knowledge constitutes a secret to the conception. In this connection, the author utilizes CSp-IDCM as an automobile exterior design concept tool for developing strategic product design by the "SQC Technical Methods".

Actually, bridging is attempted to span the research-oriented analysis as the event analysis to the exterior design in Steps 1 to 3 in Figure 3 of Chapter 8 below.

Step 1 analyzes relationships between images of vehicles desirable to customers and those actually selected to research, and it actualizes apparent relevancy whereby a vehicle type can be specified by the desirable image.

Step 2 grasps what part of a vehicle customers observe to evaluate it. By coming down from the overall assessment, partial assessment and detailed assessment, this report clarifies which design factor should better be given priority to satisfy customers.

By thinking that true customer-in is to propose a desirable thing before it is desired. In Step 3, the designers research the excellent exterior design by grasping the relevance of vehicle images and profile design (called proportion) data.

CSp-IDCM improves the designing process (work) for creation of exterior design image involving the matching profile design, form, and color. By accumulating the improvement processes, this study intends to improve the quality of designing - "creation of conception" as the dotted line indicates in Figure 3 of Chapter 8 (See to Chapter 8).

STUDIES ON AUTOMOBILE EXTERIO DESIGN MODEL FOR CUSTOMER VALUE CREATION

To realize the CS-IDCM, the author conducted the "Advanced Exterior Design Project, ADS" for raising customers' worth in Toyota Motor Corp. described in the following (Nunogaki et al., 1996; Amasaka, 1999a,b; Amasaka, 2015, 2018, 202a,b, 2023).

As for the development of "ADS" above, the author as the Chief Examiner of TQM Promotion Div. (1992-2000), and the Head of Amasaka New JIT Laboratory of Aoyama Gakuiin University (2000-2017).

Specifically, the author has organized as follows; (i) Design Div. I of Vehicle Development Center I for the development of new world car "Lexus", Design Div. II of Vehicle Development Center II for other model change of various mid-size cars, and Toyota Design Research Laboratory Tokyo for Advanced design cars, (ii) Marketing Service Div., Dealer Marketing System Div., Auto-salon *"Amulux"* Tokyo, U.S. Office, Europe Office and others for the internal and external customers information gathering, and (iii) TQC Promotion Div. for developing "Science SQC" (See to Chapter 2 and 3).

Concretely, as developing CSp, the actual studies for AEDM have been applied to Toyota and others (Nunogaki et al., 1996; Nagaya et al., 1998; Amasaka et al., 1999; Okazaki et al., 2000).

In the "ADS" project, the author has developed an "Automobile Exterior Design Model with 3 Core Methods" (AEDM-3CM) as shown in Figure 7 of Chapter 8 (Amasaka, 2017, 2018) (See to Chapter 8). This model combines 3 core methods as follows; (A) Improvement of business process methods for automobile profile design, (B) Creation of automobile profile design "Psychographics" approach methods, and (C) Automobile profile design, form and color matching support methods.

In core method (A), the 1st of the aims of the author and others was to grasp the characteristic of the profile design of the famous cars of the world such as "Benz, BMW, etc.," rationally by the concrete development of AEDM utilizing CS-IDCM.

To realize the above knowledge, in core method (B), the 2nd aim was the realization of the profile design, which is the main elements of exterior design of Toyota's strategic prestige car "new-model, Lexus GS400/LS430" by utilizing "psychographics" approach methods. In core method (C), the 3rd aim was the realization of the profile design development of various mid-size cars by the application of Lexus exterior design development.

The author has expanded AEDM for raising attractiveness to consumers by employing "Customer Information Analysis and Navigation System employing Customer Science principle" (CSp-CIANS) as shown in Figure 7 of Chapter 8 (Amasaka, 2005), and describe it in the next Sections (A), (B) and (C) (See to Chapter 8).

IMPROVEMENT OF BUSINESS PROCESS METHODS FOR AUTO- PROFILE DESIGN (A)

To enhance the design planning quality, the present study aims at achieving the following:

(1) To materialize vehicle images as desired by customers,
(2) To analyze causal relations between the customer satisfaction assessment and the vehicle appearance review factors on the basis of the appearance style assessment
(3) Based on the knowledge acquired, to understand the relevancy between vehicle images and proportion data and to reflect the results to the exterior design planning (Amasaka et al., 1999; Amasaka and Nagasawa, 2002).

Understanding Correlation Between Customers' Desirable Images of the Vehicles and Most Favored Vehicles (Step 1: Figure 3 in Chapter 8)

Questionnaire was used for 157 customers (domestic panel) various in personality. First, their desirable images were determined for the vehicles in the panel without indicating the appearance shown in Table 1.

Then, the photographs of four representative vehicle models, consisting of domestic and imported models (BMW850i/1990 model, BENZ 300-24/1989 model, Legend Coupe/1991 model and Soarer 4.0GT/1991 model), without indicating vehicles' brand and name were presented to answer their most favorite vehicle as shown in Table 2.

In Tables 1 and 2, the design image words (1)-(20) reveal the customers' impressions on the four models. These image words are not independent enough to be able to reply exactly, and some similarity of words exists (Nunogaki et al., 1996; Okazaki et al., 2000; Suzuki et al., 2000; Amasaka, 2003a, 2004b, 2015, 2018, 2022a,b, 2023).

Using the desirable images and the most favored vehicle data from the questionnaire, the author conducted four-group discriminant analysis.

As a result, it was found that all four models are roughly discriminated with a discrimination ratio of around 70%, and that the desirable panel images selected for individual models are almost similar.

On the basis of the obtained data, the panel groups belonging to "BMW, BENZ, Legend and Soarer" are specifically extracted.

Figure 1 shows a scatter diagram of individual scores obtained as the result of the analysis using the quantification method of the third type to more clearly understand the correlation between the desirable images and the most favored vehicle.

It is understood from Figure 1 that a panel group in favor of Soarer has desirable vehicle images such as sophistication and / or sportiness as embodied by "Soarer".

Furthermore, the characteristics of BMW is composed of individuality, advanced, intellectual etc., while that of Benz is composed of intrepid, novel, safe, composed, etc.

Since similar results are verified with other vehicle models, it is possible to conclude that customers' liking is consistent.

It was found that the customers' words can be materialized as the concrete contour of the vehicle.

The knowledge of the above (A) corresponds to the automobile exterior designer's values and experienced intellect.

Results of these analyses become great guidelines to the stylish flag ship car as a world car for the Japanese automakers, based on "Target area BMW and Target area BENZ" in Figure 1.

Table 1. Questionnaire on "desirable images"

Table 2. Questionnaire on "desirable images" and "most favorable vehicle"

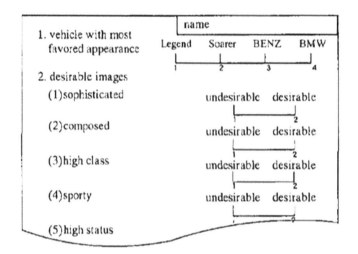

Figure 1. Correlation between desirable images and the most favored vehicle

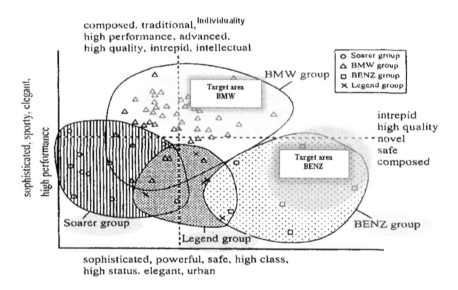

These results gave sufficient knowledge for concept creation of the exterior design of "global car - prestige Lexus" (Amasaka et al., 1999; Amasaka, 2004a,b, 2007).

Method for Exploring the Factors That Customers Emphasize in Vehicle Appearance Evaluation (Step 2: Figure 3 in Chapter 8)

The task taken up here is to objectify the above established theories (implicit knowledge) as qualitative and empirical rules for the professional designers who plan automobile profile designs. It is a well-known fact that the design of vehicle appearance has considerable weight in the customer's decision to make a purchase. To which part of vehicle appearance design do customers, domestic and overseas (regardless of age and sex), pay attention?

Professional automotive profile designers have a theory (or a rule of thumb) that, in general, Japanese customers tend to focus on the front design while North American customers look at the overall design of front, side and rear. The author's challenge here was to give an objective analysis of the theory. To the author's knowledge, no report has been made for objective verification of the theory in the academic world.

Quantitative evaluation on the sections of vehicle appearance is expected to advance a customer-in design strategy. For designing the new "Lexus GS400/LS430", 157 customers (young and old, and male and female panels) of various personality are asked to evaluate the appearance of the four major, mutually-competing models (BMW 850i/1990, Benz 300-24/1989, Legend coupe/1991 and Soarer 4.0GT/1991) and the priority of three appearance factors: front, side and rear views.

In analysis (1), the correlation between the evaluation and the priority is verified by multi-regression analysis. The three appearance factors are further divided into the design balance (profile) and detailed elements (4, 9, and 5 sections, respectively) for a similar study on their causal relationships as analysis (2).

A preliminary cluster analysis shows that the customers can be stratified, in terms of the overall liking of the vehicle appearance, into a group lower in age and annual income and a group higher in age and annual income for all four models (Amasaka, 1996, 1999a; Amasaka et al., 1999; Amasaka, 2004a,b, 2007).

Figure 2 shows an example of analytical results on a vehicle model specified to the group higher in age and annual income.

From Analysis (1), the contributory factor adjusted for the degree of freedom (R^{*2}) representing the degree of influence on the overall evaluation of vehicle appearance (X1) is 0.62, indicating a high causal relationship.

The breakdown is as follows: The influence of the front view is even higher at Bfv=0.59 while the influence of the side and rear views (X7 and X17) are relatively low at Bsv=0.18 and Brv=0.17.

In analysis (2), the analytical results are similar to those for the group with lower age and annual income, but the influence of bonnet (X4) is high on the front view (X2).

It is verified that the vehicle appearance is evaluated in a wider range; for example, the influence of the line (X19) from the rear to the trunk and the design balance (X22) of the rear as a whole are high on the side view (X7). This trend also applies to other three models.

On the other hand, in the group lower one, though not illustrated in analysis (1), the overall evaluation of vehicle appearance (X1) is 0.74, indicating a high causal relationship.

The breakdown is as follows: The influence of front view (X2) is fairly high at Bfv=0.46 while those of the side and rear views (X7 and X17) are at Bsv=0.30 and Brv=0.29, respectively, showing their positive influence.

From analysis (2), the head lamp and grille (X3) have a high degree of influence on the front view (X2). The overall side view and design (X16) exert much influence on the side view (X7).

And, the tail lamp (X20) and rear bumper design (X21) exert much influence on the rear view (X17).

A similar survey and analysis were conducted in North American market. While the front view is generally high priority in Japan, it is known that the front, side and rear views are equally valued in North America.

Another noticeable input is that Japanese customers are likely to provide individual evaluation for each front, side and rear view at a dealer.

In contrast, North American customers evaluate the front view while looking at a moving car on the opposite lane, evaluate the side view while looking at a car driving pass, and evaluate the rear view similarly on the street.

Figure 2. Causal relationships between customer satisfaction assessment and vehicle appearance assessment factors by multiple regression analysis

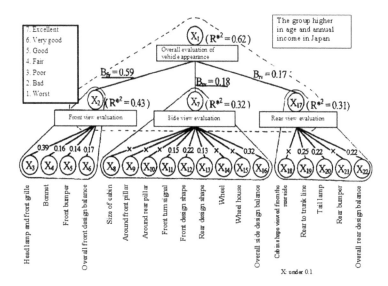

The author confirmed that, as they evaluate the three appearance factors, their focus is the total balance of the design.

Through this analytical study approach, designers understood the need for the customer-in design strategy that gives consideration to the different characteristics in each country.

These findings were the result of verification of the designer's theory (or rule of thumb), which greatly contributed to the designing of the global strategic vehicle Lexus GS400/LS430.

CREATION OF AUTOMOBILE PROFILE DESIGN "PSYCHOGRAPHICS" APPROACH METHODS (B)

Psychographics of automobile profile design refers to obtaining an appropriate outline profile (proportion), which is the skeleton of an automobile. It studies scientifically the relationships between the profile and psychological factors in the development stage so as to match the consumer' sensibility (i.e., so called "*Kansei*") (Amasaka and Nagasawa, 2002; Amasaka, 2005, 2015, 2018, 2022a,b, 2023).

As a scientific approach, Designing is applied from the viewpoint of CSp (Customer Science principle). The origin of design is explained and the process described sequentially thereafter.

During the development of a model-change vehicle, keen attention is paid to changes in the design in addition to selling points strengthened through functional improvements (Mori, 1991, 1993; Shinohara et al., 1996; Amasaka et al., 1999; Amasaka, 2007).

The appearance of a vehicle or the design is of high importance when consumers decide which one to buy. The reason consumer goods such as vehicles can sometimes be purchased on impulse is that consumers are impressed with and attracted by their designs.

It is generally considered that the factors determining good or poor design are (1) beauty, (2) freshness and (3) appearance worthy of the price. It seems that there is a universal rule on the correlation between beauty and product value to a certain extent. Freshness, changes and latest design are the main factors in selling a product by stimulating consumption.

Without these factors, there is no reason for customers to purchase new cars. Freshness of design has an important meaning for demonstrating functional improvements. The reason is that a consumer is not motivated to try unless functional improvements can be seen at a glance. For this reason, the spotlight is cast on design freshness.

The factors that make the appearance and design of an automobile feel "fresh" are largely classified into (1) proportion, (2) form and (3) surface. It seems that most people note these points in forming a visual impression. It was considered that (1) proportion has a significant weight since it can be recognized from a considerable distance, and a new proportion cannot be realized without technical innovation.

This subsection focused on freshness not as transitory fashion (fad) but as evolutionary freshness, which should be made explicit to a certain degree. The author paid attention to proportion since it typically represents this characteristic.

As in the expression of "form over substance," the form and surface generally represented by angles and roundness are considered to represent transitory fashions (Nagaya et al., 1998; Amasaka and Nagaya, 2002).

Studies on the Customer-Orientedness of Profile Design (Step 2: Figure 3 in Chapter 8)

Based on the abovementioned survey and analysis results, it can be assumed that "purchase because it is worth the price" and "purchase because it is newer than the model owned at present" can be represented by "high grade feel: price" and "freshness: year model". Here, analysis is done with the following approaches to estimate what would be the proportion of a vehicle that would sell in future.

If the author studies why customers buy merchandise, the following two factors may be listed. They buy because they get what they pay for, and they buy because the product is newer than what they have now.

By supposing that these two factors represent "class feeling being equal price" and "newness being equal model year", the following relationships between these and profile design are analyzed: ① hood ratio: hood length/overall length, ② trunk ratio: trunk length/overall length, ③ cabin ratio: cabin skirt length/overall length, ④ roof ratio: roof length/overall length, ⑤ front overhang ratio: front overhang length / overall length, ⑥ rear overhang ratio: rear overhang length/overall length, ⑦ wheel base ratio: wheel base length/overall length, ⑧ roof/cabin ratio: roof length / skirt length, overall height ratio, ⑨ overall width ratio: overall height/overall length, and ⑩ overall width ratio: overall width/overall length.

The data used for the analysis was measured with the autograph (measurement diagram). A total of 62 vehicle models, domestic and imported (sedans), are selected to measure their proportions (ratios), and a scatter diagram using principal component score is obtained as the Figures 3 and 4.

Figures 3 and 4 show the result of stratified classification by the class and the model year, respectively. In Figure 3, the class drops toward the right of the scatter diagram. In Figure 4, the model year is younger toward the right.

For example, if one of these two sheets of scatter diagrams is placed on top the other, it is understood that it is hard to realize a combination of the highest class and the latest model.

When the scatter diagrams on these two principal components are overlapped, old and high-class vehicles with a coach-type cabin (with long hood and luggage compartment length) such as Rolls Royce, Benz W123, BMW518, Jaguar X16, etc., are laid in the second quadrant. In the opposite fourth quadrant, late-model vehicles with a long cabin and shorter hood and luggage compartment are positioned.

From the results of these analyses, it is quantitatively clarified that seemingly highest class "luxury" and latest newer "newness" are mutually contradictory elements (Amasaka et al., 1999; Amasaka and Nagaya, 2002; Amasaka, 2003a, 2004a,b, 2007).

The customers wish the luxury and newness with advanced technology (fuel consumption, driving power and delightful cabin, etc.).

To create the "Lexus profile design", the author had to realize the excellent proportion ratio by combining the short hood length (small size engine) and long cabin (large wheel base and small overhang unit-axle), excellent streamline (short roof length, large slope front / rear window by enlarging the area of window glass and cabin space spreads out) and low vehicle height (thin top and floor).

Figure 3. Classification of vehicle model by the class grade "degree of luxury"

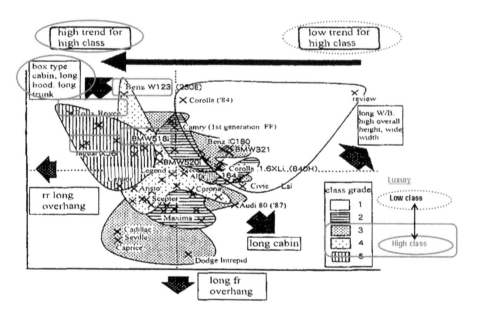

Through these realizations, being compatible with "luxury" and "newness" became possible. In other words, it can be quantitatively verified that the luxury (high class) and the newness are contrary to each other, which is a newly found knowledge.

254

Figure 4. Classification of vehicle model by the year model "degree of newness"

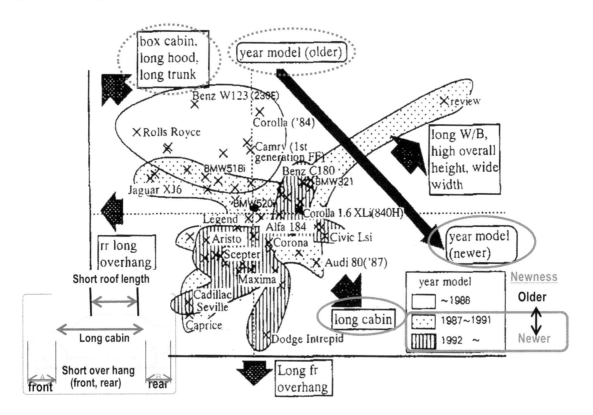

From these studies, the author has been able to obtain a desirable direction for the study to forecast the liking of future users, not of current users, by incorporating into analysis a factor containing the concept of time axis that the customers' liking can change in terms of newness.

These findings are the result of verification of the designer's theory (a rule of thumb), which greatly contributed to the designing of the global strategic vehicle "Lexus GS400/LS430".

This enabled us to propose a desirable thing to customers before they want it, thereby obtaining a guideline that helped us improve the processing of design business.

The author then identified the general rule (that was not the intentional but the natural consequence of designers' works) of the common proportional ratio (highest class) that was inherited in the world's prestigious vehicles and insusceptible to change over five or ten years.

Actual Proof Research of "Lexus GS400/LS430" Development (Step 3: Figure 3 in Chapter 8)

Upon designing a global strategic car "Lexus", the matter of primary concern was how to catch the target customer' heart. Taking advantage of the knowledge acquired in subsection 17.3.2(A), the author realized the profile design of highest class and latest car "Lexus GS400 / LS430".

Particularly, the development target of the first Lexus was the realization of the profile design of high trend/high class vehicle "BMW/BENZ/Jaguar luxury" as shown in Figures 2, 3 and 4.

The target of GS400 was the profile design surpassing middle grade class for lower in age (youth and middle age). In the same way, the target of LS430 was the profile design surpassing upper grade class for higher in age (middle and advanced age).

To verify the above Steps 1 and 2 of Figure 3 in Chapter 8, the author showed that side views with typically different proportion ratios A through F, which arrange the combination of proportion ratio of ①-⑩ in Figure 5, were selected for evaluation purposes (comparison of ratio data) (Amasaka and Nagaya, 2002).

The direction toward ratio A was judged to be fresh, as expected (plan with shorter hood, longer cabin skirt length and shorter roof).

Figure 5. Side views for survey purposes using dimensional ratios for Lexus GS400 development

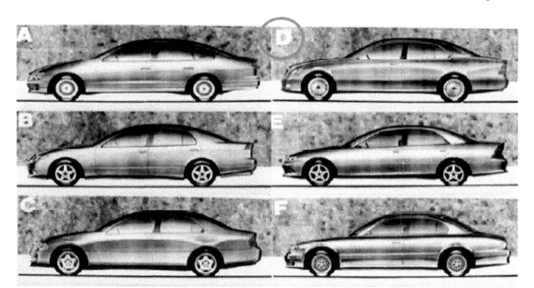

Lexus GS400 adopted proportion ratio "D"
(Source: Amasaka and Nagaya, 2002).

On the other hand, ratios D and F (ratios opposite to those of A) were judged to be "newness" (GS400 adopted a model "ratio D" by the final judgement) (Amasaka and Nagaya, 2002).

The above survey results proved that the scientific approach yields a good match with sensory evaluations by consumers. Based on the knowledge obtained using the scientific approach, GS400 was developed successfully, named Toyota "new model Aristo" in Japan (Motor Fan, 1997).

Similarly, the survey shown in Figure 6 was implemented for the development of the Lexus LS430 named Toyota "Celsior" in Japan (Motor Fan, 2000). Proportion ratios A through F of the LS430 were determined in the same way as for the GS400 (LS430 adopted a model "ratio B" by the final judgement "Newness").

Both proportion ratios definitions of GS400/LS430 are illustrated in Amasaka et al. (1999), Amasaka and Nagaya (2002) and Amasaka (2005, 2007a, 2017, 2018).

Figure 6. Side views for survey purposes using dimensional ratios for Lexus LS430 development

Lexus LS430 adopted proportion ratio "B"
(Source: Amasaka and Nagaya, 2002)

Lexus GS400/LS430 obtained satisfactory evaluation from the customers, and was successful as a flag ship car "Lexus" of Toyota sedan vehicles (Motor Fan, 1997, 2000; JD Power Associates, 1998).

ACTUAL STUDIES ON AUTOMPBILE PROFILE DESIGN, FORM, AND COLOR MACHING SUPPORT METHODS (C)

Increasingly diverse consumer values have made an automobile exterior design the most important factor influencing purchasing decisions in recent years. For this reason, identifying customers' unspoken senses (values) and scientifically presenting them from an objective standpoint have become an absolute requirement in automobile exterior design processes for mid-size car.

These processes involve successfully matching profile design (proportion), form, and color from the perspective of an objective Science SQC approach with SQC Technical Methods in AEDM (Amasaka et al., 1999; Amasaka, 2002, 2004a, b, 2014, 2017, 2018, 2022a,b, 2023).

Developing Automobile Profile Design, Form, and Color Matching Support Methods

The author has created an "Automobile Profile Design, Form and Color Matching Support Methods" (APFC-MSM) based on the customers' value elements of automobile exterior design work in developing the "AEDM-3CM" (Automobile Exterior Design Model with 3 Core Methods) described in Figure 5 of Chapter 7 and Figure 7 of Chapter 8 (Toyoda et al., 2015a; Amasaka, 2018) (See to Chapter 7 and 8).

Due to the development of "Automobile Profile Design, Form and Color Optimal Matching Model" (APFC-OMM) to mention later described in Figure 8, APFC-MSM starts with three elements (1) profile design, (2) form, and (3) color.

Recent design work strategies make it a point to optimize business processes so that they are in line with the vehicle design concept from the product planning stage. Next, each element must be matched: (4) profile design and form, (5) form and color, and (6) profile design and color.

Finally, (7) all three elements - profile design, form, and color - must be integrated harmoniously to address modern market demands.

Based on the knowledge acquired by the above-mentioned the above section "Improvement of business process methods for auto-profile design (A) and Creation of automobile profile design "Psychographics" approach methods (B)", the author developed AEDM-3CM by using APFC-MSM.

These studies were carried out in the Amasaka New JIT laboratory (in Aoyama Gakuin University) by collaboration with Toyota Motor Corp., Toyota Design Research Laboratory Tokyo, Nissan Motor Corp., Honda Motor Co., Ltd., Mazda Motor Corp., Nippon Paint Co., Ltd., Kansai Paint Co. Ltd. and others.

The synthetic optimization studies of the three elements - profile design, form and color - of exterior design are indispensable in making the knowledge of above section ""Improvement of business process methods for auto-profile design (A) and Creation of automobile profile design "Psychographics" approach methods (B)" evolving further.

Examples of actual case studies, corresponding to "(i)-(vii)" for the business process innovation that addresses optimizing the automobile exterior design for customer creation, are as follows;

(i) compatibility of profile design and package design (interior space) (Okabe et al., 2007; Yamaji and Amasaka, 2009), (ii) form design (Asami et al., 2010; Yazaki et al., 2013), (iii) color design (Muto et al., 2011; Takebuchi et al., 2012a), (iv) profile design and form matching (Takimoto et al., 2010), (v) form and color matching (Takebuchi et al., 2012b; Muto et al., 2013), (vi) profile design and color matching (Asami et al., 2011), and (vii) integrated profile design, form and color matching (Toyoda et al., 2015a; Kobayashi et al., 2016).

The author now illustrates some pioneering case studies for the development of APFC-MSM in the following.

Creation of an Automobile Exterior Form Design Optimization

In this study, by paying attention to the automobile form, the details that were previously ambiguous, such as "roundness between bonnet and fender" and "angle", are represented numerically.

With the aim of measuring the degree of influence of customer sensibility, the author developed an automobile exterior form design optimization (Asami et al., 2010).

The author created a model of a form-altered automobile to quantitatively evaluate the degree of influence of customer sensitivity by utilizing above mentioned CS-IDCM.

As a means of associating customer sensitivity (as expressed through words) with form, a car model, including several types of determined parameters, is created by design CAD, which performs a sensitivity evaluation.

The form shape parameters and corresponding customer impressions are shown in Table 3 for each car model type (ex. front view and side view). Similarly, a model for investigating nine car types (Type A-Type I) was created for the front part and side part using numerical form representation.

The purpose of the sensitivity evaluation questionnaire was to explore the differences in the impressions felt by customers, regarding nine models in which the forms of the front and side had been changed.

Accordingly, subjects were asked using the same questionnaire form for all models. These data were analyzed to determine the relationships between subjective impressions and forms. In addition, the younger age group (the 20's) was targeted for the investigation.

Table 3. Form shape parameters for nine car model types (front view and side view)

Car type / Form shape	Angle of the bonnet edge (˚)	Angle of the bonnet (˚)	Angle of the fender (˚)	Angle of the front pillar (˚)	Angle of the belt line (˚)	Angle of the character line (˚)
Type A	40	40	40	45	2	25
Type B	15	20	25	30	0	15
Type C	15	25	60	30	0	55
Type D	25	50	20	30	5	15
Type E	25	50	60	30	5	55
Type F	30	25	20	60	0	15
Type G	30	25	60	60	0	55
Type H	45	50	20	60	5	15
Type I	45	50	60	60	5	55

In the questionnaire, a seven-step evaluation (disagree=1 vs. agree=7) using sensitivity words was performed. In order to select suitable sensitivity words, a bibliographic search and an investigation hearing with the designer were conducted, and 42 sensitivity words related automobile appearance were extracted.

This set of words was then narrowed to those having a close relationship, yielding the 12 sensitivity words (sophisticated, intellectual, elegant, traditional, advanced, high class, characteristic, simple, composed, cute, powerful and sporty).

In visualization of the front/side part of relationships, covariance structural analysis was conducted to obtain an overall sensitivity evaluation by combining separate sensitivity evaluations for the front/side part into a synthetic variable in order to examine their mutual influence.

As an example, the shape parameter ~ sensitivity path diagram is shown in Figure 7 from the standardized partial regression coefficient for "sophisticated". In this figure, the strength of the relationship for the shape parameter with a strong weight is expressed by the thickness of the arrow, regardless of the positive and negative sign of the coefficient.

From the path diagram in Figure 7, one front part and one side part each exert large degrees of influence, and the angle of the front pillar conveys the sensitivity evaluation "sophisticated" as the shape parameter with the largest degree of influence. By conducting the same analysis for all sensitivity words, the front / side part of relationships can be comprehensively evaluated. Based on the above information, it is clear that, to the younger generation, the concept of "advanced", conveys a sense of stylishness.

As a result, it is clear that an "advanced" image can be expressed via a specific combination of 5 points (elements: (a)-(e)): (a) deep bonnet edge, (b) bonnet with swelling, (c) front pillar in an acute angle, (d) upward belt line, and (e) strengthened character line, each parameterized and materialized by design CAD. The effectiveness of this study was considered to be verified.

Creation of an Automobile Exterior Color Design Optimization

The quality of the exterior colors of automobiles has become a major aesthetic factor (preference) influencing the car buying process. In this study, the author created an automobile exterior color design optimization based on above mentioned CS-IDCM, and verified this study's effectiveness throughout the application of medium sedan car (Takebuchi et al., 2012a).

Figure 7. Path diagram for "sophisticated" using covariance structural analysis

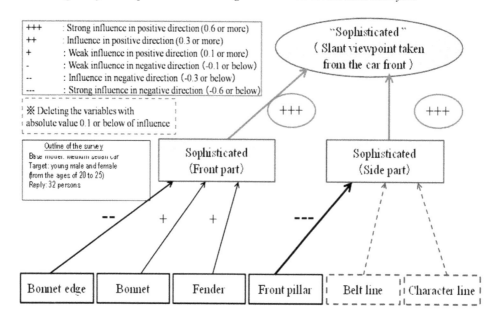

The author conducted a survey using 11 painted panels with varying color elements to determine how customers are influenced aesthetically by color variations including the textural expressions.

For this study, the panels were shaped and painted like car panels to show the relevant textural expressions as shown in Figure 12 of Chapter 8 (See to Chapter 8). Figure 12 in Chapter 8 is an example for matching of form and color optimization using front fender panels.

First, the painted panels were selected. Three panels were initially prepared for the survey, using three colors (red, black and brown) with minimal opacity and graininess.

Then, a further eight panels were prepared, varying the effect of the opacity and graininess equally to produce a total of 11 panels as shown in Figure 12(i).

The 6 elements of color (hue, luminosity, intensity, shine, opacity and graininess) of each panel were measured. The results are described in Table 4.

Then, the participants were asked to look at each painted panel and evaluate according to a seven-point scale, the extent to which each panel gave the desired impression for the above four preferences (classy, luxurious, dignified and sporty).

Aesthetic evaluation data was obtained from the results. The survey was aimed at men and women in their twenties, and the answers were obtained from a total of 94 men and women.

This path diagram of covariance structure analysis was conducted to determine the correspondence relationship between color elements and preferences as shown in Figure 12(ii) of Chapter 8 (See to Chapter 8).

The results were then used to provide information to contribute to the development of exterior colors appealing to people who value self-expression.

The coefficients of each path are standardized coefficients. The absolute values of these standardized coefficients indicate the degree of influence while the direction of influence is indicated by positive or negative numbers (Figure 7(ii) in Chapter 8).

Table 4. The six elements of color of each panel

Panel No.	Hue (H45°)	Luminosity (L45°)	Intensity (C45°)	Shine (60°)	Opacity (L15-110°)	Graininess (HG)
1	35.00	83.89	41.17	1.79	0	89.90
2	16.79	39.00	32.00	33.90	75.70	91.50
3	23.54	60.10	33.00	4.01	53.60	90.40
4	27.49	56.55	28.06	35.70	59.20	89.10
5	6.19	2.10	342.00	34.79	89.70	96.50
6	3.44	2.80	31.00	19.94	17.90	90.80
7	30.45	10.96	44.15	59.85	29.20	95.90
8	9.90	8.04	10.56	0.59	0	94.50
9	5.48	1.10	255.50	19.05	51.20	93.00
10	2.77	0.70	260.00	11.90	4.00	89.60
11	1.66	0.12	93.07	3.64	0	92.60

This model was subjected to a goodness-of-fit test once it was corrected using a modification index, resulting in reasonable scores of 0.979GFI and 0.895AGFI.

In this path diagram used for this covariance structure analysis, the "exterior color" latent variable was chosen as the latent variable that would satisfy the four vehicle design preferences of those who value self-expression.

Respondents in the self-expression group were asked to freely evaluate the three developed colors. The respondents gave positive evaluations, saying they liked how classy the colors were and that they would like a car with colors such as these.

This shows that the target users were satisfied with the exterior colors developed by employing AEDM-3CM.

Creation of an Automobile Profile Design, Form, and Color Optimal Matching Model

Moreover, the author has created an "Automobile Profile Design, Form and Color Optimal Matching Model" (APFC-OMM) as the investigation of "Optimum AEDM-3CM" as shown in Figure 8 (Toyoda et al., 2015a; Amasaka, 2018).

The development of the AOPFC-MM was applied to young people in their 20s to see whether the model selected by the method would match their preferences. The effectiveness of the research was thus confirmed as follows;

In Step 1, the authors began by conducting interviews with key automakers and dealers as well as with customers. They also consulted prior research to see which issues had already been addressed and which had yet to be clearly identified.

In Step 2, the authors conducted a customer preference survey based on the key issues identified above. In this step, customers were given a questionnaire to find out which vehicles they were interested in, and to pinpoint the main factors they considered when purchasing a vehicle.

Once the data was collected, the authors subjected it to a principal component analysis and a cluster analysis in order to quantitatively determine the relationships between different customer senses.

In Step 3, the authors used the insights gained from the customer preference survey in Step 2 to recruit test subjects that resembled target customers. Once the test subjects were selected, they were each fitted with an eye camera to analyze line-of-sight information.

Figure 8. Automobile Profile design, form, and color optimal matching model (APFC-OMM)

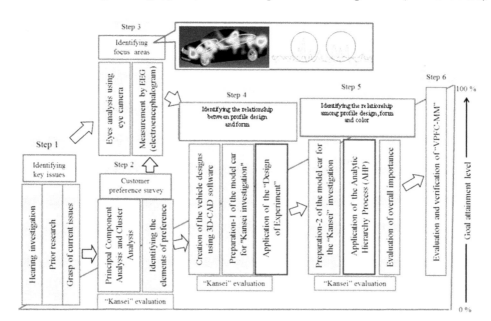

This told the author where the subjects placed their attention when looking at vehicles.

At the same time, an electroencephalogram (EEG) was used to measure brain waves. This device ascertained how the subjects were feeling when they looked at certain parts of the car.

In Step 4, the authors continued to build upon the information gained in the previous steps in order to identify the relationship between the profile design and the form, two of the critical elements in exterior automotive design.

To do this, they used 3D-CAD software to actually convert vehicle characteristics into numerical specifications, thus actually creating the vehicle designs for realizing "Customers' wants".

In Step 5, the models designed in Step 4 were then analyzed using statistics (Design of Experiments (DOE) and Analytic Hierarchy Process (AHP)) to select the model that had the optimum combination of design elements.

In Step 6, the authors evaluated the success of the model selected in Step 5. Test subjects were again fitted with EEG equipment as the authors compared their reactions to the selected model and other vehicle designs.

Concretely in Step 3, the authors used 3D-CAD software to actually design a vehicle exterior that reflected the critical proportion and form element as indicated in Figure 9.

The study used a mid-size sedan and selected standard proportion and form dimensions based on collections of dimensional body drawings as shown in Table 5, and then numerically represented them at three levels based on the knowledge of the Steps 1- 2.

The author then created models that combined different front part (length of the front windshield glass, edge and groove lines), side part (hood bulge and strength of character lines), and rear part (rear pillar angle and rear bumper width) form elements.

Figure 9. Designing a vehicle exterior design (outline of proportion and form)

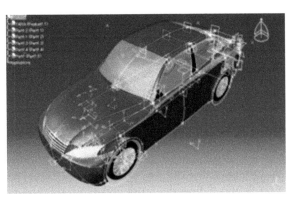

Table 5. Selected standard proportion and form dimensions

Model	Length of the front windshield (mm)	Hood bulge (°)	
A	730	40	40
B	655	25	15
C	805	25	15
D	655	50	15
E	805	50	15
F	655	25	45
G	805	25	45
H	655	50	45
I	805	50	45

The nine models were analyzed by DOE to determine the optimum combination of design elements. As shown in Table 6, the front part analysis revealed that factor A (length of the front windshield), factor B (hood bulge), factor C (edge and groove lines), and the interplay between factors B and C had a 5% significance. It indicates that shallow edge and groove lines were more significant than differences in hood bulge.

These results allowed the authors to identify two front patterns, one side pattern, and one rear pattern.

Next, in Step 4, Three colors (black, white, red) were added to the model to create six alternatives (Type A, Type B, Type C, Type D, Type E and Type F). They were then subjected to an AHP in order to identify the model that optimally matched the sensory words that represent the demands of the exterior design group.

Table 6. Analysis result (front part)

Factor	Sum of Squares	Flexibility	Variance	Variance Ratio	Test
A	20.816	1	20.816	21.575	**
B	11.391	1	11.391	11.806	**
C	5.348	1	5.348	5.543	*
AB	0.660	1	0.660	0.684	
AC	1.266	1	1.266	1.312	
BC	7.223	1	7.223	7.486	*
ABC	2.250	1	2.250	2.332	
Error	7.719	8	0.965	0.558	

Models created using 3D-CAD software were given to subjects to visually compare using a paired comparison method in order to assign priority levels to the different standards and alternatives. The nine models were analyzed by DOE to determine the optimum combination of design elements. The test subjects' subjective responses were gathered using a questionnaire.

Table 7 shows the relative priority of each standard (sensory word) and alternative (CAD model). This information was then comprehensively evaluated to identify the model with the best proportion, form, and color combination.

The analysis indicated that, among young men in their 20s within the exterior design group (the target of the study), two models created using this method (Type B, Type F) received the highest evaluation (Toyoda et al., 2014).

The results indicate that the Type B model, with a front windshield length of 805mm, a hood bulge of 50°, an edge and groove lines of 45°, emphasized strong character lines, a front pillar angle of 30°, and black paint, has the "cool" sensory word as its main focus. The Type F model had the same proportions and form as Type B in red, and it had the luxurious‖ sensory word as its main focus.

The models identified in Step 4 were verified in Step 5, and the authors were then able to develop the APFC-OMM. Here, EEG equipment was used to measure the subjects' alpha waves as they looked in random order at the models that they were asked about in the Step 4 questionnaire.

Table 7. The relative priority (standard)

Subject	Alternative 1	Alternative 2	Alternative 3	Alternative 4	Alternative 5	Alternative 6
1	0.13	0.17	0.12	0.10	0.16	0.32
2	0.13	0.15	0.10	0.13	0.18	0.30
3	0.09	0.10	0.26	0.39	0.06	0.11
4	0.25	0.29	0.03	0.04	0.15	0.24
5	0.33	0.31	0.19	0.09	0.02	0.06
6	0.31	0.34	0.16	0.10	0.06	0.03
7	0.31	0.34	0.16	0.10	0.06	0.03
8	0.06	0.11	0.08	0.07	0.21	0.48
9	0.17	0.16	0.06	0.08	0.31	0.21
10	0.19	0.23	0.09	0.25	0.11	0.13

The results of this test; namely, that the models triggering the highest alpha wave values were the same models that the study predicted through the APFC-OMM. The effectiveness of the model was thus confirmed (Toyoda et al., 2014, 2015a).

Expanding CSp for Raising Attractiveness to Consumers

Furthermore, the author has been advancing the new deployments of AEDA using CSp (Customer Science principle) for "Mid-size and Small-size cars" as follows;

The 1st research area is the free answer questionnaire analysis for consumer-needs visualization: The example of the products project support of the vehicle (Namikata and Yano, 2003), 2nd is the development of automobile exterior color and interior color matching (Koizumi et al., 2013a; Shinogi et al., 2014) and auto-instrumentation design (Yazaki et al., 2012), 3rd is the application to automobile exterior design for the women customers (Asami et al., 2011), and 4th is the development of exterior design research of a motor scooter and a bicycle (Nakamura et al., 2008; Koizumi et al., 2013b; Toyoda et al., 2015b). These have been developed in Toyota Motor Corp., and others.

CONCLUSION

The author verified the validity of AEDM as the key to implementation of Design SQC for the innovation of business process for designing attractive vehicles using CSp. As definite deployment of AEDM with CS-IDCM using SQC Technical Methods, the author focused on 3 vehicle exterior design elements that were handled separately in the past (profile design, form, and color) and attempted to match them all while pinpointing their correlations to customer senses. As actual examples for customer value creation, this model was applied in the development of Toyota Lexus and others. Furthermore, the application of ADEM is currently advancing new areas such as "exterior color and interior color matching, package design, auto instrumentation design, motor scooter and bicycle".

REFERENCES

Amasaka, K. (1995). A construction of SQC Intelligence System for quick registration and retrieval library: A visualized SQC report for technical wealth. *Lecture Notes in Economics and Mathematical Systems, Springer, 445*, 318–336. doi:10.1007/978-3-642-59105-1_24

Amasaka, K. (1996). Application of classification and related method to the SQC Renaissance in Toyota Motor. Data Science, Classification, and Related Methods. Springer.

Amasaka, K. (1999a). A demonstrative study of a new SQC concept and procedure in the manufacturing industry: Establishment of a New Technical Method for conducting Scientific SQC. *An International Journal of Mathematical & Computer Modeling, 31*(10-12), 1–10.

Amasaka, K. (1999b). A study on Science SQC by utilizing Management SQC: A demonstrative study on a new SQC concept and procedure in the manufacturing industry. *International Journal of Production Economics, 60-61*, 591–598. doi:10.1016/S0925-5273(98)00143-1

Amasaka, K. (2002). New JIT, a new management technology principle at Toyota. *International Journal of Production Economics*, *80*(2), 135–144. doi:10.1016/S0925-5273(02)00313-4

Amasaka, K. (2003a). *Development of "New JIT", Key to the excellence design "LEXUS": The validity of "TDS-DTM", a strategic methodology of merchandise.* Proceedings of the Production and Operations Management Society, Hyatt Regency, Savannah, Georgia.

Amasaka, K. (2003b). Proposal and implementation of the Science SQC, quality control principle. *Mathematical and Computer Modelling*, *38*(11-13), 1125–1136. doi:10.1016/S0895-7177(03)90113-0

Amasaka, K. (2004a). *Customer Science: Studying consumer values.* Japan Journal of Behavior Metrics Society, The 32nd Annual Conference, Aoyama Gakuin University, Sagamihara, Kanagawa.

Amasaka, K. (2004b). *Science SQC, new quality control principle: The quality strategy of Toyota.* Springer-Verlag. doi:10.1007/978-4-431-53969-8

Amasaka, K. (2005). Constructing a *Customer Science* application system "CS-CIANS": Development of a global strategic vehicle "*Lexus*" utilizing *New JIT, WSEAS (World Scientific and Engineering Academy and Society). Transformations in Business & Economics*, *2*(3), 135–142.

Amasaka, K. (2007). The validity of "*TDS-DTM*", a strategic methodology of merchandise: Development of *New JIT*, Key to the excellence design "*LEXUS*". *The International Business & Economics Research Journal*, *6*(11), 105–115.

Amasaka, K. (Ed.). (2007). New Japan Model: Science TQM: Theory and practice for strategic quality management, The quality management of the manufacturing industry, Maruzen, Tokyo.

Amasaka, K. (2015). New JIT, New Management Technology Principle, Taylor and Francis Group, CRC Press, Boca Raton, London, New York.

Amasaka, K. (2017). Studies on Automobile Exterior Design Model for customer value creation utilizing Customer Science Principle, *The 7th International Symposium on Operations Management Strategy, Tokyo Metropolitan University, Tokyo,* 27-28.

Amasaka, K. (2018). Automobile Exterior Design Model: Framework development and supporting case studies. *The Journal of Japanese Operations and Strategy*, *8*(1), 67–89.

Amasaka, K. (2022a). New Manufacturing Theory: Surpassing JIT (2nd Edition). Lambert Academic Publishing, Germany.

Amasaka, K. (2022b). *Examining a New Automobile Global Manufacturing System.* IGI Global Publisher. doi:10.4018/978-1-7998-8746-1

Amasaka, K. (2023). New Lecture-Surpassing JIT: Toyota Production System-From JIT to New JIT-, Lambert Academic Publishing.

Amasaka, K., & Nagasawa, S. (2002). *Fundamentals and application of sensory evaluation: For Kansei Engineering in the vehicle.* Japanese Standards Association.

Amasaka, K., & Nagaya, A. (2002). *Engineering of the new sensitivity in the vehicle: Psychographics of LEXUS design profile", Development of articles over the sensitivity: The method and practice.* Edited by Japan Society of Kansei Engineering, Nihon Shuppan Service Press.

Amasaka, K., Nagaya, A., & Shibata, W. (1999). Studies on Design SQC with the application of Science SQC: Improving of business process method for automotive profile design. *Japanese Journal of Sensory Evaluations, 3*(1), 21–29.

Asami, H., Ando, T., Yamaji, M., & Amasaka, K. (2010). A study on Automobile Form Design Support Method "AFD-SM". *Journal of Business & Economics Research, 8*(11), 13–19. doi:10.19030/jber.v8i11.44

Asami, H., Owada, H., Murata, Y., Takebuchi, S., & Amasaka, K. (2011). The A-VEDAM for approaching vehicle exterior design. *Journal of Business Case Studies, 7*(5), 1–8. doi:10.19030/jbcs.v7i5.5598

Evans, R. J., & Lindsay, M. W. (2004). *The management and control of quality.* South-Western College Publishing.

Kobayashi, T., Yoshida, R., Amasaka, K., & Ouchi, N. (2016). A statistical and scientific approach to deriving an attractive exterior vehicle design concept for indifferent customers. *Journals International Organization of Scientific Research, 18*(12), 74–79.

Koizumi, K., Kanke, R., & Amasaka, K. (2013a). Research on automobile exterior color and interior color matching. *International Journal of Engineering Research and Applications, 4*(8), 45–53.

Koizumi, K., Kawahara, S., Kizu, Y., & Amasaka, K. (2013b). A Bicycle Design Model based on young women's fashion combined with CAD and Statistical Science. *Journal of China-USA Business Review, 12*(4), 266–277.

Mori, N. (1991). *The design plan of the engineering soft system of a designing.* Asakura-Shoten.

Mori, N. (1993). *The research of the scientific method of a left brain designing.* Kaibundou.

Moriya, A., & Sugiura, K. (1999). *Collage Therapy: Esprit of Today.* Shibundou.

Muto, M., Miyake, R., & Amasaka, K. (2011). Constructing an Automobile Body Color Development Approach Model. *Journal of Management Science, 2,* 175–183.

Muto, M., Takebuchi, S., & Amasaka, K. (2013). Creating a New Automotive Exterior Design Approach Model: The relationship between form and body color qualities. *Journal of Business Case Studies, 9*(5), 367–374. doi:10.19030/jbcs.v9i5.8061

Nagaya, A., Matsubara, K., & Amasaka, K. (1998). *A study on the customer tastes of automobile profile design (Special Lecture).* Union of Japanese Scientists and Engineers, The 28th Sensory Evaluation Sympojium, Tokyo.

Nakamura, M., Kuniyoshi, M., Yamaji, M., & Amasaka, K. (2008). Proposal and validity of the product planning business model "A-POST": The application of text mining method to scooter exterior design. *Journal of Business Case Studies, 4*(9), 61–71. doi:10.19030/jbcs.v4i9.4808

Namikata, A., & Yano, Y. (2003). *Free answer questionnaire analysis for consumer-needs visualization: The example of the products project support of the vehicle* [Bachelor thesis, Aoyama Gakuin University].

Nunogaki, N., Shibata, K., Nagaya, A., Ohashi, T., & Amasaka, K. (1996). *A study of customers' direction about designing vehicle's profile: A deployment of "Design SQC" for design business process.* The Japanese Society for Quality Control.

Okabe, Y., Yamaji, M., & Amasaka, K. (2007). Research on the Automobile Package Design Concept Support Methods "CS-APDM": Customer Science approach to achieve CS for vehicle exteriors and package design [in Japanese]. *Journal of Japan Society for Production Management, 13*(2), 51–56.

Okazaki, R., Suzuki, M., & Amasaka, K. (2000). *Study on the sense of values by age using Design SQC.* The Japan Society for Production Management, The 11th Annual Technical Conference, Okayama University, Okayama.

Shinogi, T., Aihara, S., & Amasaka, K. (2014). Constructing an Automobile Color Matching Model (ACMM). *IOSR Journal of Business and Management, 16*(7), 7–14. doi:10.9790/487X-16730714

Shinohara, A., Sakamoto, H., & Shimizu, Y. (1996). *Invitation to Kansei Engineering.* Morikita-Shuppan.

Suzuki, M., Okazaki, R., & Amasaka, K. (2000). *The research studies of the value by the generation by employing Design SQC: The development of Customer Science by utilizing Science SQC.*, The Japan Society for Production Management.

Takebuchi, S., Asami, H., & Amasaka, K. (2012b). An Automobile Exterior Design Approach Model linking form and color. *Journal of China-USA Business Review, 11*(8), 1113–1123.

Takebuchi, S., Nakamura, T., Asami, H., & Amasaka, K. (2012a). The Automobile Exterior Color Design Approach Model. *Journal of Japan Industrial Management Association, 62*(6E), 303–310.

Takimoto, H., Ando, T., Yamaji, M., & Amasaka, K. (2010). The proposal and validity of the Customer Science Dual System, *China-USA. Business Review (Federal Reserve Bank of Philadelphia), 9*(3), 29–38.

Toyoda, S., Koizumi, K., & Amasaka, K. (2015b). Creating a Bicycle Design Approach Model based on fashion styles, *IOSR (International Organization of Scientific Research). Journal of Computational Engineering, 17*(3), 1–8.

Toyoda, S., Nishio, Y., & Amasaka, K. (2014). *Matching methods of the car proportion, form, color based on the customer sensitivity approaches and that validity.* Japanese Operations Management and Strategy Association, The 6th Annual Conference on Takushoku University, Tokyo.

Toyoda, S., Nishio, Y., & Amasaka, K. (2015a). Creating a Vehicle Proportion, Form, and Color Matching Model. *Journals International Organization of Scientific Research, 17*(3), 9–16.

Yamaji, M., & Amasaka, K. (2009). An Intelligence Design Concept Method utilizing Customer Science. *The Open Industrial and Manufacturing Engineering Journal, 2*(1), 10–15. doi:10.2174/1874152500902010021

Yazaki, K., Takimoto, H., & Amasaka, K. (2013). Designing vehicle form based on subjective customer impressions. *Journal of China-USA Business Review, 12*(7), 728–734.

Yazaki, K., Tanitsu, H., Hayashi, H., & Amasaka, K. (2012). A model for design auto instrumentation to appeal to young male customers. *Journal of Business Case Studies, 8*(4), 417–426. doi:10.19030/jbcs.v8i4.7035

Chapter 13
Automobile Exterior and Interior Color Matching Model Using Design SQC

ABSTRACT

In previous chapters, the author has developed the AEDM (automobile exterior design model) for realizing profile design, form, and color matching. In this chapter, furthermore, the author has created the automobile exterior and interior color matching model using design SQC (AEDM-EICMM). Specifically, then, the author has developed the Design SQC to Toyota vehicle Passo using CSp-IDCM (intelligence design concept method using customer science principle) as follows: (1) Preference surveys were conducted to determine which "exterior colors, interior colors, and front panel colors" suit the preferences of women in their twenties (20's); (2) Color combinations were created based on the data obtained from the preference surveys; (3) Effectiveness of the AEDM-EICMM was confirmed by conducting surveys to determine whether the color combinations created were suitable for women in their 20's, and (4) Similar applications in Toyota and others.

OUTLINE OF DESIGN SQC FOR AUTO-EXTERIOR DESIGN

Design SQC in Toyota's Auto-Exterior Design Strategy

It is quite important for mapping up design strategies to study on "what style of vehicles would sell in the future?". To enhance the automobile profile design planning quality, the author has created the "Design SQC in Toyota's strategic auto-profile design" through the application of "Customer Science principle" using "Science SQC, new Quality Control principle" (Amasaka et al., 1999; Amasaka, 2004a,b, 2005) (See to Chapter 2, 3, 7 and 8).

In Design SQC strategy, the author has developed the "Automobile Intelligence Design Concept Method using Customer Science principle" (CSp-IDCM) for realizing an automobile exterior design for customer value creation based on research of psychographics by the viewpoint of customers' life stage and lifestyle (See to Chapter 8).

DOI: 10.4018/978-1-6684-8301-5.ch013

Specifically, the author has established the "AEDM-3CM" (Automobile Exterior Design Model with 3 Core Methods); (A) Improvement of "Business Process Methods for Automobile Profile Design" (BPM-APD), (B) Creation of "Automobile Profile Design using "Psychographics Approach Methods" (APD-PAM), and (C) "Automobile profile design, form and color matching support methods" (APFC-MSM) (Amasaka, 2015, 2018).

Design SQC, Conception, and Analysis in Auto-Exterior Design

A concept of "Design SQC" by utilizing *"Kansei Engineering"* has been in recognition for long time (Shinohara, et al., 1996; Amasaka et al., 1999; Amasaka and Nagasawa, 2000; Amasaka and Nagaya, 2002).

For example, the author has developed variable data out of sensory elements such as the digitization of sensory inspection of vehicles and / or mechanization of hunch and knack work, which are being applied to actual development (Amasaka, 1972, 1976, 1983; Shimizu & Amasaka, 1975).

On the other hand, examples of study are introduced where it is possible to apply a statistical method (analysis) to the development business of "product development design", which can be interpreted as sense itself (Amasaka, 1996; Mori, 1991, 1993).

When it comes to the application to actual business, however, there are few concrete examples of analysis worthy of introduction to a design development story for the presentation of new vehicles model. This is attributable to the graphics that the design business (hereinafter referred to as "Designing") often ends up paying more importance to the end result, and that proposing good designs in the design planning or development process is based on the conception, which is not closely related with the analysis (Nunogaki et al., 1996).

It is apparent that the higher analysis develops, the more important the conception becomes. But the key point to the matter is how to yield a new conception, and the process of developing conception is important. In the advanced information society where identical conditions and/or data are shared, it hardly occurs to have remarkable difference in the environment. Under the circumstances, the method for proposing a leading conception cannot be the same as before the advancement of the information society (Nagaya et al., 1998; Okazaki et al., 2000).

The author may be prone to use human power for what a machine can handle and may commit an error of misunderstanding that a conception has been gained. The high-quality creative activities can be carried out by determining and studying the fields assigned to the conception adapted to the times (Amasaka, 2004a, 2005, 2018).

Positioning of Design SQC

Thus, it is the objective of this study to find the guideline for establishing a method for scientifically supporting "designing" so as to establish it expressly as a more creative activity from the state of tacit knowledge. It is considered that the very analysis process for establishing it as an activity would be the key to the successful conception making. It is necessary to create a live particular solution that catches the liking of the next generation, not the object teleology that seeks a general solution as the "automotive design" (Nunogaki et al, 1996; Amasaka et al., 1999; Amasaka and Nagaya, 2002; Amasaka, 2004b; Yamaji and Amasaka, 2009).

In this connection, "Science SQC" is applied to the flow of designing to actually enhance the quality of the designer's job. This is defined as "Design SQC" and applied as the activity guideline for "ADS" (Advanced Design by Utilizing "Science SQC" in Toyota) project. In this project, "SQC Technical Methods" will be applied as the "Mountain-climbing for problem-solving", and by analyzing the "bridge portion" between analysis (research) and conception (creation of contour images), "Design SQC" will be applied as the concrete conception support tool (Amasaka et al., 1999; Okazaki et al., 2000; Amasaka, 2004b, 2005, 2018, 2022).

PRELIMINARY RESEARCH

Automobile Exterior Colors

Since the beginning of the 21ˢᵗ century, automobiles have evolved significantly in many ways and "auto-exterior colors" have also diversified. The auto-industry has a incorporated colors that women tend to prefer. With user-friendly functionality as well as "fashionable colors and designs" through the "Profile design x color matching and form x color matching", the automobiles' image is likely to develop significantly (Nagaya et al., 1998; Fujieda et al., 2007; Asami et al., 2010, 2011; Takebuchi et al., 2010, 2012a,b; Muto et al., 2011, 2013: Koizumi et al., 2014; Toyoda et al., 2015a; Amasaka, 2017).

Colors are also entering the next stage of their evolution. Automobile exterior colors are influenced by a range of factors such as the "current needs and sense of values of customers" and "colors of other industries' products, trend information, styles and designs, colored materials, paints and painting techniques". It is necessary to have a sufficient understanding of these factors when undertaking color. This is because more people in their 20s tend to purchase a car for personal use than any other age group, making them the ideal targets in terms of understanding the needs of customers.

Furthermore, women also tend to be more interested in design than men. Specifically, "Passo" made by Toyota Motor Corp. has been chosen as the vehicle for this research as it was created from the perspective of women customers, and the developers took into account feedback from women right from the development stage (Koizumi et al., 2014a).

Automobile Interior Colors

When designing automobiles, the interior design has become just as important as the exterior design. Unlike the exterior, the image sketches for the interior design portray the interior as viewed from the inside. For the interior design, designers do not produce large numbers of image sketches like they do for the exterior (Koizumi et al., 2014a).

This is because a design will not be adopted if, for example, it impedes the operation of functions or the visibility of the meters. Although individuals' color-related interests and preferences vary significantly, there are clear trends from the consumer perspective when viewed statistically. Color trends change through a "10-to-12-year cycle".

These color trends also vary slightly depending on the region such as Japan, US and Europe. In addition to trends, consumers' color preferences also depend on cultural factors and the significance of a color may vary slightly in each market. "Color designers" need a good sense of color, but individual designers also have their own preferences and biases. Designers tend to believe that other users will like the same colors as they do. It is necessary to consider consumers overall when developing colors and materials.

Therefore, designers must select the most appealing colors based on an understanding of the market environment and trends, gained through a statistical interpretation of consumers' preferred colors (Asakura et al., 2011; Aihara and Shinogi, 2012)

The Target Vehicle "Passo"

There are 9 exterior colors available for Toyota "Passo", enabling customers to choose colors according to their preferences as shown in Figure 1. However, only one interior color is available (Asakura et al., 2011; Aihara and Shinogi, 2012; Koizumi et al., 2014a).

Figure 1. Summary of Toyota Passo
(Toyota (2010) and Koizumi et al. (2014a))

Seat in cabin

Front panel n cabin

Size:
Total length 3,640 mm / Width 1645 mm / Overall height 1535 mm
Riding capacity 5 people
Total emission INR-FE 1.329L / 1KR-FE 0.996L
Price 1,000,000 yen – 1,470,000 yen

DEVELOPMENT OF AUTOMOBILE EXTERIOR AND INTERIOR COLOR MACHING MODEL

In this section, the author has created the an "Automobile Exterior and Interior Color Matching Model" (AEDM-EICMM) as shown in Figure 2 (Refer to Asakura et al. (2011), Aihara and Suzuki, (2012), Shinogi et al., (2014) and Koizumi et al., (2014a) in detail).

Specifically, the author gas been adopted the "Design SQC approach" based on the knowledge and the validity of "AEDM-3CM" (Automobile Exterior Design Model with 3 Core Methods) for the "Auto-profile design, form and color optimal matching" in Chapter 13 above.

Concretely, AEDM-EICMM expands the methodology of "CSp-IDCM" (Automobile Intelligence Design Concept Method using Customer Science principle) with "Step 1-Step 4" using "SQC Technical Methods" as shown below.

- In Step 1, "Market Research" contains the (i) "Market research" and (ii) "Hearing survey".
- In Step 2, "Preference Survey" contains the (iii) "Determination of "Chroma" and value", (iv) "Preference survey of women in their 20's", (v) "Classification of preference", and (vi) "Extraction

of preference elements" by using "Cluster Analysis (CA) and Principal Component Analysis (PCA)" as the "Preference survey".

- In Step 3, "Creating Colors" contains the (vii) "Extraction of color combinations", (viii) "Analyzing color combinations", (ix) "Analyzing relationship between fashion and color combinations", and (x) "Creating colors" using "AHP (Analytic Hierarchy Process), Multiple Regression Analysis and Covariance Structure Analysis".
- In Step 4, "Application" contains the (xi) "Actual study" and (xii) "Verification".

Figure 2. Automobile exterior and interior color matching model (AEDM-EICMM)

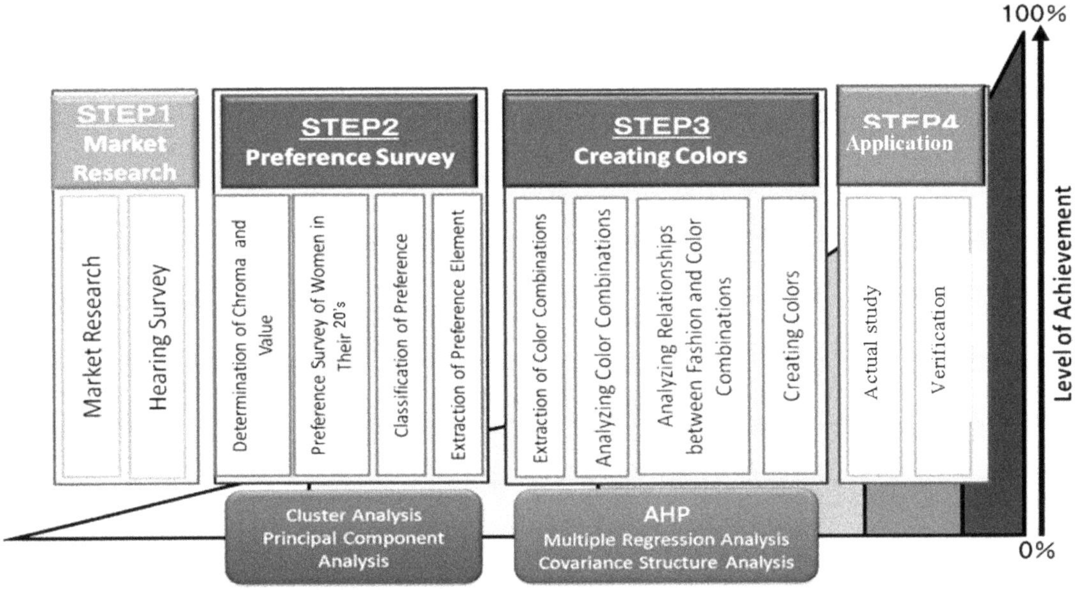

MARKET RESEARCH

From Step 1 in Figure 2, "Questionnaire surveys" were conducted to clarify what kind of cars women in their twenties prefer, and what they require from an automobile. The question items are as shown below (Refer to Harada and Ozawa (2008), Asakura et al. (2011) and Aihara and Suzuki (2012) in detail).

(1) Question items for determining the "automobile-related preferences of women in their 20s": (a) Automobile-related interests / awareness, (b) Preferred automobile "exterior colors and interior colors", and (c) Priorities when choosing a car.

(2) Question items for determining the "lifestyle-related preferences of women in their 20s": (d) Regularly read "fashion magazines", and (e) "Usual fashion / style".

PREFERENCE SURVEY

From Step 2 in Figure 2, based on the results of the questionnaire surveys, CA was used to separate the car-related requirements into 5 groups: Group 1: Design color group, Group 2: Visual importance group, Group 3: Price importance group, Group 4: Function importance group and Group 5: Ride importance group.

Moreover, PCA based on the result of CA was then used to derive the (1) "factor loadings", and was indicated to create (2) "positioning map" as an example of "Categorization by Preferences" as shown in Figure 3 (Koizumi et al., 2014a).

In Figure 3(2), because the research involved varying the "exterior colors, seat colors, and front panel colors", the author decided to target "Group 1: Design color group" which prioritized the "interior design and colors" as well as the "exterior design and colors".

Figure 3. An example of "categorization by preferences" using PCA

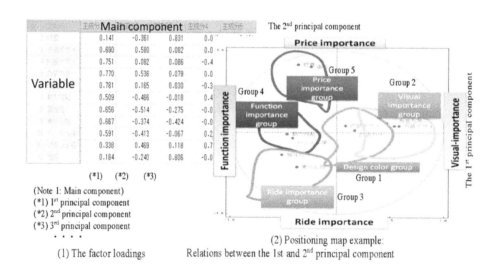

(1) The factor loadings

(2) Positioning map example:
Relations between the 1st and 2nd principal component

For example, the results of the analysis for "Group 1" are as shown below.

Group1: Design Color Group

In Figure 3, the author guessed the characteristic of "Group 1: Design color group" based on the "factor loading" of the "1st, 2nd and 3rd principal component" as follows (Refer to Asakura et al. (2011), Aihara and Suzuki (2012) in detail).

Specifically, "Design color group" is positioned strongly towards the positive direction, indicating that "appearance" is an important requirement for an automobile from 2nd principal component, and is distributed around the center, indicating a tendency to "pay little attention to ride comfort and price" from 3rd principal component. It is also clear that the "interior design and color" are important as well as the "exterior design and color".

CREATING COLORS

In Step 3 in Figure 3, the author shows the "Creating colors" based on the Step 1 and 2.

Producing Colors

"Market research" (Step 1) indicates that "bright and pale colors" (bright colors that are positioned comparatively close to the top of the "color solid" in the diagram) are popular as the "automobile exterior colors for women in 20s", who are the target for this research. Observation in urban areas also found that many women in their 20s tend to "drive bright and pale" colored vehicles (Japan Color Research Institute, 1996).

Thus, 10 light toned colors (five basic color hues and five intermediate color "hues") were selected to use when varying the "exterior colors, seat colors, and front panel colors".

In terms of "PCCS" (Practical Color Co-ordinate System) [*4] numbers, the "colors" had uniform saturation (6S), and luminosity as indicated (Red: 7, yellow red: 7.75, yellow: 8.5, yellow green: 8, green: 7.5, blue green: 6.75, blue: 6, purple blue: 6, violet: 6, red purple: 6.5) (Refer to Asakura et al. (2011) and Aihara and Suzuki (2012) in detail).

Due to the nature of color hues, "brighter colors" tend to look "yellower" while "darker colors" tend to look more "purple blue". Thus, the "luminosity of a particular tone" will differ depending on the "hue". Based on the selected colors, the "exterior colors, seat colors, and front panel colors" were varied using "GIMP" (GNU Image Manipulation Program)[*5] image editing software. The "hue" was changed while keeping the "luminosity and saturation constant".

(Note 2: [*4] https://ja.wikipedia.org/wiki/PCCS and [*5] GIMP: https://ja.wikipedia.org/wiki/GIMP)

A total of 10 colors were created, with 5 basic colors (red, yellow, green, blue, purple) and 5 complementary colors (Yellow red, yellow green, blue green, purple blue, red purple) as shown in Table 1. These 10 colors were used for the "exterior colors, seat colors, and front panel colors" respectively to produce a lineup of (10×10×10=) 1000 color combinations. These color combinations are analyzed in the next subsection.

Color Analysis Using AHP

Then, first, "AHP" was used to identify three important colors each for the "exterior colors, seat colors, and front panel colors" respectively. The "exterior colors, seat colors and front panel colors" thus identified as shown in Table 2. Weighting (importance) in AHP was calculated for each of the 10 resulting "exterior colors".

As a result, in "alternative and weight: ", "yellow: 0.13, red purple: 0.12, and blue green: 0.11" were identified as the important "exterior colors".

The same analysis was then conducted for the "seat colors and front panel colors". "Red, yellow red, and purple blue" were identified as the important "seat colors" while "yellow red, green, and purple" were identified as the important "front panel colors".

Table 1. Evaluation of color parts

Sample	Picture 1	Yellow exterior color	Yellow red exterior color	Yellow green exterior color	...
1	5	5	4	5	...
2	4	6	3	4	...
3	3	4	5	2	...
4	2	6	4	5	...
5	5	7	4	3	...
6	4	5	2	3	...
7	6	5	1	2	...
:	:	:	:	:	...

Table. 2. Result of AHP (exterior color example)

Alternative	Weight	Alternative	Weight
Yellow	0.13	Purple blue	0.10
Yellow red	0.09	Blue green	0.11
Yellow green	0.08	Red	0.10
Purple	0.10	Red purple	0.12
Blue	0.10	Green	0.10

Analyzing Color Combination With Color Palette and Triangle

Second, the results of the color analysis identified "3 colors each for the exterior colors, seat colors, and front panel colors". These colors were combined to produce a "total of 27 combinations", which were then evaluated as shown in Table. 3.

Averages were calculated for each of the evaluations. The average for the "exterior color (red purple), seat color (yellow red), and front panel color (purple) (hereafter, Red purple + yellow red + purple)" was the 1st highest average at 4.705.

The 2nd highest average was 4.647 for the combination of "blue green + purple blue + green".

These combinations look very different when seen in photographs. However, when represented on the "color pallet and triangle", they appear as shown in Figure 4.

The 3 colors of "exterior color, seat color, and front panel color" are plotted as the points on the "color pallet", and the points are joined with lines to create a "triangle".

Seen in the diagram, the balance (triangle) of the "exterior color, seat color, and front panel color" has a similar shape. Looking at the triangle separately, the difference in "hue between the "exterior color and seat color" is 5, while the difference in "hue between the exterior color and front panel color" is 2 to 3, and the difference in "hue between the front panel color and seat color" is 7 to 8. Thus, they are clearly related.

Furthermore, "4 other triangles" were extracted from the "color pallet" (① Exterior: red purple, Seat: yellow red, and front panel: purple, ② Exterior: blue green, Seat: purple blue, and front panel: green, ③ Exterior: yellow green, Seat: blue green, and front panel: yellow, ④ Exterior: yellow red, Seat: yellow green, and front panel: red).

Then, "triangles ① and ②"are the original triangles, while "triangles ③ and ④" are newly-found triangles. Of the 24 colors on the color pallet, this research uses "5 basic colors and 5 intermediate colors", making a total of 10 colors. These 10 colors are represented on the color pallet by the numbers 2, 5, 8, 10, 12, 15, 18, 20, 22 and 24.

Table 3. Combination of "exterior colors, seat colors and front panel colors"

Exterior color Yellow (Y)	Exterior color Blue green (BG)	Exterior color Red purple (RP)
Y/ Yellow red / Green	BG/ Yellow red / Green	RP/ Yellow red / Green
Y/ Yellow red / Yellow red	BG/ Yellow red / Yellow red	RP/ Yellow red / Yellow red
Y/ Yellow red / Purple	BG/ Yellow red / Purple	RP/ Yellow red / Purple
Y/ Purple bule / Green	BG/ Purple bule / Green	RP/ Purple bule / Green
Y/ Purple bule / Yellow red	BG/ Purple bule / Yellow red	RP/ Purple bule / Yellow red
Y/ Purple bule / Purple	BG/ Purple bule / Purple	RP/ Purple bule / Purple
Y/ Red / Green	BG/ Red / Green	RP/ Red / Green
Y/ Red / Yellow red	BG/ Red / Yellow red	RP/ Red / Yellow red
Y/ Red / Purple	BG/ Red / Purple	RP/ Red / Purple

Note 3: (Ex.).   Exterior color (Yellow) + Seat color (Yellow red) + Front panel color (Green)

Figure 4. Color expressed in the color palette and triangle that is extracted

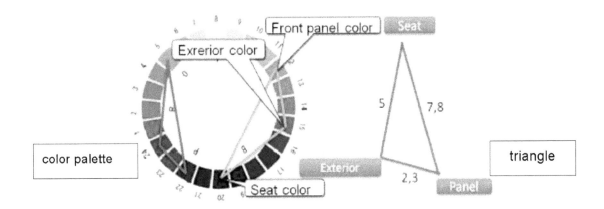

Identifying Similar Color Combinations Using Multiple Regression Analysis

Third, "Multiple regression analysis" was used to analyze the 27 color combinations identified through color analysis. For this analysis of the color combinations, the combination of colors with the highest average "Red purple + yellow red + purple" was used as the objective variables while the other combinations were used as explanatory variables.

When the variables were freely varied using the stepwise method it was found that, with respect to the objective variables "Red purple, yellow red and purple" described in Table 4, the "partial regression coefficient" for the combination of "Blue green + red + yellow red " was comparatively high at 0.617. Similarly, the partial regression coefficient for the combination of "Red purple + red + purple " was also comparatively high at 0.522.

The combination of "Blue green + red + yellow red" is represented on a "color pallet and triangle" (PCCS hue rings), as indicated in Table 4.

The difference in hue between the "exterior color and seat color" is 11, while the difference in hue between the "exterior color and front panel color" is 10 and the difference in hue between the "front panel color and seat color" is 3.

Analyzing Relationships Between Fashion and Color Combinations

Fourth, to investigate the relationships between the "9 color combinations" as shown in Table 4, and the "8 types of fashion" defined as shown in Figure 5, "Multiple regression analysis" was conducted using the "27 color combinations" identified through the "color analysis". In Figure 5(1), the author gives an example of "collage" representing "8 types of fashion". Specifically, for example, it is clear from the diagram that "feminine" styles directly contrast with "masculine" styles, and "country basis" styles directly contrast with "urban (city)" styles.

Table 4. Color combination of nine were extracted from the analysis

Exterior	Seat	Front panel	Color pallet and triangle (PCCS hue rings)
Red purple	Yellow red	Purple	
Blue green	Purple blue	Green	
Yellow Green	Blue Green	Yellow	
Yellow red	Yellow green	Red	
Blue green	Red	Yellow red	
Blue	Yellow red	Yellow	
Red purple	Red	Purple	
Purple blue	Purple	Blue	
Yellow green	Green	Yellow	

Figure 5. Relationship diagram of eight fashion and fashion collage (ex. romantic, elegance, sophisticate, modern, mannish, active, country, and ethnic)

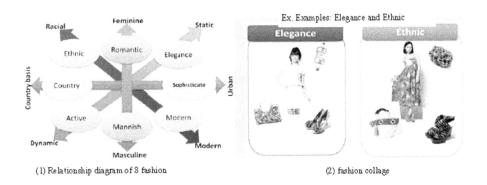

(1) Relationship diagram of 8 fashion   (2) fashion collage

In Figure 5(2), the "fashion collage" was constructed from the results of "questionnaires targeting women in their 20's".

The surveys investigated the kind of "magazines" the respondents read and the kind of brands they like, and "collages" were created for each.

"Pass Diagram" by "Multiple regression analysis" as shown in Figure 6 was repeated "nine times" using "each of the color combinations" as the objective variables and "eight types of fashion collage" as the explanatory variables (Table 4 and Figure 5(1)).

With respect to the "Blue-green, Blue-violet, and Green" objective variables, the "partial regression coefficient for "romantic fashion" was comparatively high at 0.723 while the other explanatory variables showed minus values described in Figure 6.

In Figure 6, with respect to the "Blue green + red + yellow red", objective variables, the "partial regression coefficient" was 0.710, and "active" and "country" showed plus values. Thus, when the variables were added, each of "partial regression coefficient values" became lower. However, if only the country explanatory variable was used, the partial regression coefficient was high at 0.718. This indicates that there is a relationship between "country" and "combination of "Blue green + purple blue + green"".

Results were also calculated for the other color combinations and summarized in a path diagram as shown in Figure 6. In Figure 6, the author shows that there is a relationship between "Blue green + red + yellow red" and two types of fashion "Ethnic and Country".

Similarly, there is a relationship between "Red purple + red + purple" and "elegance and sophisticated", as well as between "Blue green +purple + blue", "Yellow green + green + yellow" and "modern and mannish. However, the relationship between the two is unclear.

Then, "Covariance structure analysis of relationship of the fashion (1)" was conducted in order to clarify this relationship as shown in Figure 7.

In Figure 7, for example, the regression coefficient in the "elegance and sophisticated" was 0.76 which clarified the relationship. In the same way as the "fashion (2)", the analysis revealed that the regression coefficient for "sophisticated and elegant" was 0.81, indicating that there is a relationship as shown in Figure 8.

Figure 6. Pass diagram

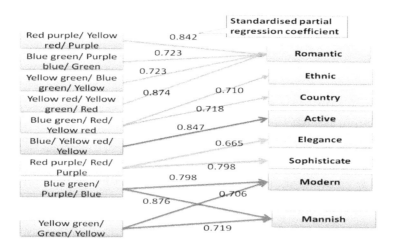

For example, in the same way as the Fashion (1), the regression coefficients for each of the "modern and mannish" and "Ethnic and country" were high at 0.88 and 0.83 respectively (Figure 7). In the same way as the "fashion (2)", the regression coefficients for "mannish and modern" and "country and ethnic" were high at 0.84 and 0.82 respectively (Figure 8). The other regression coefficients have minus (-) values, or low values. This indicates that there is a weak relationship for fashion styles other than "sophisticated and elegance", "mannish and modern" and "country and ethnic".

When the "fashion collages" were reviewed based on these results, it was found that there were definite similarities between each of these pairs. These fashion styles are also positioned similarly described in Figure 5(1) and Figure 5(2) examples.

In the "sophisticated and elegance collage", there is an overall white theme. In the "manish and modern collage", similar kinds of "beige or blacks" are used for a relaxing impression. In the "ethnic and country collage", even similar words are used.

Based on the above results, "fashion collages" were reconstructed for the pairs "sophisticated and elegant", "mannish and modern", and "ethnic and country" as shown in Figure 9 (Refer to Asakura et al. (2011) and Aihara and Suzuki (2012) in detail).

The "sophisticated and elegance" collage gives a mature impression overall. The "ethnic and country collage" incorporates both "ethnic style and country style elements" (Figure 9).

The "mannish and modern collage" has a "beige and black theme", giving a relaxing impression. Conducting analysis in this way has enabled previously unnoticed relationships to be discovered between those fashion styles described in Figure 9.

Discussion of Color and Fashion

When reviewing the color combinations and fashion styles for which relationships were discovered during analysis, it becomes evident that they are connected to a certain degree by using "PCCS hue rings" (color pallet and triangle) above.

Figure 7. Covariance structure analysis of the relationship of the fashion (1)

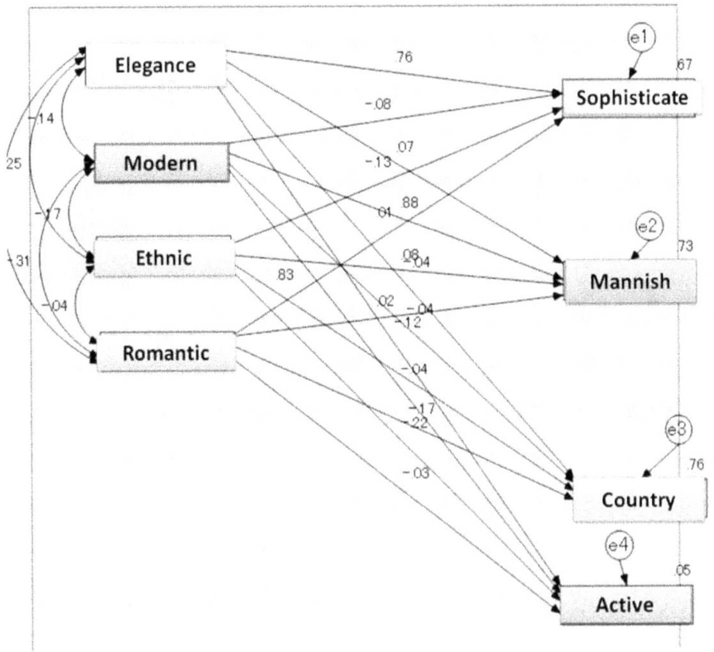

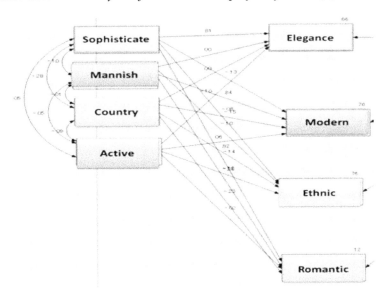

Figure 8. Covariance structure analysis of the relationship of the fashion (2)

Romantic Example

Overall, this is an intermediate color combination with only slight differences between the hues, giving rise to indistinct color combinations as shown in Figure 10.

Figure 9. New fashion collage

Women who prefer "romantic fashion" also tend to be interested in slight color differences. Indistinct color combinations are often used, and it is clear that this triangle also applies to fashion.

Figure 10. Triangles of romantic

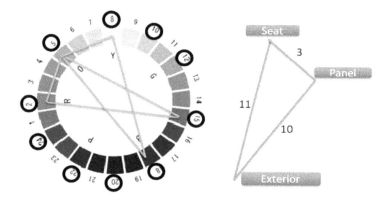

Ethnic x Country Example

Both the seat colors and the front panel colors are complementary colors of the exterior colors, making this a contrasting color combination as shown in Figure 11.

As an energetic and stimulating color combination, the stimulating colors relate to the "country and ethnic" fashion styles while the energetic colors relate to "active".

Figure 11. Triangles of ethnic × country

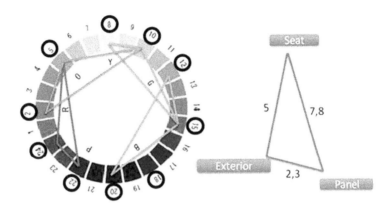

Modern x Mannish and Sophisticate

Overall, there are only slight differences between the hues, making this a triangle of analogous colors as shown in Figure 12.

The overall unity of the hues gives a relaxing impression. The related fashion styles also have a mature image and a relaxing impression. The color combinations therefore have a similar relationship to the styles.

Figure 12. Triangles of modern × mannish and sophisticate

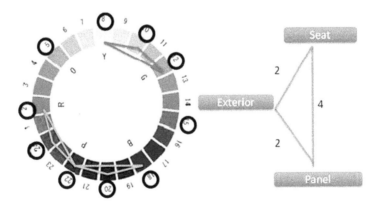

APPLICATION OF AEDM-EICMM

In Step 4 in Figure 3, the author describes the "Actual study and verification through the application of "AEDM-EICMM" to Toyota vehicle "Passo" and other cases based on the knowledge of Step 1, 2 and 3 above, and shows by using the following references (Refer to Asakura et al. (2011), Aihara and Suzuki (2012) and Shinogi et al. (2012) in detail).

Actual Study for AEDM-EICMM Application

Determining the "Targeted parts for exterior and interior color matching"

To identify actual parts of automobiles observed by consumers, and changes in "brainwaves" during observation, data for analysis was collected from "12 females in their 20's", using an "eye camera", for the analysis of "human sight", and an "electroencephalograph", for the "measurement of brainwaves".

Firstly, an "eye camera" was used to analyze the time that "eye sight" resided on a particular part, and the "parts of an automobile" that were most focused on were identified as shown in Figure 13.

"Colors" in the graph closer to "red" indicate parts that were looked at more closely. The "body, seats, steering wheel, and front panel" were identified as the parts of an automobile that were most focused on.

Figure 13. The result of analyzing the time that "eye sight" resided on a particular part in eye camera: Exterior and interior of Toyota vehicle "Passo"

Subsequently, brainwaves were measured to clarify brain reactions to the identified parts (body, seats, steering wheel, and front panel). The "brainwaves" were measured whenever the subject looked at each part for 5 seconds. The measurement results are indicated in Figure 14.

When the subjects observed the "body, seats or front panel", "brain patterns" substantially changed in many cases. "Brainwaves" did not show substantial changes only when the subjects looked at the steering wheel. This suggests that, although the subjects frequently looked at the steering wheel, they potentially paid little attention to it. Therefore, the "body, seats, and front panel" were selected as the "parts targeted for color matching".

Finally, the priority of each targeted part was identified through "counting the number of times the subjects" looked at the targeted part area using the data collected with the "eye camera". The priority was the largest for the ① "body", followed by the ② "seats" and then the ③"front panel" as shown in Figure 15.

Consumer Preference Survey

A consumer preference survey was conducted to stratify "consumer preferences", and to clarify what consumers wanted from "automobile exterior and interior colors". A questionnaire was used as a means of the survey. The survey was conducted over 15 females in their 20's, asking 1) what automobile colors they liked, and 2) what magazines they were reading.

Figure 14. The result of brainwaves

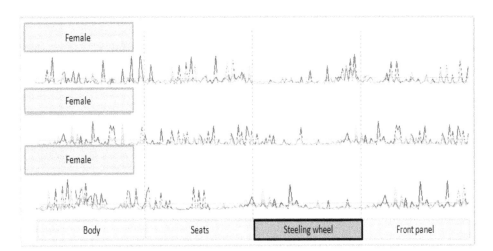

Figure 15. The priority of the parts in color matching

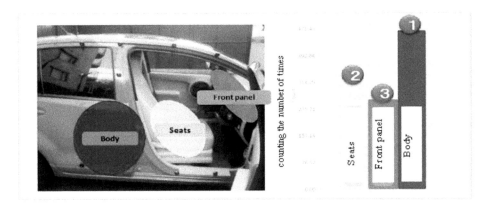

In Question 1), "20 keywords by the perception" were provided the "cute, cool, unique, refined, orthodox, chic, novel, sophisticated, futuristic, girly, classy, simple, traditional, fresh, fashionable, urban, conspicuous, clean, gentle, sporty", to represent images that the subjects wanted from "automobile exterior and interior colors" as listed in Figure 16. Respondents were asked to answer rating each keyword on a scale of 7.

In Question 2), respondents were supposed to freely describe their answers.

Classification of Preferences and Identification of Perceptional Keywords

"Cluster analysis" was conducted to classify the subjects by their preferences. Responses to Question 1) were used as data for analysis. The analysis method applied sample classification, standardization, "Wards method, and Squared Euclidean distance" as shown in Figure 17. In Figure 17(a), the author indicates the dendrogram for this analysis, suggesting that females in their 20s could be classified into

Groups A through D. Freely described responses to Question 2) indicate that subjects in the same group prefer magazines for similar fashion styles.

Furthermore, based on the "results of fashion survey" by using "various magazines", "Group A was defined as Girlish", "B as Elegant", "C as Boyish", and "D as "Thrifty" from "classify fashion" indicated in Figure 17(b).

Figure 16. The perceptional keywords

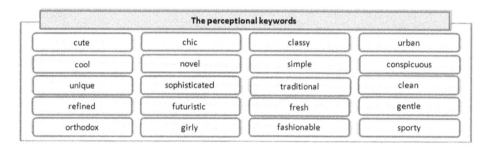

Figure 17. Classifying preferences and fashion "Group A- D" using cluster analysis

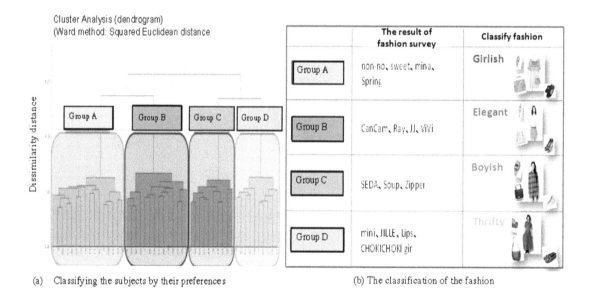

Subsequently, PCA (Principal Component Analysis) was conducted using data collected in Question 1), in order to clarify relationships between "consumer preferences and perceptional keywords".

For the purpose of verifying the "significance of each principal component" by using "factor loading", which indicates correlation between the component and the original variant, was identified as shown in Figure 18.

Then a "factor loading scatter diagram" was formulated for "Principal component 1 (horizontal axis) and Principal component 2 (vertical axis)".

According to "Principal component 1, "conspicuous, novel, unique and similar perceptional keywords" indicated "strong positive values, while "chic, refined, clean", etc. showed negative values.

According to "Principal component 2", "sporty and cool" had positive values, in contrast to negative values for "cute and girly".

Figure 18. The principal component analysis with factor loading scatter diagram

Principal Component analysis

Cluster analysis

The "principal component score" scatter diagram

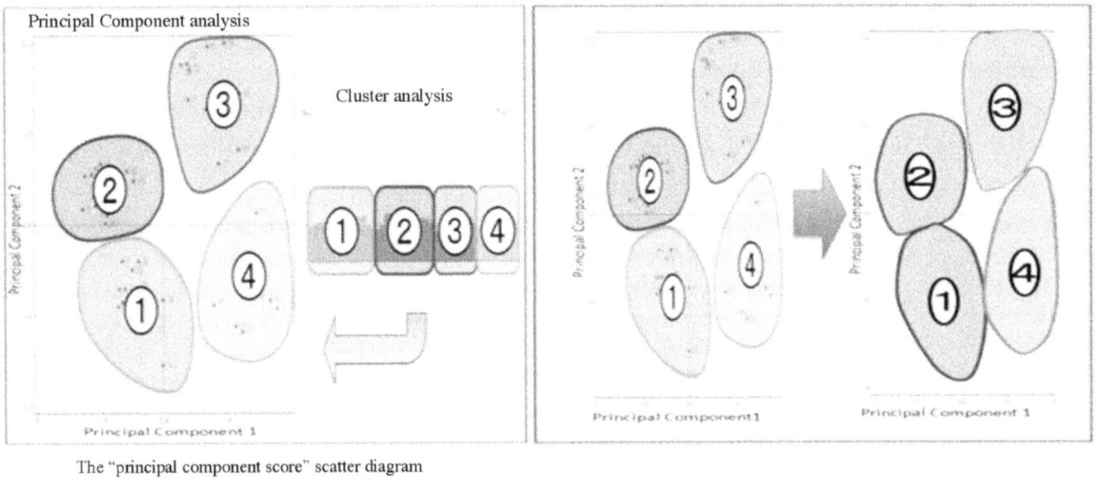

Based on the "Cluster Analysis" in Figure 17

These results can be interrupted as suggesting that, in the context of "automobile exterior and interior colors", Principal component 1 is an axis (horizontal axis) for preference to flamboyant or chic, while preference to cool or cute is indicated by an axis (vertical axis) of Principal component 2.

Thus, relationships between perceptional keywords and significances of principal components could be clarified.

Finally, the results of cluster analysis were compared with the results of PCA, in order to clarify the characteristics of each group, and identify perceptional keywords by preference.

Subjects were classified according to a scatter diagram formulated based on Principal Component 1(horizontal axis) and Principal Component 2 (vertical axis), and were compared to the groups defined in cluster analysis as described in Figure 18.

The 4 groups (A, B, C and D) based on cluster analysis could also be delineated on the "principal component "score" scatter diagram (see to Group ①,②,③ and ④)"based on the "similar analysis and it's knowledge of "factor" score scatter diagram, perceptional keywords for each group could be picked up from its distribution area in the diagram.

The identified keywords for ① "Girlish" were "cute, girly, gentle and urban". For ② "Elegant", the keywords were "refined", "classy", "chic" and "clean". For ③ "Boyish", the keywords were "cool, sporty, simple and orthodox"; and for ④ "Thrifty", the keyword were "unique, novel, conspicuous, and fashionable".

Color Matching

Subsequently, preferred "automobile exterior and interior colors" were identified for each of the 4 groups "A, B, C and D" as classified in the above subsection ("Girlish, Elegant, Boyish and Thrifty"), and "color matching for exterior and interior" was undertaken based on the AHP.

1st, to identify preferred automobile colors, subjects were shown "PCCS hue rings" as shown in Figure 19. The alternative "Subjects (No. 1 - 12)" were asked to answer "1" if they liked the hue, and "0" if they disliked it indicated in Figure 19(a). Subsequently, a "PCSS hue ring" liked by each group was identified using the "Quantification Method Type III".

Figure 19(b) indicates the "Category score" scatter diagram using "Scores of respective variants". To the "vertical axis", the positive area was defined as "high saturation", because tones with "high saturation" were concentrated in this area. According to the "horizontal axis", the positive area was defined as "low brightness", because tones with "low brightness" were concentrated in this area. Similarly, the negative area was defined as "high brightness".

Then, the author shows the "Color of the scope of matching for 4 groups (Girlish, Elegant, Boyish and Thrifty)" as shown in Figure 20. Referring to the grouping based on the "consumer preference survey" in previous subsection, the "tones" (composite of saturation and brightness) preferred by each group could also be identified as indicated in Figure 20(a) as follows;

The ① "Girlish" group preferred "light grayish, light and soft tones", ② "Elegant" group liked "dull, deep and grayish tones", ③ "Boyish" group liked "vivid, soft and dull tones", and ④ "Thrift" group preferred "vivid, bright and strong colors". These results also seem reasonable from the viewpoint of their respective preferred "fashion styles".

Figure 19. Color matching for exterior and interior for 4 groups (girlish, elegant, boyish, and thrifty) using AHP and quantification method type III

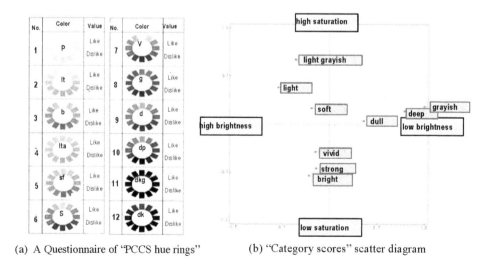

(a) A Questionnaire of "PCCS hue rings"   (b) "Category scores" scatter diagram

Figure 20. The color of the scope of matching for girlish, elegant, boyish, and thrifty

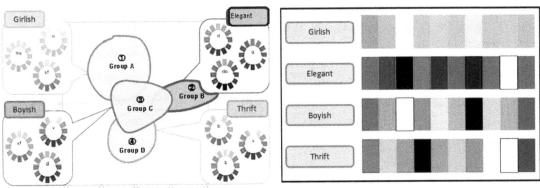

(a) The classification of "PCCS hue rings" by each group based on the Figure 19

(b) The color of the scope of matching: For 4 groups "*Girlish, Elegant, Boyish and Thrifty*"

Because the ① "Girlish" group likes "soft and cute clothes", ② "Elegant" group likes "chic and classy clothes", ③ "Boyish" group prefers "sporty and / or standard items", and ④ Thrift Group Loves Unique Items, The Classification Above Can Be Deemed Appropriate

Subsequently, "a questionnaire survey by the scale of seven" was conducted concerning colors contained in the identified "PCCS hue rings", and colors with high mean values were Included in the Scope of Matching. Figure 20 (b) Indicates the Selected Colors

Finally, AHP analysis for matching was conducted for the "colors selected for each group", in the order of priority of parts (: "exterior (body) and interior (seats, and front panel) in this order"). The aim for AHP was the matching of automobile colors. The "perceptional keywords for each group", as identified in "Consumer preference survey" above, were used as the relevant parameters. Colors selected for each group were used as alternatives.

The following analysis process is described taking the ① "Girlish" group as an example. 1st, the "matching for body and seat color" was determined as shown in Figure 21 below.

As Figure 21(a) indicates, "Alternative subject No.2" was selected as the "body color", because its significance was the highest. In the subsequent process, the "selected body color" was set while the "seat color" was determined for the ① "Girlish" group.

As Figure 21(b) indicates, "Alternative subject No.5" was selected as the "seat color", because its significance was the highest for the ① "Girlish" group.

As Figure 21(c) indicates, "Alternative subject No.2" was selected as the "front panel color", because it indicated the largest value. Through these steps, "automobile design matching" has been determined for the ① "Girlish" group.

Finally, an example of selected combination of "body color, seat color and front panel" was determined as shown in Figure 22. In Figure 22, the "each of the color matching" for 4 groups "① "Girlish", ② Elegant, ③ Boyish and ④ Thrift" was analyzed and determined.

Figure 21. Girlish group for body, seat, and front panel color matching

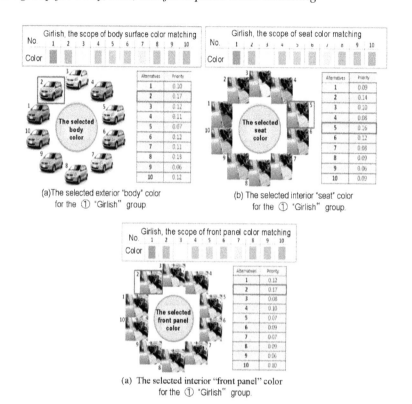

(a)The selected exterior "body" color
for the ① "Girlish" group

(b) The selected interior "seat" color
for the ① "Girlish" group.

(a) The selected interior "front panel" color
for the ① "Girlish" group.

Figure 22. An example of exterior (body) and interior (seat and front panel) color matching

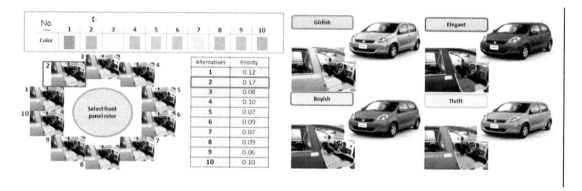

Verification

Verification was undertaken to check that the each of the 4 groups "color matching" in "auto-exterior (body) and interior (seat and front panel)" as identified through analysis, and its influence on consumer perception, would be practically useful.

The "color matching" for verification was compared to the existing color matching of the Passo. A questionnaire survey by the scale of seven was conducted to identify the color matching with the highest mean value among respondents.

The following process is described by using the "Girlish" group as an example. Respondents were asked to evaluate the existing car color line and the "color matching" developed through above "AEDM-EICMM" by ranking them on a scale of seven, to see how effectively the intended perceptions were attained.

As an example of "Girlish", Figure 23 indicated the "color matching" based on "AEDM-EICMM", and obtained high evaluations as the "cute, girly, gentle and urban" (using the questionnaire survey).

In the same way, colors for the other groups "Elegant, Boyish and Thrift" were also verified in the same manner, all leading to high evaluations.

Thus, it was verified that the intended perceptions were attained through color matching based on "AEDM-EICMM".

This model and the resulting color matching, as created by the authors, were acknowledged by the design institute of automobile manufacturer "Toyota" to a certain extent.

Figure 23. The result of verification

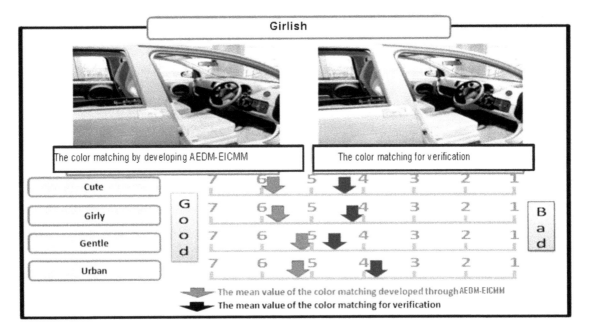

Expanding Similar Cases

With its effectiveness verified, the author was able to apply the new developments of Design SQC as follows;

(i) Research on the Automobile Package Design Concept Support Methods "CS-APDM": Customer Science approach to achieve CS for vehicle exteriors and package design (Okabe et al., 2007), (ii) A model for design auto-instrumentation to appeal to young male customers (Yazaki et al., 2012), (iii) A statistical and scientific approach to deriving an attractive exterior vehicle design concept for indifferent customers (Kobayashi et a., 2016).

Furthermore, as the other applications, (iv) Customer behavior analysis using motion pictures: Research on Attractive Flier Design Method (Koyama et al., 2010), (v) A Bicycle Design Model based on young women's fashion combined with CAD and statistics (Koizumi et al., 2013) and (vi) Creating Automobile Pamphlet Design Methods: Utilizing both biometric testing and statistical science (Koizumi et al., 2014b).

CONCLUSION

In this study, the author has created the "AEDM-EICMM" (Automobile Exterior and Interior Color Matching Model" using Design SQC. Specifically, then, the author has developed the Specifically, then, the author has developed the AEDM-EICMM for the "Auto-profile design, form and color optimal matching" to Toyota vehicle "Passo" using "CSp-IDCM" (Intelligence Design Concept Method using Customer Science principle. The effectiveness of the AEDM-EICMM was confirmed by conducting surveys to determine whether the "color combinations created were suitable for women in their 20s, and similar applications in Toyota and others.

REFERENCES

Aihara, S., & Shinogi, T. (2012). An *establishment of Automobile Exterior Color and Interior Color Matching Model: The 20th woman example*. [Thesis, School of Science and Engineering, Aoyama Gakuin University]. .

Amasaka, K. (1972). Measurement of "gear's noise control for inspection of differential gear assembly, *The 2nd Sensory Symposium in JUSE, Tokyo,* 5-12.

Amasaka, K. (1976). Sensory characteristic of "differential gear noise occurrence. *Quality Control (an extra edition), 26*(11), 5-12.

Amasaka, K. (1983). *The mechanization of "Kan and Kotsu in experience" work: Distortion modification of "Rear axle shaft."* The 26th Technical Conference in JSQC (Journal of the Japanese Society for Quality Control), Nagoya.

Amasaka, K. (1996). *A demonstrative study of a new SQC concept and procedure in the manufacturing industry: Establishment of a New Technical Method for conducting Scientific SQC.* Proceedings of the 2nd Australia-Japan Workshop on Stochastic Models in Engineering, Technology and Management, The University Queensland, Australia.

Amasaka, K. (2004a). Customer Science: Studying consumer values. *Japan Journal of Behavior Metrics Society,* 196-199.

Amasaka, K. (2004b). *Science SQC, new quality control principle—The quality strategy of Toyota.* Springer-Verlag Tokyo. doi:10.1007/978-4-431-53969-8

Amasaka, K. (2005). Constructing a Customer Science Application System "CS-CIANS": Development of a Global Strategic Vehicle "Lexus" utilizing New JIT. *WSEAS Transactions on Business and Economics*, *2*(3), 135–142.

Amasaka, K. (2015). New JIT, new management technology principle. Taylor & Francis Group, CRC Press, Boca Raton, London, New York.

Amasaka, K. (2017). *Studies on Automobile Exterior Design Model for customer value creation utilizing Customer Science Principle*. The 7th International Symposium on Operations Management Strategy, Tokyo Metropolitan University, Tokyo.

Amasaka, K. (2018). Automobile Exterior Design Model: Framework development and supporting case studies. *Journal of Japanese Operations Management and Strategy*, *8*(1), 67–89.

Amasaka, K. (2022). A New Automobile Product Development Design Model: Using a Dual Corporate Engineering Strategy. *Journal of Economics and Technology Research*, *4*(1), 1–22. doi:10.22158/jetr.v4n1p1

Amasaka, K., & Nagasawa, S. (2000). Automobile's Kansei Engineering. Fundamentals and applications of Sensory Evaluation: For Kansei Engineering in automobile, Japanese Standards Association.

Amasaka, K., & Nagaya, A. (2002). New Kansei Engineering in Automobile: Psychographics of "Lexus" design profile. Product development in Kansei Engineering, Nihon-Shuppan Service, 55-72.

Amasaka, K., Nagaya, A., & Shibata, W. (1999). Studies on Design SQC with the application of Science SQC - Improving of Business Process Method for automotive profile design. *Japanese Journal of Sensory Evaluations*, *3*(1), 21–29.

Asakura, S., Kanke, R., & Tobimatsu, K. (2011). A study on Automobile Exterior Color and Interior Color Matching Model: The 20th woman example (senior thesis), *School of Science and Engineering, Aoyama Gakuin University, in Amasaka New JIT Laboratory, Sagamihara, Kanagawa, Japan.* .

Asami, H., Ando, T., Yamaji, M., & Amasaka, K. (2010). A study on Automobile Form Design Support Method "AFD-SM". *Journal of Business & Economics Research*, *8*(11), 13–19. doi:10.19030/jber.v8i11.44

Asami, H., Owada, H., Murata, Y., Takebuchi, S., & Amasaka, K. (2011). The A-VEDAM for approaching vehicle exterior design. *Journal of Business Case Studies*, *7*(5), 1–8. doi:10.19030/jbcs.v7i5.5598

Fujieda, S., Masuda, Y., & Nakahata, A. (2007). Development of automotive color designing process. *Journal of Society of Automotive Engineers of Japan*, *61*(6), 79–84.

Harada, O., & Ozawa, T. (2008). Color trends and popularity in Auto China 2008 and Chinese urban area. *Research of Paints*, (151), 58–63.

Kobayashi, T., Yoshida, R., Amasaka, K., & Ouchi, N. (2016). A statistical and scientific approach to deriving an attractive exterior vehicle design concept for indifferent customers. *Journals International Organization of Scientific Research*, *18*(12), 74–79.

Koizumi, K., Kanke, R., & Amasaka, K. (2014a). Research on Automobile Exterior Color and Interior Color Matching. *International Journal of Engineering Research and Applications*, *4*(8), 45–53.

Koizumi, K., Kawahara, S., Kizu, Y., & Amasaka, K. (2013). A Bicycle Design Model based on young women's fashion combined with CAD and statistics. *China-USA Business Review, 12*(4), 266–277.

Koizumi, K., Muto, M., & Amasaka, K. (2014b). Creating Automobile Pamphlet Design Methods: Utilizing both biometric testing and statistical science. *Journal of Management, 6*(1), 81–94.

Koyama, H., Okajima, R., Todokoro, T., Yamaji, M., & Amasaka, K. (2010). Customer behavior analysis using motion pictures: Research on Attractive Flier Design Method, *China-USA. Business Review (Federal Reserve Bank of Philadelphia), 9*(10), 58–66.

Mori, N. (1991). *Engineering of design: Software System of design engineering.* Asakura-Shoten.

Mori, N. (1996). *Designing with Left Brain: The research of the scientific methodology.* Kaibun-Dou.

Muto, M., Miyake, R., & Amasaka, K. (2011). Constructing an Automobile Body Color Development Approach Model. *Journal of Management Science, 2*(2), 175–183.

Muto, M., Takebuchi, S., & Amasaka, K. (2013). Creating a New Automotive Exterior Design Approach Model: The relationship between form and body color qualities. *Journal of Business Case Studies, 9*(5), 367–374. doi:10.19030/jbcs.v9i5.8061

Nagaya, A., Matsubara, K., & Amasaka, K. (1998). *A Study on the customer tastes of automobile profile design (Special Lecture).* Union of Japanese Scientists and Engineers, The 28th Sensory Evaluation Sympojium, Tokyo .

Nunogaki, N., Shibata, K., Nagaya, A., Ohashi, T., & Amasaka, K. (1996). *A study on the "customers' preference" of "Automobile profile design": Development of Design SQC which is useful for Design business process.* The 26th Annual Conference of JSQC, Gifu, Japan.

Okabe, Y., Yamaji, M., & Amasaka, K. (2007). Research on the Automobile Package Design Concept Support Methods "CS-APDM": Customer Science approach to achieve CS for vehicle exteriors and package design [in Japanese]. *Journal of Japan Society for Production Management, 13*(2), 51–56.

Okazaki, R., Suzuki, M., & Amasaka, K. (2000). *Study on the sense of values by age using Design SQC.* The Japan Society for Production Management, The 11th Annual Technical Conference, Okayama University, Okayama, Japan.

Shimizu, H., & Amasaka, K. (1975). Quality assurance of low-speed steering force. *Quality Control* (an extra edition), in *JUSE* (*Union of Japanese Scientists and Engineers), 26* (11), 42-46.

Shinogi, T., Aihara, S., & Amasak, K. (2014). Constructing an Automobile Color Matching Model (ACMM). *IOSR Journal of Business and Management, 16*(7), 7–14. doi:10.9790/487X-16730714

Shinohara, A., Yoshio, O., & Sakamoto, H. (1996). *Invitation to Kansei Engineering.* Morikita-Shuppan.

Takebuchi, S., Asami, H., & Amasaka, K. (2012b). An Automobile Exterior Design Approach Model linking form and color. *China-USA Business Review, 11*(8), 1113–1123.

Takebuchi, S., Asami, H., Nakamura, T., & Amasaka, K. (2010). *Creation of Automobile Exterior Color Design Approach Model "A-ACAM".* The 40th International Conference on Computers & Industrial Engineering, Awaji Island, Japan.

Takebuchi, S., Nakamura, T., Asami, H., & Amasaka, K. (2012a). The Automobile Exterior Color Design Approach Model. *Journal of Japan Industrial Management Association, 62,* 303–310.

Toyoda, S., Koizumi, K., & Amasaka, K. (2015b). Creating a Bicycle Design Approach Model based on fashion styles, *IOSR (International Organization of Scientific Research). Journal of Computational Engineering, 17*(3), 1–8.

Toyoda, S., Nishio, Y., & Amasaka, K. (2015a). Creating a Vehicle Proportion, Form & Color Matching Model. *Journals International Organization of Scientific Research, 17*(3), 9–16.

Yamaji, M., & Amasaka, K. (2009). Intelligence Design Concept Method using Customer Science. *The Open Industrial and Manufacturing Engineering Journal, 2*(1), 21–25. doi:10.2174/1874152500902010021

Yazaki, K., Tanitsu, H., Hayashi, H., & Amasaka, K. (2012). A model for design auto- instrumentation to appeal to young male customers. *Journal of Business Case Studies, 8*(4), 417–426. doi:10.19030/jbcs.v8i4.7035

Chapter 14
Highly–Accurate CAE Analysis Model for Bolt–Nut Loosening Solution

ABSTRACT

In this chapter, the author has established the highly-accurate CAE analysis model for bolt-nut loosening solution as a developing new JIT strategy. To enable the high quality assurance - based research aimed at the innovation of product design processes, it is necessary to initiate the transition from the conventional prototype testing method to predictive evaluation method by the combination of various experiments and CAE. This model's validity is verified with application to study on loosening mechanism of bolt-nut tightening as the worldwide auto-manufacturers. bottleneck technology. Specifically, moreover, the author has created an excellent "New design nut" for the prevention of bolt-nut looseness in excellent cost performance as the viewpoint of bottleneck solution in the vehicle market claim "looseness of bolt-nut tightening".

CREATION OF THE HIGHLY-ACCURATE CAE ANALYSIS MODEL FOR BOLT-NUT LOOSENING SOLUTION

The number of bolts-nuts exceeds 3000 in an automobile. For many years, about 15% of automobiles' quality recall is the looseness of bolt-nut tightening around powertrain system (MLIT, 2005).

In this study, therefore, the author has established the "Highly-Accurate CAE Analysis Model" to explain undiscovered technological mechanisms using "SSTTM" (Strategic Stratified Task Team Model) activity with "Science SQC" based on the "Advanced TDS" in New JIT strategy (Amasaka, 2004, 2007, 2008, 2010, 2012, 2017, 2019a,b,c; Amasaka et al., 2012, 2014) (See to Chapter 2, 3, 7, 8 and 11).

Specifically, this created CAE model is no gap between its results and the results of actual machine tests, allows the realization of intelligent product development design described in "Step 1- Step 5" using "Mountain-climbing for Problem-solving Methods" as shown in Figure 1 (Ueno et al., 2009; Takahashi et al., 2010; Yamaji and Amasaka, 2011; Onodera and Amasaka, 2012; Kozaki et al., 2012; Hashimoto et al., 2014; Hashimoto, 2015; Nomura et al., 2015, 2016; Shimura and Sakurai, 2015; Amasaka, Ed., 2007, 2012; Amasaka, 2015a,b, 2022a,b, 2023a,b) as follows;

DOI: 10.4018/978-1-6684-8301-5.ch014

Figure 1. Highly-accurate CAE analysis model

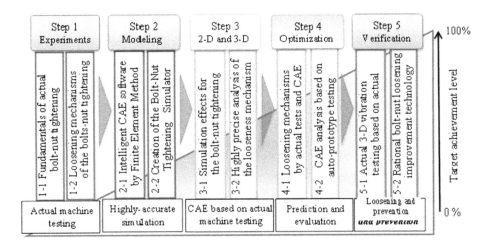

1.   Experiments - Actual machine testing (Step 1)

In Step 1, to visualize the loosening mechanism of the bolt-nut tightening, it is necessary to make clear the dynamic behavior that accompanies occurrence of the problem as follows; The aim of "Step 1-1 is to catch the fundamentals of actual bolt-nut tightening, and Step 1-2 is to verify the loosening mechanism of the bolts-nut tightening.

2.   Modeling - Highly accurate simulation (Step 2)

In Step 2, to draw the information obtained in Steps 1, the author lists the specific components which make up this modelling as follows; The aim of Step 2-1 is to develop the intelligent CAE software by "Finite Element Method" (FEM), and Step 2-2 is to create of the "Bolt-Nut Tightening Simulator" for the "Optimization of highly accurate CAE analysis and comprehensive manner".

3.   2D and 3D analysis - CAE based on actual machine testing (Step 3)

In Step 3, to perform a highly accurate numerical simulation, the author develops the combination of 2D and 3D in Step 3-1 and Step 3-2 based on the investigation gained in Step 2, and articulates implicit knowledge about the bolt-nut loosening mechanism integrated on a qualitative level where the visualization produced through actual testing can be reproduced.

4.   Optimization - Prediction and evaluation (Step 4)

In Step 4, the author conducts a "highly reliable numerical simulation" that enables the absolute values. Step 4-1 focuses on visualization of internal stress distribution of bolt-nut tightening by combination of actual tests and CAE analysis using "photo-elastic experiment". Moreover, Step 4-2 develops the CAE analysis by comparing results of numbers of vibration cycles of axial force reduction between prototype testing and CAE analysis.

5.   Verification – Loosening and prevention (Step 5)

Finally, Step 5-1 illustrates the actual 3D vibration testing to observe the actual moment of "loosening behavior of bolt-nut" using a visualization device. In Step 5-2, furthermore, the author creates the "New design nut" having high-cost performance which gave priority to mass-productivity by employing "rational bolt-nut loosening improvement technology", and verifies the validity of developing.

APPLICATIONS OF HIGHLY-ACCURATE CAE ANALYSIS MODEL

Experiments–Actual Machine Testing (Step 1)

Fundamentals of Actual Bolt-Nut Tightening Experiments (Step 1-1)

The test piece is a "M12 x 1.25 9T hexagon bolt and nut with flange" that is often used to tighten parts of power-train system in automotive applications, and an experiment are conducted in order to confirm an important parameter of the friction coefficient based on the "dynamic behavior of angle x torque x axial force" of the (a) "threaded portion" and (b) "nut bearing surface" as shown in Figure 2.

Then, the author showed the relationship between "angle and torque" for 60KN of tensile force as well as compression load described in Figure 3.

In Figure 3(a), the friction coefficient of the (a) threaded portion is measured using a 2-axial fatigue testing machine, which is capable of "applying axial force and twist force" at the same time. The "bolt head" was pulled with the force of 20kN, 40kN and 60kN, conducting the experiment for each tensile force five times in total, while torque and axial force were measured.

Figure 2. Measuring the friction coefficient to tighten bolt-nut

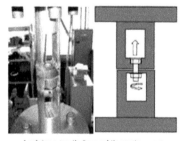  

Applying a tensile force while rotating a nut

(a) Measuring the friction coefficient of the threaded portion

Applying a compression load while rotating a nut

(b) Measuring the friction coefficient of the nut bearing surface

Object of experiment

Hexagon bolt and nut with flange (JIS B 1189/1190)
*Material: SCM435
*Heat treatment: HRC32-39

In Figure 3(b), the friction measurement for coefficient of the (b) "flange (bearing) surface" was similarly conducted using the "2-axial fatigue testing machine", applying a compression load of 20kN, 40kN and 60kN onto the substrate while "rotating the nut so as to confirm the torque and axial force".

Based on the results of the "friction coefficient measurement" of the (a) "threaded portion" and (b) "nut bearing surface", torque and axial force were obtained.

Applying theoretical equations to these experiment results, the friction coefficients were calculated.

Figure 3. The relationship between angle and torque

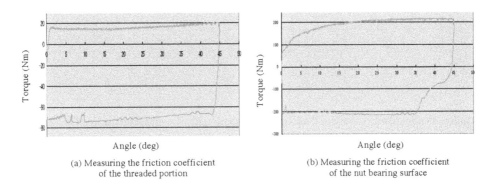

(a) Measuring the friction coefficient
of the threaded portion

(b) Measuring the friction coefficient
of the nut bearing surface

When calculating the friction coefficient of the threaded portion, μ_s, equation (1) below was used.

Also, the friction coefficient of the nut bearing surface μ_w was calculated using equation (2) (Tanaka et al., 1981; Suzuki etc., Eds., 2005).

The bolt and nut dimensions necessary for calculation of these coefficients were given in Table 1.

$$\mu_s = \frac{\dfrac{2T_s}{d_2} - F \tan \beta}{\dfrac{2T_s}{d_2} \tan \beta + F} \cos \alpha \tag{1}$$

$$\mu_w = \frac{2T_w}{Fd_w} \tag{2}$$

where μs: The friction coefficient of the threaded portion, Ts: The torque of the threaded portion, d2: Pitch diameter, F: Axial force, β: Lead angle, α: Thread angle, μw: The friction coefficient of the nut bearing surface, Tw: The torque of the nut bearing surface, dw: The equivalent diameter of torque on nut bearing surfaces.

The dimensions of the bolt are actually measured here, and the bolt and nut (units) are in millimeters.

However, pitch and other dimensions that are difficult to measure are taken from the ISO standard (ISO 4161/15071/5072/10663, 1999).

Table 1. The bolt and nut measurements

	Bolt	Nut
Major diameter of external thread (mm)	12	12
Pitch diameter (mm)	11.19	11.19
Minor diameter of external thread (mm)	10.65	10.65
Pitch of threads (mm)	1.25	1.25
Width across flat (mm)	16.91	16.91
Flange diameter (mm)	26	26
Lead angle (deg)	2.03	2.03
Thread angle (deg)	30	30

*1 Source: ISO15071:1999, ISO15072:1999, ISO4161:1999 and ISO10663:1999
*2 Note: the author measured or calculated

Using equations and dimensions such as those mentioned above, the friction coefficient of the threaded portion and bearing surface was then calculated.

As an example, the friction coefficients of the threaded portion and bearing surface when the tensile force and compression load in axial direction were both 60kN at the angle of 40° are shown in Table 2.

Using these calculated friction coefficients, bolt tightening simulation was conducted and the relationship between simulative axial force and torque was confirmed.

Table 2. The friction coefficients

The friction coefficient of the threaded portion			The friction coefficient of the nut bearing surface		
angle (deg)	torque (Nm)	friction coefficient	angle (deg)	torque (Nm)	friction coefficient
40	18.63	0.02	40	215.26	0.35
40	23.05	0.03	40	221.63	0.37
40	35.30	0.06	40	198.09	0.33
40	28.44	0.04	40	208.39	0.34
40	32.85	0.05	40	226.53	0.37

Loosening mechanisms of bolts-nut tightening (Step 1-2)

An external force was added in the "bolt-nut vibration test to gauge changes to axial force!, which was the "bolt-nut tightening force", and in the "stress distribution of the nut bearing surface" and contacted surface of the base material as shown in Figure 4 ((1) Test-A and (2) Test-B).

Here, "oscillation amplitude" was applied to the "bolt-nut joint", and the" reduction in axial force of the joint and stress distribution" at the "nut bearing surface" were measured by using the "strain gauge equipped test bolt" and the "nut bearing surface pressure sheets for measuring the stress distribution".

The "stress distribution" is measured at the "bolt and nut bearing surfaces" using pressure sheets with 20kN, 35kN and 50kN at oscillation load of ±40%, ±60% and ±90% (±1.8kN, ±2.8kN and ±4.0kN) of the "measured static release load (fall-off load of axial force) that pith of threads (hereinafter referred to as the pitch)" is 1.25mm, respectively.

Figure 4. The bolt-nut vibration test

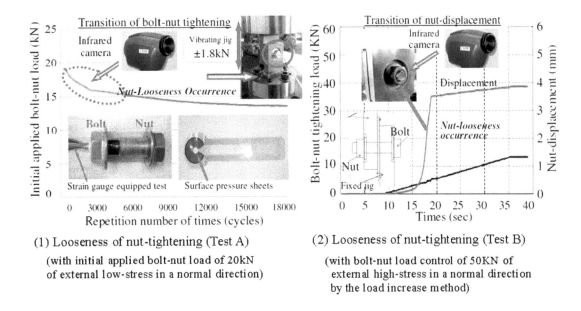

(1) Looseness of nut-tightening (Test A)

(with initial applied bolt-nut load of 20kN
of external low-stress in a normal direction)

(2) Looseness of nut-tightening (Test B)

(with bolt-nut load control of 50KN of
external high-stress in a normal direction
by the load increase method)

Here, the author visualizes the "looseness behavior of bolt-nut tightening" by using "Infrared camera". An example of (1) "Test-A" showed the "transition of bolt-nut load" with the initial applied bolt-nut load of 20kN of external stress in a normal direction described in Figure 4(1).

Here, each measurement result was plotted with the bolt-nut load on the vertical axis, and the sampling number (5Hz) at ±1.8kN and ±2.8kN of "oscillation load as the repetition number of times".

From (1) "Test-A", "rotation looseness of nut-tightening" was confirmed at a stage of the beginning described in Figure 4(1).

In same way, an example of (2) "Test-B" showed the "transition of nut-displacement" with the bolt-nut load control of 50KN of external stress in a "normal direction using load control by increasing load employing tensile load jig".

Here, the "looseness of nut-tightening" was confirmed at a stage after 4.4kN of the "bolt-nut tightening" load described in Figure 4(2).

Next, the author checked the loosening mechanism of bolts-nut tightening in detail by the following tests ("Test-C, -D and -E").

An example of "Test-C" shows the "comparison of flange surface stress distribution as the bolt and nut" with the initial applied bolt-nut load of 20kN of external low-stress in a normal direction, and the sampling number (5Hz) at ±1.8kN and ±2.8kN of "oscillation load on the vertical axis vibration" as shown in Figure 5.

In Figure 5(a), the flange (bearing) surface stress distribution of bolt was the "uniform stress all-around the base material" named "substrate" using coloring type pressure gauge.

In Figure 5(b), the bearing surface stress distribution of nut was the "un-uniform stress all-around the base metal".

Figure 5. Comparison of flange surface stress distribution as the bolt and nut (Test-C) (with the initial applied bolt-nut load of 20kN of external low-stress in a normal direction)

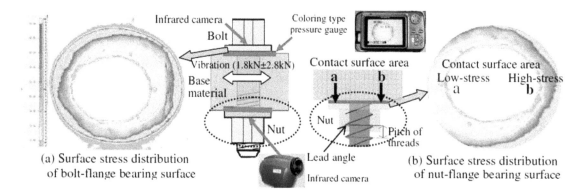

The contact of around "b" surface area with "Nut-flange bearing surface and base material" showed the high-stress distribution under the "influence of strong contact of spiral lead angle". Then, as described in the above-mentioned Figure 4(1) and 4(2), the author checked by "Infrared camera" that nut-looseness occurs.

However, as these acquired knowledge above, the "looseness of bolt-flange surface" was not checked. Then, the author has reached the "Relation between (1) Pitch and Nut bearing surface stress (Test-D) and (2) Pitch and Axial force reduction rate (Test-E)" as shown in Figure 6.

In Figure 6(1), an example of (1) "Test-D" showed the relation between "pitch diameter and nut bearing surface stress" with initial axial forth of bolt-nut load 35kN of external stress.

When the pitch was long, the lead angle grew big, and nut bearing surface stress became high value. In an example of pitch diameter "0.50mm" on the vertical axis vibration, initial axis forth was decreasing rapidly, and this example's reduction rate was 12.5%.

In addition, from the knowledge of the above "Test-A, -B, -C and -D" the author has researched the (2) Test-E. as the "Relation between "pitch diameter and axial force reduction rate" with the "initial axial forth of bolt-nut load" of 35kN of external stress as shown in Figure 6(2).

As the looseness of bolt-nut tightening, an example of "Test-E" showed comparison of axial force reduction rate between vertical axis vibration and horizontal axis vibration.

Modeling–Highly accurate simulation (Step 2)

Intelligent CAE Software by Finite Element Method (Step 2-1)

The main issues in creating CAE software are grasping the actual state of contact between the threaded portion and grooves, and the friction coefficient of threaded portion and nut bearing surface based on the bolt and nut measurements (Table 1) and friction coefficients (Table 2).

For structural analysis to function properly, there are two minimum requirements. These are necessary to simulate the dynamic behavior of angle, torque, and axial force for both the threaded portion and nut bearing surface during bolt-nut tightening.

Figure 6. Relation between "(1) "Pitch" and "Nut bearing surface stress" (Test-D) and (2) "Pitch" and "Axial force reduction rate" (Test-E)"

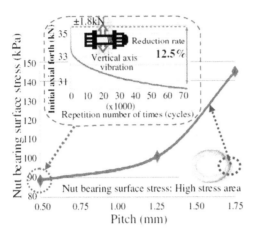 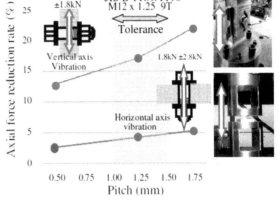

(1) Relation between "Pitch" and "Nut bearing surface stress" (Test-D)
(with the relation between "Repetition number of times (cycles)" and "Initial axial forth" of bolt-nut load 35kN of external stress)

(2) Relation between "Pitch" and "Axial force reduction rate" (Test-E)
(with the initial axial forth of bolt-nut load 35kN of external stress)

These require the contact between the threaded portion and groove as well as contact between the threaded portion and bearing surface.

These should be done on the basis of elastic static analysis and elastic dynamic analysis results, calculated using the "Finite Element Method" (FEM).

Creation of "Bolt-Nut Tightening Simulator" (Step 2-2)

Then, the author creates a "Bolt-Nut Tightening Simulator" using "Highly Precise CAE Technology Element Model" as shown in Figure 7 as follows; (Baggerly, 1996; Gamboa and Atrens, 2003; Amasaka, Ed., 2007)

(i)   Relevant issues for conducting "Numerical simulation of physical chemical phenomena" contain the (1) Relationship between axial force and torque", (2) "Dynamic friction coefficient and static friction coefficient" and (3) "Contact surface pressure on the threaded portion and nut bearing surface".

(ii) "Structuring models" for solving these issues contain the (1) "Dynamic elemental model", (2) Elastic-plastic model", (3) "Contact model" and (4) "Material component rule model".

(iii) Useful algorithms contain the (1) "Finite element method" and (2) "Dynamic analysis method".

(iv) Suitable theoretical equations contain the "(1) structural mechanics and (2) equation of equilibrium".

(v)   Creative solutions for ensuring precision and realistic time calculation contain the (1) "Temporal integration method", (2) "Matrix function", (3) "Penalty method" and (4) "Normal Lagrange Multiplier Method".

To facilitate highly precise CAE analysis, the author has developed the "bolt-nut tightening technical element model" in order to allow accurate reproduction of actual machine tests.

This model addresses the bolt-nut loosening mechanism through steps (i) to (v) below.

(i)   Problem-The author assessed the phenomenon by means of actual machine tests employing the "2nd Law of Thermodynamics", interpreting temperature changes detected using high speed camera and infrared camera as stress changes.

(ii) Modeling-To convert the entire nut being targeted in "Numerical analysis" into a model that can work with "CAE analysis", the author broke it down into "elements and modeled the elements' material constitutive principle (component rule) model" as an equation.

Figure 7. Bolt-nut tightening simulator using highly precise CAE technology element model

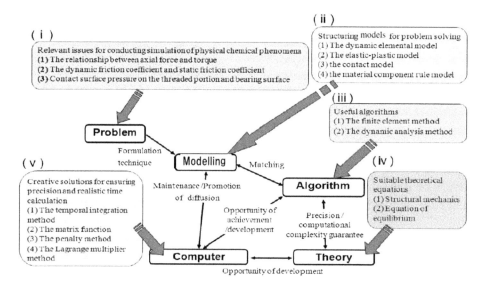

(iii) Algorithms-The author synchronized important algorithms where the effects of external force were observed and eliminated gaps with actual machine tests so that the accuracy of "FEM analysis" and "Nonlinear Analysis" (Newton-Rraphson Method) increase.

(iv)   Theory-Since the validity, application scope, and performance of the algorithms themselves can be derived from theory, the author calculated "coefficients of friction" using "experimental values and a theoretical equation" which vary with temperature, accurately by holding the "temperature constant using visualization technology".

(v)   Computer-The author changed from the penalty method to the "Augmented Lagrange Algorithm" to enhance the "engagement precision of the bolt and nut", which is characterized by numerical stability and a tendency to converge on solutions.

2D and 3D Analysis– CAE Based on Actual Tests (Step 3)

Simulation effects for "bolt-nut tightening" (Step 3-1)

First, the author utilizes the axis symmetry 2D model (using FEM) for "contact surface of bolt-nut tightening" using "Bolt-Nut Tightening Simulator" as shown in Figure 8.

Regarding the analysis parameters, axial force is applied to part of the "upper substrate and bolt bounded" from all directions in the same way as it is applied to part of the bolt head and lower substrate.

Concretely, the "friction coefficient" was calculated by conducting experiments on parts of the "contact bearing surfaces and the contact surface of the threaded portions".

Figure 8. The axis symmetry 2D model for the "contact surface of bolt-nut tightening": Using "bolt-nut tightening simulator"

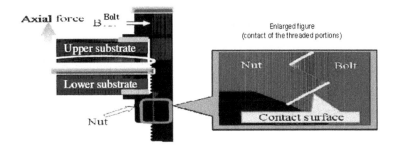

Second, the author utilizes 3D model (using FEM) simulating the helical structure of the threaded portion as shown in Figure 9.

In the diagrams of Figure 9(a) and (b), the analysis process involves the following steps:

① Place parts to be clamped using 2 substrates between the bolt and nut.
② Apply axial force to the bolt and nut determine the distribution of stress during contact.
③ Fix the edge of the lower substrate and apply upward perpendicular force (axial force) to the upper substrate.
④ Determine the pressure on the contact nut bearing surface.

Furthermore, the results of 3D analysis showed the examples of stress distribution of the nut surface and nut bearing surface in contact with lower substrate in the initial axial force of load 25kN and 35kN of external stress as shown in Figures 10.

The diagrams of "Mises stress" in Figure 10(a) showed the "stress distribution of the nut surface in contact with lower substrate".

In Figure 10(a), the "stress value of initial axial forth" of load 35kN was a little bigger than the stress value of initial axial forth of load 25kN.

Figure 9. 3D finite element models simulating "helical structure of the threaded portion"

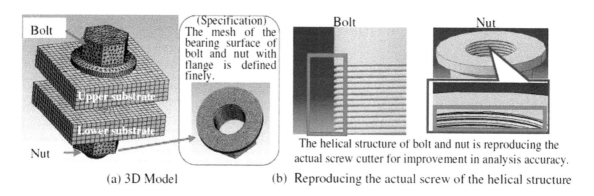

(a) 3D Model   (b) Reproducing the actual screw of the helical structure

In same way, Figure 10(b) showed the "stress distribution of the nut bearing surface in contact with lower substrate".

In Figure 10(b), the "stress value of initial axial forth" of load 35kN was much bigger than the stress value of initial axial forth of load 25kN.

Figure 10. Stress distribution of the (a) nut surface and (b) nut bearing surface in contact with lower substrate (with the initial axial forth of bolt-nut load 25kN and 35kN of external stress)

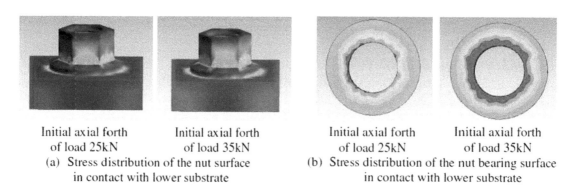

Initial axial forth of load 25kN   Initial axial forth of load 35kN   Initial axial forth of load 25kN   Initial axial forth of load 35kN

(a) Stress distribution of the nut surface in contact with lower substrate   (b) Stress distribution of the nut bearing surface in contact with lower substrate

Then, the author illustrates the "comparison results of experiment and 3D analysis" through the "relationship between torque and axial force of bolt-nut tightening" as shown in Figure 11.

The results of experiment were represented by the dotted line (experiment results), and the results of the "numerical simulation by CAE" were represented by the "solid line (CAE results)".

This diagram showed that the "results of the CAE analysis closely match the results of the experiment".

Figure 11. Comparison result between experiment and CAE simulation

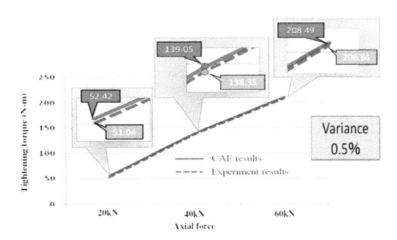

Highly Precise Analyses of Bolt-Nut Loosening Mechanism (Step 3-2)

Figures 12 showed the results of the 2D analyses conducted as a stepping stone to conducting the 3D analysis. The following "stress distribution diagram" indicated the presence of higher stress in the inner portion of the nut" as well as in the base material.

Because the purpose of the axially symmetrical 2D model was to understand the boundary conditions and model integrity, surfaces will be discussed in the 3D model with stress values increasing as the colors change.

Figure 12. The results of axis symmetry 2D model

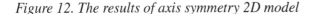

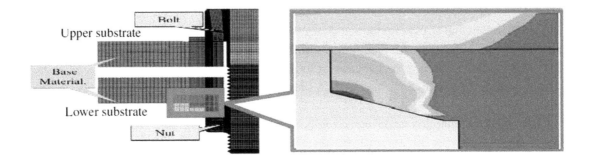

Moreover, the author will discuss the 3D analysis results at the "bolt-nut bearing surface" when an axial force of 20kN was applied to the bolted joint along with the sampling number (5Hz) at ±1.8kN of oscillation load applied to the base material on the oscillation side (vertical axis vibration).

Figure 13 shows the change in stress distribution at the nut bearing surface due to oscillation load. "High stress" was observed at the "nut bearing surface", particularly at the "inner portion (starting point and end point) of the helix structure (① in Figure 13)".

This is similar to the observation made during the "tightening analysis" discussed earlier. Also, it is evident that the "outer part of the nut bearing surfaces experience less stress (②, ③ and ④ in Figure 13)".

Figure 13. An example of transition of stress distribution at the nut bearing surfaces

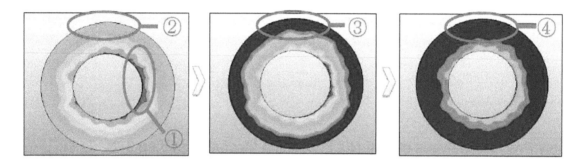

This result indicates that a "reduction in contact force" at the "nut bearing surface of the area closer to the outer edge of a nut leads to slippage at the bearing surface when a bolt's axial force reduces with externally applied force.

Here, the author will compare the "stress distribution of nut bearing surfaces" during the "vibration test" and "stress distribution of nut bearing surfaces from the 2D and 3D analyses".

The stress at the "nut bearing surface" was higher in the "area closer to nut helix", and showed a "stress distribution similar to the CAE simulation result".

Although there are differences in the values that can be attributed to the "distortion in the base materials, wear and roughness of the contact surfaces", the "simulation" was successful in "reproducing similar stress values and stress distribution changes".

Therefore, by conducting CAE analysis according to the "Highly-Accurate CAE Analysis Approach Model" created in this subsection "Creation of the Highly-Accurate CAE Analysis Model" in bolt-nut loosening solution", the author was able to show the effectiveness of this model by reproducing an actual phenomenon.

It is also plausible that this model can contribute to "bolt loosening" and "bolt fracture analysis" by understanding the changes in stress distribution at the "nut bearing surfaces".

The 3D analysis procedure is conducted in the next step.

1ˢᵗ, the 2 bolted workpieces are placed between the bolt and the nut using pitch 0.50 mm and 1.75 mm.

Furthermore, the author will discuss the 3D analysis results at the bolt bearing surface when an axial force of 35kN was applied to the bolted joint along with ±1.8kN of oscillation load applied to the base materials on the oscillation side (vertical axis vibration) as shown in Figure 14.

First, axial force is applied to the bolt (securing the edge of the lower substrate) and perpendicular to the upper substrate.

Figure 14. 3D analysis of pitch differences between two kinds (Ex. Stress distribution of nut bearing surface with initial axial forth of bolt-nut load 35kN of external stress by the sampling number (5Hz) at ±1.8kN of oscillation load employing nonlinear structural analysis (elastic analysis))

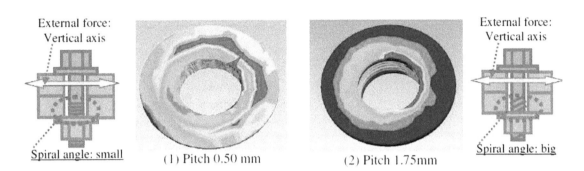

Second, the analysis looks at the pressure on the contact surface between the bolt-nut bearing surface and the base materials (lower substrate and upper substrate, and between the bolt threads and the nut).

Figure 14 showed a lack of uniformity in terms of nut bearing surface response (stress distribution) and a localized strong response where the nut thread structure begins.

In looking at the differences between the 2 pitch values, the "response distribution was more concentrated around the initial thread structure on the bolt with the longer pitch (1.75 mm) with stress values increasing as the "colors change from blue to red".

These results suggested that "differences in pitch also lead to differences in non-uniformity on nut bearing surfaces and a more pronounced gap between areas of strong and weak response".

In terms of new information, the 3D analysis revealed that the "length of bolt and nut pitch impacts the reduction in contact force on contact surfaces".

Then, comparing the "axial force reduction behavior measurements" obtained through "actual testing and those obtained through CAE analysis" verified the "precision of the CAE analysis results".

Moreover, Figure 15 compares the results of actual testing and CAE analysis for the bolt and nut with 1.75 mm pitch under initial axial forth of 35 kN.

The dashed line shows a margin of error of 3%, indicating that a high-quality CAE analysis was achieved.

Similar results were achieved for the 0.50 mm-pitch diameter bolt and nut as well.

Optimization–Prediction and Evaluation (Step 4)

Loosening Mechanisms by Actual Test and CAE Analysis (Step 4-1)

Generally, internal stress distribution of screw engagement part made of the metal can't be measured as shown in Figure 16. In actual test of Step 4-1, the author measured the stress distribution of bolt-nut tightening made from polycarbonate by interference fringes employing photo-elastic experimental device as described in Figure 16(a).

Figure 15. Comparing the results of actual force reduction between actual testing and CAE

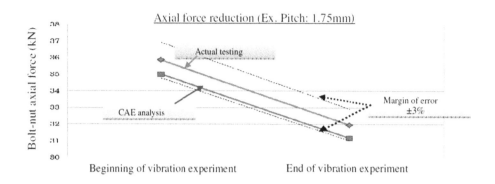

Figure 16. Visualization of the internal stress of bolt-nut tightening by actual test and CAE

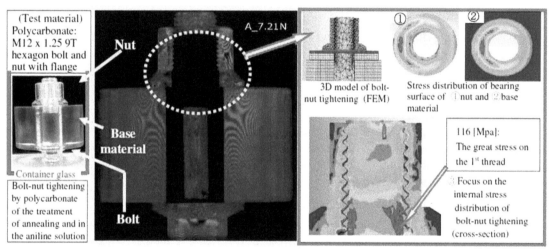

(a) Actual test-Stress distribution of bolt-nut tightening by interference fringes employing photo-elastic experiment device

(b) 3D analysis - Stress distribution of bolt-nut tightening with initial axial forth load 35kN (Pitch: 1.25mm)

In 3D analysis, furthermore, the author reproduced the result of "photo-elastic experiment in highly-accurate CAE analysis" as the internal stress distribution of bolt-nut tightening made from "SCM435" as described in Figure 16(b).

In Figure 16(a), the author chose the "polycarbonate" from the following reasons; the (i) characteristic of polycarbonate had 1/20 intensity of SCM435, (ii) stress-strain curve of these 2 materials "SCM435 and polycarbonate" are similar in shape, (iii) cutting processing of "polycarbonate" is good, and (iv) "highly transparent polycarbonate" which can visualize internal stress distribution by having the "isotropic homogeneity and property of causing the stress distribution" and by having the "isotropic homogeneity and property of causing the birefringence".

To make a "photo-elastic experiment" succeed, the author implemented (v) "annealing treatment for removal of residual strain" and (vi) "aniline solution of the same refractive index as the polycarbonate", and (vii) "glass containers not having the properties of optical birefringence".

In Figure 16(b), the author showed the "stress distribution of bearing surface of ① nut and ② base material", and focuses on the "③ internal stress distribution of bolt-nut tightening (cross-section)" by 3D analysis with initial axial forth load 35kN (Pitch: 1.25mm).

From Figure 16(b)-③, the author checked the "transitions of internal stress distribution of bolt-nut tightening". Particularly, the author caught the "existence of great stress on the 1st thread of engagement between the bolt and nut in detail".

Moreover, the author understood the "state of change of internal stress distribution with sufficient accuracy through CAE analysis" as the difference of initial axial forth load (20kN, 35kN and 50kN) of "external stress", the difference of pitches (0.5mm, 1.25mm and 1.75mm), etc.

CAE Analysis Based on Automobile Prototype Testing (Step 4-2)

In Step 4-2, the author implemented the "highly-accurate CAE analysis" with "no error (gap) compared to tests for ensure the optimization of prediction and evaluation of bolt-nut loosening mechanism". As the prototype testing and CAE analysis", Figure 17 showed a case of "Comparing the results of numbers vibration cycles of actual force reduction" between prototype testing and CAE analysis.

The author chose "M12 9T flanged hexagonal bolts and nuts" with 3 pitches (0.50mm, 1.25mm, 1.75mm) for "automobile prototype testing using rear suspension arm with mount rubber-bush" as follows;

(1)   Accelerated test using 3D vibration testing machine

Concerning the equipment used in bolt-nut looseness tests, the author reproduced a working chassis designed to simulate "prototype (bench) testing of a typical actualvehicle". These bolt-nut tightening with the mount and suspension were the automotive parts most prone to loosening.

The author combined these parts to conduct tests using 3D vibration. In accelerated tests, it is possible to trigger a failure in a shorter amount of time (smaller number of vibration cycles) than usual by applying a large-stress amplitude.

A typical accelerated stress procedure consisted of creating the "S-N curve" based on the results of past tests conducted by the Amasaka-Laboratory, and then using "Miner's rule" to calculate the acceleration coefficient "4.4" employing test condition" in Figure 17.

(2)   Result of axial force reduction by prototype testing and CAE analysis

(i)   1st, the author illustrated the overview of 2D model and stress distribution of 2D analysis with stress values increasing as the "colors change from blue to red".

(ii)   2nd, the author showed a "typical result of prototype testing and 3D analysis" from the dual standpoints of time sequence and accuracy.

As in the time sequence comparison in Figure 17, axial force is shown on the vertical axis, and number of 3D vibration cycles on the horizontal axis.

Figure 17. Comparing results of numbers vibration cycles of axial force reduction between prototype testing and 3D analysis (Pitch: 1.25 mm)

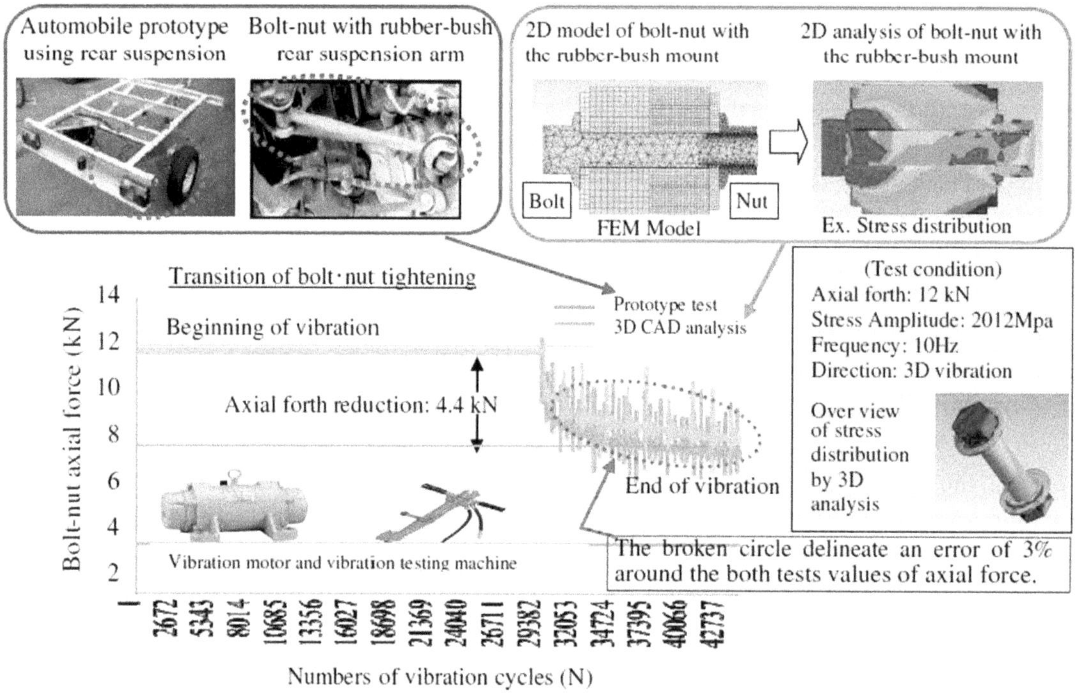

(3)   Actual machine test results are shown in blue, and CAE values in green

This figure reveals that the author achieved an analysis with a "good level of accuracy in terms" of both the "timing at which the bolt-nut loosened and in the extent of the decline in axial force".

Moreover, to verify accuracy, the author confirmed that the broken circle delineates an error (gap) of 3% around the both test values of axial forth reduction "4.4kN" line in Figure 17.

Verification–Loosening and Prevention (Step 5)

Actual 3D Vibration Testing Based on Actual Vehicle Test (Step 5-1)

In Step 5-1, the author conducts 3D oscillating experiment and CAE analysis. To clarify the "loosening movement of bolt-nut joint", the author gave "oscillating load" to the "bolting part for the vibration under actual market running" with 4 kinds of combination of the "flat road, winding road, intense gravel road of ups and downs", and "intense bad road of ups and downs" in the "Japan National Route 16" over "Tokyo and Kanagawa-Prefecture".

Figure 18 illustrated an example of "actual 3D vibration testing" using "actual vehicle test and CAE analysis" as follows;

(i)   Figure 18(a) shows an example of actual vehicle vibration test using "Intense gravel road" based on "Oscillatory movement of rear suspension arm with shock absorber of Toyota Prius". In acceleration data of Figure 17(a), each of X-axis, Y-axis and Z-axis wireless vibration recorder.

(ii)   Figure 18(b) shows an equipment of bolt-nut tightening using actual 3D vibration system based on actual vehicle test in Figure 17(a). This 3D vibration system contains ①"Step up vibration test" named "Incremental vibration method, 0G→1G→3G→5G" and ②"Rating vibration test" (Frequency: 3.3Hz, Acceleration: Max 5G, Exciting force: 4.9kN) and others.

(iii) In Figure 18(c), the author shows an example of "looseness of bolt-nut tightening" by "Step up vibration test" increasing 0.5G per 30 seconds about the "acceleration of each axis of x, y and z".

Figure 18(c) expresses the "movement of looseness of bolt-nut tightening" as the relation between "Bolt-nut axial force and Time" in initial tightening torque 40N-m by using "High- speed camera as the visualization technology".

From "transition of bolt-nut axial force", in this example, in Stage 0→①→②→③, initial loosening occurred in 1 minute after "vibration beginning of stage ①", and "complete loosening of bolt-nut" has occurred 5 minutes after "stage ③". In stage ①, especially, the author could observe the "moment of nut loosening (: initial nut-loosening is 10 degrees) at the same time".

(iv) As a result of the experiment of 3-D vibration testing, the author illustrates an example of "stress distribution of bolt-nut tightening by 3-D analysis based on Stage 0→①→②→③ by Step up vibration test. As a concrete instance, Figure 18(d) shows the stress distribution of ① Nut bearing surface and cross-section of cross-section of the ② Nut / ③ Bolt-nut tightening.

In Stage 0, the stress distribution of "❶ Nut bearing surface is uniform distribution", and internal stress distribution of "❷ Nut and ❸ Bolt-nut-base material (cress-section)" is uniform distribution.

In Stage ①, ② and ③, the stress distribution of "❶ Nut bearing surface, internal stress distribution of ❷ Nut and ❸ Bolt-Nut-base material" are "un-uniformity distribution with the inclination".

Especially, the author has confirmed that "internal stress distribution of base material had large distortion".

Rational bolt-nut loosening improvement technology (Step 5-2)

Then, the author discerns the "loosening mechanism of a bolt-nut tightening" obtained in the "various experiments and CAE analyses" which were mentioned above, and tries the "design improvement of the nut form which is the main factor of looseness".

Specifically, the author creates the "high-cost performance nut" which gave priority to mass-productivity by utilizing "rational bolt-nut loosening improvement technology", and verifies the "validity of developing "Advanced TDS" in New JIT strategy" as the viewpoint of looseness prevention" as shown in Figure 19 (See to Chapter 7).

Figure 19(a) illustrates a validity of "New design nut" preventing bolt-nut looseness by actual test and CAE analysis.

In Figure 19(a), first, the author showed the comparison of "Form and stress distribution of nut bearing surface" of bolt-nut tightening between "Normal nut and New design nut" based on 3-D model.

Figure 18. An example analysis of actual 3D vibration testing using actual vehicle test and CAE based on oscillatory movement of rear suspension arm with shock absorber of Toyota Prius

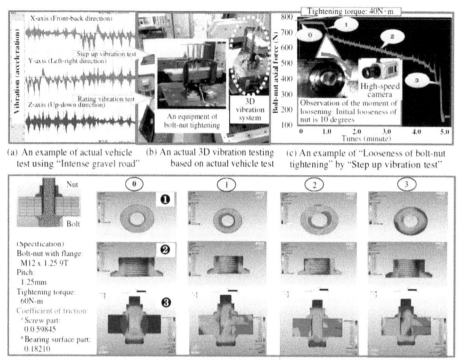

(a) An example of actual vehicle test using "Intense gravel road"   (b) An actual 3D vibration testing based on actual vehicle test   (c) An example of "Looseness of bolt-nut tightening" by "Step up vibration test"

(d) An example of stress distribution of ① Nut bearing surface and ② Nut / ③ Bolt-nut tightening by CAE analysis

In the design of "New design nut", by "Boring" (Depth: 3.5-4.0mm / Inner diameter: φ14.0), the author deleted 3 pitches of the screw thread of bearing surface portion of "Normal nut" used by a mass-production vehicle, and lengthened 3 pitches for the full length of the "Normal nut" for "securing of screw strength" as shown in Figure 19(a)-left.

In Figure 19(a)-right, furthermore, the author showed the "Normal nut" and "New design nut" which were actually manufactured, and the examples of both "stress distribution of bearing surface" and "cross section by 3D analysis".

Compared with the "Normal nut", the author checked that the bearing surface stress of "New design nut" was small by the effect of boring.

In Figure 19(b), 2nd, the author showed the comparison of "Time to the looseness occurrence" between "Normal nut and New design nut" based on the "Step up vibration test".

Especially, the author could check the "sharp decline phenomenon of bolt-nut axial force stress" through the "Transition of final loosening of bolt-nut tightening" as follows;

(i)   In normal nut, elapsed time from initial loosening to final loosening is 5 minutes.
(ii)   In New design nut, elapsed time from initial loosening to final loosening is 35 minutes.

From this study, the author could observe that the "New design nut" preventing bolt-nut looseness is 7 times compared with "Normal nut".

Figure 19. A validity of new design nut preventing bolt-nut looseness

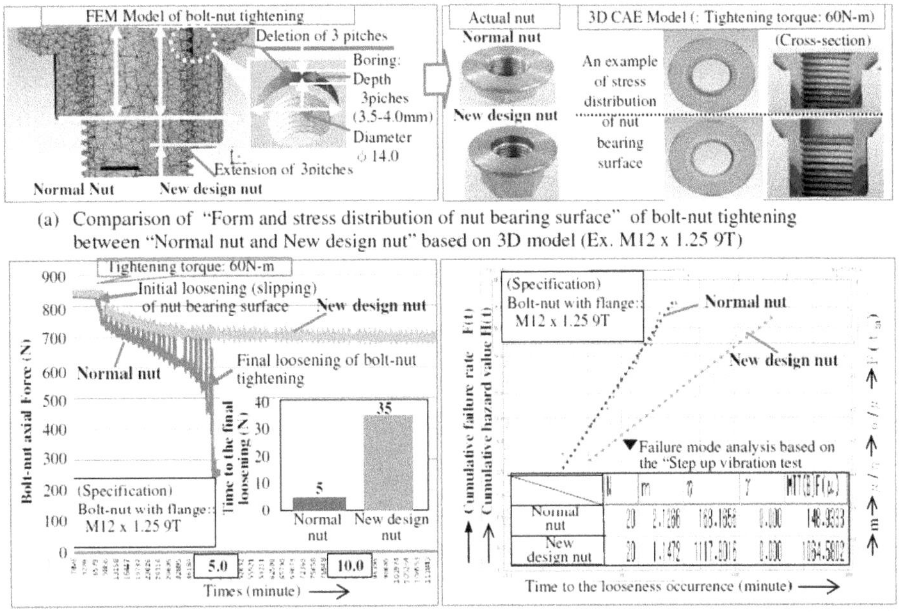

(a) Comparison of "Form and stress distribution of nut bearing surface" of bolt-nut tightening between "Normal nut and New design nut" based on 3D model (Ex. M12 x 1.25 9T)

(b) Comparison of "Time to the looseness occurrence" between "Normal nut and New design nut" based on the "Step up vibration test"

(c) Comparison of "Failure mode of bolt-nut loosening" between "Normal nut and New design nut" by using "Weibull type – Cumulative hazard paper"

In Figure 19(c), 3rd, the author showed the comparison of "Failure mode of bolt-nut loosening" between the "Normal nut" and "New design nut" by using "Weibull type – Cumulative hazard paper" based on the "Step up vibration test".

In this "Failure mode analysis", the author could calculate "m" (shape parameter), "η" (scale parameter), "γ" (location parameter) and "MTTF" (Mean Time to Failure) by "Step up vibration test" of N=20 cases to each of "Normal nut" and "New design nut".

From the comparison of MTTF, the author could check that the "New design nut" preventing bolt-nut looseness is 7 time compared with Normal nut.

Although a detailed explanation is omitted, furthermore, the author checked that the well-known the "Looseness prevent nut" named "Hard-lock nut, U-nut and Self-locking nut" which are marketed (E.g., Hardlock Industry Co., Ltd., 2014 and Fuji Seimitsu Co., Ltd., 2014), and "New Design Nut" were equivalent performances as the preventing bolt-nut looseness.

From the author's trial calculation, the manufacture cost of "New design nut" in the mass production is 25 yen/1peice, and is 1.2 times of "Normal nut" (21yen/1piece), However, is 1/6 of Hard-lock nut (150yen/1piece) (Hashimoto, 2015).

From the viewpoint of bottleneck solution of the market claim "looseness of bolt-nut tightening", the author concludes that "New design nut" has an excellent cost performance in the diffusion expansion to the mass production vehicle (At present, it is progressing).

CONCLUSION

In this chapter, the author has established the Highly-Accurate CAE Analysis Model by experiments and CAE employing Advanced TDS in New JIT strategy as part of principle-based research aimed at the evolution of product design processes. This model's effectiveness for predictive evaluation method in design work has been demonstrated with various actual tests and precision CAE analysis to application to study on "automotive bolt-nut loosening mechanism" with no discrepancy between the experiment and CAE. Furthermore, the author has created the excellent "New design nut" prevention bolt-nut looseness in high-cost performance from the viewpoint of bottleneck solution of the vehicle market claim "looseness of bolt-nut tightening".

REFERENCES

Amasaka, K. (2004). *Science SQC, New Quality Control Principle: The Quality Strategy of Toyota.* Springer-Verlag. doi:10.1007/978-4-431-53969-8

Amasaka, K. (2007). Highly Reliable CAE Model, The key to strategic development of *Advanced TDS.* *Journal of Advanced Manufacturing Systems, 6*(2), 159–176. doi:10.1142/S0219686707000930

Amasaka, K. (Ed.). (2007). Establishment of a needed design quality assurance: Framework for numerical simulation in automobile production. Japanese Society for Quality.

Amasaka, K. (2008). An Integrated Intelligence Development Design CAE Model utilizing New JIT. *Journal of Advanced Manufacturing Systems, 7*(2), 221–241. doi:10.1142/S0219686708001589

Amasaka, K. (2010). Proposal and effectiveness of a High-Quality Assurance CAE Analysis Model: Innovation of design and development in automotive industry, *Current Development in Theory and Applications of Computer Science. Engineering and Technology, 2*(1/2), 23–48.

Amasaka, K. (2012). Constructing Optimal Design Approach Model: Application on the Advanced TDS. *Journal of Communication and Computer, 9*(7), 774–786.

Amasaka, K. (Ed.). (2012). Science TQM, New Quality Management Principle: The quality management strategy of Toyota. Bentham Science Publisher.

Amasaka, K. (2015a), New JIT, new management technology principle, Taylor & Francis Groups.

Amasaka, K. (2015b). Constructing a New Japanese Development Design Model "NJ-DDM": Intellectual evolution of an automobile Product Design. *TEM Journal Technology Education Management Informatics, 4*(4), 336–345.

Amasaka, K. (2017). Strategic Stratified Task Team Model for Realizing Simultaneous QCD Fulfilment: Two Case Studies. *Journal of Japanese Operations Management and Strategy, 7*(1), 14–35.

Amasaka, K. (2019a). Studies On New Manufacturing Theory, *Noble. International Journal of Scientific Research, 3*(1), 1–20.

Amasaka, K. (2019b). *Developing Automobile Optimal Product Design Model.* Japanese Operations Management and Strategy Association, The 9th International Symposium on Operations Management and Strategy 2019, Tokyo Keizai University, Tokyo.

Amasaka, K. (2019c). Establishment of an Automobile Optimal Product Design Model: Application to study on bolt-nut loosening Mechanism, *Noble. International Journal of Scientific Research, 3*(9), 79–102.

Amasaka, K. (2022a). *Examining a New Automobile Global Manufacturing System.* IGI Global Publisher. doi:10.4018/978-1-7998-8746-1

Amasaka, K. (2022b). New Manufacturing Theory: Surpassing JIT (2nd Edition), Lambert Academic Publishers.

Amasaka, K. (2023a). A New Automobile Product Development Design Model: Using a Dual Corporate Engineering Strategy. *Journal of Economics and Technology Research, 4*(1), 1–22. doi:10.22158/jetr. v4n1p1

Amasaka, K. (2023b). New Lecture-Surpassing JIT: Toyota Production System-From JIT to New JIT-. Lambert Academic Publishing.

Amasaka, K., Ito, T., & Nozawa, Y. (2012). A New Development Design CAE Employment Model. *The Journal of Japanese Operations Management and Strategy, 3*(1), 18–37.

Amasaka, K., Onodera, T., & Kozaki, T. (2014). Developing a Higher-Cycled Product Design CAE Model: The evolution of automotive product design and CAE. *International Journal of Technical Research and Application, 2*(1), 17–28.

Baggerly, R. (1996). Hydrogen-Assisted stress cracking of high-strength wheel bolts. *Engineering Failure Analysis, 3*(4), 231–240. doi:10.1016/S1350-6307(96)00023-4

Fuji Seimitsu Co. Ltd. (2014). *U-Nut.* FUN. https://www.fun.co.jp/u_nut/u_nut.html/

Gamboa, E., & Atrens, A. (2003). Environmental influence on the stress corrosion cracking of rock bolts. *Engineering Failure Analysis, 10*(5), 521–558. doi:10.1016/S1350-6307(03)00036-0

Hardlock Industry Co. Ltd. (2014). Hard-lock nut. https://www.hardlock.co.jp/products/ hln/

Hashimoto, H. (2005). Automotive Technological Handbook (3), Design and Body, Chapter 6, CAE. Society of Automotive Engineers of Japan, Tosho-Shuppan-Sha.

Hashimoto, K. (2015). *Internal stress analysis of bolt with experiment and CAE analysis* [Master's thesis, Graduate School of Science and Engineering, Aoyama Gakuin University in Amasaka New JIT laboratory]

Hashimoto, K., Onodera, T., & Amasaka, K. (2014). Developing a Highly Reliable CAE Analysis Model of the mechanisms that cause bolt loosening in automobiles. *American Journal of Engineering Research, 3*(10), 178–187.

Kozaki, T., Yamada, H., & Amasaka, K. (2012). A Highly Reliable Development Design CAE Analysis Model: A Precision CAE Analysis Approach to automotive bolt tightening. *The International Business & Economics Research Journal, 10*(1), 1–9.

Ministry of Land. (2015). *Infrastructure, Transport and Tourism*. MLit. https://www.mlit.go.jp/ jidosha/ carinf/rcl/data.html

Nomura, R., Hori, K., & Amasaka, K. (2015). Problem Prevention Method for product design based on predictive evaluation: A study of bolt-loosening mechanisms in automobile. *American Journal of Engineering Research*, *4*(6), 174–178.

Nomura, R., Shimura, T., Sakurai, Y., & Amasaka, K. (2016). *The elucidation of the loosening mechanism of automobile bolt nut conclusion by using experiments and CAE*. Japanese Operations Management and Strategy Association, The 6th Annual Technical Conference, Kobe University, Kobe.

Onodera, T., & Amasaka, K. (2012). Automotive bolts tightening analysis using contact stress simulation: Developing an Optimal CAE Design Approach Model. *Journal of Business & Economics Research*, *10*(7), 435–442. doi:10.19030/jber.v10i7.7148

Shimura, T., & Sakurai, Y. (2015). *The elucidation of the nut loosening mechanism: Combined use of the three-dimensional system experiment by actual vehicle data and CAE analysis* [Bachelor's degree thesis, School of Science & Engineering, Aoyama Gakuin University]

Suzuki, M., Arafune, S., & Wadachi, M. (Eds.). (2005). *Encyclopedia of Physics*. Asakura-Shoten. (in Japanese)

Takahashi, T., Ueno, T., Yamaji, M., & Amasaka, K. (2010). Establishment of Highly Precise CAE Analysis Model using automotive bolts. *The International Business & Economics Research Journal*, *9*(5), 103–113. doi:10.19030/iber.v9i5.574

Tanaka, M., Miyagawa, H., Asaba, E., & Hongo, K. (1981). Application of the finite element method of bolt-nut joints: Fundamental studies on analysis of bolt-nut joints using the finite element method. *Bulletin of the JSME*, *24*(192), 1064–1071. doi:10.1299/jsme1958.24.1064

Ueno, T., Yamaji, M., Tsubaki, H., & Amasaka, K. (2009). Establishment of bolt tightening simulation system for automotive industry: Application of the Highly Reliable CAE Model. *The International Business & Economics Research Journal*, *8*(5), 57–67.

Yamada, H., & Amasaka, K. (2011). Highly-reliable CAE analysis approach-Application in automotive bolt analysis. *China-USA Business Review*, *10*(3), 199–205.

Chapter 15
Epoch-Making Innovation in Work Quality for Auto Global Production

ABSTRACT

To strengthen corporate management technology, the author has recently worked out the epoch-making innovation in work qualityy for auto-global production strategy. Specifically, to realize the high-productivity in New JIT strategy, the author has organized the strategic global production activity named AWD-6P/J (aging working development six projects) for evolution of the work environment, and has verified its validity at an advanced car manufacturer Toyota. While many vehicle assembly shops depend on a young male and female workforce, innovation in optimizing an aging workforce is a necessary prerequisite of TPS (JIT). Elements necessary for enhancing work value and motivation, and work energy, including working conditions and work environment (amenities and ergonomics), were investigated through objective survey, and analyzed from labor science perspectives.

PREREQUISITE OF STRATEGIC GLOBAL PRODUCTION

INNOVATION IN WORK QUALITY

Today's challenge for business management lies in providing customers with products of excellent QCD (quality, cost and delivery) performance in the pursuit of customer satisfaction (CS) and staying ahead of competitors through market creation activities.

To do this, New JIT was established as the new management technology principle for 21st century manufacturing in order to realize the high-productivity employing both "Science SQC and SSTTM (Strategic Stratified Task Team Model) (Amasaka, 2000a, 2002, 2003, 2004a, 2007a,b,c, 2013a,b, 2014, 2015, 2017a,b, 2018, 2020a,b, 2021, 2022a,b,c, 2023; Amasaka and Sakai, 2010, 2011) (See to Chapter 7 and 9).

The author believes that the key to a company's prosperity is a global production strategy that enables supply of leading products with high quality assurance and simultaneous global production start-up (the same quality and production at optimal locations) in both developed and developing countries. Innovation for optimizing an aging workforce is a necessary prerequisite of TPS (Ebioka et al., 2007a,b; Amasaka, 2007a).

DOI: 10.4018/978-1-6684-8301-5.ch015

It is essential to identify elements necessary for enhancing work value, motivation and work energy, as well as an optimum work environment (amenities and ergonomics), through objective survey and analysis from labor science perspectives (Sakai and Amasaka, 2006, 2007, 2008a,b).

In Today, major manufacturing companies are facing the strong need to innovate their businesses for global production. The current and future health of manufacturing performance in Japan, as well as the possibility of simultaneously attaining the same quality level in overseas plants, remains unanswered.

Depending on the situation in each country, including product specifications, production volume and market conditions, manufacturing may be fully automated or require manual labor. If so, the success of global production is highly dependent upon the quality of workers, indicating the necessity of work skill innovations.

Therefore, to achieve long-term growth, companies should also undertake drastic improvement of the work environment (Amasaka, 2004b,c, 2007a,b,c).

Background of the Concepts Regarding Consideration for Older Workers

Japan, with the fame of its big manufacturing businesses, faces industrial changes-expansion of overseas production, stagnant domestic demand, and diminished recognition of manufacturing due to changing preferences of young people.

This has resulted in reduced employment of new, young workers. In the case of Toyota, the average and oldest age of workers on vehicle assembly lines has been on the rise for the past decade (1995-2005), as shown in Figure 1 (Suzumura et al., 1988; Eri et al., 1999; Amasaka, 2000b, 2000b, 2007a).

This figure also indicates an increase of female workers. To cope with this trend more extensive consideration of worker motivation and physical condition is essential.

In other words, manufacturers should shift from work-oriented shop designs to people- oriented shop designs that put more focus on the work environment.

Figure 1. Variation of aging workers

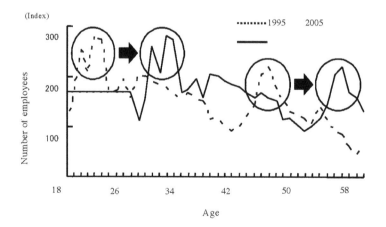

As an example of manufacturing innovation in overseas countries, various governmental actions for older workers have been taken in Scandinavian countries that have aging workforces.

Though the necessity of such actions has been advocated in Japan, action has been relatively modest compared with that in Scandinavian countries [*1] (Amasaka, 2000b, 2007a: Niemela et al., 2002: Vecchio et al., 2003).

At Toyota, improvements have been based mainly on TVAL (Toyota-verification Assembly Load) [*2] (Toyota Motor Corp. and Toyota Motor Kyushu, 1994) for quantitative evaluation of workloads.

At its Kyushu Plant, Toyota has also implemented a worker-oriented line for the assembly process based on the new concept shown in Table 1 (Amasaka., 2000b). Further enhancement of these activities for aging workers will be indispensable in the future.

(*1) An example, Finish Institute of Occupational Health (Finland) and Institute for Gerontechnology (Eindhoven University of Technology (Netherlands) etc. (Japan Machinery Federation and Japan Society of Industrial Machinery Manufactures, 1995a, 1995b).

(*2) Quantitative evaluation of assembly workloads, using electromyographic values.

Table 1. New concept for auto-assembly lines

New Concept
1. Increase worker motivation
2. Reduce workload
3. Automation that people want to work with comfortable work environment

Consideration for Older Workers According to New JIT Strategy

To create a workplace that is friendly to aged workers, 4 steps were planned as shown in Figure 2 (Suzumura et al., 1998; Amasaka, 2007a, 2015; Eri et al. 1999).

Step (1) was to interview middle-aged and elderly workers engaged in the assembly process. The interview brought to light positive and negative aspects.

The first-hand information obtained from middle-aged and older workers was then classified according to body functions.

To ensure accurate interpretation of responses collected during the interviews, the study members participated in actual line operations for 5 weeks.

In step (2), objective data concerning body function were collected by investigation and from various documents in order to clarify the implications of the interview results.

Step (3) was an investigation and evaluation of existing measures for improving conditions related to the physical attributes classified in Step 1.

In step (4), areas were identified in which countermeasures are necessary: (i) workers, (ii) car and equipment, and (iii) management.

For example, to complement insufficient muscular strength due to unnatural working posture and heavy work, assisting devices were introduced that remarkably improved the situation.

Figure 2. Method of setting concept of consideration for aged workers

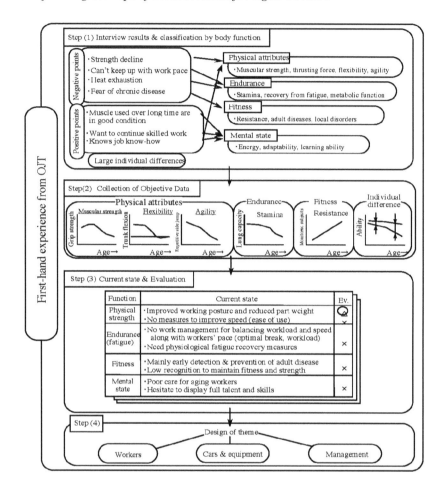

However, the workshop still had a problem in terms of work speed and there were no measures taken to assist work speed.

In terms of endurance, there was no study for assigning jobs according to the work speed of individual operators and establishing effective breaks.

In terms of basic physical strength, early detection and prevention of adult diseases had been the primary consideration, while maintaining and improving physical strength had not been emphasized much.

From the perspective of mental health, care for new employees was sufficient, but the care for older employees was not.

From these analyses, the following imperatives were "selected-I. boost morale", "II. study work standards to reduce fatigue", "III. build up physical strength for assembly work", "IV. alleviate heavy work by employing easy-to-use tools and devices", "V. ensure temperature conditions suited to assembly work characteristics", and "VI. Reinforce preventative measures against illness and injury".

DEFINITE PLAN FOR CONCEPT ACTUALIZATION

Formation of Project Team, AWD-6P/J, and Activity Optimization

It was concluded that tackling these six themes separately is not effective because they are strongly interrelated.

As specialized investigations are necessary, a project team named "AWD-6P/J (Aging and Work Development 6 Programs Project) was formed within Toyota as shown in Figure 13 in Chapter 9 (Refer to Eri et al. (1999) and Amasaka (2000b, 2007a, 2015) in detail).

And it shows the project themes of AWD-6P/J based on the AWD-COS (Aging & Work Development–Comfortable Operating System) in "NJ-GPM" (New Japan Global Production Model) described in Figure 4 in Chapter 9. (Suzumura et al., 1998; Amasaka, 2000b, 2007a; Amasaka and Sakai, 2011).

The relation diagram of Figure 13 in Chapter 9 shows the inter-relationship of each theme. The diagram was created to emphasize team unity during the project.

AWD-6P/J team structure is also shown in Figure 3 (Amasaka, 2000b, 2007a, 2015).

A division specializing in a theme acted as the leader and members including the vehicle assembly division and related divisions acted as a total task management team. The project was mainly promoted by the Vehicle Production Engineering Div., Safety & Health Promotion Div. and Human Resources Development Div.

Figure 3. AWD6P/J structures

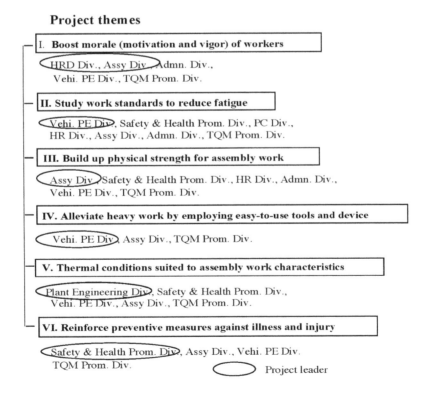

Project themes

I. **Boost morale (motivation and vigor) of workers**
   HRD Div., Assy Div., Admn. Div.,
   Vehi. PE Div., TQM Prom. Div.

II. **Study work standards to reduce fatigue**
   Vehi. PE Div., Safety & Health Prom. Div., PC Div.,
   HR Div., Assy Div., Admn. Div., TQM Prom. Div.

III. **Build up physical strength for assembly work**
   Assy Div., Safety & Health Prom. Div., HR Div., Admn. Div.,
   Vehi. PE Div., TQM Prom. Div.

IV. **Alleviate heavy work by employing easy-to-use tools and device**
   Vehi. PE Div., Assy Div., TQM Prom. Div.

V. **Thermal conditions suited to assembly work characteristics**
   Plant Engineering Div., Safety & Health Prom. Div.,
   Vehi. PE Div., Assy Div., TQM Prom. Div.

VI. **Reinforce preventive measures against illness and injury**
   Safety & Health Prom. Div., Assy Div., Vehi. PE Div.
   TQM Prom. Div.   ⬭ Project leader

The TQM Promotion Div. coordinated the overall project. Also, by having directors (vice president, senior managing director and managing director) as project advisors, systematic implementation throughout the organization became possible (Eri et al., 1999; Amasaka, 2000b, 2007a, 2015).

Total Task Management Team Activity by Applying Science SQC

Each team activity took the form of *total task management* (Suzumura et al., 1999; Eri et al., 199; Amasaka, 2000b). By applying Science SQC, team activities by managers and staff members ensured the rotation of the PDCA cycle (plan, do, check and action) (Amasaka, 2003).

The relation diagram of Figure 13 in Chapter 9 and the mountain climbing chart for problem solving in Figure 4 were made for the management of the overall activity so that all teams share the same milestones and steps for attaining goals and recognize the inter-relationship between individual teams and the direction of each activity.

As the main players in the assembly line are workers, a worker-oriented approach is the key to problem solving in each project team.

Figure 4. Diagram of "climbing mountain of problem-solving" for all projects

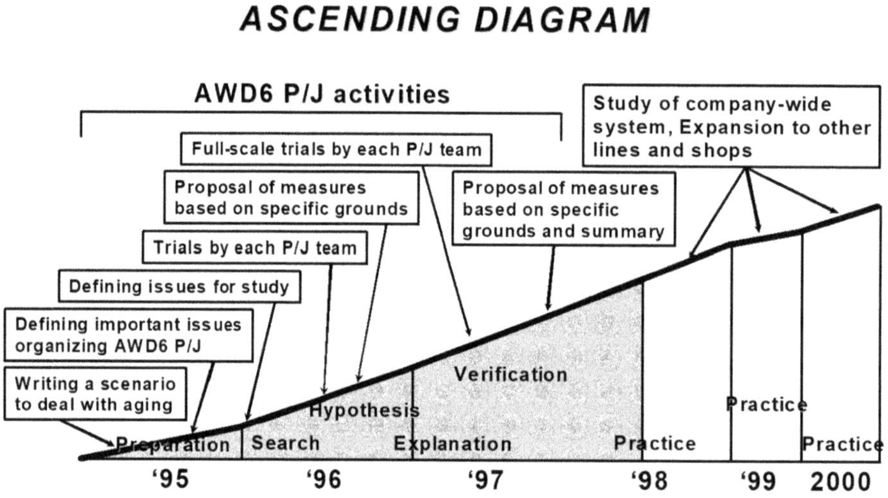

Specifically, Figure 5 shows the steps for making implicit knowledge explicit via the practical application of Science SQC. First, implicit knowledge (ambiguous, subjective information) such as opinions, intuition and worker sense should be quantified with objective and subjective indicators.

These quantitative data can then be scientifically analyzed to identify causal relationships within the given phenomena, and these indicators make possible objective, universal evaluation and make implicit knowledge explicit.

The next section presents examples of AWD6-P/J activities.

Figure 5. Science SQC approach

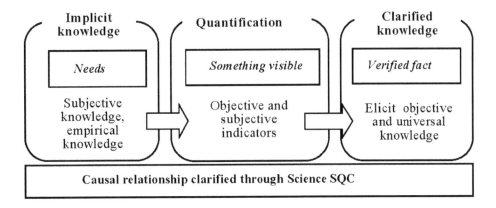

ACTUAL DEVELOPMENT OF AWD6-P/J ACTIVITIES IN TOYOTA

Study Work Standards to Reduce Fatigue (P/J II)

The objective of this project was to achieve work standards that minimize fatigue and maximize its recovery. The following three actions were taken: (1) Establishing technologies for evaluating the fatigue of assembly line workers, (2) Identifying the types of fatigue experienced by aging workers and (3) Setting and testing methods for reducing fatigue (Suzumura et al., 1998; Eri et al., 1999: Amasaka, 2000b, 2007a, 2015).

Fatigue Evaluation and Types of Fatigue Experienced by Aging Workers

Continuous assembly work results in fatigue. Fatigued workers have the desire to rest. In other words, fatigue involves changes in physical and mental state. These changes first appear as symptoms such as declining productivity and health and increased operational errors as physical function decreases.

Fatigue evaluation, therefore, requires the analysis of these changes. Effective indicators in analyzing the fatigue of workers have been confirmed in both subjective and objective terms. Through evaluation, it was concluded that aging workers experience chronic fatigue rather than acute fatigue.

Experiment by Changing the Rest Pattern and Testing the Obtained Knowledge on Model Lines

In this subsection, the author described the "Project II example -Study work of changing the rest pattern to reduce fatigue" as shown in Figure 14 of Chapter 9.

As a way to minimize fatigue, rest patterns were studied. Two rest patterns (varying the time of continuous work and breaks) were tested experimentally to analyze differences in fatigue level.

In (a) "Changed break time and effect" as shown in Figure 14(a) of Chapter 9, fatigue during operations gradually increases with time and decreases after each break. It was confirmed that fatigue increases as a whole with ups and downs.

When the number of breaks was increased and the length of continuous working time was reduced in the afternoon (when fatigue tends to increase), the fatigue level at the end of the operation became lower, according to both subjective and objective indicators (e.g., physiological data).

To verify the above-mentioned finding, two assembly lines at the "Toyota Motomachi -Plant", where the *CROWN* and *IPSUM* car model at an advanced car manufacturer A are manufactured, were selected as experimental lines for a two-month trial.

Two patterns show the "(b) Brake pattern comparison: "Normal and Trial "New pattern A and B)" described in Figure 14(b) of Chapter 9.

Figure 14(b) of Chapter 9 was set for the trial, with continuous 90-minute operation as the base. Pattern A adopts 5 minutes for the second break in the afternoon while pattern B adopts 10 minutes.

Pattern A follows pattern B on the No. 1 line and pattern B follows pattern A on the No. 2 line. The trial was conducted with about 500 workers on the No. 1 and No. 2 lines.

The effect perceived by workers and their opinions were used as subjective indicators, while the effects on productivity were used as objective indicators to provide the major basis for evaluation.

Perceived Effect and Free Opinions (Subjective Indicators)

For both patterns A and B, most workers found that fatigue was reduced at the end of daily operations. It was also confirmed that pattern B produced a greater effect than pattern A. A possible reason for this result is that workers felt psychological stress from a 5-minute break, because they were used to 10-minute breaks.

Furthermore, answers to questions about expected retirement age based on confidence in physical strength showed a rise, indicating a reduced physical load. Free opinions on changes in physical, mental and operating conditions during the trial were collected, classified and summarized.

About 70% responded that breaks were "good for health" and that they felt "less load to parts of the body" in terms of physical condition.

As for the mental aspect, about 60% said that they "felt more relaxed with work time reduced to 90 minutes" and were able to "concentrate more on the operation". Lastly, regarding operations, about 50% declared that such changes resulted in "less operational delay" and "fewer errors".

These responses indicated improved operational quality resulting from the synergy of physical and mental effects.

Effect on Productivity (Objective Indicators)

The line stops time data between "12:25 and 13:25" when delay is most likely to occur was sampled, as it was assumed that "line stop time" is closely related to delays in "operation on the No. 2 assembly line" described the "(c) Example of line stop time between 12:25 and 13:25 that causes work delays" as shown in Figure 14(c) of Chapter 9.

In Figure 14(c) of Chapter 9, the line stop time decreased by about 2 minutes on average and productivity increased by about 3% during the trial period of December and January compared to November.

Free opinions were collected from foremen who constantly watch assembly lines to check operation rates and product quality. About 50% of foremen said that there were fewer operational delays. With regard to product quality, about a 40% declared improvement. These results demonstrate the effectiveness of the rest pattern change in decreasing worker fatigue and improving productivity.

Temperature Conditions Suited to Work Characteristics (P/J V)

The project team aimed to realize an air conditioning system that considers individual differences in temperature preferences so that temperature conditions do not adversely affect fatigue levels. The focuses of their activities were as follows: (i) Clarifying the relationships between temperature and fatigue (ii) Analyzing problems of current air conditioning for assembly lines and (iii) Development of an air conditioning system suitable for assembly processes.

Suitable Temperature Conditions for Minimizing Fatigue

Activities revealed that a temperature of 28°C to 31°C and airflow of 1 meter / second are desirable for suppressing fatigue. When this environment was created using line-flow air conditioning and synchronous fans, suppression of fatigue was observed from both subjective indicators and physiological data.

Also, it was also found that non-breathable, sweaty work clothes have the "various adverse feelings (affects)" on body heat control. Therefore, development of comfortable work clothes with excellent moisture absorption and drying properties was promoted.

Development of Comfortable Work Clothes and Test on Model Lines

The present punch-knit work clothes made of 100% cotton offer good moisture (sweat) absorption performance but poor heat radiation, ventilation and drying, resulting in sticking. As a result of a technical survey and development efforts, stitch-knit work clothes made of a special fiber (porous hollow-section polyester) and cotton were made as prototype comfortable work clothes with 2.6 times the ventilation and drying capability of present work clothes, and similar moisture absorption properties.

An awareness survey of actual workers on model lines (Motomachi-Plant assembly lines: Nos. 1 and 2) was conducted to test the following: (a) Any difference in the feeling of workers with and without synchronous fans (with new wide-area exposure function) and (b) Differences between comfortable work clothes and conventional work clothes

Since the fatigue reduction effect of line-flow air conditioning had been made clear through past activities, we assumed line-flow air conditioning for the model lines. Since the temperature was felt to be 1 to 2 °C lower when wearing comfortable work clothes, the air conditioner outlet temperature was raised by 1°C from that at the time of the last evaluation.

Figure 6 shows the scattered diagram obtained by principal component analysis (by use of correlation matrix) of the survey results regarding the awareness of model line workers. The results show that workers felt that comfortable work clothes were better than conventional work clothes.

Also, the installation of a synchronous fan was evaluated highly in terms of both air conditioning and work clothes. As the result of checking actual opinions, many workers commented that the comfortable work clothes were less stuffy and sticky and allowed for easy work movement.

Comfortable work clothes were also evaluated favorably by workers involved in processes that have insufficient exposure to airflow. While the initial prototype clothes (100% special fiber) were favored for processes exposed to airflow at 1 meter/second or more, they ware unpopular for processes exposed to less airflow because they allowed for little heat radiation (due to the heat insulation effect of the pores in the special fiber).

Figure 6. Principal component analysis (of awareness survey results)

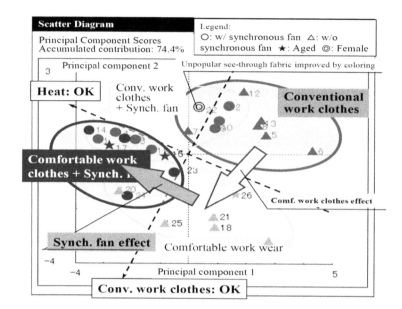

This problem was solved by mixing the special fiber with cotton. Stitch knitting may have reduced stickiness to the skin (contributing to easy motion) thanks to the added surface roughness. In the actual application test, evaluation was made by raising the air conditioner outlet temperature by 1°C from that of the previous evaluation, resulting in reduced steam consumption by the condenser of about 13%.

This yielded energy savings in addition to improving the work environment.

Outline of Other Projects

1. P/J I: Implemented a new system for line workers (New Life Action Program) and promoted work development for skilled workers.
2. P/J III: Verified the effectiveness of stretch exercise in fatigue recovery at a model workshop successfully promoted stretch exercises and achieved fatigue reduction.
3. P/J IV: Improved high load work by providing easy to use tools and devices.
4. P/J VI: Developed methods to evaluate load to fingers for disease prevention and successfully conducted disease prevention activities at a model workplace.

APPLICATION EXAMPLES USING SUPERIOR PRODUCTION TECHNOLOGIES IN TOYOTA

To strengthen the "Epoch-making innovation in work quality for auto-global production strategy" more, it is important to accelerate the validity of AWD-6P/J.

Then, as mentioned with Chapter 6 "Evolution of TPS employing Advanced TPS", the author develops the "DGEM" (Dual Global Engineering Model) with both of the "NJ-GPM and NJ-GMM" (New

Japan Global Production Model and New Japan Global Manufacturing Model) employing "SCTTM" (Strategic Cooperative Task Team Model based on the SSTTM) in Toyota.

Specifically, to reform of production planning and preparation, the author focuses on the (1) "Digital engineering by Human Digital Pipeline System" (HDP) using HI-POS (Human Intelligence Production Operating System), and (2) "Strengthen the highly skilled operators" using "ATTS" (Accelerated Technical Training System) as follows (Amasaka, 2015, 2017a,b, 2018, 2020a,b, 2021, 2022a,b,c, 2023).

Digital Engineering by HDPS Using HI-POS

In typical vehicle assembly line using "HI-POS, strengthening of the intelligent operator", to conduct the simulation as to whether operators can complete their entire operation of one process in the set "*Takt-Time* and *Heijunka*", the author has developed the "Digital engineering by HDPS (Human Digital Pipeline System)" employing the combination of "PPOS and WCOS" ("Production Process Optimization Simulation" and "Work Operation Combination Simulation") described in Figure 7, 8, 10 and 11 described in Chapter 9 (Sakai and Amasaka, 2006, 2007a,b, 2008a,b, 2011; Amasaka, 2007a; Yamaji et al., 2007; Yamaji and Amasaka, 2007a,b, 2009; Amasaka and Sakai, 2010) (See to "Dual Global Engineering Model to advanced NJ-PMM" and "Application examples" in Chapter 9).

If it says more, this system carries out the image training of production processes in the actual order of assembly operations.

This is done at the preparatory stage before beginning mass production, and without having a real product on hand.

Also, this system verifies the line process setup in advance in order to make sure whether or not the production planning for manufacturing a number of models on one line can be carried out in the real production takt time.

This HDPS promotes the leveling of the workload of the operators in each process, and then completes the building up of the production line even before launching it by developing "(i) Hardware configuration, (2) Software configuration" and "Simulation algorithm".

As a result, the improvement in the understanding and acquisition level of technique / skill processes before the start of mass production is possible.

Hardware configuration of HDPS

The contents of such visual manuals should be revised as needed to reflect clearer ideas. The "Hardware" configuring "HDPS" is depicted in Figure 7.

The conventional work procedure manuals in which hand-written letters and drawings were used are to be done away with. Instead, intelligent, user-friendly operation manuals are prepared, which clearly present the items listed and instructed in an easily understandable manner, and offered to production operators.

More specifically, the CAD data as well as CAE data used from development to production engineering are stored in ① Product D/B through the digital pipeline.

Next, ② Production Information D/B, containing the production management information, such as production scale and volume as well as the parts arrangement information regarding procurement status and locally procured parts is connected to ③ Work Procedure D/B, which accumulates typical examples of past work procedures, completing a total linkage.

In such a procedure, a work procedure manual is prepared from work data and parts data in advance and offered to production operators. Also, at the same time, based on this work procedure manual, the routes which can be taken by the operators when moving in the production line are prepared.

Figure 7. Hardware, software and simulation configuration of HDPS

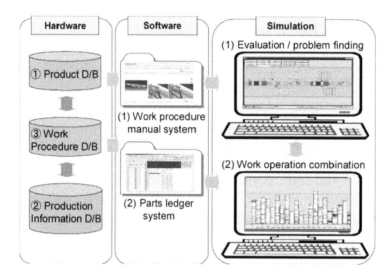

Then, from among these routes, the optimal combination of production operations is selected and arranged through simulation.

As a result, the process setup which rearranges the work processes to correspond to multi-model production will be verified before the start of mass production.

In addition, the workload of each process is totaled for comparison, so that an uneven distribution of the workload (uneven distribution of the work amount among differing vehicle models on the production line) is leveled out.

The workload as well as the work posture of the operators are confirmed beforehand, which then is subjected to evaluation and fault finding.

Software Configuration of HDPS

The "Software" configuring HDPS is depicted in Figure 7. The work procedure manual mentioned above is prepared using the (1) Work Procedure Manual System.

This system contains a wide range of information, such as the work data consisting of work names, times, locations, the specification data consisting of the specification, number, and quantity of parts, in addition to quality, work posture, instruments, safety, intuitive knacks and know-how, etc. Based on all this, the work procedure manual mentioned above is prepared.

Next, the visual data of parts is generated by using the (2) "Parts Ledger System" (PLS). Concretely, target parts are extracted using the information about their number, name, model, or quantity, and the associated 3-D shape data (CAD data) or verification data (CAE data) are searched. Based on the above systems of (1) and (2), the linkage between work and parts is made, and the elemental instruction sheet is prepared for each of the parts in the order of the steps in which the parts are being assembled.

For the types of work operations that cannot be fully instructed through explanations and photographs, video images are added to the "Visual Manual" (Sakai and Amasaka, 2006) to describe the acquired knacks and know-how.

This is to ensure more accurate work operations by instructing the procedure to be followed and the things not to be performed by providing animation.

Simulation Algorithm of HDPS

(1) Evaluation / Problem finding

The "Simulation" configuring HDPS is depicted in Figure 8. To arrange for a production operator to be able to move from his current work point to another work point in a straight line described in Figure 8, in addition to the speed of the conveyor, the following 3 factors are to be taken into consideration, for the calculation of the coordinate locations below:

V; walking speed
V_G; takt speed
M_G; moving distance from the start point
V_X; X coordinate of the work point of V
V_Y; Y coordinate of the work point of V
M_{GX}; maximum moving time per unit time of V_X
M_{GY}; maximum moving time per unit time of V_Y
T; walking time

"Walking Time Calculation Formula"

$$T = (-b + \sqrt{b^2 - ac}) / a$$

$$a = V^2 - V_{GX}^2 - V_{GY}^2$$

$$b = -M_{GX} \times V_{GX} - M_{GY} \times V_{GY}$$

$$c = -M_{GX}^2 - M_{GY}^2$$

(a) For simulating the avoidance route necessary to avert contact between an operator and a body (vehicle), an advance setting is programmed for the avoidance points for each move from one point to another, as shown in Figure 9. For the work operation inside the vehicle body, 4 avoidance points, namely; body front (engine compartment), rear (luggage), and each point on the right and left (center pillars) when viewing the whole body as a rectangle, are set. This program is designed to avoid any contact between a production operator and a vehicle body).

Figure 8. Calculation of coordinate locations

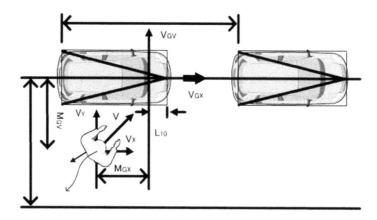

Figure 9. Advance setting for avoidance points

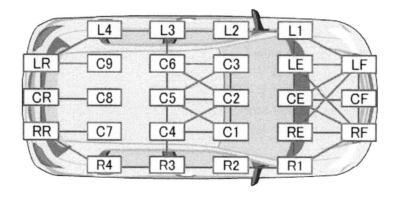

(b) Using Dijkastra method (Dijkastra, 1959; Dijkastra and Feijen, 1988), (calculation method for the shortest route) for calculating the shortest route and shortest distance, the next optimal work points for an operator to move to is calculated. (After determining the pattern of routes that a process operator can take for each vehicle model as described in Figure 7, the points are connected with lines, for example, R3-C4-C2 described in Figure 9, and the shortest route of these patterns is calculated using the Dijkastra method).

(c) As part of the condition, the next movement coordinate is calculated using the time needed for each step. (The initial setting is 1 sec.). The shorter this step time is, the more precisely the simulation can be carried out. (However, as the calculation amount increases, the load on the PC increases, too).

Operator Interference is verified through simulation to determine which combinations of vehicle models cause overlapping of operators in the time series. More specifically, the distance from the center coordinate of each operator is calculated, and if the results are smaller than the radius of the operators' movement, such will be defined as "operators' interference."

Since it is possible to decide freely which models are put on the line in which order, simulation based on the aforementioned hardware information (② Production Information D/B) is possible (see to above Figure 7).

Through simulation, verification will be conducted through simulation as to whether an operator can complete the entire operation of one process within the set takt time, or if the line conveyor needs to be stopped.

Based on the results of this simulation, advance verification is possible about which process, on which vehicle model, caused the operator to stop the line conveyor.

(2)   Work operation combination

Using the parameters of work hours, and designated walking time, entered by means of the work procedure system mentioned earlier while discussing the software ((1) Work Procedure Manual System), as well as the width and length of the vehicle body, an accumulative calculation is conducted described in Figure 7 above.

With regard to the walking time of an operator, the data computed through the above-mentioned (1) Evaluation / Problem Finding can be used as well.

For the calculation of the averaged weight, the data imported by the software ((1) Work Procedure Manual System) is used by setting as parameters in advance the production volume of a model, as well as the production volume of vehicle options.

Accumulative simulation will be carried out for the production workload for each vehicle model, which appears at random according to the ratio of the production volume of each model.

Based on this, the fluctuations in workload for each model can be verified beforehand, so that the process layout can be re-examined to realize the optimal load distribution.

It is also capable of displaying graphs for each vehicle model, process by process, which enables the verification of the work hours for each vehicle model, and furthermore, singles out in advance the priority items for shortening the work hours in view of the process distribution.

Optimization of Logistics in Toyota's Application Example

In recent years in overseas factories, work-related accidents by workers engaged in distribution occur frequently. The ratio of temporary employees is rising for domestic factories, and this problem has become more pronounced.

For example, truck stopping areas are small and tractors may protrude and interfere with the surrounding environment. Or parts cabbies come in contact with people in narrow routes.

Because many countries have introduced legislation towards security in industry, companies have increased safety precautions and stepped-up checks on work environments. Therefore, optimizing distribution routes is an important matter.

In the Figure 10, the authors digitized distribution and planned the "optimization by digital simulation of logistics".

This simulation optimizes placement and the i) "logistic route" around ii) "main body line" so that iii) "parts cabbies" do not interfere with workers and facilities having iv) "Finishing: interference".

The facilities were re-designed. Workers can prevent the "accidents and mental workloads of workers" are reduced by realizing this.

Figure 10. Optimization by digital simulation of logistics

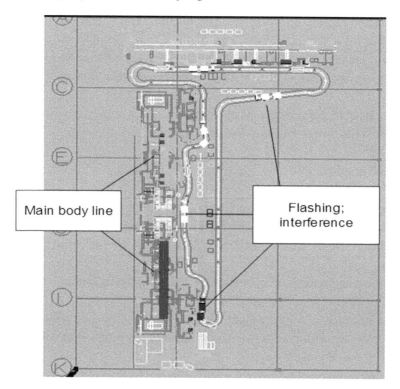

Strengthen the Highly Skilled Operators Using ATTS

In Chapter 9, to realize the evolution of auto-manufacturing, the developing ATTS become indispensable necessary conditions for strengthening of the highly skilled operators in global manufacturing strategy using "HI-POS" with (a) "HIAS" and (b) "HIDS" (Human-integrated Assist system and Human Intelligence Diagnosis System) (Sakai and Amasaka, 2006, 2007, 2008a,b).

Shortened Training for the Advancing the Work Procedure in Toyota

To improve the skills of new (inexperienced) production operators both in Japan and overseas, "ATTS" is designed for the accurate transmission of operation skills and repetitive training until the trainee fully understands the work procedure employing "V-IOS" (Virtual–Intelligent Operator System) described in Figure 12 of Chapter 9, and "HIAS" (Human

Integrated Asist System) using "Visual Manual" (See Figures in the Appendix).

Specifically, this system provides the footage of work operations performed after the repetitive image training via the visual manual mentioned above (i).

As shown in Figure 11, a series of work operations demonstrated by a "highly skilled trainer" are video-recorded and the same operations performed by a newly recruited production operator are also recorded and uploaded as video data.

These two files are processed to be played side by side on a PC display. In this way, production operators can view their weak points objectively, and work on correcting them repeatedly so as to improve their skills to the desired level in a short period of time.

Figure 11. Image comparisons of a highly skilled trainer and a new operator

Newly employed operator   Highly skilled trainer

Assessment of Aptitude / Inaptitude by Aptitude Test

It has been determined that, of the production operators assigned to assembling processes, those who are assessed in advance as "apt" have a tendency to improve their skill in a short time after the assignment.

The correlation between fundamental skill training items in the assembling process and skill acquisition on the site has also been identified.

Based on these notions, the aptitude test sheet shown in Figure 12 was devised for the advanced assessment of aptitude and inaptitude.

This illustrates the example of an assembling process in which a test is carried out for eight fundamental skill items, and the total points for each item are used for the assessment.

Based on the results, those who are assessed as "apt" immediately start receiving training at the actual production line, whereas those who proved to be "inapt" go through repeated "dynamic training", which simulates a similar movement to the actual operations at the production line, to accelerate their skill development.

Figure 12. Aptitude test sheets (ex. assembly process)

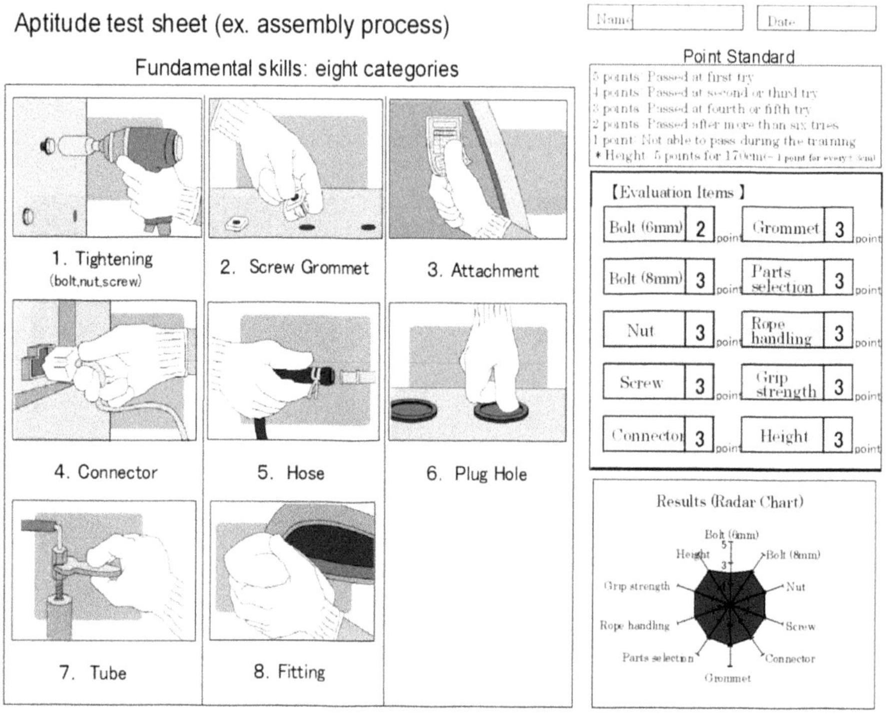

Optimization of Training Steps

Figure 13 presents the flow chart outlining the training steps below.

Repetitive training using the visual manual mentioned in the figure (1) will be repeated along with the self-study video described in the figure (2) to make the trainee understand and recognize their weak points until they reach the skill level designated for each process on the specified evaluation sheet.

Next, in accordance with the results of the aptitude test conducted earlier consisting of the eight fundamental skills discussed in the figure (3), assessment of aptitude and inaptitude will be made.

Those who are assessed as "apt" immediately start receiving training at the actual production line.

Those who proved to be "inapt" go through repeated "dynamic training", which simulates a similar movement until they reach the skill level designated on the evaluation sheet, after which training at the production line is given.

ATTS Application Example

The author explains the specific examples of the implementation of the "ATTS system" using "HIA system" and the perceived effectiveness of this system at Toyota.

(1)   Application case of skill training for newly employed production operator

Figure 13. Flow chart of training program

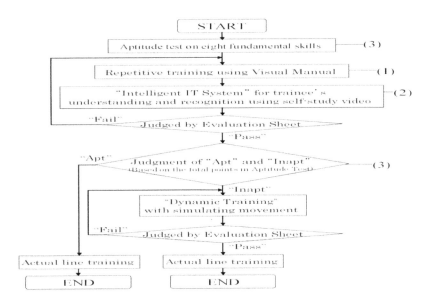

The skill training for newly employed production operators in Japan and abroad was conducted in the production preparations stage for startup with a stabilized regular production system. Skill training for assembly jobs is explained as an example here.

Skill training is conducted for production operators so that they are able to operate correctly at the time of a new plant start-up or model change. Especially in the case of assembly jobs, accurate work completion within the specified time is required for target attainment at an early stage before start-up or changeover. Since the conventional training used a group of operations for judgment of the training result based on completion in the specified cycle time, it brought about a disparity in the quality of final products.

To be more definite, an example of a trimming process is shown in Figure 13. Fundamental skills are divided into eight items: bolt (6 Mm tightening), bolt (8 Mm tightening), screw tightening, connector installation, screw grommet installation, parts selection and rope routing. The training is conducted with the stress placed on bolt (6 Mm tightening) and nut / screw tightening.

This radar chart shows the difference between the requirements and the personal diagnosis so as to make each person aware of, and able to overcome their weak points through training.

The training is conducted repeatedly until attainment of the target level is determined by evaluation using the specified evaluation sheet.

In Figure 14, the author shows an example of the "Visual Manual" concerning the bolt feeding operation. First, the operational procedure is broken down into smaller operational movements. Accurate motions are clearly indicated visually using still pictures, movie images, and animation. The "Visual Manual" is in a standardized form. The screen consists of a procedure block, description block, and a display image block, which has a main and a sub-screen.

This manual is read by turning the page using the forwarding button according to the procedure. Particularly in the description block, key points representing the know-how of each intelligence operator are written and visual information not displayed on the main screen is shown in detail in the sub-screen.

The explanatory text under each image describes why the posture is needed, what role it plays in quality assurance, or other information from the intelligence operator so as to share the best practices in the world.

The Figure 14 displays a movie showing the nut feeding procedure for assembling product parts on the main center screen. The sub-screen on the right side displays a still picture showing a key point of good posture by maintaining a right angle between the arm and the fingers.

Figure 14. An example of skill levels between requirements and personal diagnosis

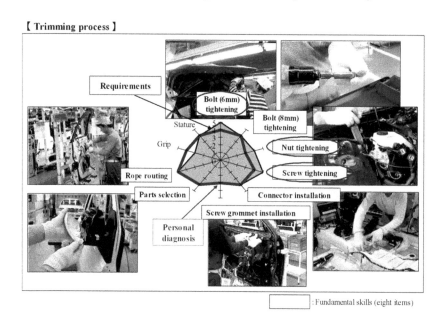

(2) Learning evaluation for new employees assigned to trimming process

Then, Figure 15 shows the learning evaluation conducted for new employees assigned this time to the trimming process. The learning curve for conventional training mainly consisting of OJT using the actual vehicle is compared with the new training using the "Visual Manual".

The degree of learning indicated in time series for the assigned trimming process job, according to the individual evaluation sheet (details are omitted), show that it took four weeks until satisfaction of the specified accuracy within the specified work time in the case of conventional training. However, this was reduced to one half (two weeks) with the new method.

The analytical results are as follows:

1) Training with the visual manual to detect primary individual weak points based on personal diagnosis has improved the understanding of the assigned job, and achieved faster learning compared to the conventional method.

Figure 15. Learning evaluation for new employees assigned to trimming process

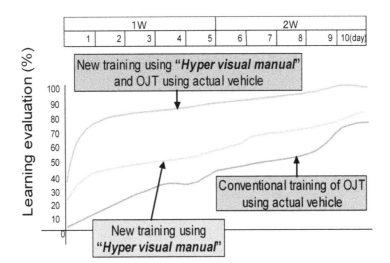

2) When training with the visual manual is combined with OJT on the actual vehicle, it has been confirmed that the learning speed can be increased through repetition of training that places an emphasis on personal weak points.

3) Efficient training was attained without disparity in the degree of learning by teaching the same contents in the same manner according to the clarified teaching process and because the training procedure was not dependent on the quality of the trainers.

(4) Application of shortened training for new overseas production operators

Below is an application case involving the overseas production operators at a leading auto-manufacturer, Toyota Motor Corporation. The deployment of created "ATTS application" for training newly employed production operators at domestic and overseas manufacturing plants reduced the conventional training period by more than half, from two weeks to five days, leading the full-scale production to a good start.

In Figure 15, specifically, the author indicates the skill training curriculum for new overseas production operators. After conducting the aptitude test for fundamental skills, skill training utilizing the "Visual Manual & Intelligent IT system" was repeatedly carried out until the standard set out in the "evaluation sheet" was met (See to Subsection "Innovation of automobile manufacturing engineering based on the NJ-GGM" in Chapter 9).

Those whose results on the aptitude test proved to be "inapt" went through "dynamic training" using the simulation training chart to simulate the movement repeatedly until the standard set out in the evaluation sheet was similarly met. Afterwards, actual line training was finally given, and the designated skill level was reached in a period less than half compared to before.

In this case, the launch of an overseas production plant, the target operating rate was achieved in four months after the start-up as shown in Figure 16.

Figure 16. Transition of operation rate after launch of production line

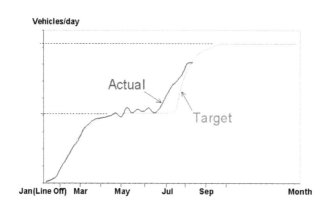

CONCLUSION

In this chapter, to realize the high-productivity in the driving force "New JIT strategy", the author has organized the "AWD-6P/J" (Aging Working Development 6 Projects) for evolution of the work environment, and has verified its validity at an advanced car manufacturer Toyota. Specifically, in the progress of digital engineering for superior quality assurance, to make the results of AWD-6P/J develop, the author has carried out both of the "AWD-COS and HDP system". Concretely, 1st, in "AWD-COS", the author has developed the comfortable workplace for older workers through objective analysis from behavior science perspectives. Moreover, 2nd, in "HDP system", the author has developed the total linkage of intellectual production information from designing to manufacturing through a digital pipeline for the reform of production preparation. These systems are greatly contributing to global production strategies of Toyota.

REFERENCES

Amasaka, K. (2000a). *New JIT, A New Principle for Management Technology 21^C: Proposal and demonstration of "TQM-S" in Toyota (Special Lecture).* Proceedings of the I World Conference on Production and Operations Management, Sevilla, Spain.

Amasaka K. (2000b). *AWD-6P/J report of first term activity 1996-1999: Creation of 21st century production line in which people over 60's can work vigorously (Project leader of AWD-6P/J).* Toyota Motor Corporation.

Amasaka, K. (2002). *New JIT*, a new management technology principle at Toyota. *International Journal of Production Economics, 80*(2), 135–144. doi:10.1016/S0925-5273(02)00313-4

Amasaka, K. (2003). Proposal and implementation of the "Science SQC" quality control Principle. *Mathematical and Computer Modelling, 38*(11-13), 1125–1136. doi:10.1016/S0895-7177(03)90113-0

Amasaka, K. (2004a). Development of "Science TQM", a new principle of quality management: Effectiveness of Strategic Stratified Task Team at Toyota. *International Journal of Production Research*, *42*(17), 3691–3706. doi:10.1080/0020754042000203867

Amasaka, K. (2004b). Global production and establishment of production system with high quality assurance, Toward the next-generation quality management technology (Series 1). *Quality Management*, *55*(1), 44–57.

Amasaka K. (2004c). New development of high-quality manufacturing in global production, Toward the next-generation quality management technology (Series 4). *Quality management, 55*(4), 44-58.

Amasaka, K. (2007a). *Applying New JIT-* Toyota's global production strategy: Epoch-making Innovation in the work environment. *Robotics and Computer-integrated Manufacturing*, *23*(3), 285–293. doi:10.1016/j.rcim.2006.02.001

Amasaka, K. (2007b). High Linkage Model "*Advanced TDS, TPS & TMS*": Strategic Development of "New JIT" at Toyota. *International Journal of Operations and Quantitative Management, 13*(3), 101–121.

Amasaka, K. (2007c). New Japan Production Model, an advanced production management principle: Key to strategic implementation of New JIT. *The International Business & Economics Research Journal*, *6*(7), 67–79.

Amasaka, K. (2013a). *The strategic development of Advanced TPS based on the New Manufacturing Theory*. Recent Advances in Industrial & Manufacturing Technologies The 1st WSEAS International Conference on Industrial and Manufacturing Technologies (INMAT13), Vouliagmeni, Greece.

Amasaka, K. (2013b). The development of a Total Quality Management System for transforming technology into effective management strategy. *International Journal of Management*, *30*(2), 610–630.

Amasaka, K. (2014). New JIT, new management technology principle: Surpassing JIT, *ScienceDirect*. *Procedia Technology*, *16*(special issues), 1135–1145. doi:10.1016/j.protcy.2014.10.128

Amasaka, K. (2015). New JIT, new management technology principle, Taylor & Francis Group, CRC Press, Boca Raton, London, New York.

Amasaka, K. (2017a). Strategic Stratified Task Team Model for realizing simultaneous QCD fulfilment: Two case studies. *Journal of Japanese Operations Management and Strategy*, *7*(1), 14–35.

Amasaka, K. (2017b). *Toyota: Production system, safety analysis and future directions*. Nova Science Publishers.

Amasaka, K. (2018). Innovation of automobile manufacturing fundamentals employing New JIT: Developing Advanced Toyota Production System, *International Journal of Research in Business. Economics and Management*, *2*(1), 1–15.

Amasaka, K. (2020a). Evolution of Japan manufacturing foundation: Dual Global Engineering Model-Surpassing JIT. *International Journal of Operations and Quantitative Management*, *26*(2), 101–126. doi:10.46970/2020.26.2.3

Amasaka, K. (2020b). Studies on New Japan Global Manufacturing Model: The innovation of manufacturing engineering. *Journal of Economics and Technology Research*, *1*(1), 42–64. doi:10.22158/jetr.v1n1p42

Amasaka, K. (2021). New Japan Automobile Global Manufacturing Model: Using Advanced TDS, TPS, TMS, TIS & TJS. *Journal of Business Management and Economic Research*, *6*(6), 499–523.

Amasaka, K. (2022a). A New Automobile Global Manufacturing System: Utilizing a Dual Methodology. *Business and Management and Economics Research*, *8*(4), 41–58. doi:10.32861/sr.83.51.58

Amasaka, K. (2022b). *Examining a New Automobile Global Manufacturing System*. IGI Global Publisher. doi:10.4018/978-1-7998-8746-1

Amasaka, K. (2022c). New Manufacturing Theory: Surpassing JIT (2nd Edition), Lambert Academic Publishing.

Amasaka, K. (2023). New Lecture-Surpassing JIT: Toyota Production System-From JIT to New JIT, Lambert Academic Publishing.

Amasaka, K., & Sakai, H. (2010). Evolution of TPS fundamentals utilizing New JIT strategy–Proposal and validity of Advanced TPS at Toyota. *Journal of Advanced Manufacturing Systems*, *9*(2), 85–99. doi:10.1142/S0219686710001831

Amasaka, K., & Sakai, H. (2011). The New Japan Global Production Model "NJ-GPM": Strategic development of *Advanced TPS*. *The Journal of Japanese Operations Management and Strategy*, *2*(1), 1–15.

Aspy, P. P., & Wai, K. M. (2001). An online evaluation of the Compete Online Decision Entry System (CODES). *Developments in Business Simulation & Experimental Leaning*, *28*, 188–191.

Aspy, P. P., Wai, K. M., & Dean, S. R. (2000). Facilitating learning in the new millennium with the Complete Online Decision Entry System (CODES). *Developments in Business Simulation & Experimental Leaning*, *27*, 250–251.

Chawla, R., & Banerjee, A. (2002). An automated 3D facilities planning and operations model: Generator for synthesizing generic manufacturing operations in virtual reality. *Journal of Advanced Manufacturing Systems*, *1*(1), 5–17. doi:10.1142/S0219686702000039

Dijkastra, E. W. (1959). *Communication with an Automatic Computer*. Excelsior.

Dijkastra, E. W., & Feijen, W. H. J. (1988). *A method of programming*. Addison-Wesley Longman Publishing Co., Inc.

Ebioka, K., Sakai, H., Yamaji, M., & Amasaka, K. (2007b). A New Global Partnering Production Model "NGP-PM" utilizing "Advanced TPS". *Journal of Business & Economics Research*, *5*(9), 1–8.

Ebioka, K., Yamaji, M., Sakai, H., & Amasaka, K. (2007a). Strategic development of *Advanced TPS* to bring overseas manufacturing to Japan standards: Proposal of a New Global Partnering Production Model and its effectiveness, *Production Management*. *Transaction of the Japan Society for Production Management*, *13*(2), 51–56.

Eri, Y., Asaji, K., Furugori, N., & Amasaka, K. (1999). *The development of working conditions for aging worker on assembly line (#2).* The Japan Society for Production Management, The 10th Annual Technical Conference; Nagoya University, Japan.

Niemela, R., Rautio, S., Hannula, M., & Reijula, K. (2002). Work environment effects on labor productivity: An intervention study in a storage building. *American Journal of Industrial Medicine, 42*(4), 328–335. doi:10.1002/ajim.10119 PMID:12271480

Sakai, H., & Amasaka, K. (2005). V-MICS, Advanced TPS for strategic production administration: Innovative maintenance combining DB and CG. *Journal of Advanced Manufacturing Systems, 4*(6), 5–20. doi:10.1142/S0219686705000540

Sakai, H., & Amasaka, K. (2006). Strategic *HI-POS*, intelligence production operating system: Applying *Advanced TPS* to Toyota's global production strategy. *WSEAS Transactions on Advances in Engineering Education, 3*(3), 223–230.

Sakai, H., & Amasaka, K. (2007a). Human Digital Pipeline Method using Total Linkage through design to manufacturing. *Journal of Advanced Manufacturing Systems, 6*(2), 101–113. doi:10.1142/S0219686707000929

Sakai, H., & Amasaka, K. (2007b). The robot reliability design and improvement method and Advanced Toyota Production System. *The Industrial Robot, 34*(4), 310–316. doi:10.1108/01439910710749636

Sakai, H., & Amasaka, K. (2008a). Demonstrative verification study for the next generation production model: Application of the Advanced Toyota Production System. *Journal of Advanced Manufacturing Systems, 7*(2), 195–219. doi:10.1142/S0219686708001577

Sakai, H., & Amasaka, K. (2008b). Human-Integrated Assist System for Intelligence Operators, Encyclopedia of Networked and Virtual Organization, II(G-Pr), 678-687.

Suzumura, H., Sugimoto, Y., Furusawa, N., Amasaka, K., Eri, Y., Asaji, K., Furugori, N., & Fukumoto, K. (1998). *The development of working conditions for aging worker on assembly line (#1).* The Japan Society for Production Management, The 8th Annual Technical Conference; 1998, Kyushu-Sangyo University, Japan.

The Japan Machinery Federation., & Japan Society of Industrial Machinery Manufacturers. (1995b). Research Report: Production System Model Considering Aged Workers, Tokyo, 1-2.

The Japan Machinery Federation and the Japan Society of Industrial Machinery Manufacturers. (1995a). *Research Report: Advanced Technology Introduction in Machinery Industry.* The Japan Machinery Federation and the Japan Society of Industrial Machinery Manufacturers.

Toyota Motor Corp., & Toyota Motor Kyushu Corp. (1994). Development of a new automobile assembly line. *Business Report Awarded with Okochi Prize, 1993*(40th), 377–381.

Vecchio, D., Sasco, A. Jr, & Cann, I. C. (2003). Occupational risk in health care and research. *American Journal of Industrial Medicine, 43*(4), 369–397. doi:10.1002/ajim.10191 PMID:12645094

Yamaji, M., & Amasaka, K. (2009). Strategic Productivity Improvement Model for white-collar workers employing Science TQM, *JOMS. The Japanese Operations Management and Strategy Association*, *1*(1), 30–43.

Yamaji, M., Sakai, H., & Amasaka, K. (2007). Evolution of technology and skill in production workplaces utilizing Advanced TPS. *Journal of Business & Economics Research*, *5*(6), 61–68.

APPENDIX

Human Integrated Assist System (HIAS)

This provides for creative, meaningful working, based on a production philosophy as the engineering nucleus of the Advanced TPS (Aspy et. al., 2000; Aspy and Wai, 2001; Sakai and Amasaka, 2005, 2006, 2007).

Three basic requirements of production are as follows:

1) Establishing a production system that ensures overall line reliability, maintainability, and improves "advanced production equipment operation technologies" (fault diagnosis, maintenance and preventive maintenance of production equipment) to prevent availability degradation and quality defects.

2) Upgrading engineering and other skill levels to levels comparable with Japanese staff in a wide range of areas.

This includes manual lines for small-lot production in automotive and other general assembly industries involving many hard-to-automate jobs, especially in developing countries.

3) Attaining "mastered skills" equivalent to those of production operators in advanced countries equipped with sophisticated mass-production type equipment that allows intelligent diagnosis by individuals in the quality incorporating stage.

For systematization to satisfy these three requirements, it is necessary to improve the "sophisticated production equipment operating skills" and "mastered skills" of production operators themselves.

Developing the intelligence operators for upgrading and unifying the production capabilities of worldwide operators is required before the "advanced production process", which is continuously driven by IT and digitization, creates a black box that depends on personal discretion (Chawla, 2002; Sakai and Amasaka, 2007).

The author, therefore, develops the "Human-integrated Assist system" (HIAS) that realistically copes with these requirements (Sakai and Amasaka, 2008b).

Figure 17A shows the HIAS construction. As will be explained from the next section, the HIAS covers system (tools, execution plan, skill requirements, and evaluation necessary) and methodology (formulating and establishing shortened training methods) for early fostering of operators with "mastered skills" as one of the two requirements explained earlier for a global production strategy.

Figure 17. HIAS construction in ATTS

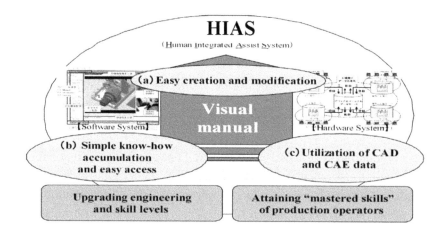

Visual Manual in ATTS

The author has developed the "Visual Manual" as a new communication tool as a key technology in ATTS (Accelerated Technical Training System) below.

(1) Easy creation and modification

Conventional, hard-to-understand operation manuals written only in character text should be replaced with user-friendly intelligent manuals allowing easy understanding of the content by intelligence operators. The contents of such visual manuals should be revised as needed to present new ideas clearly.

Figure 18. "Hyper visual manual" creation

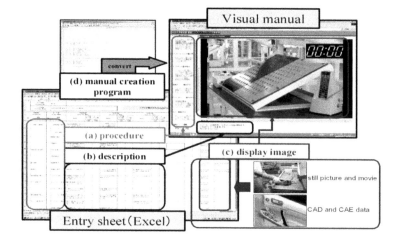

Figure 16B-(1) shows the "visual manual creation process". The entry sheet is created by modifying a world-standard Microsoft Excel sheet. This sheet is for the input of (a) procedure, (b) description and (c) display image, and the input information is then converted to a visual manual using (d) the "manual creation program. Anybody can easily revise the visual manual by using this sheet to modify the display image and explanatory text.

Through this method, manuals that have conventionally been used only for basic knowledge learning can be improved to facilitate accumulation of specialized knowledge.

(2) Simple know-how accumulation and easy access

It is necessary for intelligence operators at worldwide production sites to be able to have access to the necessary visual manual content for immediate use. Furthermore, revision of and additions to the visual manual requires simultaneous data storage and distribution.

Figure 16B-(2) shows the system for visual manual selection. Depending on the selection method, a series of screens matching each particular situation can be selected.

Method (a) clearly indicates the contents (title of each item is in a list) for the search of necessary content. On the other hand, method (b) employs a search program using a key word for selection of content from the search result.

The visual manual data is handled in the general HTML (Hyper Text Markup Language) format, enabling it to be used anywhere in the world, and the data can be delivered locally and globally through the internet.

Figure 19. System for visual manual selection

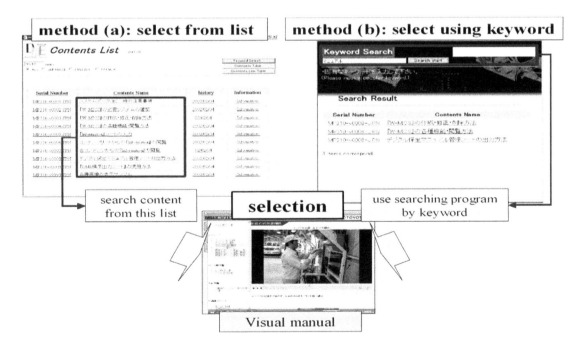

With regard to the hardware system, a plant server is installed at each plant to establish a server and client system. It is used to view the data on each client system (PC) at each office or work-site through the internet, and a note can be written as required on the spot.

Then, each plant server is controlled in synch with the central server so as to make the system simultaneously update and distribute any modifications.

Thus, the intelligence operators at local and overseas plants can virtually experience the production method related to each production process and also obtain the related know-how, knowledge, and information concerning the same production process.

(3) Utilization of CAD and CAE data

Explanation of work instructions by character text only creates problem such as lack of clarity and difficulty accessing the required item. On the other hand, visually appealing work instructions provide a perfect description of the scene so as to enable members that use different languages to obtain unified understanding of the same material.

As the skill level and training method may vary from trainer to trainer, use of the still picture and movie images of the CAD / CAE data in the visual manual will convey consistent information at a higher level.

Figure 19 shows the flows of CAD and CAE data. The product data is mainly provided in still images for use by the (a) design division. Movie images, on the other hand, are used as visual manual data in the (b) R&D and (c) production engineering divisions and in the manufacturing shop.

Thus, CAD and CAE data information has been made available on a network to production engineering and up to the manufacturing process.

It has become possible to instantaneously create procedures for the manufacturing process based on the design data, which in turn has made it possible to shrink the lead times from design to production.

Figure 20. Flows of CAD and CAE data

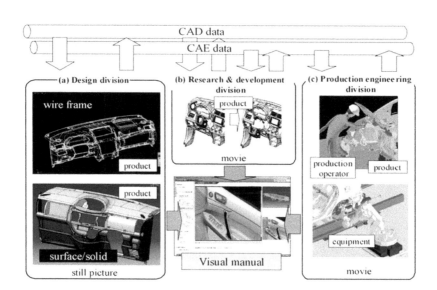

Chapter 16
New Global Partnering Production Model for Overseas Manufacturing

ABSTRACT

In this chapter, the author describes the new global partnering production model (NGP-PM) for Japan's expanding overseas manufacturing strategy. Recently, leading companies aimed to succeed in localizing production as a global production strategy; the key to this is success in global production in order to get ahead in the worldwide quality competition. Specifically, to improve quality at leading manufacturers' overseas production bases from the perspective of global production, the new global partnering production model (NGP-PM)—the strategic development of the new global production model (NJ-PM) for Toyota's expanding overseas manufacturing—is established. In the enforcement stage, to development of New JIT strategy, strategic global deployment of created strategic productivity improvement model (SPIM) has proved especially the validity of these models.

PROBLEM WITH SUCCESS IN AUTO-GLOBAL PRODUCTION

Recently leading companies aim to succeed in localizing production as a global production strategy; the key to this is success in global production. However, it has been observed that, despite the fact that overseas plants have the relevant production systems, facilities, and materials equivalent to those that have made Japan the world leader in manufacturing, the "building up of quality–assuring of process capability (Cp)" has not reached a sufficient level due to the lack of skills of the production operators at the manufacturing sites (Amasaka et al., 1999; Amasaka. 2002, 2004a,b).

Under such a circumstance, there are many studies abroad for globalization (Lagrosen, 2004; Ljungström, 2005) and TQM (Burke et al., 2005; Hoogervorst et al., 2005). As a countermeasure to such a problem, and in order not to lag behind the "evolution of digital engineering–the transition to advanced production systems at production plants", the Japanese manufacturers expect the production plants in Japan to serve as "mother plants".

They would welcome overseas production operators to these plants, and promote a "local production program – transplanting the know-how of Japanese manufacturing" (Amasaka, 2002; Sakai and Amasaka, 2006).

DOI: 10.4018/978-1-6684-8301-5.ch016

However, it is by no means easy to transfer the "know-how of Japanese manufacturing" directly to overseas production bases as mentioned above.

In other words, there is always "an obstacle to overcome - a suitable production system for each production base", due to the difference in ability (level of skill and education) or national characteristics between the local production site and Japan.

Therefore, to cope with this situation, an environment in which the creation of labor values – ES (employee satisfaction), advanced skills, a sense of achievement, and self-development can be realized must be urgently considered (Amasaka et al., 2006; Yamaji et al., 2006, 2007a).

To accomplish the above, the author surmises that it is necessary to develop a type of manufacturing which fits the local circumstances of various overseas production bases, and to advance from "Japanese mother plants" to "global mother plants" (Amasaka, 2013, 2019, 2015, 2016, 2017a, 2022a,b, 2023a).

NEW GLOBAL PARTNERING PRODUCTION MODEL FOR JAPAN'S OVERSEAS MANUFACTURING STANDARDS

The Significance of Global Partnering Research

The author has defined "global partnering" as knowledge sharing in order to promote continual evolution of the production plants in Japan and overseas, as well as greater cooperation among production operators (Amasaka, 2004a).

In other words, global partnering means that Japanese companies no longer stick to a one-sided promotion of Japanese concepts and systems onto overseas plants as they have in the past, but that Japanese and overseas plants exchange opinions, accept each other, and share knowledge in order to continue developing the best-suited manufacturing system for each environment (Yamaji and Amasaka, 2006, 2007b; Yamaji et al., 2006, 2007b).

Therefore, for expedient achievement of globally consistent levels of quality and simultaneous global launch (production at optimal locations), the authors consider global partnering through cooperation from fresh standpoints to be the key to "achieving worldwide quality competitiveness—the simultaneous achievement of QCD" (Ebioka et al., 2007a,b; Amasaka, 2008a, 2017b).

New Japan Production Model "NJ-PM" for Overseas Strategy

To develop the New JIT strategy, the author shows a new concept "New Japan Production Model" (NJ-PM) for overseas strategy as shown in Figure 1 (Amasaka, 2005) based on the "Advanced TPS, strategic production management model" called "NJ-PMM" (New Japan Production Management Model) (Amasaka, 2021, 2023b) (See to Section "Developing Advanced TDS, TPS, TMS, TIS & TJS in New JIT strategy" in Figure 3 of Chapter 7 and Section "Developing New Japan Production Management Model (NJ-PMM) surpassing JIT" in Figure 1 of Chapter 9).

In Figure 1, as the "Key to strategic application of Advanced TPS overseas strategically", the aims of "NJ-PM" are realization of "digitized production, renewal of production management system, increasing older and female workers and creating attractive workshop environment" in the embodiment of 4 core elements: (a) Intelligent quality control system, (b) Highly reliable production system, (c) Renovating work environment System, and (d) Intellectual operator training system.

Figure 1. New Japan production model (NJ-PM) for overseas strategy

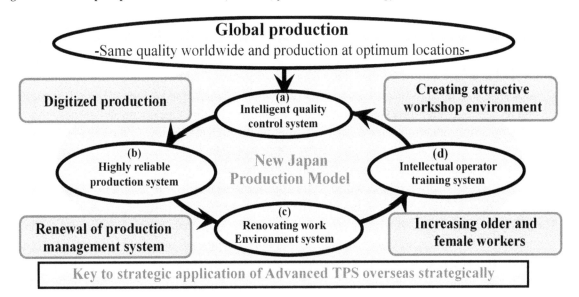

New Global Partnering Production Model "NGP-PM" for Overseas Manufacturing

For advanced Japanese companies in the process of developing global production, the most important issue is how to bring "overseas manufacturing to Japan standards" at overseas plants for expanding above "NJ-PM".

Therefore, the author has established the "NGP-PM" (New Global Partnering Production Model) which generates a synergetic effect that organically connects and promotes continual evolution of the production plants in Japan and overseas, as well as greater cooperation among production operators described in Figure 5 and 21 of Chapter 9 (Ebioka et al., 2007a,b) (See to Subsection "New Global Partnering Production Model (NGP-PM) for expanding overseas manufacturing" in Chapter 9).

The mission of "NGP-PM" is the simultaneous achievement of QCD in order to realize high quality assurance. The essential strategic policies include the following three items:

First of all, (A) the establishment of a foundation for global production, "realization of global mother plants-advancement of Japanese production sites";

Second, (B) achieving the "independence of local production sites" through the incorporation of the unique characteristics (production systems, facilities, and materials) of both developing countries (Asia) and industrialized countries (US, Europe);

Third, (C) the necessity of a "Developing Intelligence Operators by employing HIAS" to promote knowledge sharing among the production operators in Japan and overseas as well as for the promotion of higher skills and enhanced intelligence (Sakai and Amasaka, 2008).

To realize NGP-PM strategy, with above "NJ-PM" as the foundation, it is essential to create a spiraling increase in the four core elements by increasing their comprehensiveness and high cyclization.

Specifically, in "realizing global mother plants", if Japanese and overseas manufacturing sites are to share knowledge from their respective viewpoints, the core elements must be advanced.

To achieve this, a necessary measure is to design separate approaches suited to developing and industrialized countries.

First, in "developing countries (1)", the most important issue is increasing the autonomy of local manufacturing sites. At these sites, "training for highly skilled operators (focus on manual laborers)" that is suited to the manual-labor-based manufacturing sites is the key to simultaneous achievement of QCD.

Second, similarly, in "industrialized countries (2)", where manufacturing sites are based on automatization and increasingly high-precision equipment, "training of intelligence operators" resulting in "realizing highly reliable production control systems and ensuring high efficiency" is the key to simultaneous achievement of QCD (Sakai and Amasaka, 2005, 2006; Amasaka and Sakai, 2010).

Third, moreover, production operators trained at "global mother plants (3)" can cooperate with operators at overseas production bases, and in order to generate synergistic results, can work to "localize global mother plants" in a way that is suited to the overseas production bases.

The HIAS can then be effectively utilized to ensure that this contribution continues indefinitely. Furthermore, the HIAS is critical in ensuring the smooth exchange of essential information in order to realize overseas manufacturing at Japan standards.

This essential information includes quality control information for each production base as well as facilities planning information, kaizen information, and information on the level of skill of human resources.

Developing Strategic Productivity Improvement Model "SPIM" for Global Production

Global Production Strategy for Expanding Manufacturing

The management values shared by so-called winning companies are shifting from emphasis on materials to human resources (Amasaka and Sakai, 2010). The companies have amassed human resources, materials, and finances.

It is easy to procure materials as well as finances. However, human resources take time to develop, and therefore, is not as easy as the foregoing to procure.

The companies are endeavoring to grasp the information on human resources, take hold of the work, and formulate the vision in order to compete at a higher level.

It has been increasingly difficult to differentiate companies only in terms of high product quality, cost performance, and superiority in the business process.

It is imperative therefore to improve the value of human resources, but only a few companies have actually constructed a mechanism for improving human resources.

Up until now, each department has acquired information on human resources from the personnel affairs division, and systems for offering such information have been insufficient. By improving the system for sharing information on human resources, similar to sharing information on materials and finances, the business assets of companies can be effectively utilized.

Moreover, the information of in-company systems has not been completely updated until the end of fiscal terms. Therefore, the accuracy of the information provided has been inferior, and the judgments regarding management tend to rely on personal experience or inspiration. In other words, information sharing has not been speedy enough (Amasaka and Sakai, 2011).

From now on, it is necessary to offer information with high precision based on the PDCA (plan, do, check and action) cycle so that decision-making on management matters will be based on facts.

To that end, a system capable of analyzing all data, including human resources, from a variety of angles must be prepared. Under such a circumstance, there are many studies for globalization (Prajogo, 2006), TQM and information sharing (Evans and Lindsay, 1995; Goto, 1999; Lagrosen, 2004; Burke et al., 2005; Hoogervorst et al., 2005; Ljungström, 2005; Grundspenkis, 2007; Halevi and Wang, 2007, Kakkar and Narag, 2007; Yamaji and Amasaka, 2008; Amasaka, 2016).

In this application, information system, engineering control, production control and purchasing control are defined as management layer of functional division.

The management layers are the core of corporate activity and therefore it is vital for it to reinforce the functions of "business management technology" so as to strengthen and enrich both internal and external management.

This is also done in order to create a business linkage with the general administration department and cooperate with the on-site departments, such as development designing, production, and sales departments, as well as with business partners.

Total planning, personal division, etc. are defined as administrative divisions. It is increasingly important for the general administration-related department to advance corporate management by grasping the changing domestic and overseas environment surrounding the industry. It should also cooperate with the management layers so as to strengthen internal and external management.

To achieve this, it is urgently necessary to position human resource development at the core of management policy in order to enhance corporate, organizational, and human reliability as the basis of strategic quality management.

It is also necessary, at the stage of utilizing human resources, to strengthen the function of improving intelligent productivity through cooperation with all the related departments. Administrative and management layers are defined as white-collar sections, and the human resources working within the divisions are defined as white-collar workers.

Additionally, the results produced by these divisions are referred to as intellectual productivity (Yamaji and Amasaka, 2009).

Establishment of the Strategic Productivity Improvement Model (SPIM)

In recent years, Japanese companies have introduced the Western-style division of labor. Such a western labor division system is designed for easy replacement of labor forces and is achievement-oriented, it evaluates the degree of personal contribution to target achievement.

Therefore, when a problem arises, it is dealt with not as a personal problem, but only within their responsibility range. Consequently, many activities and actions based on such a principle have been criticized by society.

On the other hand, in the former Japanese system, problems were handled by all members across departments, which was a strong point of the Japanese way.

In this sub-section, an investigation has been carried out on problems and administrative issues currently facing a number of manufacturers, including automobile manufacturers.

A questionnaire survey yielded feedback from 12 division managers at 10 companies. The survey was carried out from May to June in 2008. The responses to the survey questions have been summarized and categorized into 238 issues.

These issues have been further categorized into 10 groups using an "Affinity Diagram", and Cluster Analysis (Ward's Method, normalized data) has been carried out as shown in Figure 2 (Yamaji and Amasaka, 2007b).

Figure 2 Result of the cluster analysis (dendrogram) (Ward's method: Squared Euclidean distance)

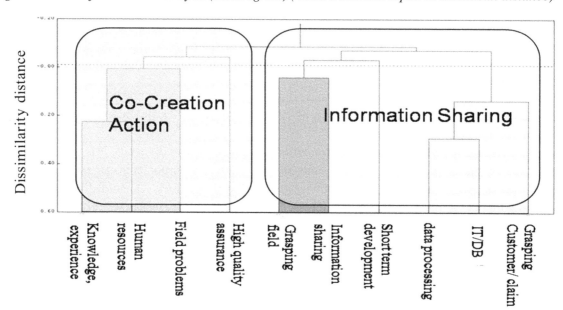

Therefore, the author has established the "Strategic Productivity Improvement Model" (SPIM) so called Toyota's NJ-QMM (New Japan Quality Management Model) for global production strategy as shown in Figure 10 of Chapter 7 (Yamaji and Amasaka, 2008) (See to subsection "Developing Strategic Productivity Improvement Model" in Chapter 7).

In Figure10 of Chapter 7, the following are the functions of the white-collar sections as corporate environment factors for succeeding in "global marketing", customer-first, 1) high quality product for 2) CS, ES (Employee Satisfaction), SS (Social Satisfaction) and as a strategic factor to realize it, in order to 3) high white-collar productivity and human resource development. Then, 4) Global production is realized by these. So, the same quality and production at optimal locations are achieved.

For that purpose, the highest priority was given to the i) Strategic information sharing and ii) Strategic co-creative action so that the "Strategic Intelligence Application System and Business Process High Linkage System" can effectively function.

When implementing a high cyclization of a business flow that consists of the setup of management policy, creation of a business plan, budget establishment, business deployment, optimal workforce distribution, task management, and evaluation, the relevant information needs to be shared among many departments and they need to also grasp the numerical values that show company-wide trends.

In that way, upon confirming abnormal numerical indications, the problem can be identified and solved at an early stage.

For example, if overtime labor cost shows an unusual figure, the project manager should find out the cause by checking which division, as well as which position or which process, is showing such a trend. In the administrative division, consideration needs to be given to possible deterioration of the work environment or lack of labor force, and in the sales division, care must be exercised to not delay delivery. Each division exchanges information with the other divisions to solve the problem as a whole company.

The solutions shall be evaluated so that the information sharing of both problems and solutions can become a preventive measure and food for thought for the next plan. Such a partnership among divisions, involves creation of a system to visualize information flow as well as its effective and practical application.

To that end, a leader who links human resources is indispensable, and therefore, the cornerstone of corporate management is to foster leaders who have the understanding of the vision of management directors, a broad view of world trends, and communication skills to create a network of personnel inside the company.

Simply putting, the cultivation of an entrepreneurial mind or professional mind is what the authors are intending by proposing this business model of strategic productivity improvement for the white-collar sections.

In the following this chapter, the effectiveness of the created "SPIM" is verified through application cases at the successful companies.

Effectiveness of SPIM

The effectiveness of the established SPIM is verified through application cases at the successful companies by employing "NJ-GPM" above as follows;

Strategic Information Sharing

"V-MICS" (Virtual-Maintenance Innovated Computer System) for a global maintenance network system, and "TPS-QAS" (TPS-Quality Assurance System) for a quality control system will be introduced as successful applications of "strategic information sharing", one of the core technologies of the established "SPIM" (Sakai and Amasaka 2005; Amasaka and Sakai 2009).

Both cases embody the successful sharing of the kind of information that cannot be expressed in words in order to achieve advances in global quality management.

Additionally, similar activities at other companies will be introduced. Administrative divisions play the role of taking a proactive leadership role in putting together all the related divisions and built these systems below.

(i)   V-MICS, a highly reliable production system by combining "DB and CG"

Japanese automotive industries are grappling with the global marketing, simultaneous new vehicle start-up globally and other major propositions under our policy of production where customers need the product and the quantity needed.

Accordingly, it is imperative for production plants to carry out production without breakdown so as to provide customers promptly with high quality products. The author has established the "V-MICS, a highly reliable production system", which has been systematized and practiced by using data base (DB) and computer graphics (CG), which form the nucleus of this advanced maintenance system.

(ii)   TPS-QAS, an intelligent quality control system by utilizing T-QCIS

The application line of "TPS-QAS" as the new production quality management model using "Toyota's Quality Control Information System" (T-QCIS) is the automated assembly line, which assembles a part that transmits engine-driven power to the tires.

This software shows the necessary control characteristics hierarchically specified as item, detailed items, and extraction conditions to improve operability and provide an expansion function.

Strategic Co-Creative Action

The key to success in global production is the reinforcement of product power, or the "simultaneous achievement of QCD" (Amasaka, 2004a, 2006).

To realize this, it is vital to reinforce Japanese-style partnering, or "partnering of competition and collaboration—Japan Supply Chain Management" between automobile manufacturers (hereafter termed Assembly Makers) and affiliated or non-affiliated parts manufacturers (hereafter termed Suppliers) (Amasaka 2004a, 2005, 2006, 2008a). This is also called the Japan Supply System (JSS).

For further advancement in this area, Assembly Makers should not only reinforce internal partnering with their own operational departments (such as engineering, production, and sales), but must also strengthen external partnering.

This means establishing cooperative relationships with other companies while advancing and establishing global partnering through strategic collaboration with both foreign and domestic suppliers.

The author presents global development of SPIM employing NGP-PM between Toyota and a group of cooperation companies as case of internal partnering, and "Improvement in Painting Quality of Chassis Parts of Automobiles" as case of external partnering in Toyota and suppliers.

In these examples, management layers play the role of taking a proactive leadership role in putting together all the related divisions and solve problems below.

(i)   Global development of "Science TQM" between "Toyota and Toyota's group of cooperation companies"

In first example, Toyota's quality management develops the Science SQC (Amasaka 2004b), which is the administrative staff's activity for improving quality management technology by using Science TQM (Amasaka, Ed., 2012), and became popularized and expanded through joint task team activities with "affiliated and non-affiliated suppliers" (Amasaka, 2008a, 2017a,b).

(ii)   Improvement in painting quality of chassis parts of automobiles: cooperative activities with affiliated and non-affiliated companies

The second example is a case where the quality of appearance and paint corrosion resistance (resistance to Salt Spray Test (SST)) were improved without increasing cost, in order to improve the market strength of automotive chassis parts (front and rear axles).

Making a global initiative to achieve simultaneous fulfillment of QCD, Toyota formed joint task teams with Aisin Kako Co., an affiliate, and Tokyo Paint Co., a non-affiliate.

For example, the author shows an example where the joint task team of Toyota and Tokyo Paint raised the product value (VA=performance/cost) of the front axle (Amasaka, 2008a, 2017b).

APPLICATION EXAMPLES

New Turkish Production System (NTPS), Integration Production of Japan and Turkish

Amidst the rapid progress of globalization, the Republic of Turkey, one of the most prominent of Europe's newly emerging economies, is expected to achieve significant growth, centering on its manufacturing industries.

Given its strong international competitiveness in terms of labor and raw materials, the country's automobile industry has particular potential for growth.

Therefore, as an opening move in Turkey's new global production strategy, the author has created a "New Turkish Production System" (NTPS). The "NTPS" has been developed by integrating "Toyota Production System" (traditional TPS and Advanced TPS), itself an evolved model of the current "Toyota manufacturing strategy", the "leading Japanese production system", with "Advanced Turkish Production System", which is an evolved model of current "Traditional Turkish Production System" (TTPS) cultivated to date (Sakalsiz, 2009; Siang et al., 2010; Amasaka, 2015, 2017a, 2022a,b, 2023a).

Current Status of Turkey's Automobile Industry

Turkey currently ranks 16th in the world for production of automobiles, one of the highest rankings among the newly industrialized countries. The strategic advantages of its location in terms of production and distribution have enabled the Turkish automobile industry to establish its own unique production system of which reflects local industries, known as the TTPS (Traditional Turkish Production System) above, while achieving steady growth at the same time.

In recent years, in an opening move in the global production strategy of the Turkish automobile industry, leading overseas companies have become increasingly active in local production. For example, since 1994, "Toyota Motor Manufacturing Turkey" (TMMT) has been conducting training in "The Toyota Way" (Jeffrey, 2004) at Toyota in Japan, and has introduced the "Traditional Toyota Production System" called JIT (See to Chapter 3).

In the background to this is an undercurrent that Turkey has an environment that is amenable to the introduction of a Japanese style production system, given the similarities in character between the Turkish and Japanese people and their belief that success will come if one works hard.

Creation of a New Turkish Production System (NTPS)

Having investigated the actual situation in Turkey, the author creates the NTPS (New Turkish Production System) with the objective of integration and evolution of the TTPS (Traditional Turkish Production System) and the aforementioned Toyota Production System (Traditional TPS and Advanced TPS) for the growth of the next-generation automobile industry in the Republic of Turkey (Sakalsiz, 2009; Yeap et al., 2010; Siang et al., 2010).

(1) Concept of NTPS

Based on the research of Ramarapu et al (1995) and Amasaka et al. (Amasaka, 2004a, 2007a,b,c,d, 2008a,b, 2009; Yamaji and Amasaka, 2006, 2008; Sakai and Amasaka, 2006, 2007, 2008, 2009; Ebioka et al., 2007a,b; Yamaji et al., 2007; Amasaka and Sakai, 2010), the author creates a concept for "NTPS" as shown in Figure 3.

Figure 3. Concept of a new Turkish production system (NTPS)

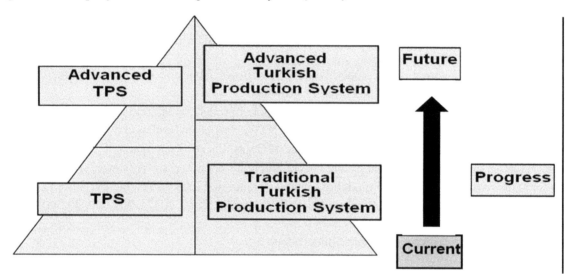

(2) Extraction of important keywords for NTPS

To extract the common factors (elements) and the Turkey-specific unique factors (elements) required for NTPS, the author conducts field surveys of Japanese (Toyota, Honda, Denso, Central Motors, Nihon Spring, and others) and Turkish local manufacturers (TOFAS, OYAK-Renault, FORD OTOSAN, TOYOTA Turkey, and DENSO Turkey).

In these surveys, approximately 500 language data were obtained from interviews, findings obtained from manufacturing plant tours, on-site case studies and reference literature. An affinity diagram of the data thus obtained, based on 5M-E (man, machine, material, manufacturing, measuring, and environment), was used to investigate the relationships between the data based on empirical technologies.

As shown in Figure 4, the author was able to extract, as important keywords for the NTPS, ten common factors - four unique factors specific to Turkey, and two unique factors specific to Japan as follows.

(i) The factors common to Turkey and Japan are the "1. quality assurance", "2. SQC education", "3. QCD activities", "4. kaizen activities", "5. creative proposal programs", "6. improvements through environmental regulations", "7. automatic management methods", "8. global partnering", "9. Safety", and "10. definition, awareness and elimination of *"muda"* (waste).

Figure 4. Important NTPS keywords obtained from affinity diagram

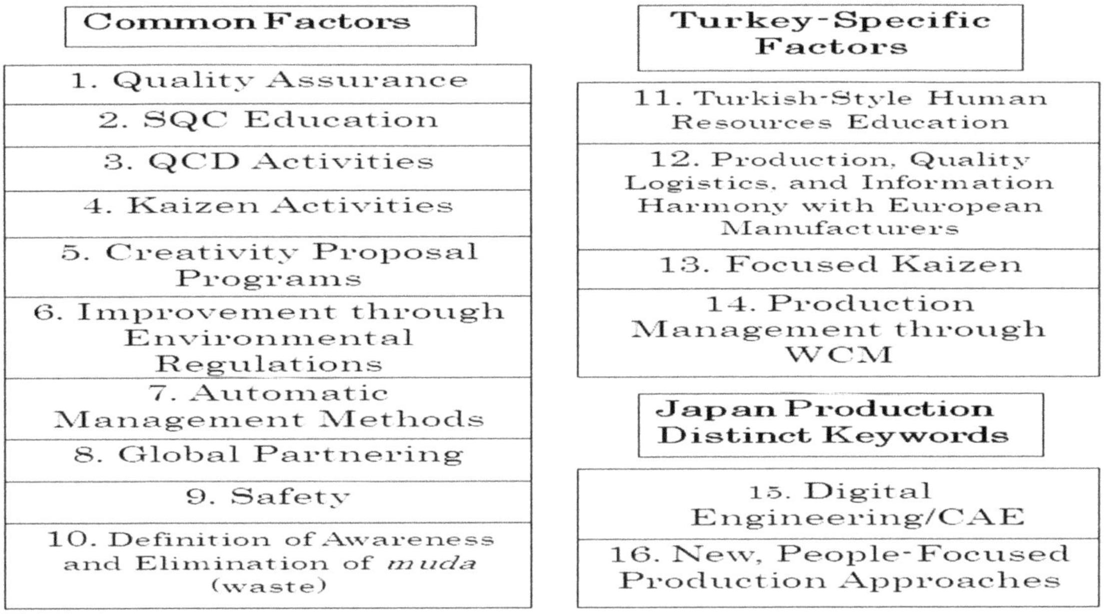

Common Factors
1. Quality Assurance
2. SQC Education
3. QCD Activities
4. Kaizen Activities
5. Creativity Proposal Programs
6. Improvement through Environmental Regulations
7. Automatic Management Methods
8. Global Partnering
9. Safety
10. Definition of Awareness and Elimination of *muda* (waste)

Turkey-Specific Factors
11. Turkish-Style Human Resources Education
12. Production, Quality Logistics, and Information Harmony with European Manufacturers
13. Focused Kaizen
14. Production Management through WCM

Japan Production Distinct Keywords
15. Digital Engineering/CAE
16. New, People-Focused Production Approaches

(ii)   The Turkey-specific factors are the "11. Turkish-style human resources education", "12. production, quality, logistics and informational harmony with European manufacturers", "13. focused kaizen", and "14. production management through world-class manufacturing (WCM)".

(iii) The Japan-specific factors are the "15. digital engineering / CAE (computer aided engineering)", and "16. new, people-focused production approaches".

(3)   Text mining analysis

The author further used text mining analysis to explore in more detail the relationships between the language data obtained based on Figure 4. Examples of the results of analysis of the 5M-E keywords for the NTPS are explained below.

First, the relationships between TPS and Advanced TPS (necessary keywords that each should possess) are shown clearly in Figure 5. Furthermore, in a similar manner, the relationships between the TTPS (Traditional Turkish Production System) and ATPS (Advanced Turkish Production System) are shown clearly in Figure 6.

Figure 5 shows that "TPS, JIT, workplace environment, SQC (statistical quality control), and standard work orientation" have a high degree of relationship and that the key elements (factors) that should be present in "Advanced TPS" include worker education, assurance of high quality, DE (design of experiment), CG (computer graphics), TQM (total quality management), Global Production, Virtual Plant, simulation and partnering.

Similarly, Figure 6 shows that in the TTPS, "safety, cost and quality control" essential to global production, line supervision, environmental improvement and problem-solving methods" essential to "Turkish style human resources development" and "utilization have a high degree of relationship".

Figure 5. Relationships between TPS and advanced TPS

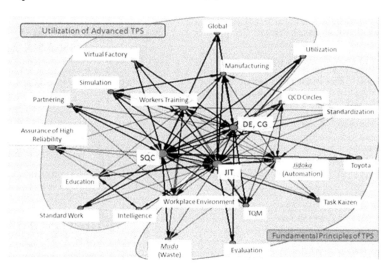

Figure 6. Connections of traditional Turkish production system and advanced Turkish production system

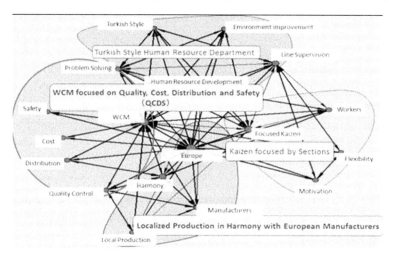

The key elements (factors) that should be present in the ATPS are the "motivated and flexible workers and line", and "kaizen activities" have a "high degree of relationship with focused kaizen" by sections, and "reinforcement of the "influence and collaboration between Turkish and European manufacturers", which is essential to localized production that is in harmony with European manufacturers.

(4) Outline of NTPS

Based on the results of analysis obtained, as described above, and the findings from same, the author has created the NTPS (New Turkish Production System) as shown in Figure 7.

This system represents the integration and evolution of the Advanced TPS, which is itself an evolved model of the current TPS (Toyota Production System), and ATPS (Advanced Turkish Production System), which is an evolved model of "Traditional Turkish Production System" cultivated to date. From the perspective of 5M-E, factors that are common to Japan and Turkey, Turkey-specific factors, and Japan-specific factors have been considered as the technological elements (factors) required for production.

Not only will this enable the Turkish automobile industry to enhance its technological capabilities and increase production, but it will also lead to expectations of the production of high-quality products and the development of flexible production systems in the future.

Figure 7. Outline of a new Turkish production system (NTPS)

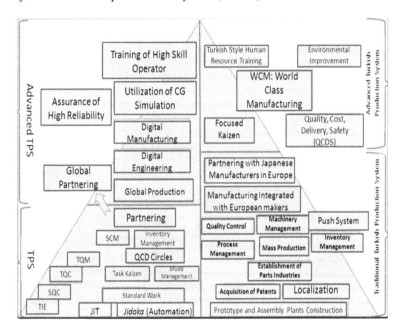

(5)   Development of NTPS

The measures taken by a leading manufacturing Toyota, to achieve successful global production will be discussed. Toyota is taking measures through its "Global Production Center" (GPC) to realize global production.

GPC offers Japanese plants and its human resources education division to the "Global Mother Plant in NGP-PM" (See to Figure 4 in Chapter 9).

In development of NTPS, the author adapts the "Human Intelligence - Production Operating System" (HI-POS) as the "creation of a person-centered new production system" to increasingly realize improvements in skills and proficiency (Sakai and Amasaka, 2006) (See to Figure 4 and 8 in Chapter 9).

An important part of this system is the "Virtual–Intelligent Operator System" (V-IOS) and "Human Integrated Assist System" (HIAS) (See to Appendix 17(a)) which allows human wisdom to be translated into increased skills and oral tradition (Sakai and Amasaka, 2003, 2008) (See to Figure 4, 8 and 12 in Chapter 9).

Within this, the skill training curriculum for overseas production operators and the important tools will be discussed.

(5-1) Operator Training Process for Assembly Works

The skill training for newly hired domestic and overseas production operators, which used to take more than two weeks, has been shortened to just 5 days—less than half the time—and has ensured a stable mass production system. Specific content of the skill training curriculum for overseas production operators is shown in Figure 8. This figure shows an example of operator training process for automobile assembly works.

After everything from (1) Class room training (basic skills training), (2) Skill training to a competency evaluation test are completed and (3) Off-line training (utilizing tools such as the Visual Manual and the HIA-Intelligent IT system) is conducted repeatedly until a certain standard is achieved on an evaluation sheet. Finally, (4) actual training on the line is conducted, and achievement of a set skill level was shown to be achieved in half the time previously needed.

Figure 8. Operator training processes for assembly works

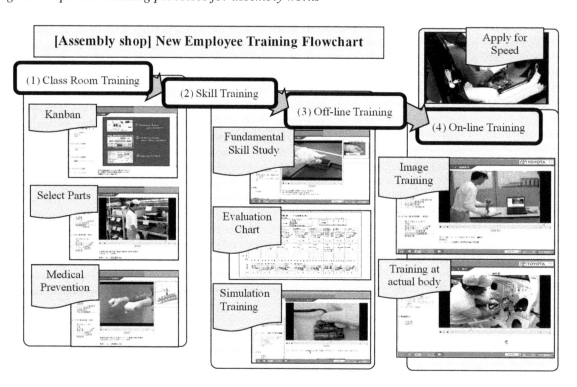

As an example, Figure 9 shows the screen setup of the "Hyper Visual Manual" (HVM) a communication tool that allows for increased exchange of knowledge and information regarding production processes as follows;

(i)   shows the process (Procedure of tighten training), while the section in (ii) shows video, animation, or pictures (Display image-main screen) and (iii) describes the process in words (Description, know-how) (See to Appendix 16A and 16B(1)(2)(3) in Chapter 16).

In Figure 9, HVM allows the operator to follow along, clicking the continue button to turn the pages as needed. With this tool, production operators can be guided using shared content worldwide, and operators can also engage in self-study between actual skill training sessions.

Through repeated image training, skill level can drastically improve at the initial stages, contributing significantly to proficiency improvement in highly skilled operators.

Figure 9. An example of "hyper visual manual" (concerning bolt feeding operation)

(5-2) Bringing Up Intelligent Operators

As mentioned earlier, the author has established the "Virtual–Maintenance Innovated Computer System–for Educational Management" (V-MICS) (See to Figure 4 and 9 in Chapter 9), which improves production operators' operational technology abilities in regard to integrated equipment, in preparation for global production (Sakai and Amasaka, 2005).

This system works to make possible support for improvements in production operators' operating skills and techniques, such as equipment availability administration, defect analysis, and other intelligence-based functions (Amasaka and Sakai, 1996, 1998).

The author, therefore, has established the V-MICS-EM (–Educational Management), an educational management system for the robot operation and maintenance by utilizing a "Visual Manual Format" (VMF) exemplified in Figure 10, which improves levels of mastered skills among production operators (Sakai and Amasaka, 2009).

Manuals describing the operation procedure have so far been kept by each operator or shop station, and the know-how obtained through use is a personal asset.

To make manufacturing operators strong, the equipment (manufacturing) manual itself should allow simple revision by each handling person. The screen consists of paging block, procedure explanation block, visual information display block, and explanation display block.

The manual is to be read by turning the page using the forwarding button according to the procedure. Especially in the explanation display block, the key point representing the know-how of each operator is written and the visual information not indicated on the main screen is shown in details in the sub screen.

The contribution these systems make to the evolution and dissemination of production operators' mastered skills lead to improved productivity among production operators when setting up a new overseas plant, and the given benefits have already been acknowledged.

Deployment to production lines in Japan and abroad has been promoted with satisfactory results.

Figure 10. An example of the visual manual format

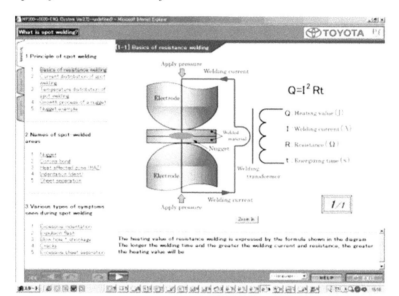

New Vietnam Production System, Hybrid Production of Japan and Vietnam

Summary of the New Vietnam Production System

Automakers in Vietnam are currently expanding production with the intent of becoming a global industry. In addition to increasing production volumes with the aim of achieving globalization, the Vietnamese automobile industry is now required to implement production strategies that will enable survival in the "global quality competition" (Ebioka et al., 2007a,b; Amasaka, Ed., 2012; Amasaka, 2015, 2017a, 2022a,b, 2023a; Miyashita and Amasaka, 2014).

At the same time, these manufacturers need a production strategy to beat out the global quality competition and overcome associated new challenges, including the management of quality, production, and personnel and strategic collaboration with overseas automakers (Amasaka, 2007c). This will involve overcoming new challenges such as quality management, production management, human resources management, and strategic link-ups with overseas automobile manufacturers (Amasaka and Sakai, 2010).

Above all, the production systems employed by the automobile industry are lagging behind compared to advanced companies in terms of people's awareness, standardization, and efficiency improvements by employing NG-GPM (New Japan Global Production Model) above (Amasaka and Sakai, 2011).

Thus, seeing the need to improve the quality of Vietnamese automobile manufacturing in the future, the authors have created a "New Vietnam Production System", (NVPS) by developing hybrid production of Japan and Vietnam (Miyashita and Amasaka, 2014).

This model integrates the TPS (Toyota Production System) (See to Chapter 4 and 5), the leading Japanese production system, and the current Vietnamese production system from a new perspective.

Specifically, the authors' research examines (i) the successful "Advanced TPS" (Amasaka, 2007a,c, 2009), which is an advanced model of the TPS (See to Chapter 4,5,7 and 9), and (ii) the "Traditional Vietnamese Production System" (VPS-T), which is currently established in Vietnam, as well as the "localized Vietnamese Production System" (VPS-L) (Miyashita and Amasaka, 2014).

Research Background and Field Survey

Japanese companies are looking to countries such as China, Thailand, Vietnam, and India as viable prospects for overseas bases. Currently, under the national "industrialization" policy Vietnam aims to enter the ranks of industrial nations by 2020. Thus, the author focuses on the automobile sector of the general assembly industry (Amasaka, 2013).

Vietnam now has an urgent need for a new production system to enhance the country's competitive strength. Looking at the industry composition ratio of the country's actual domestic general production, the proportion covered by the manufacturing industry was 41.0% in 2005 and further growth is expected in this industry area.

Above all, the production systems employed by the automobile industry are lagging behind compared to advanced companies in terms of people's awareness, standardization, and efficiency improvements. Thus, the authors saw a need to create a new Vietnamese production model focusing on the automobile industry.

The VPS-T (Traditional Vietnamese Production System) is often performed manually assembling and inspection.

Recent years, Vietnams companies depend on the import from the foreign countries for the high facilities, but the simple work employs a person. Currently, there are many companies introducing production systems, such as Japan which is a developed country.

Creation of the New Vietnam Production System (NVPS)

(1) Outline of the NVPS

Having investigated the current situation of the automobile industry in Vietnam, the author has created a concept model for the "New Vietnam Production System" (NVPS) by developing hybrid production of Japan and Vietnam as shown in Figure 11 (Miyashita and Amasaka, 2014).

This model integrates the "Traditional Vietnamese Production System" (VPS-T) and "Localized Vietnamese Production System" (VPS-L), based on the core technology of the "Toyota Production System" (TPS) called "traditional TPS", the leading Japanese production system, and "Advanced TPS" which is an advanced model of the TPS that enables global production. In creating this concept model, the author conducts field surveys of automobile suppliers and assembly companies that are implementing and developing VPS-T and VP-L.

Similarly, the authors also conducted field surveys of several companies that are implementing and developing TPS and Advanced TPS, such as Toyota Vietnam. The surveys focused mainly on aspects such as 5M-E (man, machine, material, manufacturing, measuring, and environment), management perspectives, employee awareness, the Vietnamese national character, quality management at plants, and human resources management.

Figure 11. Outline of the new vietnam production system (NVPS)

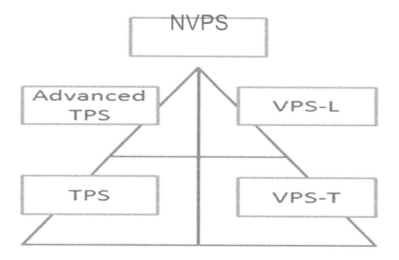

(2) Field survey

Automakers in Vietnam are currently expanding production with the intent of becoming a global industry.

At the same time, these manufacturers need a production strategy to beat out the global quality competition and overcome associated new challenges, including the management of quality, production, and personnel and strategic collaboration with overseas automakers.

The author made appointments with Japanese companies that have bases in Vietnam, and conducted field surveys to determine the actual situation there. While working alongside the workers in Vietnam, the authors noticed differences from Japan.

In particular, they noticed that workers tend to do the jobs that they are given exactly as instructed, and prefer simple tasks. Interview questions were created based on the results of feedback, and a large majority of the responses were related to human resources development and procurement.

Rather than developing human resources from scratch, there is a tendency to acquire talented employees through head-hunting. However, there are costs involved and it is not necessary to employ large numbers of people, so it is considered important to find a small number of highly talented prospects.

Vietnamese people also tend to dislike doing dirty work and strongly prefer to be in charge, which means that talented people often start their own business.

Furthermore, it costs less to use people rather than investing in machinery and equipment, and can sometimes lead to improvements in quality. Actually, companies such as Toyota Vietnam are doing similar things.

(3)   Identifying the requisite elements for NVPS

The author took the information obtained from company surveys conducted in Japan and Vietnam and categorized the information according to 5M-E. Having done so, they then organized the information using sticky notes as shown in Figure 12.

Figure 12. Extraction of the factors of NVPS

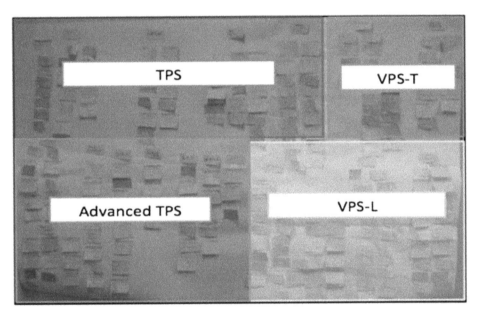

Next, they conducted text mining using the categorized key words. The use of text mining enabled the authors to investigate the relationship between the production models in Japan and Vietnam, and then summarize the characteristics common to the Japanese model and the "Vietnamese model", as well as the distinctive characteristics of each model in order to objectify the various elements.

The results of the "Text mining" are summarized in Figure 13 (NTT DATA). The distinctive characteristics of the Japanese model and the "Vietnamese model" were identified, and then the characteristics common to both models were identified.

First, the author will explain the elements categorized by the text mining, considering the four items organized by using the sticky notes from the perspective of 5M-E. The obtained data was organized as shown in Figure 14.

Arrows are used to show the associations between words, and the arrows are concentrated around words that have strong associations.

Figure 13. "Text mining" of TPS and advanced TPS

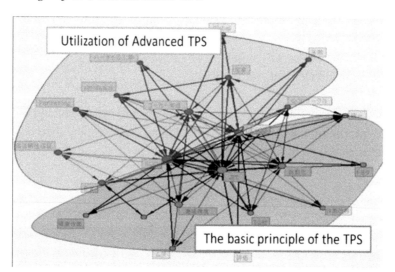

Figure 14. "Text mining" of VPS-T and VPS-L

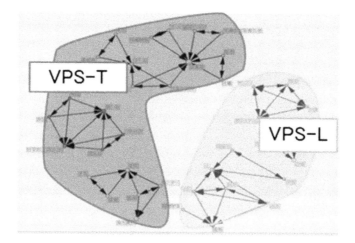

(4)   Creating NVPS

The use of text mining enabled TPS, Advanced TPS, VPS-T, and VPS-L to be identified as the relevant elements. This system utilizes TPS and Advanced TPS, and encompasses VPS-T and VPS-L.

The left side of the model is comprised of TPS and Advanced TPS while, similarly, the right side is comprised of VPS-T and VPS-L. As far as the author is aware, there has been no other research on efficient production systems in Vietnam up to now.

Thus, utilizing their production-related findings from research in China and other parts of Asia, including Japan, the author aims to develop local production to ensure survival amidst global competition.

As Vietnam has not yet become an advanced country, the author considers the need for production capabilities equivalent to those of advanced countries.

In creating this model, the author anticipates improvements in productivity to be achieved by creating links with all production systems, possessing advanced technological capabilities, and implementing thorough kaizen of local production in Vietnam.

NVPS has created by the consumers is as shown in Figure 15. To evaluate the effectiveness of the proposed NVPS, the author conducted questionnaire surveys targeting company employees that have had the experience of being based in Vietnam, and collated the results.

A 7-level evaluation system was used for the questionnaires, and the questions were devised based on 18 items derived from feedback during the field surveys.

Figure 15. Outline of a new Vietnam production system, NVPS

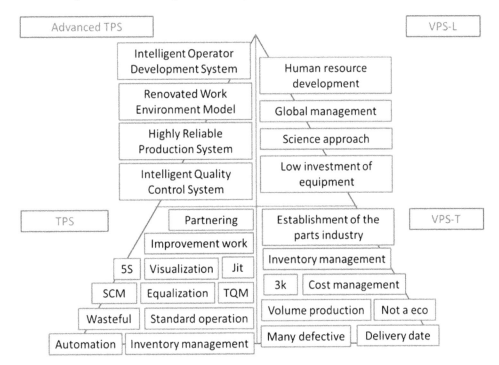

The questionnaire mainly covered elements considered necessary to develop workers at Vietnamese plants into highly skilled human resources. Of the 41 people asked to participate in the questionnaire survey, responses were obtained from 36 people.

The contents of the questionnaire are enumerated as follows.

- Same facilities as Japan
- Facilities management for overseas production bases
- Understanding of local problems
- Education regarding Japanese culture and customs
- Shared information management
- Clarification of roles
- Rule-making

Table 1. The results of multiple regression analysis

Variable name	Partial regression coefficient
Guidance that incorporates new knowledge	0.914
Human resources development and training system for local employees	0.523
Flexibility in response to changes	0.493
Understanding of local problems	0.364
Development of leaders for overseas bases	0.344
Communication	0.157
Rule-making	0.125

- Communication
- Trust-based relationships
- Visualization of know-how
- Clarification of cause-effect relationships using a scientific approach
- Study programs in Japan
- Development of leaders for overseas bases
- Thorough implementation of non-compliance countermeasure reports
- Flexibility in response to changes
- Differentiation of managerial positions
- Human resources development and training system for local employees
- Development of equipment suitable for local people
- Guidance that incorporates new knowledge
- Management goals and visions for employees

The items of questionnaire content were extracted from raw data obtained during field surveys and interviews conducted in Vietnam. Most of these items relate to people.

This is probably due to differences in culture and regional characteristics. Multiple regression analysis was conducted using the collated results to analyze the degree of importance.

The results of the multiple regression analysis are as shown in Table 1. Based on the relative size of the partial regression coefficients described in Table 1, the author identified "Guidance that incorporates new knowledge", "Human resources development and training system for local employees", "Flexibility in response to changes", and "Development of leaders for overseas bases".

While actually working at a local site, the authors were told that "Vietnamese workers are not particularly sensitive to new things in the world around them", which reaffirmed the importance of gaining new knowledge for Vietnamese people.

(3-5) Verification

To verify the validity of NVPS, the authors conducted the questionnaire survey. The survey targeted workers either with the experience of working, or currently working, at overseas bases.

The aim of this questionnaire survey was to receive raw feedback (free opinions) concerning the proposed model in order to lend credibility to the verification (evaluation).

The main opinions received are listed below.

- If a model existed that was suitable for both advanced countries and developing countries, it would be applied a lot more.
- This is the first time we have ever considered the matter objectively, so it's very interesting.
- The practicality would be much greater if it was possible to search related cases in real-time.

Based on the above opinions, it is possible to ascertain that the model will be both necessary and effective at actual production sites. Through this research, the author has created a NVPS with the aim of contributing to the development of the manufacturing industry in Vietnam.

The ultimate objective of Japanese companies is to enhance the locally-based production technology without placing a burden on Japan, in order to effectively and efficiently maintain a level of quality that is equivalent to or above that in Japan.

As a future view, the author will research the elements of the model deeply and the elements' details taken by Vietnam seriously in the production.

OTHER CASES EXPANDING NJP-PM

In the same way, to develop those results and validities, the author has created the "New Malaysia Production Model" (NMPM) (Shan et al., 2011), and "China Local Automobile Manufacturer's Production System" (Shan, 2012).

More recently, as a strengthening of the productivity in "NJP-PM", the author is "developed the "Robot Reliability Design and Improvement Methods" (RRD-IM) in "NJ-GPM" (New Japan Global Production Model) (See to Figure 4 in Chapter 9), and obtains the successes as expected as shown in Figure 16 (Sakai and Amasaka, 2007a).

This model proceeds through the following stages, (i) Initial Failure (-IF), (ii) Random Failure (-RF) and (iii) Wear Failure (-WF). (Refer to Amasaka (2015, 2017a, 2022a,b, 2023a) in detail).

As representative examples, specifically, to develop the "RRD-IM" as the "Advanced TPS" strategy, the author has established the (i) "Robot control method for curved seal extrusion for high productivity in an advanced Toyota production system", (ii) "Body Auto Fitting Model "BAFM" using "NJ-GPM" at Toyota" and (iii) ""High Quality Assurance Production System and Production Support Automated System" (Refer to Sakai and Amasaka (2007b, 2013 and 2014) in detail).

Figure 16. Robot reliability design and improvement method (RRD-IM)

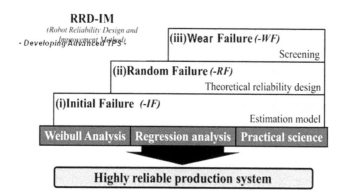

CONCLUSION

The author has established a model and conducts verification for the strategic development of Advanced TPS through the strategic use of NGP-PM (New Global Partnering Production Model) with the aim of establishing basic principles and realizing a Japan standard of quality in overseas manufacturing. As application examples, the author has established the SPIM (Strategic Productivity Improvement Model) for global production strategy, which improves the intellectual productivity. The effectiveness of SPIM was verified by going over the application results observed at Toyota and others. As the concrete examples of NGP-PM deployment, the author describes some case studies of the "New Turkish Production Model, Integration production of Japan and Turkish" (NTPM), "New Vietnam Production Model, Developing hybrid production of Japan and Vietnam" (NVPS), and others ("New Malaysia Production Model" (NMPM), a new integrated production system of Japan and Malaysia", "China Local Automobile Manufacturer's Production System".

REFERENCES

Amasaka, K. (2002). New JIT, a new management technology principle at Toyota. *International Journal of Production Economics, 80*(2), 135–144. doi:10.1016/S0925-5273(02)00313-4

Amasaka, K. (2004a). Development of "Science TQM", a new principle of quality management: Effectiveness of strategic stratified task team at Toyota. *International Journal of Production Research, 42*(17), 3691–3706. doi:10.1080/00207540420002003867

Amasaka, K. (2004b). *Science SQC, New Quality Control Principle: The Quality Strategy of Toyota.* Springer-Verlag. doi:10.1007/978-4-431-53969-8

Amasaka, K. (2005). New Japan Production Method, an innovative production management principle. *Proceedings of the 16th Annual Conference of the Production and Operations Management Society, Michigan, Chicago.*

Amasaka, K. (2006). *Evolution of TPS fundamentals utilizing New JIT strategy: Toyota's Simultaneous Realization of QCD Fulfillment Models*. Proceedings of the International Manufacturing Leaders Forum, Taipei, Taiwan.

Amasaka, K. (2007a). High Linkage Model *"Advanced TDS, TPS & TMS"*: Strategic development of *"New JIT"* at Toyota. *International Journal of Operations and Quantitative Management, 13*(3), 101–121.

Amasaka, K. (2007b). Applying New JIT - Toyota's global production strategy: Epoch-making innovation in the work environment. *Robotics and Computer-integrated Manufacturing, 23*(3), 285–293. doi:10.1016/j.rcim.2006.02.001

Amasaka, K. (2007c). New Japan Production Model, an advanced production management principle: Key to strategic implementation of New JIT. *The International Business & Economics Research Journal, 6*(7), 67–79.

Amasaka, K. (2007d). *New Japan Model, Science TQM: Theory and application of strategic quality management*. Maruzen publishing. (in Japanese)

Amasaka, K. (2008a). Strategic QCD studies with affiliated and non-affiliated suppliers utilizing New JIT, Encyclopedia of Networked and Virtual Organizations, III(PU-Z), 1516-1527.

Amasaka, K. (2008b). Science TQM, a new quality management principle: The quality management strategy of Toyota. *The Journal of Management & Engineering Integration, 1*(1), 7–22.

Amasaka, K. (2009). The foundation for advancing the Toyota Production System utilizing New JIT. *Journal of Advanced Manufacturing Systems, 8*(1), 5–26. doi:10.1142/S0219686709001614

Amasaka, K. (Ed.). (2012). Science TQM, New Quality Management Principle: The quality management strategy of Toyota. Bentham Science Publisher, Shariah, UAE, USA.

Amasaka, K. (2013). The development of a Total Quality Management System for transforming technology into effective management strategy. *International Journal of Management, 30*(2), 610–630.

Amasaka, K. (2015). New JIT, New Management Technology Principle, Taylor & Francis Group, CRC Press, Boca Raton, London, New York.

Amasaka, K. (2016). *Innovation of automobile manufacturing fundamentals employing New JIT: Developing Advanced Toyota Production System at Toyota Manufacturing USA*. Proceedings of the 5th Conference on Production and Operations Management, P&OM Habana, Cuba.

Amasaka, K. (2017a). *Toyota: Production System, safety analysis and future directions*. Nova Science Publishers.

Amasaka, K. (2017b). Strategic Stratified Task Team Model for realizing simultaneous QCD fulfilment: Two case studies. *Journal of Japanese Operations Management and Strategy, 7*(1), 14–35.

Amasaka, K. (2019). Studies on New Manufacturing Theory, *Noble. International Journal of Scientific Research, 3*(1), 1–20.

Amasaka, K. (2021). New Japan Automobile Global Manufacturing Model: Using Advanced TDS, TPS, TMS, TIS & TJS. *Journal of Business Management and Economic Research, 6*(6), 499–523.

Amasaka, K. (2022a). New Manufacturing Theory: Surpassing JIT (2nd Edition), Lambert Academic Publishing.

Amasaka, K. (2022b). *Examining a New Automobile Global Manufacturing System.* IGI Global Publisher. doi:10.4018/978-1-7998-8746-1

Amasaka, K. (2023a). New Lecture-Surpassing JIT: Toyota Production System-From JIT to New JIT. Lambert Academic Publishing, Germany.

Amasaka, K. (2023b). A New Automobile Product Development Design Model: Using a Dual Corporate Engineering Strategy. *Journal of Economics and Technology Research*, *4*(1), 1–22. doi:10.22158/jetr. v4n1p1

Amasaka, K., Baba, J., Kanuma, Y., Sakai, H., & Okada, S. (2006). *The evolution of technology and skill - The Key to success in global production": Latest implementation of New Japan Model "Science TQM*. The Japan Society for Production Management.

Amasaka, K., & Sakai, H. (1996). Improving the Reliability of Body Assembly Line Equipment. *International Journal of Reliability Quality and Safety Engineering*, *3*(1), 11–24. doi:10.1142/S021853939600003X

Amasaka, K., & Sakai, H. (1998). Availability and Reliability Information Administration System "ARIM-BL" by methodology in Inline-Online SQC. *International Journal of Reliability Quality and Safety Engineering*, *5*(1), 55–63. doi:10.1142/S0218539398000078

Amasaka, K., & Sakai, H. (2009). TPS-QAS, New Production Quality management Model: Key to New JIT-Toyota's global production strategy. *International Journal of Manufacturing Technology and Management*, *18*(4), 409–426. doi:10.1504/IJMTM.2009.027774

Amasaka, K., & Sakai, H. (2010). Evolution of TPS fundamentals utilizing New JIT strategy-Proposal and validity of Advanced TPS at Toyota. *Journal of Advanced Manufacturing Systems*, *9*(2), 85–99. doi:10.1142/S0219686710001831

Amasaka, K., & Sakai, H. (2011). The New Japan Global Production Model "NJ-GPM": Strategic development of *Advanced TPS. The Journal of Japanese Operations Management and Strategy*, *2*(1), 1–15.

Burke, R. J., Graham, J., & Smith, F. (2005). Effects of reengineering on the employee satisfaction-customer satisfaction relationship. *The TQM Magazine*, *17*(4), 358–363. doi:10.1108/09544780510603198

Ebioka, K., Sakai, H., Yamaji, M., & Amasaka, K. (2007b). A New Global Partnering Production Model "NGP-PM" utilizing Advanced TPS. *Journal of Business & Economics Research*, *5*(9), 1–8.

Ebioka, K., Yamaji, M., Sakai, H., & Amasaka, K. (2007a). Strategic development of *Advanced TPS* to bring overseas manufacturing to Japan standards: Proposal of a new global partnering production model and its effectiveness. *Transaction of the Japan Society for Production Management*, *13*(2), 51–56.

Evans, J. R., & Lindsay, W. M. (1995). *The Management and Control of Quality*. South-Western.

Goto, T. (1999). *Forgotten origin of management-Management quality taught by G.H.Q, CCS management lecture*. Seisansei-Shuppan. (in Japanese)

Grundspenkis, J. (2007). Agent based approach for organization and personal knowledge modeling: Knowledge management perspective. *Journal of Intelligent Manufacturing*, *18*(4), 451–457. doi:10.1007/s10845-007-0052-6

Halevi, G., & Wang, K. (2007). Knowledge based manufacturing system (KBMS). *Journal of Intelligent Manufacturing*, *18*(4), 467–474. doi:10.1007/s10845-007-0049-1

Hoogervorst, J. A. P., Koopman, P. L., & Flier, H. (2005). Total quality management: The need for an employee-centered coherent approach. *The TQM Magazine*, *17*(1), 92–106. doi:10.1108/09544780510573084

Jeffrey, K. L. (2004). *The Toyota Way*. McGraw-Hill.

Kakkar, S., & Narag, A. S. (2007). Recommending a TQM model for Indian organizations. *The TQM Magazine*, *19*(4), 328–353. doi:10.1108/09544780710756232

Lagrosen, S. (2004). Quality management in global firms. *The TQM Magazine*, *16*(6), 396–402. doi:10.1108/09544780410563310

Ljungström, M. (2005). A model for starting up and implementing continuous improvements and work development in practice. *The TQM Magazine*, *17*(5), 385–405. doi:10.1108/09544780510615915

Miyashita, S., & Amasaka, K. (2014). Proposal of a "New Vietnam Production Model (NVPM), a new integrated production system of Japan and Vietnam. *IOSR Journal of Business and Management*, *16*(12), 18–25. doi:10.9790/487X-161211825

NTT DATA Mathematical Systems Inc. (2006). Text Mining Studio Analysis. NTT DATA. https://www.msi.co.jp/solution/tmstudio/index.html

Prajogo, D. (2006). Progress of quality management practices in Australian manufacturing firms. *The TQM Magazine*, *18*(5), 501–513. doi:10.1108/09544780610685476

Ramarapu, N. K., Mehra, S., & Frolick, M. N. (1995). A comparative analysis and review of JIT implementation research. *International Journal of Operations & Production Management*, *15*(1), 38–49. doi:10.1108/01443579510077188

Sakai, H., & Amasaka, K. (2003). *Construction of "V-IOS" for promoting intelligence operator: Development and effectiveness for visual manual format*. The Japan Society for Production Management.

Sakai, H., & Amasaka, K. (2005). V-MICS, Advanced TPS for strategic production administration: Innovative maintenance combining DB and CG. *Journal of Advanced Manufacturing Systems*, *4*(6), 5–20. doi:10.1142/S0219686705000540

Sakai, H., & Amasaka, K. (2006). Strategic HI-POS, Intelligence Production Operating System: Applying Advanced TPS to Toyota's Global Production Strategy, WSEAS (World Scientific and Engineering and Society). *Transactions on Advances in Engineering Education*, *3*(3), 223–230.

Sakai, H., & Amasaka, K. (2007a). The Robot Reliability Design and Improvement Method and Advanced Toyota Production System. *The Industrial Robot*, *34*(4), 310–316. doi:10.1108/01439910710749636

Sakai, H., & Amasaka, K. (2007b). Development of a robot control method for curved seal extrusion for high productivity in an advanced Toyota production system. *International Journal of Computer Integrated Manufacturing, 20*(5), 486–496. doi:10.1080/09511920601160262

Sakai, H., & Amasaka, K. (2008). Human-Integrated Assist Systems for intelligence operators, Encyclopedia of Networked and Virtual Iorganizations, II(G-Pr), 678-687.

Sakai, H., & Amasaka, K. (2009). Proposal and demonstration of V-MICS-EM by digital Engineering: Robot operation and maintenance by utilizing Visual Manual. *International Journal of Manufacturing Technology and Management, 18*(4), 344–355. doi:10.1504/IJMTM.2009.027769

Sakalsiz, M. M. (2009). *The proposal of New Turkish Production System utilizing Advanced TPS.* [Master's thesis, Graduate School of Science and Engineering, Aoyama Gakuin University*].*

Shan, H. (2012). *The proposal of China Local Automobile Manufacturer's Production System*, [Thesis, Graduate School of Science and Engineering, Aoyama Gakuin University].

Shan, H., Yeap, Y. S., & Amasaka, K. (2011). *Proposal of a New Malaysia Production Model "NMPM": a new integrated production system of Japan and Malaysia.* Proceedings of International Conference on Business Management 2011, Miyazaki Sangyo-Keiei University, Japan.

Siang, Y. Y., Sakalsiz, M. M., & Amasaka, K. (2010). Proposal of New Turkish Production System (NTPS), Integration and evolution of Japanese and Turkish production system. *Journal of Business Case Study, 6*(6), 69–76. doi:10.19030/jbcs.v6i6.260

Yamaji, M., & Amasaka, K. (2006). New Japan Quality Management Model, Hyper-cycle model "QA & TQM Dual System": Implementation of New JIT for strategic Management technology. *Proceedings of the International Manufacturing Leaders Forum.* Springer.

Yamaji, M., & Amasaka, K. (2007a). Proposal and validity of Global Intelligence Partnering Model for Corporate Strategy, GIPM-CS. *International IFIP TC 5.7 Conference on Advanced in Production Management System.* Springer.

Yamaji, M., & Amasaka, K. (2008). New Japan Quality Management Model: Implementation of New JIT for strategic management technology. *The International Business & Economics Research Journal, 7*(3), 107–114.

Yamaji, M., & Amasaka, K. (2009). Strategic Productivity Improvement Model for white-collar workers employing Science TQM. *The Journal of Japanese Operations Management and Strategy, 1*(1), 30–46.

Yamaji, M., Sakai, H., & Amasaka, K. (2006). Intellectual working Value Improvement Model utilizing Advanced TPS: Applying New JIT, Toyota's global production strategy. *Proceedings of the International Applied Business Research Conference, Cancun, Mexico.*

Yamaji, M., Sakai, H., & Amasaka, K. (2007b). Evolution of technology and skills in production workplaces *ss &. Economic Research Journal, 5*(6), 61–68.

Yeap, Y. S., Sakalsiz, M. M., & Amasaka, K. (2010). Proposal of New Turkish Production System, NTPS, Integration and Evolution of Japanese and Turkish Production System. *Journal of Business Case Study, 6,* 69–76.

Chapter 17
CS and CL to Boost Marketing Effectiveness in Auto Dealerships

ABSTRACT

Faced with a sluggish economy, car sales have been disappointing in recent years. Given this situation, it is critical that auto-dealerships shift the focus of their sales and marketing activities from attracting new customers to keeping the customers they already have. To develop advanced TMS, this move can be expected not only to reduce sales costs, but also contribute to healthy profits. To strengthen auto-dealerships, therefore, the author proceeds with researching customer satisfaction (CS) and customer loyalty (CL) as a way of boosting marketing effectiveness by using covariance structure analysis/structural equation modeling (SEM), clarifying the key factors that comprise CL, and help improve the marketing strategy. Then, as the application examples, the author focuses on the development and effectiveness for CS and CL by employing video unites customer behavior and maker's designing intentions (VUCMIN) and total direct mail management model (TDMMM).

SITUATION AND PROBLEM OF JAPAN'S AUTO-SALES

In developing New JIT strategy, the author shows 5 customer lifecycle stages (Prospect - First time buyer - Initial repeater - Core customer - Withdrawal) as they apply to auto-sales (dealerships) based the success of Chapter 7, 10 and 19 as shown in Figure 1 (Okutomi, 2011; Okutomi and Amasaka, 2012, 2013).

Generally speaking, the Japanese auto industry has now maximized new vehicle sales—meaning that the market has fully matured when it comes to traditional auto-sales methods. Faced with a sluggish economy, car sales have been disappointing in recent years.

Given this situation, it is critical that dealerships shift the focus of their sales and marketing activities from attracting new customers to keeping the customers they already have.

This move can be expected not only to reduce sales costs, but also contribute to healthy profits.

A dealership's best (core) customers took on that designation through a process of satisfactory purchases with that company.

They are familiar with the features of their dealership's products and services, know how purchases are made, and know the best way to take advantage of how the dealership runs.

DOI: 10.4018/978-1-6684-8301-5.ch017

For this reason, marketing activities that target these core customers will help create a customer base that is superior to that of competitors working in the same industry.

Core customers are an auto-dealer's most priceless asset.

Figure 1. Customer lifecycle

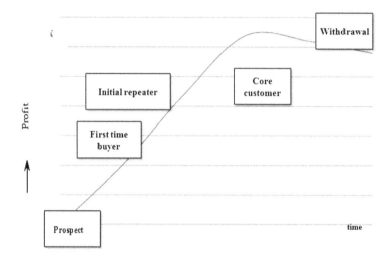

CS AND CL TO BOOST MARKETING EFFECTIVENESS AT JAPAN'S AUTO-DEALERSHIPS

Sales and Marketing at Auto-Dealerships

The author conducted interviews with six auto-dealerships, each of whom represent major automakers - 4 Japanese (A, B, C, and D) and 2 foreign (E and F) (Okutomi, 2011; Okutomi and Amasaka, 2012, 2013).

Based on these interviews and the resulting analysis, it was found that sales and marketing activities at auto dealerships were not particularly focused on boosting satisfaction and loyalty among core customers - and an unexpected majority had no awareness of these concepts.

At the same time, even if auto-dealerships were aware of the need for improvements in these areas, it was common for sales staff to simply rely on their personal knowledge (experiential data), so that related activities tended to be completely different for each person.

As a result, when it came to "Customer Satisfaction" (CS) and "Customer Loyalty" (CL), problems (issues) were arising from the mismatch between managers' intentions, salesperson activities, and customer experiences (preferences) at auto-dealers.

CS and CL to Boost Marketing Effectiveness for CR Activities

To help resolve the sales and marketing issues outlined in the previous section and simultaneously achieve higher satisfaction among core customers, the author employed statistical methods to make explicit (objectify) the operations currently being implicitly carried out at auto-dealerships based on the "employees' personal knowledge" for the "Customer Retention (CR) activities using Customer Science principle (CSp) and Science SQC, new quality control principle" (Amasaka, 2004, 2007a,b, 2005, 2011) (See to Chapter 2 and 3).

The goal was to make sales and marketing activities more effective by having them accurately reflect the needs of core customers employing CSp described in Figure 2 of Chapter 3 (See to section "Creation of the "Scientific quality management" employing CSp in Chapter 3), and visualization of "experienced - based implicit knowledge" described in Figure 3 of Chapter 2 (See to subsection "Reinforcement of the corporate management function for realizing customers' wants" in Chapter 2).

The specific procedure shown in Figure 1 used to do this was as follows;

1. In Step 1, the author specified the key factors that make up CS and CL
2. In Step 2, the author collected customer information and subjected it to an analysis of "covariance structure". The insights gained from this analysis were then used to determine the degree to which each key factor impacts CS and CL (Okutomi, 2011; Okutomi and Amasaka, 2012, 2013).
3. In Step 3, concretely, the key factors comprising satisfaction and loyalty among core customers at the six target dealerships - 4 Japanese (A, B, C, and D) and 2 foreign (E and F) were identified in order to determine the level of impact each carry.

Specify the Key Factors (Step One)

In Step 1, the author collected and analyzed sales information from core customers to identify the components of satisfaction and loyalty among them.

One way they did this was by actually going to dealerships to participate in new car (vehicle) presentations and collecting the required information from customers.

The author put together their own questionnaire for the "information collection process", with items that not only has reflected the personal knowledge of salespeople and managers, but also information has gained from direct interviews with customers.

As a result of these information collection efforts, the author was able to add survey items that accurately reflected how important service was to customers - both before and after a new car purchase.

Working from these insights as a basis, the author has added in the "SERVQUAL framework" (a service quality evaluation measurement tool) to create a new evaluation and measurement scale that was ultimately used to generate the questionnaire (Ex. Parasuraman et al., 1988).

Because the SERVQUAL method is not limited to a specific industry, it has some issues in terms of applicability to individual business activities (including auto-sales).

To eliminate these issues, the author used the aforementioned interviews to ensure that they had a complete grasp of the factors specific to the auto-industry. (Okutomi, 2011; Okutomi and Amasaka, 2012, 2013).

Collecting and Analyzing the Customer Information (Step Two)

Once the items extracted from these interviews were integrated with the SERVQUAL framework, the authors were able to specify 29 individual factors as shown in Table 1.

These included the "17 product factors, 3 employee factors, 6 dealership factors, and 3 corporate factors".

In Step 2, the 29 factors listed in Table 1 "Survey (Questionnaire form)" were rated on a seven-point scale along with CS and CL to identify the kinds of things that customers were looking for.

As the table indicates, core customers from six auto-dealers were asked to complete a survey (questionnaire form) covering four key topics (factors): "products, dealership, employees, and company".

A total of 226 valid responses were received from 138 men and 88 women. After collecting the survey data, the author subjected it to an analysis of covariance structure in order to find out the degree of impact each factor had on CS and CL.

Table 1. Survey (questionnaire form)

factor	Question items	Rating scales						
		Very dissatisfied	dissatisfied	A little dissatisfied	neutral	A little satisfied	satisfied	Very satisfied
product	exterior						○	
	interior						○	
	Safety device				○			
	Handling					○		
	Cornering						○	
	Straight-line stability				○			
	High-speed stability				○			
	Durability				○			
	pedal			○				
	seat							○
	Engine displacement			○				
	fuel efficiency		○					
	Interior noise					○		
	Body shaking				○			
	Car navigation system				○			
	audio				○			
	price				○			
employee	polite						○	
	knowledge					○		
	Prompt customer service						○	
dealership	Appearance				○			
	Opening Hours			○				
	location			○				
	mandatory inspection services					○		
	Emergency measure							○
	Periodic contact					○		
corporate	pamphlet · Website				○			
	TV commercials				○			
	corporate image					○		
	Customer Satisfaction						○	
	Customer Loyalty					○		

Finding out the degree of impact each factor (Step 3)

The author showed the cause-and-effect model created by the author based on the "4 key factors: product, employee, dealership, and corporate" specified in the above subsection "Step 1 and Step 2" as shown in Figure 2.

This diagram was used as a "structural model" of "Customer satisfaction (CS) and Customer loyalty (CL)" at auto-dealerships.

Figure 2. Cause-and-effect model of key factor

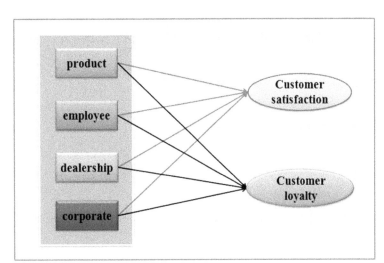

The 4 key factors (first-order compositional factors) represented described in Figure 2 are comprised of individual, or second-order compositional factors (Question items), which impact "CS and CL". In the cause-and-effect model, "CS and CL" was assumed to independent relationship as shown in Figure 3, 4 and 5.

Each figure with "products, dealership, employees and corporate" as follows;

(a) Cause-and-effect model of individual factor: Example 1.
(b) Cause-and-effect model of individual factor: Example 2.
(c) Cause-and-effect model of individual factor: Example 3.

Then, Table 2 showed the results of the "Covariance Structure Analysis / *Structural Equation Modeling (SEM)*" on each individual factor that makes up the 4 key factors as they relate to "CS and CL". Under product, this table showed that the individual factors with the greatest impact on "CS and CL" were aspects of driving performance quality (engine displacement, fuel efficiency, handling stability, etc.) as well as subjective quality factors like exterior features and seats.

Figure 3a. Cause-and-effect model of individual factor: Example 1 "Product 1, 2 and 3" x "CS and CL"

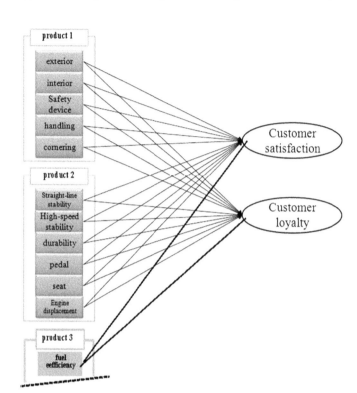

Figure 3b. Cause-and-effect model of individual factor: Example 2 "Product and dealerships x employee and corporate" x "CS and CL"

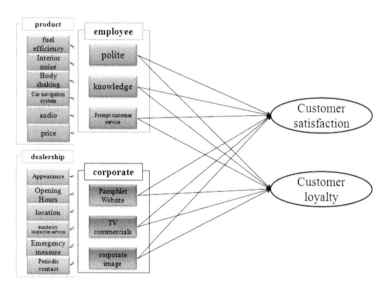

Figure 3c. Cause-and-effect model of individual factor: Example 3 "Employee and corporate x CS and CL"

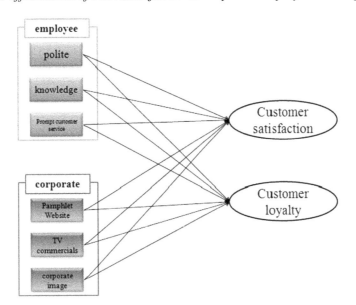

The analysis results indicated that if auto manufacturers are to prioritize "CS and CL" during their sales and marketing activities, they need to enhance product development so that it focuses on performance quality, design quality (including exterior design quality like body shape and paint color), and subjective qualities like seat comfort.

At the same time, affiliated dealerships need to work to enhance customer handling during emergencies as well as mandatory inspection services.

Similarly, it can be inferred that because corporate image and TV (television) commercials also have a powerful effect on customers, vehicles are an important factor in generating trust towards a company.

Under the topic of employees, prompt, polite customer service was found to have the greatest impact on CL.

This point is to a customer mindset whereby those looking to replace an existing vehicle have a desire to go back to the dealership where they purchased their old car because they have a lasting impression of the courteous manner with which a salesperson treated them before.

Similarly, it can be inferred that because corporate image and TV commercials also have a powerful effect on customers, vehicles are an important factor in generating trust towards a company (Refer to Okutomi (2011) and Okutomi and Amasaka (2012, 2013) in detail). The analysis results obtained above provide clues as to how more logical sales and marketing activities might be carried out in the future.

By making explicit underlying employee knowledge on the degree to which certain factors affect CS and CL that makes buyers want to come back for repeat purchases, the author was able to visually present a structural model of impact outlining the relative strength of both the key and individual factors that can help boost CS and CL.

Then, in the following section, the effectiveness of the knowledge acquired by this research is described.

Table 2. Results of the covariance structure analysis

product	Customer Satisfaction	Customer Loyalty
exterior	.612	.470
interior	.198	.217
Safety device	-.034	-.109
Handling	.087	.026
Cornering	.190	.149
Straight-line stability	.207	.213
High-speed stability	.106	-.077
Durability	-.045	-.064
pedal	-.458	-.257
seat	.519	.694
Engine displacement	.600	.343
fuel efficiency	.475	.270
Interior noise	.136	.188
Body shaking	.254	.506
Car navigation system	-.045	-.120
audio	.181	-.099
price	-.024	.119
dealership	**Customer Satisfaction**	**Customer Loyalty**
Appearance	.115	-.041
Opening Hours	.024	-.048
location	-.098	.169
mandatory inspection services	.203	.086
Emergency measure	.659	.559
Periodic contact	-.294	.278
employee	**Customer Satisfaction**	**Customer Loyalty**
polite	.291	.552
knowledge	.154	.316
Prompt customer service	.464	.077
corporate	**Customer Satisfaction**	**Customer Loyalty**
pamphlet · Website	.205	.057
TV commercials	.128	.502
corporate image	.776	.511

APPLICATION EXAMPLES

In this subsection, as application examples using those knowledge acquired above, the author focuses on the development and effectiveness for CS and CL describing "Video Unites Customer Behavior and Maker's Designing Intentions" (VUCMIN) and *Total* "Direct Mail Management Model" TDMMM in CR activities (Hifumi, 2006; Murato et al., 2008; Yamaji et al., 2010; Kawasaki and Yoshizawa, 2011; Okutomi, 2011; Okutomi and Amasaka, 2012, 2013; Ishiguro et al., 2010, Ishiguro and Amasaka, 2012a,b; Amasaka, Ed., 2012; Amasaka et al., 2013; Amasaka, 2015, 2019, 2022a,b, 2023a,b) (See to Section "Constructing a Scientific Mixed Media Model for boosting auto-Dealer visits" in Chapter 19).

A New Strategic Advertisement Model "VUCMIN" Designed to Enhance the Desire to Visit Auto-Dealers

Outline

In the 21st century, one of the important issues in the industrial world is how to create desirable and influential products. The created method "VUCMIN" uses video advertisement, and was developed based on scientific approaches and analyses that focus on the standard behavioral movements of customers who visit dealers when choosing an automobile (Hifumi, 2006; Murat et al., 2008; Yamaji et al., 2010; Amasaka et al., 2013).

Specifically, to strengthen "customer market creation", the author develops the "New JIT" employing "Advanced TDS" based on the "TMS" described in Figure 1, 2 and 3 of Chapter 7 (See to Section "Establishment of New JIT, new management technology principle" and Section "Developing Advanced TDS, TPS, TMS, TIS & TJS in New JIT" in Chapter 7).

VUCMIN, which is based on the different approaches identified in target customer profiles, aims to increase the desire of customers to visit dealers. After creating this video advertisement, customers were verified as having a positive opinion towards visiting dealers with a plan to purchase the vehicle featured in the video.

Need for a New Advertising Approach in the Automobile Industry

Advertising expenses in the Japanese automotive industry are trending to around 6 trillion Japanese Yen (Dentsu, 2003). Despite these high advertising expenses, the number of vehicle sales is stagnant (JAMA, 2005). Furthermore, it is thought that retaining existing customers and gaining new customer profiles will be difficult unless researchers construct a new advertising policy; therefore, the author creates a new advertisement approach (Amasaka, 2007a,b,c; Amasaka et al., 2013).

From the manufacturers' perspective, new designs must be created to grasp the heart of customers and find out how sale advertisements can deal with new customer intentions in the future. When looking at media-based advertising expenses, there has been an increase in internet advertising expenses recently (Dentsu, 2003). The automobile industry has also been expected to use internet video as a new advertising method.

Approach of the VUCMIN Study

The purpose of the "Video Unites Customer behavior and Maker's designing Intentions (VUCMIN) study" is to effectively generate fascination with a product, resulting in an increase in the desire of customers to visit dealers and growth in the number of vehicle sales.

The author generated the following solution to the current situation outlined in subsection "Outline and Need for a new advertising approach in the automobile industry" above, distribution of a video 1 to 2 minutes in length to a particular age group and gender segment. The video takes customer preferences and the intentions of the product planning and design departments into account.

Figure 4 represents the steps (STEP 1– 4) of the VUCMIN study, utilizing the specific concept methods outlined in "SQC Technical Methods" in Toyota's New JIT (Amasaka, 1999, 2004, 2007a,b,c, 2011, 2015, 2019, 2022a,b, 2023a,b) (See to Figure 5 in Chapter 2, Figure 3-6 in Chapter 3, Figure 6 and 9 in Chapter 10, and Figure 2 in Chapter 19).

These steps cover preparation of internet video distribution, which unites customer preferences and behaviors with the manufacturer's design intentions:

Figure 4. The process of VUCMIN study

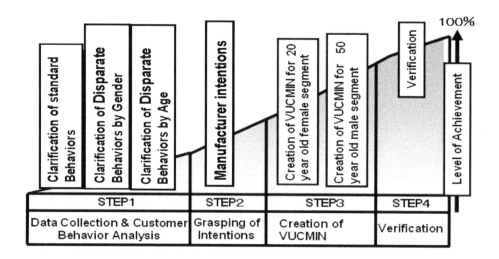

Data Collection and Customer Behavior Analysis (Step One)

This step aims to provide an up-to-date inquiry as to deep-seated customer wants in terms of customer behavior analysis (Jonker, et al., 2006). Disparate behaviors by gender and age segment will be clarified in the analysis of standard behaviors.

Grasping of Intentions (Step Two)

In this step, the intentions of the product planning department and designers are scientifically analyzed in terms of product design. Identifying which aspects of the vehicle the designers and manufacturer want to express to customers is an important step in preparing the video.

Creation of VUCKMIN (Step Three)

(1)   Framework of VUCMIN

Based on the research approach outlined in the previous chapter, the framework of the VUCMIN is created as in Figure 5.

In this figure, i) standard behaviors and ii) disparate behaviors by gender are identified and classified. After classifying the subjects by age, the details of disparate behaviors are identified mainly in terms of the front seat of the vehicle (driver's seat tools, passenger rearview mirror, etc.) and the rear seat (not shown in figure).

This knowledge of customer behaviors and knowledge of the parts that product planning and designers wish to show to customers are taken into consideration as the basis for the "VUCMIN framework".

Figure 5. Framework of VUCMIN

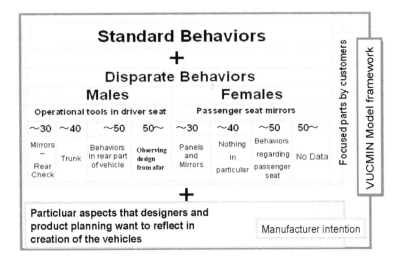

(2) The process of VUCMIN creation

Then, the author explains the process of video creation using VUCMIN.

(2-1) Data collection from "customer behaviors"

Customer behaviors are analyzed while customers are facing the vehicle from the front.

This allows collection of customer information used to create the video. It is thought that customers' desire to visit the dealer can be increased via video distribution when the customer is in the stage "prior to dealer visit".

The author therefore conducted the following survey in order to investigate customer behaviors.

Then, the author prepared the survey table in Figure 6 (: Figure 6(a) shows the outline of Sample 1, 2 and 3, and Figure 6(b) shows the enlargement of Sample 1 by indicated) (See to Figure 8(a) and Figure 8(b) in Chapter 10).

Then, the survey item categories were decided below.

In the survey table, the target car (vehicle) model is 1, gender 2, age 3, standing positions 4, and vehicle part focused on is 5. Among those items, standing positions are categorized as in Figure 7 (See to Figure 7 in Chapter 10).

Front is 1, front fender (driver seat) 2, rear fender (driver seat side) 3, trunk is 4, rear fender (passenger seat side) 5, front fender (passenger seat) 6, handle 7, shift lever 8, near passenger seat is 9.

In total, all customer behaviors (standing positions, getting in and out, operation, walking time, etc.) are categorized into 85 distinct types of behaviors.

The survey was conducted on customers visiting the "Toyota Exhibition Hall" (Tokyo) in the 2 months from August 2006 to September 2006 between 12:00 and 17:00 p.m., according to age and gender group. 316 data items were collected.

Figure 6. (a). Survey samples (Outline of Sample 1, 2, and 3); (b) Survey samples (Ex.1: The enlargement of Sample1)

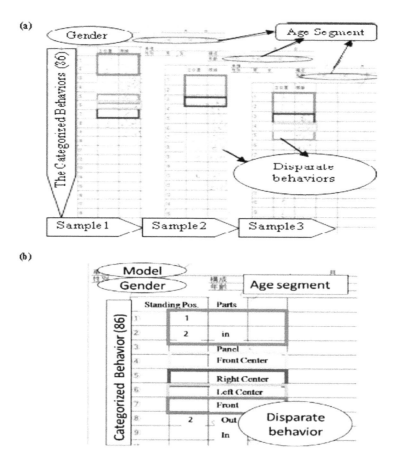

Figure 7. Samples of customer standing positions

(2-2) Data analysis of customer behaviors

As a result of the behavioral analysis of how customers observe vehicles outlined in subsection "Approach of the VUCMIN study" above, the following "(i) - (iv)" conclusions were drawn:

(i)   In general, when customers visit vehicle galleries (dealers) they pay special attention to the first vehicle and focus on some parts with interest. However, they start to loose interest by the second and third vehicles, which they observe casually and for a shorter period of time.
(ii)   When movement time is excluded, customers spend only 1-2 minutes to determine the value of each vehicle.
(iii) Regardless of gender and age, heading for the driver's seat and getting in and observing the interior, steering wheel, gauges, and other items was common standard behavior.
(iv)   Nevertheless, there were disparities in standard behaviors by both gender and age.
    (3)   Detailed analysis for creation of VUCMIN
        (3-1) Standard behaviors for creating the video

As an example, Toyota's vehicle "Mark X" was the vehicle model used in creating the video, and analyses regarding standard behavior were conducted as follows.

As seen in Figure 7, survey sheets are taken on each customer sample while they are observing the vehicle.

Sample 1 is a 20-year-old male, sample 2 is a 30-year-old female, and sample 3 is a 40-year-old male. Samples of each of the three customers are different according to their age and gender.

For further analysis, *"Text Mining Studio"* (NTT DATA, 2006) is conducted on all categorized customer behaviors and the results are shown in Figure 8, numbers 1 to 86.

The numbers in the inner circle of the figure (1, 9, 12, 16, 22, 29, 31, 32, 85) represent the "nine standard behaviors (I)" (observing front area → sitting in the driver's seat →driver operation system (control system) → leaving the driver's seat, etc.). Element resolution of the standard behaviors is also shown in Figure 8.

Next, the numbers in middle circle of the figure (17, 35, 40, 49, 53, 59, 73) represent the "seven standard behaviors (II)" (observing rear seat from driver's seat → moving to rear seat → leaving driver's seat and observing side of trunk → getting into rear seat of passenger seat → moving into passenger seat).

Moreover, the remaining 70 attached behaviors from the outer circle of Figure 11 specify individual behaviors. For example, 2 through 8 are looking at the vehicle entirely from behind, tires, and engine; 18 to 28 are passenger seat storage, side mirror, and lights: 42 to 48 are looking at vehicle diagonally from a "45-degree angle", looking under the vehicle, etc.

Finally, as a result of detailed analyzes, a pattern of standard behaviors was identified regardless of age or gender. Front → moving to driver's seat → entering driver's seat → looking out the front view → observing steering operation systems and instrument panel →looking to passenger seat side →looking at operation systems on driver's side and checking side mirrors → leaving the vehicle.

(3-2) Disparate customer behavior by "Gender and Age" for creating video

Figure 8. Text Mining Studio analysis of standard behavior (top); standard customer behaviors (bottom)

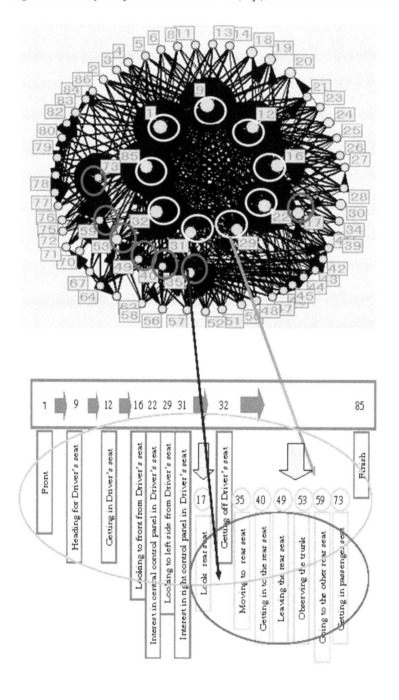

Then, the following analyses of disparate behavior are conducted regarding age and gender. On the basis of the data collected above, to identify disparities in behavior by gender, a "Correspondence Bubble Analysis" was then conducted as seen in Figure 9.

Figure 9. Disparate customer behaviors by age and gender using Text Mining Studio

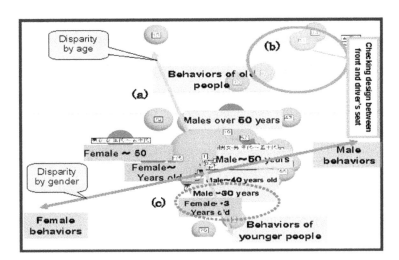

Correspondence Bubble Analysis is an analysis technique generally known as correspondence analysis. This method distributes area maps related to attributes (gender, age) and texts (customer behaviors). Items with stronger relationships are closely distributed.

In Figure 9, the difference in distributed texts (behaviors) by age and gender can be seen. (The blue line in the figure shows behaviors by gender, and the green line shows behaviors by age).

Moreover, the authors identified male and female behaviors as follows;

(i)   Female behaviors: According to the results of the Correspondence Bubble Analysis, female customer behaviors indicate that they are especially concerned with the area around the passenger seat in addition to the front fender and rear-view mirror, door opening and closing, the dashboard, sun visor (make-up mirror), etc.

(ii) Male behaviors: However, males do not show concern the with passenger seat, and instead were focused on the driver seat position, shift lever, door switches, air conditioner, operating tools, and switches around the driver seat such as audio parts.

    (4)   The creation of VUCMIN

        (4-1) Identifying of the intentions of product designers

This chapter will explain the influence of product planning and designer intentions in VUCMIN creation. Specifically, Toyota mid-size passenger car "Mark X" (named USA "Cressida") is used as a target vehicle in design inquiries.

According to common opinions from designers and product planning at Toyota Motor Corporation, the parts that are focused on to be demonstrated to customers are: (i) Front Proportions, (ii) Streamlined Side proportions, (iii) Tri-beam Headlamps (lenses), and (iv) Widened Console Box, and (v) Sharpened Rear

(4-2) The creation of VUCMIN for males in their 50s and females in their 20s/twenties

The timetable figure indicates that video time is set at 90 seconds. Video shooting order is composed specifically of scenes from 1 to 11 starting from the front, driver's seat, side, and rear of the vehicle.

Scenes are 1. direct front scene, 2. diagonally front view scene, 3. door opening scene of the driver's seat, 4. entire driver's seat view scene, 5. console box and shift lever scene, 6. Steering handle scene, 7.

Operation tools of driver's seat scene, 8. side view scene from driver's seat, 9. rear side view scene, 10. entire view of vehicle from rear, and lastly, moving back to front, 11. entire view of the vehicle scene.

In this subsection the video created for the target profile of males in their 50s will be explained using the timetable in Figure 10.

The composed scenes form the VUCMIN video on the basis of the standard and disparate behaviors of customers. Example photos representing these scenes (1 to 11) are shown in Figure 11.

Using the same approach, VUCMIN is created regarding age and gender.

Figure 10. VUCMIN creation timetable (males in their 50s, mark X)

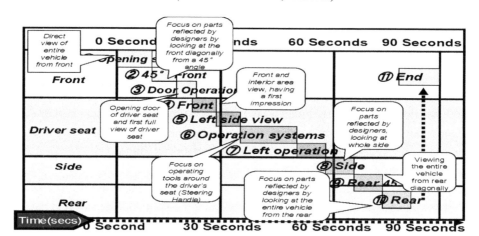

Validity of VUCMIN (Step 4)

In this subsection, customer surveys were executed in order to test the effectiveness of VUCMIN. Specifically, this was done by asking customers, after seeing the "Mark X" video, approximately when do you plan to purchase one by visiting a Toyota dealer?" in order to verify their desire to visit dealers (high, low).

According to the survey results, the desire to visit dealers (early-stage consideration of "Mark X" purchase) was not only increased for current Toyota vehicle owners but also for customers who own vehicles from other manufacturers.

The author has verified currently the effectiveness of this study as part of the strategic advertising method "VUCMIN", which utilizes internet interface with the collaboration of universities and industries (Refer to Kimura et al., 2007, Kojima et al., 2010, Murat et al., 2008 and Yamaji et al., 2010 in detail).

Figure 11. Example of representative photos for VUCMIN video

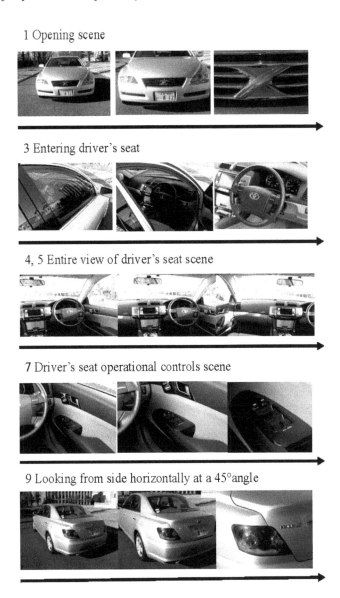

1 Opening scene

3 Entering driver's seat

4, 5 Entire view of driver's seat scene

7 Driver's seat operational controls scene

9 Looking from side horizontally at a 45°angle

Total Direct Mail Management Model (TDMMM) to Attract Customer in Auto-Dealer

Outline

Recent changes in the auto-marketing environment have made the personal relationship between businesses and their customers even more critical. Businesses are faced with the task of constructing sales schemes that are able to flexibly and accurately grasp peculiarities and trends in customer preferences.

This study looks at the effectiveness of the direct mail advertising method in bringing customers to auto dealers, based on the idea that forming personal bonds with customers is a core component of successful sales.

One of the unique features of direct mail is the advertiser's ability to select which customers will receive the mailings. This point is the focus of this study, which attempts to identify an optimum decision-making process for selecting target customers.

The way that auto-dealers currently select which customers will receive direct mail is by having individual salespeople choose them on the basis of personal experience and knowledge. Because there is no clear decision-making process, the response rate is lower than expected and dealers do not achieve the targets set forth in their sales strategies.

To address these issues, this study presents a methodology for selecting Direct Mail (DM) recipients from the dealer pool that will result in increased response (dealer visit) rates.

The proposed mechanism for selecting these optimum recipients is strategically applying a comprehensive "Total Direct Mail Management Model" (TDMMM) with the aim of bringing more customers into auto-dealerships (Murat et al., 2008; Ishiguro and Amasaka, 2012a) (See to Subsection "Development of Strategic Sales Marketing System" in Chapter 7 and Section "New Automobile Sales Marketing Model" in Chapter 10).

In developing "Advanced TMS" based on the TMS, this model uses mathematical programming and statistics to aid the decision-making process and boost the effectiveness of DM advertising in Toyota's New JIT (Amasaka, 2007b,c, 2011, 2015, 2019, 2022a,b, 2023a,b) (See to Section "Establishment of New JIT, new management technology principle" and "Developing Advanced TDS, TPS, TMS, TIS & TJS in New JIT strategy in Chapter 7).

The core systems in TDMMM used in the model are the (a) Strategically formulation: "System for analyzing customer purchase data" to increase dealer visits, (b) Direct Mail Content: "System for optimizing DM content" to target customer preferences, (c) Recipients: "System for strategically determining DM recipients" who should be sent direct mail based on who actually visits the dealership", and (d) Approach to visitors: "DM promotion system for sales staff" to increase market share.

These 4 systems are integrated and strategically applied to promote dealer visits. The author then applies this model at an actual auto dealership, where they are able to successfully increase dealer visits.

Necessity of DM in Auto Dealerships

This subsection focuses on CR (customer retention, a critical issue in auto dealerships, and discusses a methodology for preventing customer loss. The necessity of DM (direct mail) as a tool for retaining premier customers is discussed below.

(1) DM and CR

Auto-dealerships typically store customer attributes, purchase history, and other information on their clients in a database or similar storage system (Amasaka, 2004).

The database is used to identify premier customers, and pinpointing their preferences is critical if the dealership is to prevent them from taking their business elsewhere.

The author considers using a database to help find premier customers and then sending them DM to be an effective method of sales promotion. In references of DM effects, (i) Kotler and Keller (2006) identified the ability to select your recipients and the ability to offer personalized content as the advantages of DM, and (ii) Piersma and Jonker (2004) studied how often to send DM to individual customers in order to establish long-term relationships between the direct mailer and the customers.

Furthermore, (iii) Jonker et al. (2006) provided a decision support system to determine mailing frequency for active customers based on their behavioral data: their "recency, frequency, and monetary" (RFM) values, (iv) Bell et al. (2006) applied experimental designs to increase direct mail sales. (v) Beko and Jagric (2011) presented demand models for direct mail and periodicals market using time series analysis, and (vi) DM is notable as an effective way of targeting specific premier customers for sales promotions (Bult and Wansbeek, 2005).

(2) Creation of "Auto-AIDA Model" by developing DM strategy

The AIDA (attention-interest-desire-action) model is a well-known advertising information management model developed by Lewis, Elias St. Elmo (1908).

The model presents the psychology of consumer purchase behavior as a series of four steps: *attention, interest, desire*, and *action* (Ferrell and Hartline, 2005; Shimizu, 2004).

Specifically, the author has created an "Auto-AIDA Model" of customer response for use in auto sales by developing DM strategy (Amasaka, 2007b, 2009).

This model is based on the AIDA model and presented in Figure 12. The model allows dealerships to boost the percentage of customers that visit the showroom by using mass media advertising, flyers, magazine ads, and DM (direct mail) and others (DH (direct hand, Train cars ads., internet etc.).

Figure 12. Auto-AIDA model

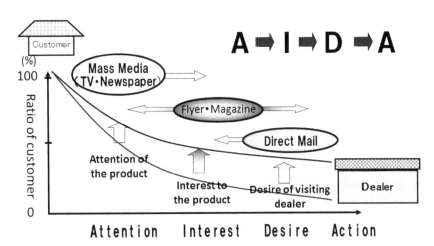

(3) Current research on DM activities

In studying the current status of direct mail activities at dealerships, the author identified the following issues:

1. Strategy formulation: target customers and sales concepts are unclear
2. Direct mail content: customer preferences are not accurately identified
3. Recipients: recipients are determined based on experiential knowledge of sales staff, which leads to a high degree of variation in the quality of work performed
4. Approach to customers who visit the dealership: purchase behavior characteristics of individual customer are not well understood.

Given these problems, direct mail cannot currently be thought of as effective in all cases. The author has demonstrated that the percentage of direct mail recipients that visit the dealers is as low as 1–5%, and that the profitability of those customers that do come in is not very high (Ishiguro et al., 2010; Ishiguro and Amasaka, 2012a,b).

It is critical that these issues be resolved so that direct mail activities are more efficient (sophisticated). This will help dealerships retain their premier customers and boost their sales. A survey of the literature did not reveal any prior research that successfully addresses these problems.

Strategically Creating a Total Direct Mail Management Model

Based on previous direct mail research and studies done on the actual state of direct mail (DM) activities, the author has created the "Strategically applying a comprehensive a comprehensive "Total Direct Mail Management Model" (TDMMM) as the strategic DM application. This model using four core functions is shown in Figure 13.

As the figure indicates, the model consists of the following 4 core functions: (a) Strategy formulation: "Scientific identification of premier customers" (Amasaka, 2011), (b) Direct mail content: "Development of content based on customer preferences" (Kimura et al., 2007), (c) Recipients (Mailing process): "Establishing a process for determining recipients based on customer information and other data" (Ishiguro et al., 2010; Ishiguro and Amasaka, 2012b), and (d) Approach to visitors (Customer handling): "Clearly identifying customer purchase behavior characteristics" (Kojima et al., 2010).

Figure 13. Creation of a total direct mail management model (TDMMM)

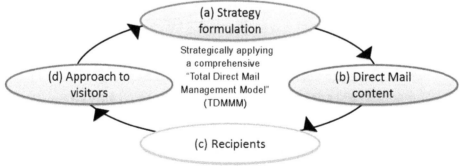

Then, Figure 14 shows the strategically "TDMMM Approach Model" to practically implementing the above strategies. The related systems for realizing four core functions are described below.

Figure 14. Strategically TDMMM approach model for realizing four core functions

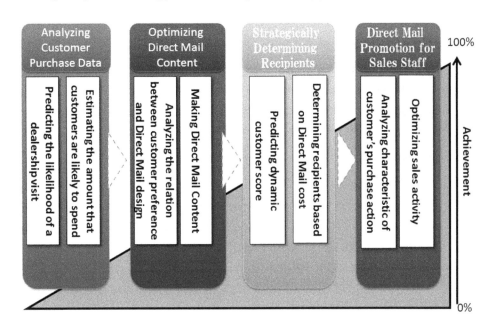

(a)   "System for analyzing customer purchase data"

As Table 3 describes, the system for analyzing customer purchase data supports dealers in formulating a direct mail strategy by identifying different premier customer segments.

(a-1) Identifying premier customer segments

Premier customers are first defined according to 2 different criteria: the (i) "likelihood that they will visit the dealership" and (ii) "amount of money that they can be expected to spend on purchases". This information is based on the purchase process (whether they buy vehicles after coming to the dealer). These 2 indicators are then combined and used to define the level of the premier customer.

Table 3. System for analyzing customer purchase data

		Likelihood of spend	
		High	**Low**
Likelihood of visit	High	Preferred customer Luxury car intention Own customer	Basic car intention Own customer
	Middle	Semi-preferred customer Luxury car intention Own/other customer	Basic car intention Own/other customer
	Low	Luxury car intention Other customer	Not preferred customer Basic car intention Other customer

(a-2) Predicting the likelihood of a dealership visit

"Discriminant analysis" is used to score each customer in terms of how likely he or she is to visit the dealership.

Past visit data is used to arrive at a discriminant score, using whether the direct mail will lead to a visit as the external criterion and customer attributes as the items.

The score is then used to carry out a logistic regression analysis (Tsujitani and Amasaka, 1993) in order to get a "Regression equation (1)" that will calculate the likelihood of a dealership visit.

$$P_i = \frac{1}{1 + \exp\{-(\alpha_0 + \sum_j \alpha_j \delta_{ij})\}}, \quad i = 1, 2, \ldots, n \tag{1}$$

i: Customer ID number (n: Number of the customers)
j: Customer attribute (J: Number of the attributes)
P_i: Likelihood that customer i will visit the dealer
α_j: Discriminant coefficient
δ_{ij}: Indicates whether or not customer i has attribute j (0 or 1)
(a-3) Estimating the amount that customers are likely to spend

Regression analysis is used to obtain a "Regression equation (2)" for this calculation, with customer purchase amount as the external criterion and customer attributes as the items. This equation is then used to estimate the amount that customers are likely to spend based on their specific attributes.

$$V_i = \beta_0 + \sum_j \beta_j \delta_{ij}, \quad i = 1, 2, \ldots, n \tag{2}$$

V_i: Amount that customer i is likely to spend
β_j: Partial regression coefficient

(a)   System for optimizing DM content

The system for optimizing direct mail content is used to analyze how customer preferences change in response to changes in direct mail design. To get this information, a "Key Graph" is used to collect free-response customer feedback on different direct mail designs.

Figure 15 shows an example of a Key Graph. In this example, the authors were able to determine that using a black background gives the design a high-end feel and causes it to stand out. By referring to concepts arrived at using these new insights as well as the system for analyzing customer purchase data, direct mail designers can now create designs that accurately reflect sales strategies and customer preferences.

Figure 15. Key graph

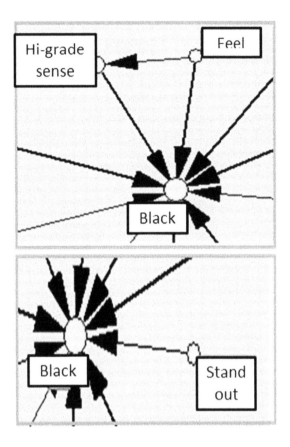

(c)   System for strategically determining DM recipients

The "System for strategically determining DM recipients" identifies who direct mail should be sent to among the premier customer categories identified using the system for analyzing customer purchase data.
"Equations (3) to (5)" are used to determine the recipients.

$$\underset{x_1,x_2,\ldots,x_n \in \{0,1\}}{Maximize} \quad \sum_i P_i V_i x_i \tag{3}$$

$$s.t. \sum_i x_i = N \tag{4}$$

$$L_j N \le \sum_i \delta_{ij} x_i \le H_j N, \quad j=1,2,\ldots,J \tag{5}$$

x_i : Whether customer i is flagged as a direct mail recipient (0 or 1)
N: Total number of direct mailings sent
L_j / H_j : Lower/Upper limit for the percentage of direct mailings sent to customers with attribute j

Mathematical programming is used to solve the above equations and determine who customer *i* should be targeted as a recipient. The values of the other parameters are determined in advance.

P_i (the likelihood of a dealership visit) and V_i (amount of money likely to be spent) are determined using the same procedure outlined for the "Customer purchase data analysis system" (See to Subsection "Necessity of DM in auto dealerships: (1) DM and CR) above, once behavioral data (in RFM: Recency Frequency Monetary analysis) is added to customer attribute data.

Table 4 shows their difference. It is used to determine who should be sent direct mail based on the current level of the premier customer (this takes into consideration different customer behaviors among different premier customer categories).

Table 4. Difference between static customer and dynamic customer

Static (Analyzing customer purchase data)	Dynamic (Strategically determining direct mail recipients)
Customer segment who like target car during the promotion term	Customer segment who like target car most now
Life stage Life style	Life stage Life style RFM

(d)   DM promotion system for sales staff

The "DM promotion system" for sales staff reveals important insights that sales staff can use in order to handle customers when they come to the dealership.

Specifically, the system surveys dealerships to determine what elements are prioritized when someone from the premier customer category purchases a vehicle.

As Figure 16 describes, survey data is then subjected to a cluster analysis and principal component analysis to identify characteristics of customer purchase behavior.

These results are then communicated to the sales staff.

Effectiveness of TDMMM

As Figure 17 shows, the author is currently using the "Strategic development of TDMMM application for auto-dealers" presented here at Company A in an attempt to increase the percentage of customers that visit the dealer as well as the amount of money they spend when they come in.

Specifically, they are using the four systems described above based on Company A's customer information and sales data as well as on customer preferences collected via surveys.

Then, to realize of TDMMM effectiveness for improving business process of auto-dealers using "Strategically TDMM Approach Model" (Figure 17), the author has proceeded the actual application examples based on the 4 core functions ((a), (b), (c) and (d) in Figure 16) as follows;

(a)   Changes in marketing process management: Establishment of "Toyota Sales Marketing System" to develop the scientific identification of premier customers (Amasaka, 2011).

Figure 16. Characteristic of customer's purchase action

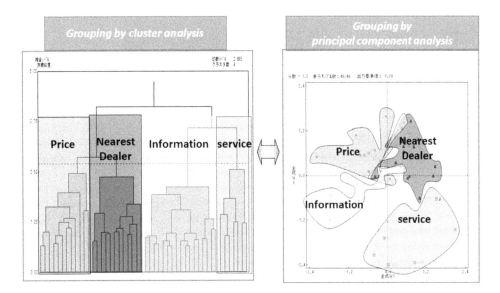

(b) Creation of "Scientific Approach Method for DM" (SAM-DM): Effectiveness of attracting customer" for the development of content based on customer preferences (Kimura, 2007),

(c) Effectiveness of a highly compelling DM method "Practical Method using Optimization and Statistics for Direct Mail" (PMOS-DM) to bring customers into auto-dealerships" to establish a process for determining recipients based on customer information (Ishiguro et al., 2010; Ishiguro and Amasaka, 2012b).

(d) Development of DM method "Practical use Model of Customer Information for Direct Mail" (PMCI-DM) for effectively attracting customers aiming clearly identifying customer purchase behavior characteristics (Kojima et al., 2010).

Developing VUCMIN and TDMMM for CS and CL to Boost Automobile Marketing

In this study, for CS and CL to boost automobile marketing, the author has developed the created "VUCMIN and TDMMM" in order to enhance the desire to visit stores in the automobile industry.

Then, taking advantage of those acquired knowledge by the validity "VUCMIN and TDMMM", the author illustrates the typical case studies for the "Progress of automobile marketing for business and sales" as follows (Amasaka, 2011; Amasaka et al., 2013; Ogura et al., 2013, 2014);

(1) Scientific Mixed Media Model for boosting auto-dealer visits.
(2) Toyota Sales Marketing System by innovating auto-dealers' sales activities.
(3) Attention-grabbing train car advertisements using auto-pamphlet.

Figure 17. Strategic development of TDMMM application for auto-dealers

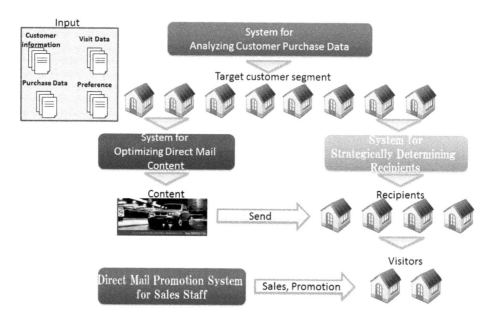

CONCLUSION

Faced with a sluggish economy, car sales have been disappointing in recent years. In this chapter, therefore, the author focused on CS as a way of boosting marketing effectiveness, clarifying the key factors that comprise CL, and helps to improve the marketing strategy employing NJ-MMM (New Japan Marketing Management Model). Specifically, the author has developed with researching CS and CL as a way of boosting marketing effectiveness, clarifying the key factors that comprise customer loyalty, and help improve the marketing strategy. Then, as application examples using those knowledges acquired above, the author illustrated the development and its effectiveness for CS and CL describing VUCMIN (Video Unites Customer Behavior and Maker's Designing Intentions) and TDMMM (Total Direct Mail Management Model).

REFERENCES

Amasaka, K. (1999). A study on Science SQC by utilizing Management SQC: A demonstrative study on a New SQC concept and procedure in the manufacturing industry. *International Journal of Production Economics*, *60-61*, 591–598. doi:10.1016/S0925-5273(98)00143-1

Amasaka, K. (2004). *Science SQC, new quality control principle: The quality strategy of Toyota*. Springer-Verlag. doi:10.1007/978-4-431-53969-8

Amasaka, K. (2005). Constructing a *Customer Science* Application System "*CS-CIANS*": Development of a Global Strategic Vehicle "*Lexus*" utilizing *New JIT*. *WSEAS Transactions on Business and Economics*, *2*(3), 135–142.

Amasaka, K. (2007a). Proposal of Marketing SQC to revolutionize dealers' sales activities-A demonstrative study on Customer Science by utilizing Science SQC. *Proceedings of the 16th International Conference Production Research. Praha Czech Public.*

Amasaka, K. (2007b). The validity of Advanced TMS, a strategic development marketing system: Toyota's Scientific Customer Creative Model utilizing New JIT. *The International Business & Economics Research Journal, 6*(8), 35–42.

Amasaka, K. (2007c). High Linkage Model *"Advanced TDS, TPS & TMS"*: Strategic development of *"New JIT"* at Toyota. *International Journal of Operations and Quantitative Management, 13*(3), 101–121.

Amasaka, K. (2009). The effectiveness of flyer advertising employing TMS: Key to scientific automobile sales innovation at Toyota, *China-USA. Business Review (Federal Reserve Bank of Philadelphia), 8*(3), 1–12.

Amasaka, K. (2011). Changes in marketing process management employing TMS: Establishment of Toyota Sales Marketing System, *China-USA. Business Review (Federal Reserve Bank of Philadelphia), 10*(6), 539–550.

Amasaka, K. (Ed.). (2012). Science TQM, new quality management principle: The quality management strategy of Toyota. Bentham Science Publishers.

Amasaka, K. (2015). *New JIT, new management technology principle.* Taylor & Francis, CRC Press.

Amasaka, K. (2019). Studies on New Manufacturing Theory, *Noble. International Journal of Scientific Research, 3*(1), 1–20.

Amasaka, K. (2022a). New Manufacturing Theory: Surpassing JIT (2nd Edition). Lambert Academic Publishing.

Amasaka, K. (2022b). *Examining a New Automobile Global Manufacturing System.* IGI Global Publisher. doi:10.4018/978-1-7998-8746-1

Amasaka, K. (2023a). New Lecture-Toyota Production System: From JIT to New JIT, Lambert Academic Publishing.

Amasaka, K. (2023b). A New Automobile Sales Marketing Model for innovating auto-dealer's sales. *Journal of Economics and Technology Research, 4*(3), 9–32. doi:10.22158/jetr.v4n3p9

Amasaka, K., Ogura, M., & Ishiguro, H. (2013). Constructing a Scientific Mixed Media Model for boosting automobile dealer visits: Evolution of market creation employing TMS. *International Journal of Engineering Research and Applications, 3*(4), 1377–1391.

Beko, J., & Jagric, T. (2011). Demand models for direct mail and periodicals delivery services: Results for a transition economy. *Applied Economics, 43*(9), 1125–1138. doi:10.1080/00036840802600244

Bell, G. H., Ledolter, J., & Swersey, A. J. (2006). Experimental design on the front lines of marketing: Testing new ideas to increase Direct Mail sales. *International Journal of Research in Marketing, 23*(3), 309–319. doi:10.1016/j.ijresmar.2006.05.002

Bult, J. R., & Wansbeek, T. (2005). Optimal selection for direct mail. *Marketing Science, 14*(4), 378–394. doi:10.1287/mksc.14.4.378

Dentsu Inc. (2003). *Dentsu Online, Total Advertisement Expenses Total of Japan*. Dentsu Inc. (http://www.dentsu.co.jp/marketing/adex/adex2005/media.html)

Elias. St. Elmo Lewis. (1908). *Financial advertising*. Levey bros. & company, Indianapolis, USA.

Ferrell, O. C., & Hartline, M. (2005). *Marketing Strategy*. Thomson South-Western.

Hifumi, S. (2006). *A study on a strategic advertising and publicity (animation): The proposal of "A-VUCMIN" for raising the volition of coming to the dealer's shop* [Master's thesis, Graduate School of Science and Engineering, Aoyama Gakuin University].

Ishiguro, H., & Amasaka, K. (2012a). Establishment of a Strategic Total Direct Mail Model to bring customers into auto-dealerships. *Journal of Business & Economics Research, 10*(8), 493–500. doi:10.19030/jber.v10i8.7177

Ishiguro, H., & Amasaka, K. (2012b). Proposal and effectiveness of a highly compelling direct mail method: Establishment and deployment of PMOS-DM. *International Journal of Management and Information Systems, 16*(1), 1–10.

Japan Automobile Manufacturers Association (JAMA). (2005). *Automobile Sale Numbers of Japan*. JAMA. (https://www.jama.or.jp/stats/product/index.html)

Jonker, J. J., Piersma, N., & Potharst, R. (2006). A Decision Support System for Direct Mailing Decision. *Decision Support Systems, 42*(2), 915–925. doi:10.1016/j.dss.2005.08.006

Kawasaki, S., & Yoshizawa, N. (2011). *Customer database marketing for the improvement in CS: The multivariate analysis of the customer information for raising commodity value* [Bachelor thesis, School of Science and Engineering, Department of Industrial and System Engineering].

Kimura, T., Uesugi, Y., Yamaji, M., & Amasaka, K. (2007). A study of Scientific Approach Method for Direct Mail, SAM-DM: Effectiveness of attracting customer utilizing Advanced TMS. *Proceedings of the 5th Asian Network for Quality Congress, Hyatt Regency*, Incheon, Korea.

Kojima, T., Kimura, T., Yamaji, M., & Amasaka, K. (2010). Proposal and development of the Direct Mail Method "PMCI-DM" for effectively attracting customers. *International Journal of Management & Information Systems, 14*(5), 15–22. doi:10.19030/ijmis.v14i5.9

Kotler, P., & Keller, K. L. (2006). *Marketing Management (12thEdition)*. Pearson Prentice Hall.

Murat, S. M., Hifumi, S., Yamaji, M., & Amasaka, K. (2008). Developing a strategic advertisement method "VUCMIN" to enhance the desire of customers for visiting dealers. *Proceedings of the International Symposium on Management Engineering*, Waseda University, Kitakyushu, Japan.

NTT DATA Mathematical Systems Inc. (2006). *"Text Mining Studio Analysis" by NTT DATA*. MSI. https://www.msi.co.jp/solution/tmstudio/index.html

Ogura, M., Hachiya, T., & Amasaka, K. (2013). A omprehensive Mixed Media Model for boosting automobile dealer visits. *The China Business Review, 12*(3), 195–203.

Ogura, M., Hachiya, T., Masubuchi, K., & Amasaka, K. (2014). Attention-grabbing train car advertisements, *International Journal of Engineering Research and Applications, 4*(1) (Version 2), 167-175.

Okutomi, H. (2011). A study on establishment of the structure model on relationship between CS and CL: An example of Japan's auto-dealerships (Bachelor thesis), *School of Science and Engineering, Department of Industrial and System Engineering, Aoyama Gakuin University, Amasaka New JIT Laboratory, Sagamihara,* Kanagawa, Japan. (in Japanese)

Okutomi, H., & Amasaka, K. (2012). Researching customer satisfaction and loyalty to boost marketing effectiveness: A look at Japan's auto dealerships. *Proceedings of the 2nd International Symposium on Operations Management Strategy.* Japanese Operations Management and Strategy Association.

Okutomi, H., & Amasaka, K. (2013). Researching customer satisfaction and loyalty to boost marketing effectiveness: A look at Japan's auto-dealerships. *International Journal of Management & Information Systems, 17*(4), 193–200. doi:10.19030/ijmis.v17i4.8093

Parasuraman, A., Zeihaml, V. A., & Berry, L. L. (1988). SERVQUAL: A multiple-item scale for measuring consumer perceptions of service quality. *Journal of Retailing, 64*(1), 12–40.

Piersma, N., & Jonker, J. J. (2004). Determining the Optimal Direct Mailing Frequency. *European Journal of Operational Research, 158*(1), 173–182. doi:10.1016/S0377-2217(03)00349-7

Shimizu, K. (2004). *Theory and Strategy of Advertisement.* Sousei-Sha. (in Japanese)

Tsujitani, M., & Amasaka, K. (1993). Analysis of Enumerated Data (1): Logit Transformation and Graph Analysis [in Japanese]. *Quality Control, 44*(4), 61–66.

Yamaji, M., Hifumi, S., Sakalsis, M. M., & Amasaka, K. (2010). Developing a Strategic Advertisement Method "VUCMIN" to enhance the desire of customers for visiting dealers. *Journal of Business Case Studies, 6*(3), 1–11. doi:10.19030/jbcs.v6i3.871

Chapter 18
Scientific Mixed Media Model for Boosting Auto Dealer Visits

ABSTRACT

In this chapter, to strengthen New JIT strategy, the author has established the scientific mixed media model (SMMM) for boosting automobile dealer visits by developing advanced TMS, strategic customer creation model based on the TMS (total marketing system) named new Japan marketing management model (NJ-MMM) in order to realize the automobile market creation. Specifically, SMMM develops and validates the effectiveness of putting together four core elements: (1) Video that unites customer behavior and manufacturer design intentions (VUCMIN), (2) Customer motion picture–flyer design method (CMP-FDM), (3) Attention-Grabbing train car advertisements (AGTCA), and (4) Practical method using optimization and statistics for direct mail (PMOS-DM) into new strategic advertisement methods designed to enhance marketing and the desire in the automotive industry. At present, SMMM was applied to a dealership representing an advanced car manufacturer Toyota, where its effectiveness was verified.

NEED FOR A MARKETING STRATEGY THAT CONSIDERS MARKET TRENDS

Today's marketing activities require more than just short-term strategies by the business and sales divisions. In a mass-consumption society, when the market was growing in an unchanging way, sales increases were achieved by means of simple mass marketing through huge corporate investments in advertising (Nikkei Business, 1999; Amasaka, 2005).

However, after the collapse of the bubble economy, the competitive market environment changed drastically. Since then, companies that have implemented strategic marketing quickly and aggressively have been the only ones enjoying continued growth (Okada et al., 2001; Amasaka, 2002, 2005).

Upon close examination, it was determined that strategic marketing activities must be conducted as company-wide, core corporate management activities that involve interactions between each division inside and outside of the company (Jeffrey and Bernard, 2005; Shimakawa et al., 2006; Amasaka, 2007).

Therefore, a marketing management model needs to be established so that business, sales, and service divisions, which are developing and designing appealing products and are also closest to customers, can organizationally learn customer tastes and desires (Amasaka et al., 2005, 2008).

DOI: 10.4018/978-1-6684-8301-5.ch018

Specifically, pursuing improvements in product quality by the continued application of objective data and scientific methodology is increasingly important (Amasaka et al., 1998; James and Mona, 2004: Amasaka, 2003, 2004, 2005).

At present, the organizational system and rational methodology that allows them to analyze data on each customer using a scientific analysis approach has not yet been fully established in these divisions; in some cases, the importance of this system has not even been widely recognized (Niiya and Matsuoka, 2001; Gary and Arvind, 2003; Ikeo, 2006; Amasaka, 2007, 2011).

EVOLUTION OF MARKET CREATION EMPLOYING ADVANCED TMS STRATEGY

Significance of Advanced TMS Based on TMS

Recently, in light of recent changes in the marketing environment, the author believes it is now necessary to develop innovative business and sales activities that adequately take into account the changing characteristics of customers who are seeking to break free from convention (Amasaka et al., 2005; Amasaka, 2011).

TMS (Total Marketeing System) as the key of the strategic development of New JIT strategy shown in Figure 1 and 2, is composed of these technological elements below (See to Figure 1 and 2 in Chapter 7 and Figure 1 in Chapter 10) below.

(a) market creation activities through collection and utilization of customer information, (b) strengthening of merchandise power based on the understanding that products are supposed to retain their value, (c) establishment of marketing systems from the viewpoint of building bonds with customers, and (d) realization of the customer focus utilizing customer information network for CS (Customer Satisfaction), CD (Customer Delight) and CR (Customer Retention) elements needed for the corporate attitude (behavior norm) to enhance customer values as discribed in Figure 2 of Chapter 2 (Amasaka, 2019) (See to Figure 7.3).

Therefore, to realize market creation with an emphasis on the customer by developing CL (Customer Royalty) to boost marketing effectiveness, the significance of Advanced TMS named "Strategic Marketing Development Model" (SMDM) based on TMS is to promote market creation and to realize quality management through scientific marketing and sales, not by sticking to conventional concepts as described in Figure 3.3 of Chapter 7 and Figure 2 of Chapter 10 (Amasaka et al., 2008; Amasaka Ed. 2012; Okutomi and Amasaka, 2013; Amasaka, 2015, 2022a,b, 2023a,b) (See to Chapter 7 and Chapter 10).

Specifically, to achieve the high cycle rate for market creation activities in Advanced TMS strategy, the author has established the "New Automobile Sales Marketing Model" (NA-SMM) for innovating auto-dealers' sales employing "Modeling of Strategic Marketing System" (MSMS) named "Scientific Customer Creative Model" (SCCM) (See to Figure 3 in Chapter 10 and Figure 9 in Chapter 7).

MSMS consists of the 4 core elements: the NSOI (New sales office image), ISINS (Intelligent Customer Information Network System), RAPS (Rational Advertisement Promotion System) and ISMS (Intelligent Sales Marketing System) as described in SCCM (Scientific Customer Creative Model) as described in Figure 9 of Chapter 7 (Amasaka, Ed., 2012; Amasaka, 2015, 2022a,b, 2023a,b)

By these elements, Advanced TMS innovates for bonding with the customer and reforms office-shop appearance and operation.

Developing Customer Science Principle for the Strengthening of Advanced TMS Based on the Science SQC

Then, the author develops the "Customer Science principle" (CSp) using Science SQC for the strengthening of Advanced TMS (Amasaka, 2002, 2003, 2004, 2005, 2019, 2022a,b, 2023a.b).

Supplying products that satisfy consumers (customers) is the ultimate goal of companies that desire continuous growth. Customers generally evaluate existing products as good or poor, but they do not generally have concrete images of products they will desire in the future.

For new product development, it is important to precisely understand the vague desires of customers. To achieve this goal, the author has created the CSp to help systematize Advanced TMS as described in Figure 2 of Chapter 3 (Amasaka, 2002, 2005) (See to Subsection "Creation of the "Scientific quality management" employing "Customer Science principle" in Chapter 3).

To plan and provide customers with attractive products is the mission of companies and the basis of their existence. This principle is particularly important to convert customer opinion (implicit knowledge) to images (linguistic knowledge) through market creation activities, and to accurately reflect this knowledge in creating products (drawings, for example) using engineering language (explicit knowledge).

This refers to the conceptual diagram that rationally objectifies subjective information (y) and subjectifies objective information ($\hat{y}$) through the application of correlation technology.

Then, the author apply the statistical science methodology "Science SQC", which has four core principles (Scientific SQC, SQC Technical Methods, Integrated SQC Network (TTIS) and Management SQC) as described in Figure 4 of Chapter 2 and which is designed to develop CSp in business and sales divisions to make changes to marketing process management (Amasaka, 2003, 2004) (See to Figure 3, 4, 5 and 6 in Chapter 3 in detail).

CONSTRUCTING A SCIENTIFIC MIXED MEDIA MODEL FOR BOOSTING AUTO-DEALER VISITS

Publicity and Advertising as Auto-Sales Promotion Activities

For many years, automobile dealers have been employing various publicities and advertising strategies in cooperation with auto-manufacturers in order to encourage customers to visit their shops. To develop the CS, CL and CR activities for the auto-market creation, the author illustrates a graphical representation of the relationship between publicity and advertising media—a relationship that helps draw customer traffic to auto-dealers as shown in Figure 8(a)(b) of Chapter 10 (Amasaka, 2009, 2011).

Area A represents multimedia advertising (the internet, CD-ROMs, etc.), Area B represents direct advertising (catalogs, direct mail, handbills (directly handed to customers), telephone calls, etc.), and Area C represents mass media advertising (TV and radio broadcasting, flyers, public transportation (train cars), newspapers, magazines, etc.).

There appear to be few cases where scientific research methods have been applied to the effect of mixed media (areas A to C) and used to study the ways in which such sales activities actually draw customers to automobile dealers.

However, the rational effects of the media-mix are insufficient as advertisement methods, and the author therefore considers the need to scientifically promote a new advertisement media mixed model (Kubomura and Murata, 1969; Melewar and Smith, 2003; Amasaka, 2007, 2011; Smith, 2009; Ogura, et al., 2013a).

Evolution of the Scientific Mixed Media Model (SMMM) for Boosting Auto-Dealer Visits

As stated with Chapter 10 and 18, to realize the "New Automobile Sales Marketing Model for innovating auto-dealer's sales", both of the "(i) Evolution of the Scientific Mixed Media Model (SMMM) for boosting auto-dealer visits and (ii) CS and CL to boost marketing effectiveness in auto-dealerships" in developing Advanced TMS is important, and is the indispensable requirement of the auto-market cultivation (Amasaka, 2011; 2015, 2022a,b, 2023a,b) (See to Chapter 7, 10 and 18).

Actually, to verify the SMMM development and its validity, the author has applied the "Intelligent Automobile Sales Marketing Model" (IASMM) described in Figure 6 of Chapter 10. and has applied SMMM to sale of the new-model car in Toyota auto-dealers, and demonstrated its effectiveness (Amasaka, 2007, 2009, 2011).

As part of an organization's market creation activities, it is important to gain a quantitative understanding of the effect of publicity and advertising, which are the principal methods involved in sales promotion and order taking, in order to aid the development of future business and sales strategies (Kobayashi and Shimamura, 1997: Kishi et al., 2000; Shimizu, 2004; Ferrell and Hartline, 2005; Amasaka et al., 2013).

After recent changes in the marketing environment, what is needed now is to develop innovative business and sales activities that are unconventional and correctly identify the characteristics of and changes in customer tastes.

There have never been a greater need for careful attention and practice in customer contact, and in order to continuously offer an appealing and customer-oriented marketing strategy, it is important to evolve current market creation activities to strengthen commercial viability and reform office/shop appearance and operations using the Customer Science approach (Amasaka, 2002, 2005).

Therefore, the author wants to construct the "Scientific Mixed Media Model" (SMMM) for boosting auto-dealer visits employing Advanced TMS for the innovation of (a) Market Creation Activities, (b) Product Value Improvement, (c) Building Ties with customers and (d) Customer Value Improvement for improving mass media and multimedia, direct advertising, internet, CD-ROM (TV, newspapers, flyers, direct mail, radio, magazines, train cars, etc.) as shown in Figure 1 (Amasaka et al., 2013).

Specifically, SMMM aims to achieve a "high cycle rate for market creation activities", and is composed of 4 core elements (1)-(4) by developing "Evolution of market creation activities" for the "Contact with customers", "Strengthening of merchandise power" and "Reform of office/shop appearance and operations" (Yamaji et al., 2010; Koyama et al., 2010; Ishiguro and Amasaka, 2012a,b; Ogura et al., 2013b; Amasaka et al., 2013; Amasaka, 2015, 2020, 2022a,b, 2023a,b) as follows;

(1)   "Video that Unites Customer behavior and Manufacturer Design Intentions" (VUCMIN)
(2)   "Customer Motion Picture–Flyer Design Method" (CMP-FDM)
(3)   "Attention-Grabbing Train Car Advertisements" (AGTCA)
(4)   "Practical Method using Optimization and Statistics for Direct Mail" (PMOS-DM)

Figure 1. Scientific mixed media model for boosting auto-dealer visits

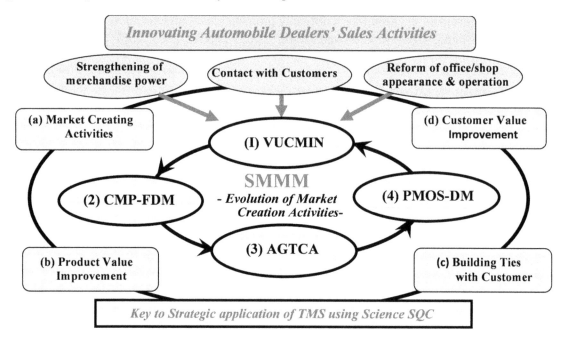

These 4 core elements aim to provide an up-to-date inquiry as to deep-seated customer wants in terms of customer behavior analysis (James et al., 2006). Disparate behaviors by gender and age segment will be clarified in the analysis of standard behaviors.

SMMM for the evolution of market creation activities improves "innovative automobile dealer sales activity knowhow" regarding repeat users of various manufacturers' vehicles. Its characteristics are described below.

(1)   VUCMIN

As stated within Chapter 10 and 18, the created "VUCMIN uses video advertisements, and is developed based on scientific approaches and analyses that focus on the standard behavioral movements of customers who visit dealers when choosing an automobile (Amasaka, 2003, 2004, 2015; Hifumi, 2006; Sakalsiz et al., 2008; Yamaji et al., 2010).

This method, which is based on the different approaches identified in target customer profiles, aims to make customers more eager to visit automobile dealers. After creating this video advertisement, customers are verified as having a positive opinion towards visiting dealers with a plan to purchase the vehicle featured in the video.

More concretely, based on the research approach outlined in the previous Chapter 7, 10 and 18, the framework of VUCMIN was established using "SQC Technical Methods" (See to Figure 9 in Chapter 10 and Figure 6 and 7 in Chapter 18).

(2)   CMP-FDM

In this study, the author establishes a method of creating "attractive flyer designs" while using customer behavior analysis with videos that help dealers attract customers and aim to reform conventional marketing activities. CMP-FDM analyzes how customers see flyers, and the author creates attractive designs that guarantee each customer's satisfaction.

Next, the author integrates the design elements into one that will satisfy all types of customers (universal type) by organizing the design features (design elements), and then validating the method (Koyama et al., 2010).

More concretely, the author shows the steps (1 to 5) for establishing CMP-FDM as shown in Figure 2. The created CMP-FDM utilizing SQC Technical Methods (Amasaka, 2003, 2004) consists roughly of two processes: research and flyer design. In the first process, the author checks the current state of flyers as preliminary research as follows;

Figure 2. Steps of establishing CMP-FDM

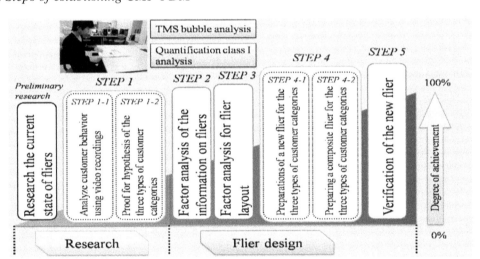

In Step 1, the author analyzes customer behavior towards the flyers using video recordings in order to understand how to the materials are actually viewed. Then, the author proves that customers can be classified into 3 types: active customers, collection-first customers and indifferent customers.

In Step 2, to address problems with the current flyer design (information appearing on the flyer, such as exterior photos, price, car name, loan information, and interior photos), the author clarifies what each customer type wants to know from the content. One problem is the provision of a lot of unnecessary information and a lack of necessary information.

In Step 3, to solve problems in flyer layout, the authors clarify what kind of layout each customer group wants. One problem is typeface that is too small to see or information that is too varied to understand.

In Step 4, based on the results of Step 2 and 3, the author incorporates the design elements into one flyer that it is attractive to all customer type.

In Step 5, the author conducts a survey to compare the composite flyer developed in Step 4.

(3)   AGTCA

This study deals with "Train car advertisements" (hanging posters, above-window posters, and sticker ads.) that have become increasingly popular in recent years.

Focusing on transit advertising, which has a good contact rate and provides long-term contact, the authors decided to examine customer relationships and how they relate to train car advertising with the aim of defining the ideal format for this type of media (Ogura et al., 2013b).

The goal was to first quantify the way passengers pay attention to "Train car advertisements", and then propose the ideal form that "in-car train advertising" should take based on a visual representation of passenger information.

More concretely, the purpose of AGTCA is to examine the correlations between passenger information and riding conditions in train car advertising in order to discover the ideal way to advertise inside passenger trains, using the same research steps as CMP-FDM using "SQC Technical Methods" (Amasaka, 2003, 2004).

Specifically, to examine and the survey data for the "Visualizing causal relationships between passenger information and riding conditions", the author analyses the procedure and goals using statistical analysis, as follows;

(i)   Look at overall trends in passenger information using a cross-tabulation method that focuses on whether passenger attention turned to hanging posters, above-window posters, or sticker advertisements and others.

(ii)   Perform a cluster analysis on riding conditions and group the results using "Quantification Theory Type III". Then, look at the relationship between (1) the riding conditions grouped in the cross tabulation and (2) whether passenger attention turned to hanging posters, above-window posters, or sticker advertisements.

(iii) Research the grouped riding condition data and basic passenger information to determine how it relates to attention rates established for the three types of advertising using a "Categorical Automatic Interaction Detector (CAID) analysis" (Murayama et al., 1982; Amasaka, et al., 1998; Amasaka, 2011).

(4)   PMOS-DM

No clear processes are used at car dealers when deciding target customers for direct mail campaigns, and individual sales representatives tend to rely on their personal experience when making such decisions (Bult and Wansbeek, 2005; Jonker et al., 2006; Bell et al.,2006; Beco and Jagric, 2011).

This means that dealer strategies lose their effectiveness and dealers fail to achieve the desired increase in customer visits. Thus, for this study, the author establishes a practical method using PMOS-DM as a method of deciding the most suitable target customers for direct mail campaigns (Ishiguro and Amasaka, 2012a,b).

Specifically, in order to both clarify the dealer's target customer types and increase the number of customer visits, the author applies mathematical programming (combinatorial optimization) using statistics to establish a model for determining the most suitable target customers for direct mail campaigns.

More concretely, this model was created using the same research steps as CMP-FDM employing SQC Technical Methods: a three-pronged approach to resolving dealers' current problems with direct mail activities, as follows;

Step 1 is to increase the response rate, or the percentage of customers who visit the dealer as a result of receiving direct mail. To achieve this, the PMOS-DM uses statistical analysis to determine which customers are most likely to respond.

Step 2 is to reflect dealer aims in the recipient selection process. This is achieved by using a simulation driven by mathematical programming to optimize the selection of target customers (refer to Appendix 3: Optimal selection using a model formula).

Step 3 is to clarify the recipient selection process by providing dealers with a model that outlines a specific approach.

Following an explicit model informed by statistics and mathematical programming keeps inconsistency among salespeople to a minimum.

The three-pronged approach proposed in this study therefore provides a direct mail method that allows dealers to both target their desired customer segment and boost response rates at the same time.

In short, PMOS-DM uses statistics and mathematical programming to create an objective decision-making process that does not rely on the current selection methods used by salespeople, which are based on personal knowledge and experience and therefore vague and implicit.

At the same time, the model aims to boost the direct mail response rate in line with dealer targets.

APPLICATION EXAMPLES

This section validates the effectiveness of SMMM for effective advertising designed to bring customers into auto dealerships by use of 4 core elements (: VUCMIN, CMP-FDM, AGTCA and PMOS-DM), and the Customer Science approach to quantitatively assess the effectiveness of various advertising (mass media, direct advertising, and multimedia) (Amasaka, 2015, 2020a,b, 2022a,b, 2023a,b).

Visualizing Causal Relationships in Customer Purchase Behavior

SMMM was developed in order to draw more attention to the vehicle, spark interest in the vehicle, and make customer want to visit the dealer. To achieve the purpose of this research, a field survey on vehicle advertising was conducted to identify the core elements of each media type and to visualize the relationship between those elements and the media as well as the causal relationships between each media type and (a) vehicle awareness, (b) vehicle interest, and (c) desire to visit dealers (sales shops).

A survey was conducted in order to better understand the causal relationships among different types of media, media elements, and customer (consumer) purchase behavior. The advertising and marketing division at an advanced car manufacturer A (Toyota Motor Corporation), Japan Toyota Dealer Y of an advanced car manufacturer A, and the Z market survey company helped to conduct an in-person survey on advertising and marketing by visiting male and female licensed drivers age 18 and older living in Tokyo, Fukuoka, and Sapporo of Japan.

A total of 318 valid responses (197 male and 121 female, generally uniform age balance) were collected. The investigation period was the five months leading up to the release of the new Q model by an advanced car manufacturer A.

Based on the author's existing research and knowledge (Amasaka, 2007, 2009, 2011), they were able to identify media mix effects in each form of media using a purchasing action model, TV ads, radio ads and newspaper ads (early June 2005), as well as internet ads (early July) and train car (transportation) ads (mid-August) before the new car sale, and flyer and magazine ads (late August), DM ads (early September) and DH (Direct Hand) ads. (mid-September) issued by Japan Toyota dealer Y.

Participants were shown TV commecials and newspaper ads promoting Japan Toyota's new Q car and then asked questions inquiring about their purchase behavior and about the media and media elements. The collected data was analyzed and the causal relationships between media, media elements, and consumer purchase behavior were outlined (Ogura et al., 2013a).

The questions that the authors used in the survey are listed in Table 1. The questionnaire was multiple choice and asked respondents to describe their opinion (item ① was yes-no, and the 5-point scale in items ③ and ④ was converted into binary data). Elements in each form of media were identified using multivariate analysis (cluster analysis, quantification theory type III) and other statistical methods shown in Figure 3.

As the example in the figure shows, the critical elements in terms of generating the distinct promotional outcomes that consumers expect are: impact, contact frequency, newsworthiness, informativeness, and memorability (Ogura et al., 2013a)

Table 1. Survey questions

①	Are you aware of the Japan Toyota Q model?
②	What media did you see advertising this vehicle?
③	Are you interested in this vehicle?
④	Do you want to or did you actually visit a dealer to inquire about this car?
⑤	What kind of influence does each type of media have on you in terms of your attention, interest, and desire?
⑥	What was your impression of the advertisement?
⑦	Which advertising elements do you consider most important?

Figure 3 also positions TV ads, transportation ads, internet ads, newspaper ads, DM in order to identify the expected advertising effectiveness of each.

Finally, the insights gained through scientific analysis were used to describe specific chracteristics of the four core elements of the proposed SMMM with "VUCMIN, AGTCA, CMP-FDM and PMOS-DM" below so that more customers would be drawn to visit auto-dealers.

Application of VUCMIN

In their previous research, the author identified behavioral patterns of customers as they focus on the exterior of a vehicle. Insights gained during this research were used to explain the influence of product planning and designer intentions in VUCMIN creation (Yamaji et al., 2010). The Toyota vehicle "Mark X" named "Cressida" in USA was used as a target vehicle in design inquiries.

According to common opinions from designers and product planning at an advanced car manufacturer A, the parts that are given the most attention when the vehicle is shown to customers are the (i) front proportions, (ii) streamlined side proportions, (iii) tri-beam headlamps (lenses), (iv) widened console box, and (v) sharpened rear.

Figure 3. A scatter diagram with a principle component scores using quantification theory type III

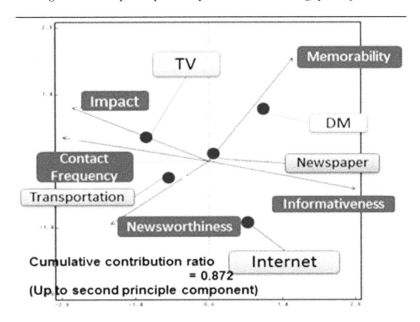

More concretely, the video created for the target profile of males in their 50s/ is explained using the timetable in Figure 13 of Chapter 18 (See to Section "A new strategic advertisement model "VUCMIN" designed to enhance the desire to visit auto-dealers" in Chapter 18).

The timetable shows video time as set to 90 seconds. Video shooting order is composed specifically of scenes from ① to ⑪ starting from the front, driver seat, side, and rear of the vehicle. Scenes are: ① direct front scene, ② diagonal front view scene, ③ driver side door opening scene, ④ entire driver seat view scene, ⑤ console box and shift lever scene, ⑥ steering wheel scene, ⑦ driver seat operational scene, ⑧ side view scene from driver's seat, ⑨ rear side view scene, ⑩ entire view of vehicle from rear, and lastly, ⑪ moving from back to front, the entire view of the vehicle scene.

The scenes that form the VUCMIN video were composed on the basis of the standard and disparate behaviors of customers. Example photos representing these scenes (① to ⑪) are described in Figure 14 of Chapter 18.

Using the same approach, VUCMIN was created for each age and gender. In this section, customer surveys were executed in order to test the validity of VUCMIN.

This was done by asking customers, "After seeing the Mark X video, when do you think you might visit an advanced car manufacturer A dealer to consider purchasing one?" to verify their desire to visit dealers (high, low).

According to the survey results, the desire to visit dealers (early consideration of Toyota Mark X purchase) not only increased for current Toyota vehicle owners but also for customers who own vehicles from other manufacturers.

The author is currently promoting the results of this research as part of the strategic advertising method VUCMIN, which utilizes an internet interface in collaboration with universities and industrial players.

Application of CMP-FDM

Flyers are a form of advertising media important for raising the customer-attraction effect. However, the results of an interviews by the author at to six dealers (national and foreign-affiliated) and two advertising agencies specializing in flyers showed that the dealers did not think that designing flyers was important—their only priority was distribution, and they outsourced the design.

Moreover, they did not understand actual customer behavior (how customers looked at flyers and what they paid attention to). Therefore, customer behavior is not reflected into the creation of flyers, and the resulting design is not attractive to customers who want to visit dealers.

The author studied how customers view flyers by analyzing browsing behavior (Koyama et al., 2010). To resolve the problems with the information contained in current flyer designs, the authors identified which information each consumer type focuses on.

In step 1 and 2, the purchase group is taken up as an example of the factor analysis results, with the Text Mining Studio corresponding bubble analysis results focusing on the purchase group shown in Figure 4.

Figure 4. Results for the information on flyers for the purchase group using Text Mining Studio "corresponding bubble analysis"

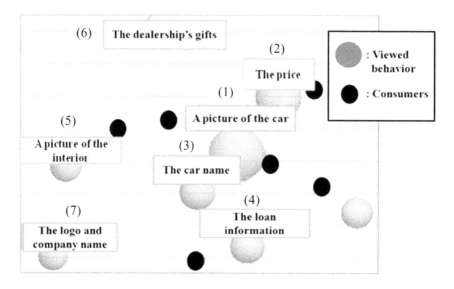

From this figure, it can be seen that the purchase group strongly correlated with viewing behaviors in the following order: (1) Looking at a picture of the car, (2) looking at the price, (3) looking at the car name, (4) looking at the loan information, (5) looking at a picture of the interior, (6) looking at dealer gifts, and (7) looking at the logo and company name. Through this analysis the authors were able to clearly reveal the information on flyers that consumers actually focus their attention on, which dealers had heretofore been unable to grasp.

In step 3, in order to resolve problems with the flyer layout, the author clarified the position and size of the information on flyers that each consumer type focuses their attention on.

In step 4 an attractive flyer design is created based on the knowledge gained in steps 1 to 3.

An example of a new attractive flyer design for Toyota's new vehicle as shown in Figure 5.

This design was intended to be appealing for a universal type of consumer. The effectiveness of the CMP-FDM method for creating the appealing flyer design described in Figure 5 was confirmed from the survey procedures and analysis obtained from steps 1 through 5, as well as the acquired results.

Figure 5. An example of new attractive flyer design

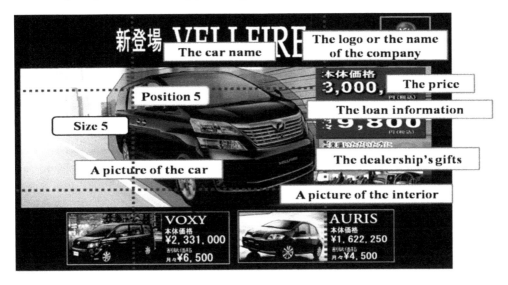

Application of AGTCA

The author researched the causal relationship between basic passenger information and ad awareness (Ogura et al., 2013a).

Firstly, in step 1, the author performed a cross tabulation on the survey data and researched the correlations between whether passengers notice each form of "Train car advertising" and "Passenger information" (age and gender) in order see how passenger information relates to attention rates.

Secondly, the author researched the causal relationship between riding conditions and ad awareness of passengers.

In step 2, the author used the survey data gathered to represent current in-train advertising conditions, and subjected it to a Cluster Analysis of group riding conditions as shown in Figure 6.

Similar results were obtained when the Quantification Theory Type III was used on the survey data shown in Figure 7.

Since the first group consisted of standing passengers who ride the train for 0–15 minutes, they were labeled "short-distance passengers".

The second group rode the train for a longer period of time and tended to sit, so this group was called the "long-distance passengers".

Thirdly, the author researched the causal relationships among basic passenger information, riding conditions and ad awareness of passengers.

Figure 6. Cluster analysis of the ride condition

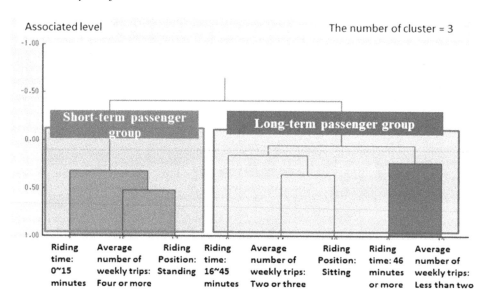

Figure 7. Quantification theory type III of the ride condition

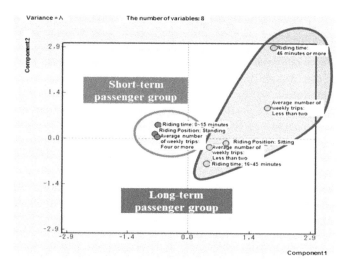

In step 3, Figure 8 shows the results of the awareness rate for hanging posters among short-term riders, which was used as a criterion variable in the CAID analysis.

The results indicate that the highest awareness rate in this group is among men in the youngest age category (15–25). Barring a few exceptions, the results indicate an overall trend where awareness rates are higher among younger people. Comparing the two groups, the author found that the short-term passengers (who tended to stand when riding) had higher awareness rates in general.

Passengers who sat, on the other hand, had more opportunities to engage in different activities during their ride, such as reading or doing work, which probably contributed to their paying less attention to advertisements than the passengers who were standing.

Figure 8. Results of the CAID analysis on the short-term passenger group

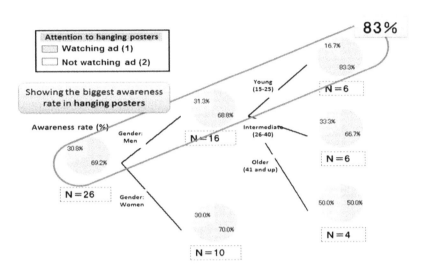

The analysis revealed that women passengers 26 and older in particular did not look at in-car advertisements. It also indicated that older passengers frequently paid attention to advertisements located above windows. Based on these conclusions, firstly, the author's recommendation is for existing train car advertising.

Because hanging posters, stickers, and other in-car advertisements are likely to attract younger riders, this space should be used to advertise weekly manga magazines, fashion magazines, sales, or other products likely to appeal to this generation. Another important consideration is using popular celebrities to catch the eye of these passengers.

Above-window advertising space, on the other hand, may be better used to appeal to those of the older generation. These passengers are more likely to be married and have children, so it may be beneficial to feature family-friendly topics.

Firstly, specifically, posters advertising events for families or travel may be ideal in this location. Also, because it was found that standing passengers tend to look at advertisements frequently, riders may pay attention not only to ads that help them pass the time while standing, but also those that stimulate their interest or desire.

Instead of showing just a picture, a magazine ad, for example, could feature headlines or other clever designs aimed at stimulating purchase behavior. It is important that other advertisements do not simply catch the eye, but encourage viewers to linger.

Secondly, the author suggests that trains adopt new forms of advertising media. The analysis results indicated that passengers who stand tend to have high awareness rates when it comes to in-car advertising, but advertisements on the floor may be easier for sitting passengers to see. Riders who sit naturally allow their eyes to fall downward, making a floor advertisement an eye-catching option. Those who sit and read are also looking downward as well, increasing the chances that they may see these advertisements.

Focusing on train car ads, which have a good contact rate and long-term contact, the author decided to examine those relationships and how they relate to train car advertising with the aim of defining the ideal format for this type of media.

Application of PMOS-DM

Putting POMS-DM to Work

The researchers teamed up with Company M to guide direct mailing efforts in conjunction with an event showcasing multiple new vehicle models (Kojima, et al., 2010; Ishiguro and Amasaka, 2012a,b).

The following steps show how optimal selection using a model formula was applied with the formulas shown in (1) to (7) (See to Appendix 19A, 19B and 19C).

(Step One) Organizing Customer Information

In Step 1, participating dealers had information on a total of 391 customers, which included data on sex (male/female), age (20s, 30s, 40s, 50s, 60+), and age of current vehicle (3-5 years, 6-8 years, 9+ years).

These were a total of 391 values assigned to j: Customer number in the formula, and a total of 10 different values assigned to m (customer attributes: e.g., sex, age, age of current vehicle). A binary code (0 or 1) was then assigned to the collected customer information in order to analyze it. This resulted in values for the f_j^m (Indicates whether or not customer j has attribute m (0 or 1)) variable. Recipients of direct mailing could now be determined based on the dealers' customer information.

(Step Two) Determining Response Likelihood

Step 2 was to conduct a survey and analyze the data to determine which customer attributes were most likely to lead customers to visit a dealer as a result of receiving DM.

The survey method used in this study was to ask customers of varying attributes (sex, age, vehicle age, etc.) whether receiving direct mail had ever caused them to visit the dealer.

Once the results were collected, they were quantified and subjected to the multivariate analysis (Quantification Theory Type II) analysis to determine which customers had the highest likelihood of responding to DM. This made it possible to analyze customer attributes in terms of whether or not they were likely to lead to a dealer visit in terms of an external standard.

These response likelihood values were then assigned the variable E^m (Effect of customer attribute (m) on the likelihood that the customer will visit the dealer).

Formula (1) below shows the results of this analysis. The discriminant ratio for the analysis results was 77.36%, indicating that they were fairly reliable.

$$y = 0x_{11} - 1.4x_{12} + 0x_{21} + 0.9x_{22} + 2.1x_{23} + 3.7x_{24} + 1.7x_{25} + 0x_{31} - 1.3x_{32} - 0.9x_{33} - 1.4 \tag{1}$$

The linear discriminant formula produced from the analysis results is as follows; If the linear discriminant is greater than 0, the customer is likely to visit the dealer as a result of receiving direct mail. If it is less than zero, it indicates that they are not likely to visit. Therefore, the coefficient produced by this formula indicates the response likelihood for the customer attribute as expressed by E^m. The results of this analysis, which allowed us to identify which customers were likely to visit the dealer, are summarized in Table 2.

Table 2. Response likelihood by attribute

	Customer attribute	Response likelihood
x11	Men	0
x12	Women	-1.4
x21	22-29 years old	0
x22	30-39 years old	0.89
x23	40-49 years old	2.1
x24	50-59 years old	3.7
x25	60+ years old	1.7
x31	Currently driving a vehicle 3–5 years old	0
x32	Currently driving a vehicle 6–8 years old	-1.3
x33	Currently driving a vehicle 9+ years old	-0.9

(Step Three) Selecting DM Recipients

In Step 3, the customer information collected in Step 1 is plugged into f_j^m, and the information on response likelihood for each customer attribute is plugged into E^m.

The number of direct mailings to be sent is plugged into C (total number of direct mailing sent). The upper and lower limits for the percentage of direct mailings to go to customers with each attribute is set at the dealer's discretion using the variables H^m (upper limit for the percentage of direct mailings sent to customers with attribute m) and L^m (lower limit for the percentage of direct mailings sent to customers with attribute m).

Once all the parameters are set, the simulation is carried out. During this process, formulas (2) through (7) are solved as a weighted constraint satisfaction problem (See to Appendix 19A, 19B and 19C).

In the weighted constraint satisfaction problem, the weighted constraints are moved to the target function as in formula (2), where they are added as a way of minimizing the level of deviation outside of the given limits. Even if a feasible solution that satisfies the constraints does not exist, the formula allows dealers to come as close as possible to meeting the constraints.

Here, in constraining the number of mailings sent to customers with the attributes defined in formula (7), it is difficult to set customer attributes L^m and H^m, ensuring that a feasible solution is more likely to exist. Therefore, when approaching the issue as weighted constraint satisfaction problem, it is best to find a solution that best satisfies formula (7).

In other words, this allows dealers to send direct mail to those customers most likely to come into the shop based on dealer strategy.

$$MIN - \sum_m \left(E^m \sum_j (f_j^m x_j) \right) + \sum_m W^m (L^m C - f_j^m x_j) + \sum_m W^m (H^m C - f_j^m x_j) \tag{2}$$

M Customer attributes (e.g. sex, age, age of current vehicle)

j Customer number

Wm Weighting for customers with customer attributes m in direct mail target group

f_j^m Indicates whether or not customer j has attribute m (0 or 1)

x_j Marks customer j for direct mailing (0 or 1)
R^m Ideal percentage with customer attributes m in direct mail target group
C Total number of direct mailings sent
L^m Lower limit for the percentage of direct mailings sent to customers with attribute m
H^m Upper limit for the percentage of direct mailings sent to customers with attribute m

(Step Four) Evaluating the Results

Once the recipients of direct mailings are selected based on the simulation, whoever is sending out the direct mail checks the simulation results to make sure that they accurately reflect the dealer's marketing strategy.

If the desired results are not achieved, the causes for the discrepancy are identified, the parameters are adjusted, and the simulation is run again.

Effectiveness of PMOS-DM

The effectiveness of the PMOS-DM was assessed by comparing the response rates (percentage of DM recipients who visited the dealer as a result) when salespeople selected DM recipients based on personal knowledge and experience and when recipients were selected using the model. 5 new models were showcased at the event held. 4 of the design concepts targeted female buyers, and one targeted male buyers.

As a result, the dealer's marketing strategy was to target women in particular throughout a wide range of age groups. This strategy was thus taken into account when verifying the effectiveness of the model. These verification results are summarized in Table 3. The response rate when DM recipients were selected on the basis of personal knowledge and experience of the sales staff was 19%.

When selection was made using the PMOS-DM model, the rate was 20.4%.

Table 3. Verification results (all)

	Dealer	**PMOS-DM**
Number of direct mailings sent	269	269
Number of resulting dealer visitors	51	59
Response rate	19.0%	20.4%

Table 4 shows the same information for female customers only (those targeted in the dealer's marketing strategy).

Table 4. Verification results (women)

	Dealer	**PMOS-DM**
Number of direct mailings sent	48	61
Number of resulting dealer visitors	2	12
Response rate	4.2%	19.8%

Salespeople generated a 4.2% response rate using their personal knowledge and experience, while the model generated a 19.8% response rate, signaling a significant improvement.

The effectiveness of the model was thereby verified in the course of this study.

Verification Results

Using the analysis results obtained in the previous section (Visualizing causal relationships in customer purchase behavior), a follow-up survey using questionnaire data from Table 1 above was then conducted to verify whether the research achieved its aim of bringing more customers into the dealers by means of raising the percentage of people affected.

In Figure 8(b) of Chapter 10: "Ratio of the number of people" in visiting an auto-dealer by using typical mixed media" in the effectiveness of "Scientific Mixed Media Model (SMMM)", the author shows the verification results from application of new mixed media by SMMM for raising the percentage of people affected (See to Chapter 10).

SMMM is effective due to its composition of four core elements (: VUCMIN, CMP-FDM, AGTCA and PMOS-DM) and use as a new strategic advertisement in nine media elements (TV, radio, newspapers, internet, train cars, flyers, magazines, DM and handbills) designed by the author (Amasaka et al., 2013). The figure shows the result of a follow-up survey using SMMM, where 16 people (percentage of people affected: 11.8%) actually visited the dealer, while 8 people signed a sales contract.

Comparative verification was done by looking at the results of "the usual experience of mixed media" when the dealer in the figure announced the old model Q4 years ago in a survey of similar size. In this case, the percentage of people affected was just 1.1%, thus validating the effectiveness of SMM.

CONCLUSION

The aim of this research study was to bring more customers into auto dealers. To achieve this, the SMMM (Scientific Mixed Media Model) was developed as a way to improve the quality of the consumer purchase behavior model in terms of vehicle awareness, vehicle interest, and desire to visit dealers. The collected research results are now being widely distributed as part of Toyota's current sales strategy.

REFERENCES

Amasaka, K. (2002). *New JIT*, a new management technology principle at Toyota. *International Journal of Production Economics*, *80*(2), 135–144. doi:10.1016/S0925-5273(02)00313-4

Amasaka, K. (2003). Proposal and implementation of the "*Science SQC*" Quality Control Principle. *Mathematical and Computer Modelling*, *38*(11-13), 1125–1136. doi:10.1016/S0895-7177(03)90113-0

Amasaka, K. (2004). *Science SQC, New Quality Control Principle: The Quality Strategy of Toyota*. Springer -Verlag. doi:10.1007/978-4-431-53969-8

Amasaka, K. (2005). Constructing a *Customer Science* Application System *"CS-CIANS"*: Development of a global strategic vehicle *"Lexus"* utilizing New JIT. *WSEAS (World Scientific and Engineering and Society Transactions on Business and Economics, 2*(3), 135-142.

Amasaka, K. (2007). The validity of *Advanced TMS,* a strategic development marketing system utilizing New JIT. *The International Business & Economics Research Journal, 6*(8), 35–42.

Amasaka, K. (2009). The effectiveness of flyer advertising employing TMS: Key to scientific automobile sales innovation at Toyota, *China-USA. Business Review (Federal Reserve Bank of Philadelphia), 8*(3), 1–12.

Amasaka, K. (2011). Changes in marketing process management employing TMS: Establishment of Toyota Sales Marketing System, *China & USA. Business Review (Federal Reserve Bank of Philadelphia), 10*(7), 539–550.

Amasaka, K. (Ed.). (2012). Science TQM, new quality management principle: The quality management strategy of Toyota, Bentham Science Publishers.

Amasaka, K. (2013). *The strategic development of Advanced TPS based on the New Manufacturing Theory (Plenary Lecture).* The 1st WSEAS International Conference on Industrial and Manufacturing Technologies, Recent Advances in Industrial & Manufacturing Technologies, Greece.

Amasaka, K. (2015). New JIT, new management technology principle. Taylor and Francis Group, CRC Press.

Amasaka, K. (2019). Studies on New Manufacturing Theory, *Noble. International Journal of Scientific Research, 3*(1), 1–20.

Amasaka, K. (2022a). New Manufacturing Theory: Surpassing JIT (2nd Edition). Lambert Academic Publishing.

Amasaka, K. (2022b). *Examining a New Automobile Global Manufacturing System.* IGI Global Publisher. doi:10.4018/978-1-7998-8746-1

Amasaka, K. (2023a). New Lecture-Toyota Production System: From JIT to New JIT. Lambert Academic Publishing.

Amasaka, K. (2023b). A New Automobile Sales Marketing Model for innovating auto-dealer's sales. *Journal of Economics and Technology Research, 4*(3), 9–32. doi:10.22158/jetr.v4n3p9

Amasaka, K., Kurosu, S., & Morita, M. (2008). *New Manufacturing Theory: Surpassing JIT - The evolution of Just-In-Time.* Maruzen.

Amasaka, K., Ogura, M., & Ishiguro, H. (2013). Constructing a Scientific Mixed Media Model for boosting automobile dealer visits: Evolution of market creation employing TMS. *International Journal of Business Research and Development, 3*(4), 1377–1391.

Amasaka, K., Watanabe, M., & Shimakawa, Y. (2005). Modeling of Strategic Marketing System to reflect latent customer needs and its effectiveness, *The Magazine of Research & Development for Cosmetics, Toiletries & Applied Industries* [in Japanese]. *Fragrance Journal, 33*(1), 72–77.

Beco, G. H., & Jagric, T. (2011). Demand models for direct mail and periodicals delivery services: Results for a transition economy. *Applied Economics*, *43*(9), 1125–1138. doi:10.1080/00036840802600244

Bell, G. H., Ledoher, J., & Swersey, A. J. (2006). Experimental design on the front lines of marketing: Testing new ideas to increase direct mail sales. *International Journal of Research in Marketing*, *23*(3), 309–319. doi:10.1016/j.ijresmar.2006.05.002

Bult, J. R., & Wansbeek, T. (2005). Optimal selection for direct mail. *Marketing Science*, *14*(4), 378–394. doi:10.1287/mksc.14.4.378

Ferrell, O. C., & Hartline, M. (2005). Marketing Strategy, Thomson South-Western, Mason, Gary, L. L., & Arvind, R. (2003). Marketing Engineering: Computer-assisted marketing analysis and planning. Pearson Education, Inc., London.

Hifumi, S. (2006). *A study on a strategic propaganda advertisement (animation): The proposal of "A-MUCMIN" Which raises the volition of coming to the auto shop* [Master's thesis, Aoyama Gakuin University].

Ikeo, K. (2006). Feature-marketing innovation. *Japan Marketing Journal*.

Ishiguro, H., & Amasaka, K. (2012a). Proposal and effectiveness of a Highly Compelling Direct Mail Method: Establishment and deployment of PMOS-DM. *International Journal of Management & Information System*, *16*(1), 1–10.

Ishiguro, H., Matsuura, S., & Amasaka, K. (2012b). Establishment of a Strategic Total Direct Mail Model to bring customers into auto dealerships. *Journal of Business & Economics Research*, *10*(8), 493–500. doi:10.19030/jber.v10i8.7177

James, A. F., & Mona, J. F. (2004). *Service Management*. McGraw-Hill Companies Inc.

James, F. E., Roger, D. B., & Paul, W. M. (2006). *Consumer Behavior*. Dryden Press Inc.

Jeffrey, F. R. and Bernard, J. J. (2005). Best face forward, *Diamond Harvard Business Review*, 62-77.

Jonker, J. J., Pielsma, N., & Potharst, R. (2006). A decision support system for direct mailing decision. *Decision Support Systems*, *42*(2), 915–925. doi:10.1016/j.dss.2005.08.006

Kishi, S., Tanaka, H., & Shimamura, K. (2000). *Theory of modern advertising*. Yuhikaku Publishers.

Kobayashi, T., & Shimamura, K. (1997). *New edition new Ad*. Dentsu.

Kojima, T., Kimura, T., Yamaji, M., & Amasaka, K. (2010). Proposal and development of the Direct Mail Method "PMCI-DM" for effectively attracting customers. *International Journal of Management & Information Systems*, *14*(5), 15–22. doi:10.19030/ijmis.v14i5.9

Koyama, H., Okajima, R., Todokoro, T., Yamaji, M., & Amasaka, K. (2010). Customer behavior analysis using motion pictures: Research on attractive flyer design method. *China-USA Business Review*, *9*(10), 58–66.

Kubimura, R., & Murata, S. (1969). *Theory of PR*. Yuhikaku.

Melewar, T. C., & Smith, N. (2003). The internet revolution: Some global marketing implications. *Marketing Intelligence & Planning*, *21*(6), 363–369. doi:10.1108/02634500310499220

Murayama, Y. (1982). *Analyzing CAID marketing review*. Japan Research Center.

Niiya, Y., & Matsuoka, F. (Eds.). (2001). *Foundation Lecture on the New Advertising Business*. Senden-Kaigi.

Nikkei Business. (1999). *Renovation of Shop, Product and Selling Method-targeting Young Customer by Nets*, 46-50. Nikkei Business.

Ogura, M., Hachiya, T., & Amasaka, K. (2013a). A Comprehensive Mixed Media Model for boosting automobile dealer visits. *The China Business Review*, *12*(3), 195–203.

Ogura, M., Hachiya, T., & Amasaka, K. (2013b). Attention-grabbing train car advertisements. *The China Business Review*, *12*(3), 195–203.

Okada, A., Kijima, M., & Moriguchi, T. (Eds.). (2001). *The mathematical model of marketing*. Sakura-Shoten.

Okutomi, H., & Amasaka, K. (2013). Researching Customer Satisfaction and Loyalty to boost marketing effectiveness: A look at Japan's auto dealerships. *International Journal of Management & Information Systems.*, *17*(4), 193–200. doi:10.19030/ijmis.v17i4.8093

Sakalsis, M. M., Hifumi, S., Yamaji, M., & Amasaka, K. (2008). Developing a Strategic Advertisement Method "VUCMIN" to enhance the desire of customers for visiting dealers. *Proceedings of the International Symposium on Management Engineering*, Waseda University, Kitakyushu, Japan.

Shimakawa, K., Katayama, K., Oshima, K., & Amasaka, K. (2006). *Proposal of Strategic marketing model for customer value maximization*. The Japan Society for Production Management, The 23th Annual Technical Conference, Osaka, Japan.

Shimamura, K., & Kobayashi, T. (1997). *New edition: New Advertisements*. Dentsu.

Shimizu, K. (2004). *Theory and strategy of advertisement*. Sousei-Sha.

Smith, D. A. (2009). Online accessibility concerns in shaping consumer relationships in the automotive industry. *Online Information Review*, *33*(1), 77–95. doi:10.1108/14684520910944409

Yamaji, M. S., Hifumi, S., Sakalsis, M. M., & Amasaka, K. (2010). Developing a Strategic Advertisement Method "VUCMIN" to enhance the desire of customers for visiting dealers. *Journal of Business Case Studies*, *6*(3), 1–11. doi:10.19030/jbcs.v6i3.871

APPENDIX

Optimal Selection Using a Model Formula

Using numerical simulation, the PMOS-DM model uses a mathematical formula to select target customers for direct mail. In coming up with a formula to determine who should be targeted by direct mail, the author referred to the formulas shown in formula (3) and (4), which were developed by Kojima et al. (2010) and Ishiguro and Amasaka (2012a).

$$\text{Min } L^m C \leq \sum_j f_j^m x_j \leq H^m C \tag{3}$$

$$\text{Subject to } \sum_m (E^m \sum_j (f_j^m x_j)) \tag{4}$$

m Customer attributes (e.g. sex, age, age of current vehicle)
j Customer number
Wm Weighting for customers with customer attributes m in direct mail target group
f_j^m Indicates whether or not customer j has attribute m (0 or 1)
x_j Marks customer j for direct mailing (0 or 1)
R^m Ideal percentage with customer attributes m in direct mail target group
C Total number of direct mailings sent
L^m Lower limit for the percentage of direct mailings sent to customers with attribute m
H^m Upper limit for the percentage of direct mailings sent to customers with attribute m

The target function of formula (3) is to minimize the gap between the ideal number of direct mailings sent to customers with attribute m (CR^m) and the number actually sent to customers with that attribute ($\Sigma f_j^m x_j$). In other words, the formula expresses the concept of setting a target value when sending out direct mail.

Accordingly, the formula can be adapted to cases where a clear, rational target value can be set. However, the formula cannot be used when it is difficult to set a logical target value for the number of direct mailings to be sent—and a dozen or so of the dealers that the authors studied did not set one.

For those dealers, the author set up a formula that would clarify the process that senior sales staff used to determine who should be targeted by a given direct mail campaign. In the process of conducting interviews, the author learned that senior sales staff use an abstract method of targeting those customers who seem like they would have an easy time coming into the dealer. The author then constructed a makeshift definition of this group of customers as follows;

(i)   Each group of customers defined by a given attribute (male, female, 20s, 30s, etc.) has different preferences that would motivate them to come into the dealer.

(ii)   Each customer's willingness to come in can be assigned a cumulative value based on that person's attributes.

Those with a high cumulative value can be considered the ones who are likely to come into the shop. With this line of thinking, the author developed a formula for calculating the total willingness for customers targeted by direct mail. They then constructed a model for optimizing those values. Finally, the author came up with a set of constraints in order to put limits on the number of mailings dealers would send, with the aim of maximizing the effectiveness of those that were sent.

Model Formula Used in the Numerical Simulation

This is the model formula used in the numerical simulation.

$$\text{Max} \sum_{m \in M} W^m \left(\sum_{j \in J} f_j^m x_j - CR^m \right)^2 \tag{5}$$

$$\text{subject to} \sum_j x_j = C \tag{6}$$

$$L^m C \le \sum_j f_j^m x_j \le H^m C \tag{7}$$

M Customer attributes (e.g., sex, age, age of current vehicle)
J Customer number
Wm Weighting for customers with customer attributes m in direct mail target group
f_j^m Indicates whether or not customer j has attribute m (0 or 1)
x_j Marks customer j for direct mailing (0 or 1)
R^m Ideal percentage with customer attributes m in direct mail target group
C Total number of direct mailings sent
L^m Lower limit for the percentage of direct mailings sent to customers with attribute m
H^m Upper limit for the percentage of direct mailings sent to customers with attribute m

This mathematical formula is designed to determine a value for the variable x_j. If the value is 1, mailings should be sent to the customer number indicated by j. If it is 0, a direct mailing should not be sent. The other variables are parameters that must be given values before solving the formula. C, L^m, and H^m are set at the discretion of whoever is sending out the direct mail.

The value f_j^m is determined based on the customer information that the dealer has. E^m is determined later via statistical analysis. The roles of the individual formulas are as follows. The objective function in formula (5) is used to maximize the customer response rate (the percentage of customers that come to the dealer as a result of the direct mail).

The constraint in formula (6) determines the number of direct mailings that are to be sent out. The constraint in formula (7) determines how many direct mailings are to be sent to each customer segment, which is how dealer aims are incorporated into the model.

The mathematical formula is designed so that the number of customer attributes it handles (m) can be increased at will. Depending on what customer information dealers have, they can limit these attributes to basic life stages or expand them to include hobbies, preferences, and other lifestyle characteristics.

Recipient Selection Process

The author describes the procedure for using the mathematical formula provided to select direct mail recipients.

First, a "response likelihood" value must be set for each customer using the variable E^m. The list of customers is then reordered with those with the highest likelihood of responding at the top. The purpose of the objective function in formula (5) is to order customers according to their likelihood of responding (visiting the dealer as a result of direct mail). Next, this list is used to select the number of customers equal to the number of direct mailings (the constraint) to be sent out, starting with those most likely to respond.

For example, if 50 direct mailings are to be sent, they would be sent to the top 50 customers most likely to respond to them. This is the basic principle behind the development of the formulas. In addition, when the dealer has a specific aim in mind (e.g., sending a large number of direct mailings to women), the constraint function in formula (7) can be used to incorporate that aim in the calculations.

For example, if the dealer wanted at least 60% of the 50 mailings to go to women, the women customers would be listed in order of response likelihood and the top 30 customers would be selected to receive direct mail.

The remaining 20 recipients would be selected from the entire pool of target customers in order of their response likelihood as well. The purpose of this function is to allow dealers to use their marketing strategies to boost response rate.

Conclusion

The Japanese administrative management technology that contributed the most to the world in the latter half of the 20th century is typified by the "Japanese Production System" represented by the "Toyota Production System".

This system was kept at a high level by a manufacturing quality management system generally called JIT. However, in Japan and overseas, a close look at recent corporate management activities reveals various situations where an advanced manufacturer, which is leading the industry, is having difficulty due to unexpected quality related problems in corporate management.

The pressing management issue, particularly for Japanese manufacturers, to enable survival in the global market is the worldwide uniform quality and production at optimal locations, which is a prerequisite for successful global production.

Therefore, to transforming Japanese corporate management strategy based on the "New JIT, new management technology principle", the author has verified the validity of the *Revolutionary Automobile Production Systems for Optimal Quality, Efficiency, and Cost.* through the Section 1, 2 and 3 in this book.

Specifically, Section 1 is the "Foundation of Automobile Production Systems: Developing Toyota Production System (TPS)", Section 2 is the "Revolutionary Automobile Production Systems: Innovation of Toyota management technology", and Section 3 is the "Optimal Quality, Efficiency, and Cost: Realizing Simultaneous QCD Fulfilment".

Concretely, as the corporate strategy in next generation, to transforming Japanese corporate management strategy, the author has created the "New JIT, Surpassing JIT" for successful global production: "worldwide uniform quality and production at optimal locations".

New JIT consists on the 5 core elements: The TDS (Total Development System), TPS (Total Production System), TMS (Total Marketing System), TIS (Total Business Intelligence Management System), and TJS (Total Job Quality Management System) based on the "Scientific total quality management" named "Science TQM, new quality management principle" by developing "Advanced TDS, TPS, TMS, TIS & TJS" for the worldwide marketing creation, and employs both of the "Customer Science principle (CSp) and statistical science named "Science SQC, new quality control principle.."

At present, New JIT has been applied at many leading Japanese companies where its effectiveness has been verified, and it is now known as strategic global management technology model in Toyota and advanced manufactures. It is hoped that this book will contribute to the evolution of management technology for expanding global manufacturing through "Automobile simultaneous QCD fulfilment for customer value creation".

Finally, it is lucky if this book is a help for the various studies of the "students, educators, researchers and business-persons in the world".

Kakuro Amasaka
Aoyama Gakuin University, Japan

Compilation of References

Aihara, S., & Shinogi, T. (2012). An *establishment of Automobile Exterior Color and Interior Color Matching Model: The 20th woman example*. [Thesis, School of Science and Engineering, Aoyama Gakuin University]. .

Akimoto, S., & Shimoda, S. (2000). Modeling of the conceptual composition of invention and right: Research on objectification of the patent value by opinion poll of engineers [Thesis, School of Science and Engineering, Amasaka New JIT Laboratory].

Amagai, M., & Amasaka, K. (2003). A study for the establishment of the quality management performance measure, *Proceedings of the 17th Asia Quality Symposium, Beijing Jiuhua Country Villa, Beijing, China*, 243-250.

Amasaka K. (2000b). *AWD-6P/J report of first term activity 1996-1999: Creation of 21ˢᵗ century production line in which people over 60's can work vigorously (Project leader of AWD-6P/J)*. Toyota Motor Corporation.

Amasaka K. (2004c). New development of high-quality manufacturing in global production, Toward the next-generation quality management technology (Series 4). *Quality management, 55*(4), 44-58.

Amasaka K. et al. (2000). *AWD6P/J Report of First Term Activity 1996-1999: Creation of 21C production line in which people over 60's can work vigorously*. Toyota Motor Corporation.

Amasaka, K. (1972). Measurement of "gear's noise control for inspection of differential gear assembly, *The 2ⁿᵈ Sensory Symposium in JUSE, Tokyo*, 5-12.

Amasaka, K. (1976). Sensory characteristic of "differential gear noise occurrence. *Quality Control (an extra edition), 26*(11), 5-12.

Amasaka, K. (1983). *Mechanization for operation depended on intuition and knack – Corrective method for distortion of rear axle shaft*. The Japanese Society for Quality Control, The 26th Technical Conference, Nagoya, Aichi, Japan.

Amasaka, K. (1983). *The mechanization of "Kan and Kotsu in experience" work: Distortion modification of "Rear axle shaft."* The 26th Technical Conference in JSQC (Journal of the Japanese Society for Quality Control), Nagoya.

Amasaka, K. (1986). *The improvement of the corrosion resistance for the front and rear axle unit of the vehicle-QCD research activities by plant production engineers*. The Japanese Society for Quality Control, The 30th Technical Conference, Nagoya, Japan.

Amasaka, K. (1988). Concept and progress of Toyota Production System (Plenary lecture). The Japan Society of Precision Engineering, The Japan Society for Design & Drafting and Hachinohe Regional Advance Technology Promotion Center Foundation and others Hachinohe, Aomori-ken, Japan.

Amasaka, K. (1988). Concept and progress of Toyota Production System (Plenary lecture). The Japan Society of Precision Engineering, The Japan Society for Technology of Plasticity, The Japan Society for Design and Drafting, and Hachinohe Regional Advance Technology Promotion Center Foundation, Hachinohe, Aomori-ken, Japan.

Amasaka, K. (1988). Concept and progress of Toyota Production System (Plenary lecture). The Japan Society of Precision Engineering.

Amasaka, K. (1988a). Electrode s of Arc welding, Surface modification technology-A dry process and its application. The Japan Society for Precision Engineering.

Amasaka, K. (1989). TQC at Toyota: Actual state of quality control activities in Japan (Special lecture). The 19th Quality Control Study Team for Europe, Union of Japanese and Engineers, Statistical Quality Control, 39, 107-112.

Amasaka, K. (1989). TQC at Toyota: Actual state of quality control activities in Japan (Special lecture). The 19th Quality Control Study Team for Europe, Union of Japanese and Engineers.

Amasaka, K. (1995). The Q.A. Network activities for prevent rusting of vehicle by using SQC, *Journal of the Japanese Society for Quality Control, The 50th Technical Conference.*

Amasaka, K. (1996). *A demonstrative study of a new SQC concept and procedure in the manufacturing industry: Establishment of a New Technical Method for conducting Scientific SQC.* Proceedings of the 2nd Australia-Japan Workshop on Stochastic Models in Engineering, Technology and Management, The University Queensland, Australia.

Amasaka, K. (1996). Application of classification and related method to the SQC Renaissance in Toyota Motor. Data Science, Classification, and Related Methods. Springer.

Amasaka, K. (1997). The development of Total QA Network using Management SQC-Cases of Quality Assurance of Brake Pads. *Journal of the Japanese Society for Quality Control, The 55th Technical Conference,* 17-20.

Amasaka, K. (1998). Application of classification and related methods to the SQC Renaissance in Toyota Motor. Hayashi, C. et al. (Eds), Data Science, Classification and Related Methods. Springer.

Amasaka, K. (1999b). *The TQM responsibilities for industrial management in Japan - The research of actual TQM activities for business management.* The Japanese Society for Production Management, The 10st Annual Technical Conference, Kyushu-Sangyo University, Fukuoka, Japan.

Amasaka, K. (1999b). *The TQM responsibilities for industrial management in Japan – The research of actual TQM activities for business management.* The Japanese Society for Production Management, The 10st Annual Technical Conference, Kyushu-Sangyo, University, Fukuoka, Japan.

Amasaka, K. (1999b). *The TQM responsibilities for industrial management in Japan - The research of actual TQM activities for business management.* The Japanese Society for Production Management.

Amasaka, K. (2000). Basic Principles of JIT–Concept and progress of Toyota Production System (Plenary lecture), *The Operations Research Society of Japan, Strategic research group-1, Tokyo.*

Amasaka, K. (2000). *Partnering chains as the platform for Quality Management in Toyota.* Proceedings. of the 1st World Conference on Production and Operations Management, Seville, Spain.

Amasaka, K. (2000a). Basic principles of JIT – Concept and progress of Toyota Production System (Plenary lecture), *The Operations Research Society of Japan.* Strategic research group-1, Tokyo.

Amasaka, K. (2000a). *Basic Principles of JIT – Concept and progress of Toyota Production System (Plenary lecture).* The Operations Research Society of Japan, Strategic research group-1, Tokyo.

Amasaka, K. (2000a). *New JIT, A New Principle for Management Technology 21^C: Proposal and demonstration of "TQM-S" in Toyota (Special Lecture).* Proceedings of the I World Conference on Production and Operations Management, Sevilla, Spain.

Amasaka, K. (2000a). *New JIT, a new principle for management technology 21^C: Proposal and demonstration of "TQM-S" in Toyota (special lecture).* The 1st World Conference on Production and Operations Management, Sevilla, Spain.

Amasaka, K. (2000b). *Partnering chains as the platform for quality management in Toyota.* Proceedings of the 1st World Conference on Production and Operations Management, Sevilla, Spain.

Amasaka, K. (2000b). *Partnering chains as the platform for Quality Management in Toyota.* Proceedings. of the 1st World Conference on Production and Operations Management, Seville, Spain.

Amasaka, K. (2001). Quality management in the automobile industry and the practice of the joint activities of the automaker and the supplier. A change in the order system such as a part in the manufacturing industry: The condition of the supplier continuance in the automobile industry. Japan Small Business Research Institute.

Amasaka, K. (2001). Quality management in the automobile industry and the practice of the joint activities of the automaker and the supplier. Japan Small Business Research.

Amasaka, K. (2002). Intelligence production and partnering for embodying a high-quality assurance (Plenary Lecture). The Japan Society of Mechanical Engineers, The 94th Course of JSME Tokai Branch, Nagoya, Japan.

Amasaka, K. (2002). *Intelligence production and partnering for realizing high quality assurance (Invited lecture).* The 94th of Japan Society of Mechanical Engineers Tokai branch, Nagoya, Aichi, Japan.

Amasaka, K. (2002a). Intelligence production and partnering for embodying a high-quality assurance (Plenary Lecture). The 94th Course of JSME Tokai Branch, Nagoya, Japan.

Amasaka, K. (2002b). Intelligence production and partnering for embodying a high-quality assurance (Plenary Lecture), The Japan Society of Mechanical Engineers, The 94th course of JSME Tokai Branch, Nagoya, Japan, 35-42.

Amasaka, K. (2002b). *Quality of engineer's work and significance of patent application.* The 2nd patent seminar, Aoyama Gakuin University, Research Institute, Shibuya, Tokyo.

Amasaka, K. (2002b). Reliability of oil seal for Transaxle–A Science SQC approach at Toyota, Case Studies in Reliability and Maintenance. John Wiley & Sons.

Amasaka, K. (2003). *New application of strategic quality management and SCM.* Proceedings of Group Technology/ Cellular Manufacturing World Symposium, Ohio.

Amasaka, K. (2003a). *Development of "New JIT", Key to the excellence design "LEXUS": The validity of "TDS-DTM", a strategic methodology of merchandise.* Proceedings of the Production and Operations Management Society, Hyatt Regency, Savannah, Georgia.

Amasaka, K. (2003a). *New application of strategic quality management and SCM-A Dual Total Task Management Team involving both Toyota and NOK.* Proceedings of the Group Technology / Cellular Manufacturing World Symposium, Columbus, Ohio.

Amasaka, K. (2003a). *The validity of "TJS-PPM", Patent Value Appraisal Method in the corporate strategy: Development of "Science TQM", a new principle for quality management (Part 3).* The Japan Society for Production Research, The 17th Annual Conference, Gakushuin University, Tokyo, 189-192.

Amasaka, K. (2003b). *A "Dual Total Task Management Team" involving both Toyota and NOK: A cooperative team approach for reliability improvement of transaxle*. Proceedings of the Group Technology / Cellular Manufacturing World Symposium, Columbus, Ohio.

Amasaka, K. (2003b). A demonstrative study on the effectiveness of "Television Ad." for the automotive sales, *Japan Society for Production Management, The 18th Annual Conference., Nagasaki, Japan*, 113-116.

Amasaka, K. (2003b). *A demonstrative study on the effectiveness of "Television Ad." for the automotive sales*. Japan Society for Production Management, The 18th Annual Conference, Nagasaki, Japan.

Amasaka, K. (2004). *The past, present, future of production management*. The Japan Society for Production Management, The 20th Annual Technical Conference, Nagoya Technical College, Aichi, Japan.

Amasaka, K. (2004a). *Applying New JIT—A Management Technology Strategy Model at Toyota-Strategic QCD studies with affiliated and non-affiliated suppliers*. Proceedings of the 2nd World Conference on Production and Operations Management Society, Cancun, Mexico.

Amasaka, K. (2004a). Customer Science: Studying consumer values. *Japan Journal of Behavior Metrics Society,* 196-199.

Amasaka, K. (2004a). Custom*er Science: Studying consumer values*. Japan Journal of Behavior Metrics Society, The 32nd Annual Conference, Aoyama Gakuin University, Sagamihara, Kanagawa.

Amasaka, K. (2004b). The past, present, future of production management (Keynote lecture), *The Japan Society for Production Management, The 20th Annual Technical Conference, Nagoya Technical College, Aichi, Japan*, 1-8.

Amasaka, K. (2004b). *The past, present, future of production management (Keynote lecture)*. The 20th Annual Technical Conference, Nagoya Technical College, Aichi, Japan. .

Amasaka, K. (2004b). *The past, present, future of production management (Keynote lecture)*. The Japan Society for Production Management, The 20th Annual Technical Conference, Nagoya Technical College, Aichi, Japan.

Amasaka, K. (2004c). *Establishment of strategic quality management–Performance Measurement Model – Key to successful implementation of Science TQM (Part 6)*. The Japan Society for Production Management.

Amasaka, K. (2005). Constructing a *Customer Science* Application System *"CS-CIANS"*: Development of a global strategic vehicle *"Lexus"* utilizing New JIT. *WSEAS (World Scientific and Engineering and Society Transactions on Business and Economics, 2*(3), 135-142.

Amasaka, K. (2005a). *New Japan Production Model*, an innovative production management principle: Strategic implementation of *New JIT*. Proceedings of the 16th Annual Conference of the Production and Operations Management Society, Michigan, Chicago IL.

Amasaka, K. (2006). *Evolution of TPS fundamentals utilizing New JIT strategy: Toyota's Simultaneous Realization of QCD Fulfillment Models*. Proceedings of the International Manufacturing Leaders Forum, Taipei, Taiwan.

Amasaka, K. (2007). *Developing the high-cycled quality management system utilizing New JIT*. Proceedings of the Seventh International Conference on Reliability and Safety, Beijing, China.

Amasaka, K. (2007). *The forefront of manufacturing technology: Evolution of Toyota Production System-The quality technological strategy by Science TQM (Special lecture)*. The Small and Medium Enterprise Agency, etc., Hachinohe intelligent plaza/Hachinohe Regional Advance Technology Promotion Center Foundation, Hachinohe, Japan.

Amasaka, K. (2007b). *Final report of WG4's studies in JSQC research activity of simulation and SQC (Part-1)-Proposal and validity of the High Quality Assurance CAE Model for automobile development design.* Transdisciplinary Science and Technology Initiative, The 2nd Annual Technical Conference, Kyoto University, Kyoto, Japan.

Amasaka, K. (2008). Strategic QCD studies with affiliated and non-affiliated suppliers utilizing New JIT, Encyclopedia of Networked and Virtual Organizations, 3(PU-Z), 1516-1527.

Amasaka, K. (2008). Strategic QCD studies with affiliated and non-affiliated suppliers utilizing New JIT., Encyclopedia of Networked and Virtual Organizations, III(PU-Z), 1516 -1527.

Amasaka, K. (2008a). Simultaneous fulfillment of QCD - Strategic collaboration with affiliated and non-affiliated suppliers. *New Theory of Manufacturing – Surpassing JIT: Evolution of Just-in-Time.* Morikita-Shuppan.

Amasaka, K. (2008a). *Simultaneous fulfillment of QCD-Strategic collaboration with affiliated and non-affiliated suppliers.* New theory of manufacturing-Surpassing JIT: Evolution of Just-in-Time, Morikita-Shuppan, Tokyo.

Amasaka, K. (2008a). Strategic QCD studies with affiliated and non-affiliated suppliers utilizing New JIT, Encyclopedia of Networked and Virtual Organizations, III(PU-Z), 1516-1527.

Amasaka, K. (2008b). Strategic QCD studies with affiliated and non-affiliated suppliers utilizing New JIT, Encyclopedia of Networked and Virtual Organizations, III(PU-Z), 1516-1527.

Amasaka, K. (2008b). Strategic QCD studies with affiliated and non-affiliated suppliers utilizing New JIT, Encyclopedia of Networked and Virtual Organizations, Information Science Reference, Hershey, New York, III(PU-Z), 1516-1527.

Amasaka, K. (2008b). Strategic QCD studies with affiliated and non-affiliated suppliers utilizing New JIT. Encyclopedia of Networked and Virtual Organizations, III(PU-Z), 1516-1527.

Amasaka, K. (2008b). Strategic QCD studies with affiliated and non-affiliated suppliers utilizing New JIT. Encyclopedia of Networked and Virtual Organizations, Information Science Reference, Hershey, New York.

Amasaka, K. (2008b). Strategic QCD studies with affiliated and non-affiliated suppliers utilizing New JIT. Encyclopedia of Networked and Virtual Organizations, VIII(PU-Z), 1516-1527.

Amasaka, K. (2008c). Science TQM, a new quality management principle: The quality management *of Management & Engineering Integration, 1*(1), 7-22.

Amasaka, K. (2010). *Manufacturing of the 21st Century: The proposal of a "New Manufacturing Theory" surpassing JIT (Plenary Lecture), The Japan Society of Production Management.* The 23th Annual Conference, Kobe University, Kobe, Japan.

Amasaka, K. (2010a). Chapter 4. Product Design, Quality Assurance Guidebook. Guide Book of Quality Assurance. The Japanese Society for Quality Control, Union of Japanese Scientists and Engineers Press.

Amasaka, K. (2010b). Chapter 4. Product Design, Quality Assurance Guidebook. Guide Book of Quality Assurance. The Japanese Society for Quality Control & Union of Japanese Scientists and Engineers Press.

Amasaka, K. (2011). *New JIT, new management technology principle (Plenary Lecture).* VIII Siberian Conference Quality Management-2011, Krasnoyarsk, Russia, Siberian Federal University.

Amasaka, K. (2012a). *Strategic development of New JIT for global transforming management technology (Plenary Lecture).* Computers and Materials, The 5th WSEAS International Conference on Sliema, Malta.

Amasaka, K. (2012b). *Prevention of the automobile development design, precaution and prevention.* Japanese standards Association Group, Tokyo.

Amasaka, K. (2012c). *Prevention of the automobile development design, precaution and prevention.* Japanese standards Association Group, Tokyo.0000000df

Amasaka, K. (2013). *The strategic development of Advanced TPS based on the New Manufacturing Theory (Plenary Lecture).* The 1st WSEAS International Conference on Industrial and Manufacturing Technologies, Recent Advances in Industrial & Manufacturing Technologies, Greece.

Amasaka, K. (2013a). *The strategic development of Advanced TPS based on the New Manufacturing Theory.* Recent Advances in Industrial & Manufacturing Technologies The 1st WSEAS International Conference on Industrial and Manufacturing Technologies (INMAT13), Vouliagmeni, Greece.

Amasaka, K. (2015). New JIT, new management technology principle, Taylor & Francis Group, CRC Press, Boca Raton, London, New York.

Amasaka, K. (2015). New JIT, New Management Technology Principle, Taylor & Francis Group, CRC Press, Boca Raton, London, New York.

Amasaka, K. (2015). New JIT, new management technology principle, Taylor and Francis Group, CRC Press, Boca Raton, London, New York.

Amasaka, K. (2015). New JIT, New Management Technology Principle, Taylor and Francis Group, CRC Press, Boca Raton, London, New York.

Amasaka, K. (2015). New JIT, new management technology principle, Taylor and Francis Group, CRC Press.

Amasaka, K. (2015). New JIT, New Management Technology Principle. Boca Raton, USA: Taylor and Francis Group, CRC Press.

Amasaka, K. (2015). New JIT, New Management Technology Principle. CRC Press, Taylor & Francis Group.

Amasaka, K. (2015). New JIT, new management technology principle. Taylor & Francis Group, CRC Press, Boca Raton, London, New York.

Amasaka, K. (2015). New JIT, New Management Technology Principle. Taylor & Francis Group.

Amasaka, K. (2015). New JIT, new management technology principle. Taylor and Francis Group, CRC Press, Boca Raton, London, New York.

Amasaka, K. (2015). New JIT, new management technology principle. Taylor and Francis Group, CRC Press.

Amasaka, K. (2015a), New JIT, new management technology principle, Taylor & Francis Groups.

Amasaka, K. (2015a). New JIT, new management technology principle. Taylor & Francis Group, CRC Press.

Amasaka, K. (2016). *Innovation of automobile manufacturing fundamentals employing New JIT: Developing advance Toyota Production System at Toyota Manufacturing USA.* Proceedings of the 5th Conference on Production and Operations Management, Habana International Conference Center, Cuba.

Amasaka, K. (2016). *Innovation of automobile manufacturing fundamentals employing New JIT: Developing Advance Toyota Production System at Toyota Manufacturing USA.* Proceedings of the 5th Conference on Production and Operations Management, Havana, Cuba.

Amasaka, K. (2016). *Innovation of automobile manufacturing fundamentals employing New JIT: Developing Advanced Toyota Production System at Toyota Manufacturing USA.* Proceedings of the 5th Conference on Production and Operations Management, P&OM Habana, Cuba.

Amasaka, K. (2017). Studies on Automobile Exterior Design Model for customer value creation utilizing Customer Science Principle, *The 7ᵗʰ International Symposium on Operations Management Strategy, Tokyo Metropolitan University, Tokyo,* 27-28.

Amasaka, K. (2017). *Studies on Automobile Exterior Design Model for customer value creation utilizing Customer Science Principle.* The 7th International Symposium on Operations Management Strategy, Tokyo Metropolitan University, Tokyo.

Amasaka, K. (2017b). *Studies on Automobile Exterior Design Model for customer value creation utilizing Customer Science Principle.* The 7th International Symposium on Operations Management Strategy, Tokyo Metropolitan University, Tokyo.

Amasaka, K. (2019b). *Developing Automobile Optimal Product Design Model.* Japanese Operations Management and Strategy Association, The 9th International Symposium on Operations Management and Strategy 2019, Tokyo Keizai University, Tokyo.

Amasaka, K. (2022a). New Manufacturing Theory: Surpassing JIT (2nd Edition), Lambert Academic Publishing, Germany, Printed by Printforce, United Kingdom.

Amasaka, K. (2022a). New Manufacturing Theory: Surpassing JIT (2nd Edition), Lambert Academic Publishing, Germany.

Amasaka, K. (2022a). New Manufacturing Theory: Surpassing JIT (2nd Edition), Lambert Academic Publishing.

Amasaka, K. (2022a). New Manufacturing Theory: Surpassing JIT (2nd Edition), Lambert Academic Publishing.

Amasaka, K. (2022a). New Manufacturing Theory: Surpassing JIT (2nd Edition). Lambert Academic Publishing, Germany.

Amasaka, K. (2022a). New Manufacturing Theory: Surpassing JIT (2nd Edition). Lambert Academic Publishing.

Amasaka, K. (2022b). New Manufacturing Theory (2nd Edition). Lambert Academic Publishing.

Amasaka, K. (2022b). New Manufacturing Theory: Surpassing JIT (2nd Edition), Lambert Academic Publishers.

Amasaka, K. (2022b). New Manufacturing Theory: Surpassing JIT (2nd Edition). Lambert Academic Publishers, Germany.

Amasaka, K. (2022b). New Manufacturing Theory: Surpassing JIT (2nd Edition). Lambert Academic Publishing.

Amasaka, K. (2022c). New Manufacturing Theory: Surpassing JIT (2nd Edition), Lambert Academic Publishing.

Amasaka, K. (2022c). New Manufacturing Theory: Surpassing JIT (2nd Edition). Lambert Academic Publishing.

Amasaka, K. (2023). *New Lecture-Surpassing JIT: Toyota Production System -From JIT to New JIT.* Lambert Academic Publishing.

Amasaka, K. (2023). New Lecture-Surpassing JIT: Toyota Production System. Lambert Academic Publishing.

Amasaka, K. (2023). New Lecture-Surpassing JIT: Toyota Production System-From JIT to New JIT, Lambert Academic Publishing.

Amasaka, K. (2023). New Lecture-Surpassing JIT: Toyota Production System-From JIT to New JIT-, Lambert Academic Publishing.

Amasaka, K. (2023). New Lecture—Surpassing JIT: Toyota Production System—From JIT to New JIT. Lambert Academic Publishing, Germany, Printed by Printforce, United Kingdom.

Amasaka, K. (2023). New Lecture-Surpassing JIT: Toyota Production System-From JIT to New JIT. Lambert Academic Publishing, Germany.

Amasaka, K. (2023). *New Lecture-Surpassing JIT: Toyota Production System-From JIT to New JIT.* Lambert Academic Publishing.

Amasaka, K. (2023). New Lecture-Surpassing JIT:Toyota Production System-From JIT to New JIT, Lambert Academic Publishers.

Amasaka, K. (2023). New Lecture-Toyota Production System: From JIT to New JIT, Lambert Academic Publishing, Germany, Printed by Printforce, United Kingdom.

Amasaka, K. (2023). New Lecture-Toyota Production System: From JIT to New JIT. Lambert Academic Publishing.

Amasaka, K. (2023a). New Lecture-Surpassing JIT: Toyota Production System-From JIT to New JIT-, Lambert Academic Publishing.

Amasaka, K. (2023a). New Lecture-Surpassing JIT: Toyota Production System-From JIT to New JIT. Lambert Academic Publishing, Germany.

Amasaka, K. (2023a). New Lecture-Surpassing JIT: Toyota Production System-From JIT to New JIT. Lambert Academic Publishing.

Amasaka, K. (2023a). New Lecture-Toyota Production System: From JIT to New JIT, Lambert Academic Publishing.

Amasaka, K. (2023a). New Lecture-Toyota Production System: From JIT to New JIT. Lambert Academic Publishing, Republic of Moldova Europe.

Amasaka, K. (2023a). New Lecture-Toyota Production System: From JIT to New JIT. Lambert Academic Publishing.

Amasaka, K. (2023b). New Lecture-Surpassing JIT: Toyota Production System-From JIT to New JIT-. Lambert Academic Publishing.

Amasaka, K. (Ed.). (2000). Science SQC: Revolution of Business Process Quality, Study group of the Nagoya QST Research, Japanese Standards Association, Tokyo.

Amasaka, K. (Ed.). (2000). Science SQC: The reform of the quality of a business process, Japanese Standards Association.

Amasaka, K. (Ed.). (2000a). Chapter 7: The aim of "Science SQC"– Deployment as a new TQM methodology from a management perspective. Japan Standards Association Nagoya.

Amasaka, K. (Ed.). (2000b). Science SQC: Revolution of Business Process Quality. Study group of Nagoya QST Research, Japanese Standards Association, Tokyo.

Amasaka, K. (Ed.). (2007). Establishment of a needed design quality assurance: Framework for numerical simulation in automobile production. Japanese Society for Quality.

Amasaka, K. (Ed.). (2007). New Japan Model "Science TQM"– Theory and practice for strategic quality management. Study group of ideal situation on quality management at manufacturing industry, Tokyo: Maruzen.

Amasaka, K. (Ed.). (2007). New Japan Model: Science TQM – Theory and practice for strategic quality management, Study group of the ideal situation the quality management of the manufacturing industry, Maruzen, Tokyo.

Amasaka, K. (Ed.). (2007). New Japan Model: Science TQM. Study Group of the Ideal Situation on the Quality Management of the Manufacturing Industry, Maruzen, Tokyo.

Amasaka, K. (Ed.). (2007). New Japan Model: Science TQM: Theory and practice for strategic quality management, The quality management of the manufacturing industry, Maruzen, Tokyo.

Amasaka, K. (Ed.). (2007a). New Japan Model – "Science TQM": Theory and practice for strategic quality management, Study group of the ideal situation the quality management of the manufacturing industry, Maruzen, Tokyo.

Amasaka, K. (Ed.). (2007a). New Japan Model: Science TQM. Study Group of the Ideal Situation on the Quality Management of the Manufacturing Industry, Maruzen, Tokyo.

Amasaka, K. (Ed.). (2007a). New Japan Model: Science TQM-Theory and practice of strategic quality management, Study group on the ideal situation on the quality management on the manufacturing, Maruzen, Tokyo.

Amasaka, K. (Ed.). (2007a). New Japan Model: Science TQM-Theory and practice of strategic quality management. Study group on the ideal situation on the quality management on the manufacturing, Maruzen, Tokyo.

Amasaka, K. (Ed.). (2007a). New Japan Model–"Science TQM": Theory and practice for strategic quality management. Study group of the ideal situation the quality management of the manufacturing industry, Maruzen, Tokyo.

Amasaka, K. (Ed.). (2007b). Establishment of a needed design quality assurance framework for numerical simulation in automobile production, Edited by Working Group No. 4 Studies in JSQC, Study group on simulation and SQC, Tokyo.

Amasaka, K. (Ed.). (2007b). Establishment of a needed design quality assurance framework for numerical simulation in automobile production, Working Group No. 4 studies "Study group on simulation and SQC", in Japanese Society for Quality Control.

Amasaka, K. (Ed.). (2007b). Establishment of a needed design quality assurance framework for numerical simulation in automobile production. Working Group No. 4 Studies in JSQC, Study group on simulation and SQC, Tokyo.

Amasaka, K. (Ed.). (2012). Science TQM, new quality management principle, The quality management strategy of Toyota, Bentham Science Publishers, Sharjah, UAE, USA, THE NETERLANDS. doi:10.2174/97816080528201120101

Amasaka, K. (Ed.). (2012). Science TQM, New Quality Management Principle: The quality management strategy of Toyota, Bentham Science Publishers, U.A.E, USA, The Nethrlands.

Amasaka, K. (Ed.). (2012). Science TQM, New Quality Management Principle: The quality management strategy of Toyota, Bentham Science Publishers, U.A.E.

Amasaka, K. (Ed.). (2012). Science TQM, new quality management principle: The quality management strategy of Toyota, Bentham Science Publishers, UAE, USA, the Netherlands.

Amasaka, K. (Ed.). (2012). Science TQM, new quality management principle: The quality management strategy of Toyota, Bentham Science Publishers.

Amasaka, K. (Ed.). (2012). Science TQM, New Quality Management Principle: The quality management strategy of Toyota. Bentham Science Publisher, Shariah, UAE, USA.

Amasaka, K. (Ed.). (2012). Science TQM, New Quality Management Principle: The Quality Management Strategy of Toyota. Bentham Science Publisher, UAE, USA, The Netherlands.

Amasaka, K. (Ed.). (2012). Science TQM, New Quality Management Principle: The quality management strategy of Toyota. Bentham Science Publisher.

Amasaka, K. (Ed.). (2012). Science TQM, new quality management principle: The quality management strategy of Toyota. Bentham Science Publishers, Sharjah, UAE, USA, THE Netherlands.

Amasaka, K. (Ed.). (2012). Science TQM, new quality management principle: The quality management strategy of Toyota. Bentham Science Publishers.

Amasaka, K. (Ed.). (2012). Science TQM, New Quality Management Principle: The quality management strategy of Toyota. Bentham Science Publishers.

Amasaka, K. (Ed.). (2012). Science TQM, new quality management principle: The quality strategy of Toyota, Bentham Science Publishers, UAE, USA, The Netherlands.

Amasaka, K. (Ed.). (2012). Science TQM, new quality management principle: The quality strategy of Toyota, Bentham Science Publishers.

Amasaka, K. (Ed.). (2013). Science TQM, New Quality Management Principle, The quality management strategy at Toyota, Bentham Science Publishers.

Amasaka, K. et al. (1999). The preliminary survey in 68QCS: Questionnaire result of TQM which creates Japan (Special Lecture), *Union of Japanese Scientists and Engineers, Hakone, Kanagawa-ken, Japan.*

Amasaka, K., & Nagasawa, S. (2000). Automobile's Kansei Engineering. Fundamentals and applications of Sensory Evaluation: For Kansei Engineering in automobile, Japanese Standards Association.

Amasaka, K., & Nagaya, A. (2002). Engineering of the new sensitivity in the vehicle: Psychographics of LEXUS design profile, Development of articles over the sensitivity-The method and practice. Japan Society of Kansei Engineering. Nihon Suppan Service Press, 55-72.

Amasaka, K., & Nagaya, A. (2002). Engineering of the new sensitivity in the vehicle: Psychographics of LEXUS design profile. Development of articles over the sensitivity-The method and practice. Japan Society of Kansei Engineering, Nihon Suppan Service Press.

Amasaka, K., & Nagaya, A. (2002). New Kansei Engineering in Automobile: Psychographics of "Lexus" design profile. Product development in Kansei Engineering, Nihon-Shuppan Service, 55-72.

Amasaka, K., & Otaki, M. (1999). *Developing New TQM using partnering – Effectiveness of "TQM-P" employing Total Task Management Team.* The Japan Society for Product Management, The 10th Annual Conference, Kyushu Sangyo University, Fukuoka, Japan.

Amasaka, K., & Saito, H. (1983). *Mechanization of an sensory inspection - An example of viewing of a damping force – "Lissajous curve" of the front shock absorber.* Central Japan Quality Control Association, Annual Quality Control Conference 1983, Nagoya, Aichi, Japan.

Amasaka, K., & Saito, H. (1983). *Mechanization of sensory evaluation: Example of the judgment of the Lissajous waveform indicating the damping force characteristics* Proceedings of the Annual Conference of Central Japan Quality Control Association 1983, Nagoya, Japan.

Amasaka, K., Baba, J., Kanuma, Y., Sakai, H., & Okada, S. (2006). *The evolution of technology and skill - The Key to success in global production": Latest implementation of New Japan Model "Science TQM.* The Japan Society for Production Management.

Amasaka, K., Igarashi, M., Yamamura, N., Fukuda, S., & Nomasa, H. (1995). The QA Network activity for prevention rusting pf vehicle by using SQC. *Journal of the Japanese Society for Quality Control,* 35-38.

Amasaka, K., Kishimoto, M., Murayama, T., & Ando, Y. (1998). The development of Marketing SQC for Dealers' Sales Operation System. *Journal of the Japanese Society for Quality Control, The 58th Technical Conference,* 76-79.

Amasaka, K., Nitta, S., & Kondo, K. (1996). An investigation of engineers' recognition and feelings about good patents by new SQC method. *Journal of the Japanese Society for Quality Control,* 17-24.

Amasaka, K., Ohmi, H., & Murai, F. (1990). The improvement of the corrosion resistance for axle unit for vehicle: Joint QCD research activity, Coatings Technology, Japan Coating Technology Association.

Amasaka, K., Tsukada, H., & Ishida, M. (1982). *Development of the tempering machine using mid-frequency of the rear axle shaft.* The Energy Conservation Center, Annual Technical Conference 1982, Nagoya, Aichi, Japan.

Amasaka, K. (1984a). Process control of the bolt tightening torque of automotive chassis parts (Part 1) . *Standardization and Quality Control, 37*(9), 97–104.

Amasaka, K. (1985). Management of the quenching process of the automotive chassis parts in case measurement takes time . *Standardization and Quality Control, 38*(3), 93–100.

Amasaka, K. (1988). *Concept and progress of Toyota Production System (Plenary lecture), Co-sponsorship: Japan Society of Precision Engineering and others.* Hachinohe, Aomori-ken.

Amasaka, K. (1988). *Concept and progress of Toyota Production System (Plenary Lecture), Co-sponsorship: The Japan Society of Precision Engineering and others.* Hachinohe, Aomori-ken.

Amasaka, K. (1988). *Concept and progress of Toyota Production System (Plenary lecture), Co-sponsorship: The Japan Society of Precision Engineering, Hachinohe Regional Advance Technology Promotion Center Foundation and others.* Hachinohe, Aomori-ken. (in Japanese)

Amasaka, K. (1988). *Concept and progress of Toyota Production System (plenary lecture), Co-sponsorship: The Japan Society of Precision Engineering, Hachinohe Regional Advance Technology Promotion Center Foundation, etc.* Hachinohe.

Amasaka, K. (1988b). *Concept and progress of Toyota Production System (Plenary lecture). Co-sponsorship: The Japan Society of Precision Engineering, and Hachinohe Regional Advance technology Promotion Center, etc.* Hachinohe.

Amasaka, K. (1991). Application of multivariate analysis in auto-production: Establishment of high-accuracy prediction formulas for the manufacture technology . *Standardization and Quality Control, 44*(11), 91–110.

Amasaka, K. (1995). A construction of SQC Intelligence System for quick registration and retrieval library: A visualized SQC report for technical wealth. *Lecture Notes in Economics and Mathematical Systems, Springer, 445,* 318–336. doi:10.1007/978-3-642-59105-1_24

Amasaka, K. (1999). *New QC circle activities at Toyota, A training of trainer's course on evidence: Based on participatory quality improvement (Special lecture), The International Health Program (TOT Course on EPQI), Tohoku University School of Medicine (WHO Collaboration Center).* Sendai-city.

Amasaka, K. (1999a). A demonstrative study of a new SQC concept and procedure in the manufacturing industry: Establishment of a New Technical Method for conducting Scientific SQC. *An International Journal of Mathematical & Computer Modeling, 31*(10-12), 1–10.

Amasaka, K. (1999a). A study on "Science SQC" by utilizing "Management SQC-A demonstrative study on a New SQC concept and procedure in the manufacturing industry-. *International Journal of Production Economics, 60-61,* 591–598. doi:10.1016/S0925-5273(98)00143-1

Amasaka, K. (1999a). *New QC Circle activities in Toyota (Special Lecture), A Training of Trainer's Course on Evidence-Based on Participatory Quality Improvement, TOT Course on EPQI, Tohoku University School of Medicine (WHO Collaboration Center).* Sendai-city.

Amasaka, K. (1999a). *TQM at Toyota-Toyota's TQM Activities: to create better car (Special Lecture), A Training of Trainer's Course on Evidence- based on Participatory Quality Improvement, International Health Program (TOT Course on EPQI), Tohoku University School of Medicine (WHO Collaboration Center).* Sendai-city.

Amasaka, K. (1999a). *TQM at Toyota-Toyota's TQM Activities: to create better car (Special Lecture). A Training of Trainer's Course on Evidence-based on Participatory Quality Improvement, International Health Program (TOT Course on EPQI), Tohoku University School of Medicine.* WHO Collaboration Center.

Amasaka, K. (2000a). Partnering chains as the platform for Quality Management in Toyota. *Proceedings of the 1st World Conference on Production and Operations Management, Seville, Spain.*

Amasaka, K. (2000b). Partnering chains as the platform for quality management in Toyota, *Proceedings of the 1st World Conference on Production and Operations Management, Sevilla, Spain.*

Amasaka, K. (2000c). A demonstrative study of a new SQC concept and procedure in the manufacturing industry: Establishment of a New Technical Method for conducting Scientific SQC. *Mathematical and Computer Modelling, 31*(10-12), 1–10. doi:10.1016/S0895-7177(00)00067-4

Amasaka, K. (2000c). Basic Principles of JIT – Concept and progress of Toyota Production System (Plenary lecture), *The Operations Research Society of Japan, Strategic research* [Tokyo.]. *Group, 1.*

Amasaka, K. (2000c). New JIT, a new principle for management technology 21C: Proposal and demonstration of TQM-S in Toyota (Special Lecture*), Proceedings of the 1st World Conference on Production and Operations Management,* Sevilla, Spain.

Amasaka, K. (2001). Proposal of Marketing SQC to revolutionize dealers' sales activities. *Proceedings of the 16th Int. Conf. on Production Research, Prague, Czech Public.* IEEE.

Amasaka, K. (2002b). New JIT, A New Management Technology Principle at Toyota. *International Journal of Production Economics, 80*(2), 135–144. doi:10.1016/S0925-5273(02)00313-4

Amasaka, K. (2003). Proposal and implementation of the "Science SQC", quality control principle. *Mathematical and Computer Modelling, 38*(11-13), 1125–1136. doi:10.1016/S0895-7177(03)90113-0

Amasaka, K. (2004a). Development of "Science TQM", a new principle of quality management: Effectiveness of Strategic Stratified Task Team at Toyota. *International Journal of Production Research, 42*(17), 3691–3706. doi:10.1080/00207540420002003867

Amasaka, K. (2004a). Development of Science TQM, a new principle of quality management: Effectiveness of strategic stratified task team at Toyota. *International Journal of Production Research, 42*(7), 3691–3706.

Amasaka, K. (2004b). *Establishment of performance measurement toward strategic quality management: Key to successful implementation of new management principle. Proceedings of the 4th International Conference on Theory and Practice in Performance Measurement,* Edinburgh International Conference Centre, UK.

Amasaka, K. (2004b). Global production and establishment of production system with high quality assurance, Toward the next-generation quality management technology (Series 1). *Quality Management, 55*(1), 44–57.

Amasaka, K. (2004c). *Science SQC, new quality control principle: The quality control principle: The quality strategy of Toyota.* Springer-Verlag Tokyo. doi:10.1007/978-4-431-53969-8

Amasaka, K. (2005). Constructing a *Customer Science* Application System *"CS-CIANS"* - Development of a global strategic vehicle *"Lexus"* utilizing *New JIT -. WSEAS Transactions on Business and Economics, 2*(3), 135–142.

Amasaka, K. (2005). Constructing a *Customer Science* Application System *"CS-CIANS"* –Development of a global strategic vehicle *"Lexus"* utilizing *New JIT–. WSEAS Transactions on Business and Economics, 2*(3), 135–142.

Amasaka, K. (2005). Constructing a *Customer Science* Application System "*CS-CIANS*"– Development of a global strategic vehicle "*Lexus*" utilizing *New JIT*–. *WSEAS Transactions on Business and Economics, 2*(3), 135–142.

Amasaka, K. (2005). Constructing a *Customer Science* Application System "*CS-CIANS*" –Development of a Global Strategic Vehicle "*Lexus*" Utilizing *New JIT*–. *WSEAS Transactions on Business and Economics, 2*(3), 135–142.

Amasaka, K. (2005). Constructing a *Customer Science* application system "CS-CIANS": Development of a global strategic vehicle "*Lexus*" utilizing *New JIT, WSEAS (World Scientific and Engineering Academy and Society). Transformations in Business & Economics, 2*(3), 135–142.

Amasaka, K. (2005). Constructing a Customer Science Application System "CS-CIANS": Development of a Global Strategic Vehicle "Lexus" utilizing New JIT. *WSEAS Transactions on Business and Economics, 2*(3), 135–142.

Amasaka, K. (2005). Constructing a Customer Science Application System "CS-CIANS"—Development of a Global Strategic Vehicle "Lexus" utilizing New JIT–. *WSEAS Transactions on Business and Economics, 2*(3), 135–142.

Amasaka, K. (2005). Constructing a *Customer Science* application system "*CS-CIANS*"-Development of a global strategic vehicle "*Lexus*" utilizing *New JIT*. *WSEAS Transactions on Business and Economics, 3*(2), 135–142.

Amasaka, K. (2005). New Japan Production Method, an innovative production management principle. *Proceedings of the 16th Annual Conference of the Production and Operations Management Society, Michigan, Chicago.*

Amasaka, K. (2005b), *Interim Report of WG4's studies in JSQC research on simulation and SQC(1) - A study of the high quality assurance CAE model for car development design, Transdisciplinary Federation of Science and Technology. 1st Technical Conference*, Nagano, Japan.

Amasaka, K. (2006). (Editorial) Development of New JIT, new management technology principle. *Production Management Transaction of the Japan Society for Production Management, 13*(1), 143–150.

Amasaka, K. (2006). A new principle, next generation management technology: Development of New JIT (Editorial), *Production Management* [in Japanese]. *Transaction of the Japan Society for Production Management, 13*(1), 143–150.

Amasaka, K. (2007a). High Linkage Model "*Advanced TDS, TPS & TMS*: Strategic development of *New JIT* at Toyota. *International Journal of Operations and Quantitative Management, 13*(3), 101–121.

Amasaka, K. (2007a). High Linkage Model "Advanced TDS, TPS & TMS": Strategic development of "New JIT" at Toyota. *International Journal of Operations and Quantitative Management, 13*(3), 101–121.

Amasaka, K. (2007a). New Japan Production Model, an advanced production management principle: Key to strategic implementation of *New JIT. The International Business & Economics Research Journal, 6*(7), 67–79.

Amasaka, K. (2007a). Proposal and validity of Patent Value Appraisal Model "TJS-PVAM": Development of "*Science TQM*" in the corporate strategy. *Proceedings of the International Conference Risk, Quality and Reliability.* Technical University of Ostrava.

Amasaka, K. (2007a). Proposal of Marketing SQC to revolutionize dealers' sales activities-A demonstrative study on Customer Science by utilizing Science SQC, *Proceedings of the 16th International Conference Production Research, Praha, Czech.*

Amasaka, K. (2007a). Proposal of Marketing SQC to revolutionize dealers' sales activities-A demonstrative study on Customer Science by utilizing Science SQC. *Proceedings of the 16th International Conference Production Research. Praha Czech Public.*

Amasaka, K. (2007a). The validity of *"TDS-DTM"*, a strategic methodology of merchandise - Development of *New JIT*, Key to the excellence design *"LEXUS"*-. *The International Business & Economics Research Journal, 6*(11), 105–115.

Amasaka, K. (2007b). High linkage model *"Advanced TDS, TPS & TMS"*: Strategic development of *"New JIT"* at Toyota. *International Journal of Operations and Quantitative Management, 13*(3), 101–121.

Amasaka, K. (2007b). High Linkage Model *"Advanced TDS, TPS & TMS"*: Strategic Development of *"New JIT"* at Toyota. *International Journal of Operations and Quantitative Management, 13*(3), 101–121.

Amasaka, K. (2007b). The validity of "TDS-DTM", a strategic methodology of merchandise: Development of New JIT, Key to the excellence design "LEXUS". *The International Business & Economics Research Journal, 6*(11), 105–115.

Amasaka, K. (2007b). The validity of *Advanced TMS*, A strategic development marketing system utilizing *New JIT –*. *The International Business & Economics Research Journal, 6*(8), 35–42.

Amasaka, K. (2007b). The Validity of *Advanced TMS*, A strategic development marketing system utilizing *New JIT –*. *The International Business & Economics Research Journal, 6*(8), 35–42.

Amasaka, K. (2007b). The validity of Advanced TMS, a strategic development marketing system: Toyota's Scientific Customer Creative Model utilizing New JIT. *The International Business & Economics Research Journal, 6*(8), 35–42.

Amasaka, K. (2007c). Applying New JIT–Toyota's global production strategy: Epoch-making innovation in the work environment. *Robotics and Computer-integrated Manufacturing, 23*(3), 285–293. doi:10.1016/j.rcim.2006.02.001

Amasaka, K. (2007c). High Linkage Model *"Advanced TDS, TPS & TMS"*: Strategic development of *New JIT* at Toyota. *International Journal of Operations and Quantitative Management, 13*(3), 101–121.

Amasaka, K. (2007c). High linkage model "Advanced TDS, TPS & TMS"-Strategic development of "New JIT" at Toyota. *International Journal of Operations and Quantitative Management, 13*(3), 101–121.

Amasaka, K. (2007c). The validity of *"TDS-DTM"*, A strategic methodology of merchandise-Development of *New JIT*, Key to the excellence design *"LEXUS"*. *The International Business & Economics Research Journal, 6*(11), 105–115.

Amasaka, K. (2007d). Highly Reliable CAE Model: The key to strategic development of *New JIT. Journal of Advanced Manufacturing Systems, 6*(2), 159–176. doi:10.1142/S0219686707000930

Amasaka, K. (2007d). *New Japan Model, Science TQM: Theory and application of strategic quality management.* Maruzen publishing. (in Japanese)

Amasaka, K. (2007e). The validity of *Advanced TMS*, a strategic development marketing system using *New JIT. The International Business & Economics Research Journal, 6*(8), 35–42.

Amasaka, K. (2007e). The validity of *Advanced TMS*, a strategic development marketing system utilizing *New JIT. The International Business & Economics Research Journal, 6*(8), 35–42.

Amasaka, K. (2008a). Science TQM, a new management principle: The quality management strategy of Toyota. *The Journal of Management and Engineering Integration, 1*(1), 7–22.

Amasaka, K. (2008a). Science TQM, a new quality management principle: The quality management strategy of Toyota. *The Journal of Management & Engineering Integration, 1*(1), 7–22.

Amasaka, K. (2008b). Science TQM, A New Quality Management Principle: The Quality Management Strategy of Toyota. *The Journal of Management and Engineering Integration, 1*(1), 7–22.

Amasaka, K. (2008c). An Integrated Intelligence Development Design CAE Model utilizing New JIT. *Journal of Advanced Manufacturing Systems, 7*(2), 221–241. doi:10.1142/S0219686708001589

Amasaka, K. (2009). Establishment of Strategic Quality Management-Performance Measurement Model "SQM-PPM: Key to successful implementation of Science TQM,". *China-USA Business Review, 8*(12), 1–11.

Amasaka, K. (2009). New JIT, advanced management technology principle: The global management strategy of Toyota (Invitation lecture*). International Conference on Intelligent Manufacturing and Logistics System, and Symposium on Group Technology and Cellular Manufacturing*, Kitakyushu, Japan.

Amasaka, K. (2009). New JIT, advanced management technology principle: The global management strategy of Toyota (special lecture). *International Conference on Intelligent Manufacturing & Logistics Systems and Symposium on Group Technology and Cellular Manufacturing*, Kitakyushu, Japan.

Amasaka, K. (2009). Proposal and Validity of Patent Value Appraisal Model "TJS-PVAM"-Development of "Science TQM" in the Corporate Strategy, *China-USA. Business Review (Federal Reserve Bank of Philadelphia), 8*(7), 45–56.

Amasaka, K. (2009). The effectiveness of flyer advertising employing TMS: Key to scientific automobile sales innovation at Toyota, *China-USA. Business Review (Federal Reserve Bank of Philadelphia), 8*(3), 1–12.

Amasaka, K. (2009a). New JIT, advanced management technology principle: The global management strategy of Toyota (Invitation Lecture*). International Conference on Intelligent Manufacturing and Logistics System, and Symposium on Group Technology and Cellular Manufacturing*, Kitakyushu, Japan.

Amasaka, K. (2009a). The Foundation for Advancing the Toyota Production System utilizing New JIT. *Journal of Advanced Manufacturing Systems, 80*(1), 5–26. doi:10.1142/S0219686709001614

Amasaka, K. (2009b). Establishment of Strategic Quality Management - Performance Measurement Model "SQM-PPM": Key to Successful Implementation of Science TQM. *The Academic Journal of China-USA Business Review, 8*(12), 1–11.

Amasaka, K. (2009c). An Intellectual Development Production Hyper-cycle Model: *New JIT* fundamentals and applications in Toyota. *International Journal of Collaborative Enterprise, 1*(1), 103–127. doi:10.1504/IJCENT.2009.026459

Amasaka, K. (2009c). Effectiveness of flyer advertising employing TMS: Key to scientific automobile sales innovation at Toyota, *China-USA. Business Review (Federal Reserve Bank of Philadelphia), 8*(3), 1–12.

Amasaka, K. (2009d). Proposal and validity of Patent Value Appraisal Model "TJS-PVAM": Development of "Science TQM" in the corporate strategy. *The China-USA Business Review, 8*(7), 45–56.

Amasaka, K. (2010). Proposal and effectiveness of a High Quality Assurance CAE Analysis Model: Innovation of design and development in automotive industry, *Current Development in Theory and Applications of Computer Science. Engineering and Technology, 2*(1/2), 23–48.

Amasaka, K. (2010). Proposal and effectiveness of a High-Quality Assurance CAE Analysis Model: Innovation of design and development in automotive industry, *Current Development in Theory and Applications of Computer Science. Engineering and Technology, 2*(1/2), 23–48.

Amasaka, K. (2010b). Proposal and effectiveness of a High-Quality Assurance CAE Analysis Model, *Current Development in Theory and Applications of Computer Science. Engineering and Technology, 2*(1/2), 23–48.

Amasaka, K. (2011). Changes in marketing process management employing TMS: Establishment of Toyota Sales Marketing System, *China & USA. Business Review (Federal Reserve Bank of Philadelphia), 10*(7), 539–550.

Amasaka, K. (2011). Changes in Marketing Process Management Employing TMS: Establishment of Toyota Sales Marketing System, *China & USA. Business Review (Federal Reserve Bank of Philadelphia), 10*(7), 539–550.

Amasaka, K. (2011). Changes in marketing process management employing TMS: Establishment of Toyota Sales Marketing System, *China-USA. Business Review (Federal Reserve Bank of Philadelphia), 10*(6), 539–550.

Amasaka, K. (2012). *Prevention of the automobile development design, Precaution and prevention.* Japan Standard Association.

Amasaka, K. (2012b). Constructing Optimal Design Approach Model: Application on the Advanced TDS. *Journal of Communication and Computer, 9*(7), 774–786.

Amasaka, K. (2013b). The development of a Total Quality Management System for transforming technology into effective management strategy. *International Journal of Management, 30*(2), 610–630.

Amasaka, K. (2014). New JIT, New Management Technology Principle: Surpassing JIT. *Procedia Technology, 16*(special issues), 1135–1145. doi:10.1016/j.protcy.2014.10.128

Amasaka, K. (2014a). New JIT, new management technology principle. *Journal of Advanced Manufacturing Systems, 13*(3), 197–222. doi:10.1142/S0219686714500127

Amasaka, K. (2015). *New JIT, new management technology principle.* Taylor & Francis, CRC Press.

Amasaka, K. (2015). *New JIT, New Management Technology Principle.* Taylor & Francis, CRC Press.

Amasaka, K. (2015). *Strategic employment of the Patent Appraisal Method, New JIT, new management technology principle, Taylor & Francis Group.* CRC Press.

Amasaka, K. (2015b). Constructing a New Japanese Development Design Model "NJ-DDM": Intellectual evolution of an automobile product design. *TEM Journal Technology Education Management Informatics, 4*(4), 336–345.

Amasaka, K. (2015b). Constructing a New Japanese Development Design Model "NJ-DDM": Intellectual evolution of an automobile Product Design. *TEM Journal Technology Education Management Informatics, 4*(4), 336–345.

Amasaka, K. (2015b). Global manufacturing strategy of New JIT: Surpassing JIT (Keynote Lecture*). International Conference on Information Science and Management Engineering, Phuket, Thailand.*

Amasaka, K. (2017). Strategic Stratified Task Team Model for realizing simultaneous QCD fulfilment: Two case studies. *Journal of Japanese Operations Management and Strategy, 7*(1), 14–35.

Amasaka, K. (2017). *Toyota: Production System, Safety Analysis and Future Directions.* NOVA Science Publishers.

Amasaka, K. (2017a). Strategic Stratified Task Team Model for realizing simultaneous QCD fulfilment: Two Case Studies. *Journal of Japanese Operations Management and Strategy, 7*(1), 14–35.

Amasaka, K. (2017a). *Toyota: Production system, safety analysis and future directions.* NOVA Science Publishers Inc.

Amasaka, K. (2017a). *Toyota: Production System, safety analysis and future directions.* Nova Science Publishers.

Amasaka, K. (2017a). *Toyota: Production System, Safety Analysis, and Future Directions.* NOVA Science Publishers.

Amasaka, K. (2017b). Strategic Stratified Task Team Model for Realizing Simultaneous QCD Fulfilment: Two Case Studies. *Journal of Japanese Operations Management and Strategy, 7*(1), 14–35.

Amasaka, K. (2017b). Strategic Stratified Task Team Model for realizing simultaneous QCD fulfilment: Two case studies. *The Journal of Japanese Operations Management and Strategy, 7*(1), 14–35.

Amasaka, K. (2017c). *Attractive automobile design development: Study on customer values (Plenary Lecture).* Executive Lecture in Rotary Clube.

Amasaka, K. (2018). Automobile Exterior Design Model: Framework development and supporting case studies. *Journal of Japanese Operations Management and Strategy, 8*(1), 67–89.

Amasaka, K. (2018). Automobile Exterior Design Model: Framework development and supporting case studies. *The Journal of Japanese Operations and Strategy, 8*(1), 67–89.

Amasaka, K. (2018). Automobile Exterior Design Model: Framework development and supporting case studies. *The Journal of Japanese Operations Management and Strategy, 8*(1), 67–89.

Amasaka, K. (2018a). Innovation of automobile manufacturing fundamentals employing New JIT: Developing Advanced Toyota Production System, *International Journal of Research in Business. Economics and Management, 2*(1), 1–15.

Amasaka, K. (2018b). Automobile Exterior Design Model: Framework development and support case studies. *Journal of Japanese Operations Management and Strategy, 8*(1), 67–89.

Amasaka, K. (2019). Establishment of an Automobile Optimal Product Design Model: Application to study on bolt-nut loosening mechanism. *Noble International Journal of Scientific Research, 3*(9), 79–10.

Amasaka, K. (2019). Studies on New Manufacturing Theory, *Noble. International Journal of Scientific Research, 3*(1), 42–79.

Amasaka, K. (2019). Study on New Manufacturing Theory, *Nobel. International Journal of Scientific Research, 3*(1), 1–20.

Amasaka, K. (2019a). Studies On New Manufacturing Theory, *Noble. International Journal of Scientific Research, 3*(1), 1–20.

Amasaka, K. (2019b). Establishment of an Automobile Optimal Product Design Model: Application to study on bolt-nut loosening Mechanism, *Noble. International Journal of Scientific Research, 3*(9), 79–102.

Amasaka, K. (2020a). Studies on New Japan Global Manufacturing Model: The innovation of manufacturing engineering. *Journal of Economics and Technology Research, 1*(1), 42–71. doi:10.22158/jetr.v1n1p42

Amasaka, K. (2020b). Evolution of Japan Manufacturing Foundation: Dual Global Engineering Model Surpassing JIT. *International Journal of Operations and Quantitative Management, 26*(2), 101–126. doi:10.46970/2020.26.2.3

Amasaka, K. (2020c). *New Manufacturing Theory: Surpassing JIT.* Lambert Academic Publishing.

Amasaka, K. (2021). New Japan Automobile Global Manufacturing Model: Using Advanced TDS, TPS, TMS, TIS & TJS. *Journal of Advanced Manufacturing Systems, 6*(6), 499–523.

Amasaka, K. (2021). New Japan Automobile Global Manufacturing Model: Using Advanced TDS, TPS, TMS, TIS & TJS. *Journal of Business Management and Economic Research, 6*(6), 499–523.

Amasaka, K. (2022a). A New Automobile Global Manufacturing System: Utilizing a Dual Methodology, *Scientific Review, Journals in Academic Research Publishing. Group, 8*(3), 51–58.

Amasaka, K. (2022a). A New Automobile Global Manufacturing System: Utilizing a Dual Methodology. *Business and Management and Economics Research, 8*(4), 41–58. doi:10.32861/sr.83.51.58

Amasaka, K. (2022b). *Examining a New Automobile Global Manufacturing System.* IGI Global Publisher. doi:10.4018/978-1-7998-8746-1

Amasaka, K. (2022c). A New Automobile Global Manufacturing System: Utilizing a dual methodology, *Scientific Review. Journals in Academic Research*, *8*(4), 41–58.

Amasaka, K. (2022d). A New Automobile Product Development Design Model: Using a dual corporate engineering strategy. *Journal of Economics and Technology Research*, *3*(3), 1–21. doi:10.22158/jetr.v4n1p1

Amasaka, K. (2023b). A New Automobile Sales Marketing Model for innovating auto-dealer's sales. *Journal of Economics and Technology Research*, *4*(3), 9–32. doi:10.22158/jetr.v4n3p9

Amasaka, K. (Ed.). (2003). *Manufacturing fundamentals: The application of Intelligence Control Charts—Digital Engineering for superior quality assurance*. Japanese Standards Association.

Amasaka, K. (Ed.). (2019). *Fundamentals of manufacturing industry management: New Manufacturing Theory – Operations management strategy 21C*. Sankei-sha.

Amasaka, K. (Ed.). (2019). *The fundamentals of the manufacturing industries management: New Manufacturing Theory- Operations Management Strategy 21C*. Shankei-sha.

Amasaka, K. (Ed.). (2019). *The fundamentals of the manufacturing management: New Manufacturing Theory-Operations Management Strategy 21C*. Sankei-Sha.

Amasaka, K., & Ishida, T. (1991). *Tempered equipment, Aichi Invention Encouragement Prize*. Aichi Japan Institute of Invention and Innovation, Aichi-ken.

Amasaka, K., Ito, T., & Nozawa, Y. (2012). A New Development Design CAE Employment Model. *The Journal of Japanese Operations Management and Strategy*, *3*(1), 18–37.

Amasaka, K., Iwata, M., & Fujii, M. (1988a). Development and the effect of the welding nozzle made from all ceramics, *Engineering Materials . Nikkan Kogyo Shinbun. Ltd.*, *36*(10), 60–64.

Amasaka, K., & Kamio, M. (1985). Process improvement and process control: An example of the automotive chassis parts process, *Quality Control. . Union Japanese Scientist and Engineers*, *36*(6), 38–47.

Amasaka, K., & Kamio, M. (1985). Process improvement and process control: The example of axle-unit parts of the automobile, *Quality Control* [in Japanese]. *Union of Japanese and Engineers*, *36*(6), 38–47.

Amasaka, K., Kurosu, S., & Morita, M. (2008). *New Manufacturing Principle: Surpassing JIT-Evolution of Just in Time*. Morikita-Shuppan.

Amasaka, K., Kurosu, S., & Morita, M. (2008). *New Manufacturing Theory: Surpassing JIT - The evolution of Just-In-Time*. Maruzen.

Amasaka, K., Kurosu, S., & Morita, M. (2008). *New Manufacturing Theory: Surpassing JIT – The evolution of Just-In-Time*. Morikita-Shuppan.

Amasaka, K., Kurosu, S., & Morita, M. (2008). *New Manufacturing Theory: Surpassing JIT-Evolution of Just in Time*. Morikita-Shuppan.

Amasaka, K., Kurosu, S., & Morita, M. (2008). *New Manufacturing Theory: Surpassing JT-Evolution of Just in Time*. Morikita-Shuppan.

Amasaka, K., Kurosu, S., & Morita, M. (2008). *New Theory of Manufacturing-Surpassing JIT: Evolution of Just-in-Time*. Morikita-Shuppan.

Amasaka, K., Mitani, Y., & Tsukamoto, H. (1993). Research into rust prevention quality assurance for plated components using SQC: Improvement of surface smoothness of rod pistons ground by centerless grinding machine, *Quality* . *Japanese Society for Quality Control*, 23(2), 90–98.

Amasaka, K., & Nagasawa, S. (2000). *Fundamentals and application of sensory evaluation: For Kansei Engineering in the vehicle*. Japanese Standards Association.

Amasaka, K., & Nagasawa, S. (2000). *The foundation of the Sensory Evaluation and application: For Kansei Engineering in automobile*. Japanese Standards Association.

Amasaka, K., & Nagaya, A. (2002). *Engineering of the new sensitivity in the vehicle: Psychographics of LEXUS design profile, Development of articles over the sensitivity—The method and practice*. Japan Society of Kansei Engineering. Nihon Shuppan Service Press.

Amasaka, K., & Nagaya, A. (2002). *Engineering of the new sensitivity in the vehicle: Psychographics of LEXUS design profile", Development of articles over the sensitivity: The method and practice*. Edited by Japan Society of Kansei Engineering, Nihon Shuppan Service Press.

Amasaka, K., Nagaya, A., & Shibata, W. (1999). Studies on Design SQC with the application of Science SQC - Improving of business process method for automotive profile design. *Japanese Journal of Sensory Evaluations*, 3(1), 21–29.

Amasaka, K., Nagaya, A., & Shibata, W. (1999). Studies on Design SQC with the application of Science SQC - Improving of Business Process Method for automotive profile design. *Japanese Journal of Sensory Evaluations*, 3(1), 21–29.

Amasaka, K., Nagaya, A., & Shibata, W. (1999). Studies on Design SQC with the application of Science SQC improving of business process method for automotive profile design [in Japanese]. *Japanese Journal of Sensory Evaluations*, 3(1), 21–29.

Amasaka, K., Nagaya, A., & Shibata, W. (1999). Studies on Design SQC with the application of Science SQC: Improving of business process method for automotive profile design. *Japanese Journal of Sensory Evaluations*, 3(1), 21–29.

Amasaka, K., Nagaya, A., & Shibata, W. (1999). Studies on Design SQC with the application of Science SQC—Improving of business process method for automotive profile design. *Japanese Journal of Sensory Evaluations*, 3(1), 21–29.

Amasaka, K., Ogura, M., & Ishiguro, H. (2013). Constructing a Scientific Mixed Media Model for boosting automobile dealer visits: Evolution of market creation employing TMS. *International Journal of Business Research and Development*, 3(4), 1377–1391.

Amasaka, K., Ogura, M., & Ishiguro, H. (2013). Constructing a Scientific Mixed Media Model for boosting automobile dealer visits: Evolution of market creation employing TMS. *International Journal of Engineering Research and Applications*, 3(4), 1377–1391.

Amasaka, K., Ohmi, H., & Murai, H. (1990). The improvement of the corrosion resistance for axle unit of the vehicle. *Coal Technology*, 25(6), 230–240.

Amasaka, K., Okumura, Y., & Tamura, N. (1988b). Improvement of paint quality for auto chassis parts . *Standardization and Quality Control*, 41(2), 53–62.

Amasaka, K., Onodera, T., & Kozaki, T. (2014). Developing a Higher-Cycled Product Design CAE Model: The evolution of automotive product design and CAE. *International Journal of Technical Research and Application*, 2(1), 17–28.

Amasaka, K., & Osaki, S. (1999). The promotion of New SQC International Education in Toyota Motor - A proposal of "Science SQC" for improving the principle of TQM -, *The European Journal of Education on Maintenance Reliability. Risk Analysis and Safety*, 5(1), 55–63.

Amasaka, K., & Osaki, S. (1999). The promotion of New Statistical Quality Control Internal Education in Toyota Motor: A proposal of "Science SQC" for improving the principle of TQM, *The European Journal of Engineering Education, Research and Education in Reliability, Maintenance, Quality Control. Risk and Safety, 24*(3), 259–276.

Amasaka, K., & Sakai, H. (1996). Improving the reliability of body assembly line equipment. *International Journal of Reliability Quality and Safety Engineering, 3*(1), 11–24. doi:10.1142/S021853939600003X

Amasaka, K., & Sakai, H. (1998). Availability and Reliability Information Administration System "ARIM-BL" by methodology of Inline-Online SQC. *International Journal of Reliability Quality and Safety Engineering, 5*(1), 55–63. doi:10.1142/S0218539398000078

Amasaka, K., & Sakai, H. (2009). TPS-QAS, new production quality management model: Key to New JIT–Toyota's global production strategy. *International Journal of Manufacturing Technology and Management, 18*(4), 409–426. doi:10.1504/IJMTM.2009.027774

Amasaka, K., & Sakai, H. (2010). Evolution of TPS fundamentals utilizing *New JIT* strategy: Proposal and validity of Advanced TPS at Toyota. *Journal of Advanced Manufacturing Systems, 9*(2), 85–99. doi:10.1142/S0219686710001831

Amasaka, K., & Sakai, H. (2011). The New Japan Global Production Model "NJ-GPM: Strategic development of Advanced TPS. *The Journal of Japanese Operations Management and Strategy, 2*(1), 1–15.

Amasaka, K., & Sakai, H. (2011). The New Japan Global Production Model "NJ-GPM": Strategic development of *Advanced TPS. The Journal of Japanese Operations Management and Strategy, 2*(1), 1–15.

Amasaka, K., & Sugawara, K. (1988). Q.C.D research activities of the engineer of the manufacture site: Improvement of paint corrosion resistance of shock absorber parts, *Quality Control . Union Japanese Scientist and Engineers, 39*(11), 337–344.

Amasaka, K., & Sugawara, K. (1988). QCD research activities by manufacture engineers: The improvement of the corrosion resistance of the shock absorber parts, *Quality Control* [Special issue] [in Japanese]. *Union of Japanese and Engineers, 39*(11), 337–344.

Amasaka, K., Watanabe, M., & Shimakawa, K. (2005). Modeling of strategic marketing system to reflect latent customer needs and its effectiveness, *Fragrance Journal, The Magazine of Research & Development for Cosmetics. Toiletries & Allied Industries, 33*(1), 72–77.

Amasaka, K., Watanabe, M., & Shimakawa, K. (2005). Modeling of strategic marketing system to reflect latent customer needs and its effectiveness, *The Magazine of Research & Development for Cosmetics. Toiletries & Allied Industries, 33*(1), 72–77.

Amasaka, K., Watanabe, M., & Shimakawa, Y. (2005). Modeling of Strategic Marketing System to reflect latent customer needs and its effectiveness, *The Magazine of Research & Development for Cosmetics, Toiletries & Applied Industries* [in Japanese]. *Fragrance Journal, 33*(1), 72–77.

Amasaka, K., & Yamada, K. (1991). Re-evaluation of the present QC concept and methodology in auto-industry: Development of SQC renaissance at Toyota . *Quality Control, 42*(4), 13–22.

Amasaka, K., & Yamada, K. (1991). Re-evaluation of the QC concept and methodology in auto-industry, *Quality Control* [in Japanese]. *Union of Japanese and Engineers, 42*(4), 13–22.

Anabuki, K., Kaneta, H., & Amasaka, K. (2011). Proposal and validity of Patent Evaluation Method "A-PPM" for corporate strategy. *International Journal of Management & Information Systems, 15*(3), 129–137. doi:10.19030/ijmis.v15i3.4649

Asakura, S., Kanke, R., & Tobimatsu, K. (2011). A study on Automobile Exterior Color and Interior Color Matching Model: The 20th woman example (senior thesis), *School of Science and Engineering, Aoyama Gakuin University, in Amasaka New JIT Laboratory, Sagamihara, Kanagawa, Japan.* .

Asakura, S., Kanke, R., & Tobimatsu, K. (2011). A study on Automobile Exterior Color and Interior Color Matching Model: The 20th woman example [Thesis, School of Science and Engineering, Aoyama Gakuin University].

Asami, H., Ando, T., Yamaji, M., & Amasaka, K. (2010). A study on Automobile Form Design Support Method "AFD-SM". *Journal of Business & Economics Research*, *8*(11), 13–19. doi:10.19030/jber.v8i11.44

Asami, H., Owada, H., Murata, Y., Takebuchi, S., & Amasaka, K. (2011). The A-VEDAM for approaching vehicle exterior design. *Journal of Business Case Studies*, *7*(5), 1–8. doi:10.19030/jbcs.v7i5.5598

Aspy, P. P., & Wai, K. M. (2001). An online evaluation of the Compete Online Decision Entry System (CODES). *Developments in Business Simulation & Experimental Leaning*, *28*, 188–191.

Aspy, P. P., Wai, K. M., & Dean, S. R. (2000). Facilitating learning in the new millennium with the Complete Online Decision Entry System (CODES). *Developments in Business Simulation & Experimental Leaning*, *27*, 250–251.

Baggerly, R. (1996). Hydrogen-Assisted stress cracking of high-strength wheel bolts. *Engineering Failure Analysis*, *3*(4), 231–240. doi:10.1016/S1350-6307(96)00023-4

Beko, J., & Jagric, T. (2011). Demand models for direct mail and periodicals delivery services: Results for a transition economy. *Applied Economics*, *43*(9), 1125–1138. doi:10.1080/00036840802600244

Bell, G. H., Ledolter, J., & Swersey, A. J. (2006). Experimental design on the front lines of marketing: Testing new ideas to increase Direct Mail sales. *International Journal of Research in Marketing*, *23*(3), 309–319. doi:10.1016/j.ijresmar.2006.05.002

Boesen, T. (2000). Creating budget-less organizations with the balanced scorecard. Harvard Business School Publishing.

Bult, J. R., & Wansbeek, T. (2005). Optimal selection for direct mail. *Marketing Science*, *14*(4), 378–394. doi:10.1287/mksc.14.4.378

Burke, R. J., Graham, J., & Smith, F. (2005). Effects of reengineering on the employee satisfaction: Customer satisfaction relationship. *The TQM Magazine*, *17*(4), 358–363. doi:10.1108/09544780510603198

Burke, W., & Trahant, W. (2000). *Business climate shift*. Butterworth Heinemann.

Burke, W., & Trahant, W. (2000). *Business Climate Shift*. Oxford Butterworth –Heinemann.

Chawla, R., & Banerjee, A. (2002). An automated 3D facilities planning and operations model: Generator for synthesizing generic manufacturing operations in virtual reality. *Journal of Advanced Manufacturing Systems*, *1*(1), 5–17. doi:10.1142/S0219686702000039

Chowdhury, S., & Zimmer, K. (1996). QS-9000 PIONEERS. Irwin Professional Pub., Chicago, and ASQC Quality Press, Milwaukee, Wisconsin.

Corporation, N. O. K. (2000). *The history of NOK's Oil Seal-Oil Seal Mechanism*. Promotion Video.

David T., & David. B. (2001). *Manufacturing operations and supply chain management: Lean approach*. Thomson Learning, London.

Deming Prize Committee. (2003). *The bookmark of Deming Prize*. Union of Japanese Scientists and Engineers Press.

Dentsu Inc. (2003). *Dentsu Online, Total Advertisement Expenses Total of Japan.* Dentsu Inc. (http://www.dentsu.co.jp/marketing/adex/adex2005/media.html)

Dijkastra, E. W. (1959). *Communication with an Automatic Computer.* Excelsior.

Dijkastra, E. W., & Feijen, W. H. J. (1988). *A method of programming.* Addison-Wesley Longman Publishing Co., Inc.

Doos, D., Womack, J. P., & Jones, D. T., (1991). *The machine that changed the world–The story of lean production.* Rawson / Harper Perennial.

Doos, D., Womack, J. P., & Jones, D. T. (1991). *The Machine that Changed the World – The Story of Lean Production.* Rawson / Harper Perennial.

Doz, Y. L., & Hamel, G. (1998). *Alliance advantage.* Harvard Business School Press, Boston.

Ebioka, E., Sakai, H., Yamaji, M., & Amasaka, K. (2007). A New Global Partnering Production Model "NGP-PM" utilizing "Advanced TPS". *Journal of Business & Economics Research, 5*(9), 1–8.

Ebioka, E., Sakai, H., Yamaji, M., & Amasaka, K. (2007). A New Global Partnering Production Model "NGP-PM" utilizing Advanced TPS. *Journal of Business & Economics Research, 5*(9), 1–8.

Ebioka, K., Yamaji, M., Sakai, H., & Amasaka, K. (2007a). Strategic development of *Advanced TPS* to bring overseas manufacturing to Japan standards: Proposal of a New Global Partnering Production Model and its effectiveness, *Production Management. Transaction of the Japan Society for Production Management, 13*(2), 51–56.

Ebioka, K., Yamaji, M., Sakai, H., & Amasaka, K. (2007a). Strategic development of *Advanced TPS* to bring overseas manufacturing to Japan standards: Proposal of a new global partnering production model and its effectiveness. *Transaction of the Japan Society for Production Management, 13*(2), 51–56.

EFQM. (2003). *The EFQM excellence.* Zeeuw. (www.EFQM.org)

Elias. St. Elmo Lewis. (1908). *Financial advertising.* Levey bros. & company, Indianapolis, USA.

Enrique, A. (2005). *Parallel, Metaheuristics: A New Class of Algorithms.* Addison Wiley.

Eri, Y., Asaji, K., Furugori, N., & Amasaka, K. (1999). *The development of working conditions for aging worker on assembly line (#2).* The Japan Society for Production Management, The 10th Annual Technical Conference; Nagoya University, Japan.

Evans, J. R., & Dean, J. W. (2003). Total quality management, organization and strategy, Thomson, South-Western, Mason, United States.

Evans, J. R., & Lindsay, W. M. (1995). *The Management and Control of Quality.* South-Western.

Evans, R. J., & Lindsay, M. W. (2004). *The management and control of quality.* South-Western College Publishing.

Ferrell, O. C., & Hartline, M. (2005). Marketing Strategy, Thomson South-Western, Mason, Gary, L. L., & Arvind, R. (2003). Marketing Engineering: Computer-assisted marketing analysis and planning. Pearson Education, Inc., London.

Ferrell, O. C., & Hartline, M. (2005). *Marketing Strategy.* Thomson South-Western.

Fuji Seimitsu Co. Ltd. (2014). *U-Nut.* FUN. https://www.fun.co.jp/u_nut/u_nut.html/

Fujieda, S., Masuda, Y., & Nakahata, A. (2007). Development of automotive color designing process. *Journal of Society of Automotive Engineers of Japan, 61*(6), 79–84.

Fukuchi, H., Arai, Y., Ono, M., Suzuki, S. T., & Amasaka, K. (1998). *A proposal TDS-D by utilizing Science SQC: An improving design quality of drive-train components.* The Japanese Society for Quality Control, The 60th Technical Conference, Nagoya, Japan.

Furuya, Y. (1999). *TQM which creates Japan (2nd report): Results of the investigation of the questionnaire analysis team-The proposal of TQM which will aim at the 21 C based on the viewpoint of management.* Union of Japanese Scientists and Engineers.

Gabor, A. (1990). *The man who discovered quality; How Deming W. E., brought the quality revolution to America.* Random House, Inc.

Gabor, A. (1990). *The Man Who Discovered Quality; How W. Edwards Deming Brought the Quality Revolution to America- The Stories of Ford, Xerox, and GM.* Random House, Inc.

Gabor, A. (1990). *The man who discovered quality; How W. Edwards Deming, brought the quality revolution-The stories of Ford, Xerox, and GM.* Random House.

Gamboa, E., & Atrens, A. (2003). Environmental influence on the stress corrosion cracking of rock bolts. *Engineering Failure Analysis*, *10*(5), 521–558. doi:10.1016/S1350-6307(03)00036-0

Gary, L. L., & Arvind, R. (2003). *Marketing Engineering: Computer-assisted marketing analysis and planning.* Pearson Education, Inc.

Goto, T. (1999). *Forgotten Management Origin.* Seisansei-Shuppan.

Goto, T. (1999). *Forgotten Management Origin–Management Quality Taught by GHQ, CCS management lecture.* Seisansei-Shuppan.

Goto, T. (1999). *Forgotten origin of management – Management quality taught by G.H.Q, CCS management lecture.* Productivity Publications.

Goto, T. (1999). *Forgotten origin of management-Management quality taught by G.H.Q, CCS management lecture.* Seisansei-Shuppan. (in Japanese)

Goto, T. (1999). *Forgotten origin of management–Management quality taught by G.H.Q, CCS management lecture.* Seisanse-Shuppan.

Grundspenkis, J. (2007). Agent based approach for organization and personal knowledge modeling: Knowledge management perspective. *Journal of Intelligent Manufacturing*, *18*(4), 451–457. doi:10.1007/s10845-007-0052-6

Halevi, G., & Wang, K. (2007). Knowledge based manufacturing system (KBMS). *Journal of Intelligent Manufacturing*, *18*(4), 467–474. doi:10.1007/s10845-007-0049-1

Hamel, G., & Prahalad, C. K. (1994). *Competing for the future.* Harvard Business School Press.

Harada, O., & Ozawa, T. (2008). Color trends and popularity in Auto China 2008 and Chinese urban area. *Research of Paints*, (151), 58–63.

Hardlock Industry Co. Ltd. (2014). Hard-lock nut. https://www.hardlock.co.jp/products/ hln/

Hashimoto, H. (2005). Automotive Technological Handbook (3), Design and Body, Chapter 6, CAE. Society of Automotive Engineers of Japan, Tosho-Shuppan-Sha.

Hashimoto, K. (2015). *Internal stress analysis of bolt with experiment and CAE analysis* [Master's thesis, Graduate School of Science and Engineering, Aoyama Gakuin University in Amasaka New JIT laboratory]

Hashimoto, K., Onodera, T., & Amasaka, K. (2014). Developing a Highly Reliable CAE Analysis Model of the mechanisms that cause bolt loosening in automobiles. *American Journal of Engineering Research, 3*(10), 178–187.

Hayashi, N., & Amasaka, K. (1990). Concept and progress of Toyota Production System: The physical distribution improvement in manufacture (Special lecture). *The Annual Conference of Japan Physical Distribution Management Association, 3.*

Hayashi, N., & Amasaka, K. (1990). *Concept and progress of Toyota Production System: The physical distribution improvement in manufacture (Special lecture).* The Annual Conference of Japan Physical Distribution Management Association, Tokyo.

Hayes, R. H., & Wheelwright, S. C. (1984). *Restoring our competitive edge: Competing through manufacturing.* Wiley.

Hayes, R. H., & Wheelwright, S. C. (1984). *Restoring Our Competitive Edge: Competing through Manufacturing.* Wiley.

Hifumi, S. (2006). *A study on a strategic advertising and publicity (animation): The proposal of "A-VUCMIN" for raising the volition of coming to the dealer's shop* [Master's thesis, Graduate School of Science and Engineering, Aoyama Gakuin University].

Hifumi, S. (2006). *A study on a strategic propaganda advertisement (animation): The proposal of "A-MUCMIN" Which raises the volition of coming to the auto shop* [Master's thesis, Aoyama Gakuin University].

Hirota, M., & Miyamoto, T. (2000). A proposal of the patent measuring model "A-PAT" (Amasakalab's patent) as the corporate strategy [*Thesis, School of Science and Engineering, Aoyama Gakuin University].*

Hoogervorst, J. A. P., Koopman, P. L., & van der Flier, H. (2005). Total Quality Management: The need for an employee-centered, Coherent approach. *The TQM Magazine, 17*(1), 92–106. doi:10.1108/09544780510573084

Ikeo, K. (2006). Feature-marketing innovation. *Japan Marketing Journal.*

Ikeo, K. E.-C. (2006). *Feature-marketing innovation.* Japan Marketing Journal.

Ishigaki, K., & Niihara, K. (2001). A study for objective evaluation of patent value – Proposal and validity of A-PPM, School of Science and Engineering [Thesis, Aoyama Gakuin University, Amasaka.]

Ishiguro, H., Kojima, T., & Matsuo, I. (2010). A Highly Compelling Direct Mail Method "PMOS-DM": Strategic applying of statistics and mathematical programming, *The 4th Spring Meeting of Japan Statistical Society, Poster Session (Student Poster Award).*

Ishiguro, H., & Amasaka, K. (2012). Proposal and effectiveness of a Highly Compelling Direct Mail Method: Establishment and deployment of PMOS-DM. *International Journal of Management & Information System, 16*(1), 1–10.

Ishiguro, H., & Amasaka, K. (2012a). Proposal and effectiveness of a Highly Compelling Direct Mail Method – Establishment and Deployment of PMOS-DM. *International Journal of Management & Information Systems, 16*(1), 1–10.

Ishiguro, H., & Amasaka, K. (2012b). Establishment of a Strategic Total Direct Mail Model to bring customers into auto-dealerships. *Journal of Business & Economics Research, 10*(8), 493–500. doi:10.19030/jber.v10i8.7177

Ishiguro, H., & Amasaka, K. (2012b). Proposal and effectiveness of a highly compelling direct mail method: Establishment and deployment of PMOS-DM. *International Journal of Management and Information Systems, 16*(1), 1–10.

ISO 9004. (2000). *Quality Management System- Guideline for performance.*

James, A. F., & Mona, J. F. (2004). *Service Management.* McGraw-Hill Companies Inc.

James, F. E., Roger, D. B., & Paul, W. M. (2006). *Consumer Behavior.* Dryden Press Inc.

Japan Automobile Manufacturers Association (JAMA). (2005). *Automobile Sale Numbers of Japan*. JAMA. (https://www.jama.or.jp/stats/product/index.html)

Japan Patent Office. (2005). *Patent Evaluation Index*. JPO. http:// www.jpo.go.jp/

Japan Quality Award Committee. (2003). *Application form & instruction*. JQA Press.

Jeffrey, F. R. and Bernard, J. J. (2005). Best face forward, *Diamond Harvard Business Review*, 62-77.

Jeffrey, K. L. (2004). *The Toyota Way*. McGraw-Hill.

Joiner, B. L. (1994). Fourth generation management: The new business consciousness. Joiner associates, Inc., Louisville, GA.

Joiner, B. L. (1994). Fourth Generation Management: The New Business Consciousness. Joiner Associates, Inc., McGraw-Hill, New York.

Joiner, B. L. (1994). *Fourth generation Management: The new business consciousness*. Joiner Associates, Inc.

Jonker, J. J., Piersma, N., & Potharst, R. (2006). A Decision Support System for Direct Mailing Decision. *Decision Support Systems*, *42*(2), 915–925. doi:10.1016/j.dss.2005.08.006

JUSE. (2006). *Quality management survey: JUSE Home page*. JUSE. https://www.juse.or.jp/

Kakkar, S., & Narag, A. S. (2007). Recommending a TQM model for Indian organizations. *The TQM Magazine*, *19*(4), 328–353. doi:10.1108/09544780710756232

Kameike, M., Ono, S., & Nakamura, K. (2000). The Helical Seal: Sealing Concept and Rib Design. *Sealing Technology International*, *77*(77), 7–11. doi:10.1016/S1350-4789(00)88559-1

Kaneta, H. (2005). *The research on construction of the patent evaluation model* [Thesis, Graduate School of Science and Engineering, Aoyama Gakuin University].

Kaneta, Y., & Kuniyoshi, M. (2003). A Study on Establishment of the Qualitative Valuation Modeling of Patent Value [Bachelor thesis, School of Science and Engineering, Aoyama Gakuin University].

Kawahara, F., Kogane, Y., Amasaka, K., & Ouchi, N. (2016). A skill map based on an analysis of experienced workers' intuition and knowhow. *IOSR Journal of Business and Management*, *18*(12), 80–85.

Kawasaki, S., & Yoshizawa, N. (2011). *Customer database marketing for the improvement in CS: The multivariate analysis of the customer information for raising commodity value* [Bachelor thesis, School of Science and Engineering, Department of Industrial and System Engineering].

Kevin, G. R., & David, K. (2001, July). *Discovering New Value in Intellectual Property*, Diamond. *Harvard Business Review*, 98–113.

Kimura, T., Uesugi, Y., Yamaji, M., & Amasaka, K. (2007). A study of Scientific Approach Method for Direct Mail, SAM-DM: Effectiveness of attracting customer utilizing Advanced TMS. *Proceedings of the 5th Asian Network for Quality Congress, Hyatt Regency*, Incheon, Korea.

Kishi, S., Tanaka, H., & Shimamura, K. (2000). *Theory of modern advertising*. Yuhikaku Publishers.

Kobayashi, T., & Shimamura, K. (1997). *New edition new Ad*. Dentsu.

Kobayashi, T., Yoshida, R., Amasaka, K., & Ouchi, N. (2015*)*. A study for creating an vehicle exterior design at teaching in different customers. *Hong Kong International Conference on Engineering and Applied Science*, Hong Kong.

Kobayashi, T., Yoshida, R., Amasaka, K., & Ouchi, N. (2016). A statistical and scientific approach to deriving an attractive exterior vehicle design concept for indifferent customers. *Journals International Organization of Scientific Research, 18*(12), 74–79.

Koizumi, K., Kanke, R., & Amasaka, K. (2013). Research on automobile exterior color and interior color matching. *International Journal of Engineering Research and Applications, 4*(8), 45–53.

Koizumi, K., Kanke, R., & Amasaka, K. (2014a). Research on Automobile Exterior Color and Interior Color Matching. *International Journal of Engineering Research and Applications, 4*(8), 45–53.

Koizumi, K., Kawahara, S., Kizu, Y., & Amasaka, K. (2013). A Bicycle Design Model based on young women's fashion combined with CAD and Statistical Science. *Journal of China-USA Business Review, 12*(4), 266–277.

Koizumi, K., Kawahara, S., Kizu, Y., & Amasaka, K. (2013). A Bicycle Design Model based on young women's fashion combined with CAD and statistics. *China-USA Business Review, 12*(4), 266–277.

Koizumi, K., Muto, M., & Amasaka, K. (2014b). Creating Automobile Pamphlet Design Methods: Utilizing both biometric testing and statistical science. *Journal of Management, 6*(1), 81–94.

Kojima, T., & Amasaka, K. (2011). The Total Quality Assurance Networking Model for preventing defects: Building an effective quality assurance system using a Total QA Network. *International Journal of Management & Information Systems., 15*(3), 1–10. doi:10.19030/ijmis.v15i3.4637

Kojima, T., Kimura, T., Yamaji, M., & Amasaka, K. (2010). Proposal and development of the Direct Mail Method "PMCI-DM" for effectively attracting customers. *International Journal of Management & Information Systems, 14*(5), 15–22. doi:10.19030/ijmis.v14i5.9

Kotler, F. (1999). Kotler marketing. The Free Press, a Division of Simon & Schuster Inc., New York.

Kotler, P., & Keller, K. L. (2006). *Marketing Management (12thEdition)*. Pearson Prentice Hall.

Koyama, H., Okajima, R., Todokoro, T., Yamaji, M., & Amasaka, K. (2010). Customer behavior analysis using motion pictures: Research on Attractive Flier Design Method, *China-USA. Business Review (Federal Reserve Bank of Philadelphia), 9*(10), 58–66.

Koyama, H., Okajima, R., Todokoro, T., Yamaji, M., & Amasaka, K. (2010). Customer behavior analysis using motion pictures: Research on attractive flyer design method. *China-USA Business Review, 9*(10), 58–66.

Kozaki, T., Oura, A., & Amasaka, K. (2012). Establishment of TQM Promotion Diagnosis Model "TQM-PDM" for strategic quality management. *China-USA Business Review, 10*(11), 811–819.

Kozaki, T., Yamada, H., & Amasaka, K. (2012). A Highly Reliable Development Design CAE Analysis Model: A Precision CAE Analysis Approach to automotive bolt tightening. *The International Business & Economics Research Journal, 10*(1), 1–9.

Kubimura, R., & Murata, S. (1969). *Theory of PR*. Yuhikaku.

Kume, H. (1999). Quality Management in Design Development. JUSE (Union of Japanese Scientists and Engineers), Tokyo.

Kusama, M. (1992). Analysis of Engineering Trend Utilizing Patent Information, *The 29th Information Science Technology Conference*, (pp. 177-182). Research Gate.

Lagrosen, S. (2004). Quality management in global firms. *The TQM Magazine, 16*(6), 396–402. doi:10.1108/09544780410563310

Leo, J. D. V., Annos, N., & Oscarsson, J. (2004). Simulation based decision support for manufacturing system Life Cycle Management. *Journal of Advanced Manufacturing Systems, 3*(2), 115–128. doi:10.1142/S0219686704000454

Ljungström, M. (2005). A model for starting up and implementing Continuous improvements and work development in practice. *The TQM Magazine, 17*(5), 385–405. doi:10.1108/09544780510615915

Lopez, A. M., Nakamura, K., & Seki, K. (1997). *A study on the sealing characteristics of lip seals with helical ribs.* Proceedings of the 15th International Conference of British Hydromechanics Research Group Ltd., Fluid Sealing.

Magoshi, R., Fujisawa, H., & Sugiura, T. (2003). Simulation technology applied to vehicle development. *Journal Society of Automotive Engineers of Japan, 53*(3), 95–100.

Manzoni, A., & Islam, S. (2007). Measuring collaboration effectiveness in globalized supply networks: A data envelopment analysis application. *International Journal of Logistics Economics and Globalization, 1*(1), 77–91. doi:10.1504/IJLEG.2007.014496

Media, M. (2006). *WRI: MRO today Home page.* Progressive Distributer. http://www.Progressivedistributor. com/mro_today_body.htm

Melewar, T. C., & Smith, N. (2003). The internet revolution: Some global marketing implications. *Marketing Intelligence & Planning, 21*(6), 363–369. doi:10.1108/02634500310499220

Miller, N. (1978). An analysis of disk brake squeal. *SAE Technical Paper 780332.*

Ministry of Land. (2015). *Infrastructure, Transport and Tourism.* MLit. https://www.mlit.go.jp/jidosha/carinf/rcl/data.html

Ministry of Land. (2015). *Ministry of Land Infrastructure, Transport and Tourism.* MLIT. https://www. mlit.go.jp/jidosha/carinf/rcl/data.html/

Miyashita, S., & Amasaka, K. (2014). Proposal of a New Vietnam Production Model (NVPM), a new integrated production system of Japan and Vietnam. *IOSR Journal of Business and Management, 16*(12), 18–25. doi:10.9790/487X-161211825

Mori, N. (1991). *Engineering of design: Software System of design engineering.* Asakura-Shoten.

Mori, N. (1991). *The design plan of the engineering soft system of a designing.* Asakura-Shoten.

Mori, N. (1993). *The research of the scientific method of a left brain designing.* Kaibundou.

Mori, N. (1996). *Designing with Left Brain: The research of the scientific methodology.* Kaibun-Dou.

Moriya, A., & Sugiura, K. (1999). *Collage Therapy: Esprit of Today.* Shibundou.

Motor Fan. (1997). *All of new-model "Aristo", a new model prompt report, 213,* 24-30. Motor Fan.

Motor Fan. (2000). *All of new-model "Celsior", a new model prompt report, 268,* 23-24. Motor Fan.

Murat, S. M., Hifumi, S., Yamaji, M., & Amasaka, K. (2008). Developing a strategic advertisement method "VUCMIN" to enhance the desire of customers for visiting dealers. *Proceedings of the International Symposium on Management Engineering,* Waseda University, Kitakyushu, Japan.

Murayama, Y. (1982). Analyzing CAID marketing review. Japan Research Center.

Murayama, Y. (1982). *Analyzing CAID marketing review.* Japan Research Center.

Muto, M., Miyake, R., & Amasaka, K. (2011). Constructing an Automobile Body Color Development Approach Model. *Journal of Management Science, 2,* 175–183.

Muto, M., Takebuchi, S., & Amasaka, K. (2013). Creating a New Automotive Exterior Design Approach Model: The relationship between form and body color qualities. *Journal of Business Case Studies, 9*(5), 367–374. doi:10.19030/jbcs.v9i5.8061

Nagaya, A., Matsubara, K., & Amasaka, K. (1998). *A Study on the customer tastes of automobile profile design (Special Lecture).* Union of Japanese Scientists and Engineers, The 28th Sensory Evaluation Sympojium, Tokyo .

Nagaya, A., Matsubara, K., & Amasaka, K. (1998). *A study on the customer tastes of automobile profile design (Special Lecture).* Union of Japanese Scientists and Engineers, The 28th Sensory Evaluation Sympojium, Tokyo.

Nagaya, A., Matsubara, K., & Amasaka, K. (1998). *A Study on the customer tastes of automobile profile design (Special Lecture).* Union of Japanese Scientists and Engineers, The 28th Sensory Evaluation Sympojium, Tokyo.

Nakamura, M., Kuniyoshi, M., Yamaji, M., & Amasaka, K. (2008). Proposal and validity of the product planning business model "A-POST": The application of text mining method to scooter exterior design. *Journal of Business Case Studies, 4*(9), 61–71. doi:10.19030/jbcs.v4i9.4808

Namikata, A., & Yano, Y. (2003). *Free answer questionnaire analysis for consumer-needs visualization: The example of the products project support of the vehicle* [Bachelor thesis, Aoyama Gakuin University].

Nezu, K. (1995). *The scenario of the U.S. manufacturing industry: Remarkable progress aimed at by CALS.* Kogyo-Chosakai Publishing.

Niemela, R., Rautio, S., Hannula, M., & Reijula, K. (2002). Work environment effects on labor productivity: An intervention study in a storage building. *American Journal of Industrial Medicine, 42*(4), 328–335. doi:10.1002/ajim.10119 PMID:12271480

Niiya, Y., & Matsuoka, F. (Eds.). (2001). *Foundation Lecture on the New Advertising Business.* Senden-Kaigi.

Nikkei Business. (1999). *Renovation of Shop, Product and Selling Method-targeting Young Customer by Nets*, 46-50. Nikkei Business.

Nikkei Business. (1999). Renovation of Shop, Product and Selling Method-targeting Young Customer by Nets. Nikkei Business.

Nikkei. (2002). *Corporate Innovation through Competitive Patent.* Nikkei.

Nomura, R., Shimura, T., Sakurai, Y., & Amasaka, K. (2016). *The elucidation of the loosening mechanism of automobile bolt nut conclusion by using experiments and CAE.* Japanese Operations Management and Strategy Association, The 6th Annual Technical Conference, Kobe University, Kobe.

Nomura, R., Hori, K., & Amasaka, K. (2015). Problem Prevention Method for product design based on predictive evaluation: A study of bolt-loosening mechanisms in automobile. *American Journal of Engineering Research, 4*(6), 174–178.

Nonobe, K., & Ibaraki, T. (2001). An improved tabu search method for the weighted constraint satisfaction problem. *INFOR, 39*(2), 131–151. doi:10.1080/03155986.2001.11732431

Nozawa, Y., Ito, T., & Amasala, K. (2013). High precision CAE analysis of automotive transaxle oil seal leakage. *China-USA Business Review, 12*(5), 363–374.

Nozawa, Y., Takahiro Ito, T., & Amasaka, K. (2013). High precision CAE analysis of Automotive transaxle oil seal leakage. *The China Business Review, 12*(5), 363–374.

NTT DATA Mathematical Systems Inc. (2006). *"Text Mining Studio Analysis" by NTT DATA.* MSI. https://www.msi.co.jp/solution/tmstudio/index.html

NTT DATA Mathematical Systems Inc. (2006). Text Mining Studio Analysis. NTT DATA. https://www.msi.co.jp/solution/tmstudio/index.html

Nunogaki, N., Shibata, K., Nagaya, A., Ohashi, T., & Amasaka, K. (1996). *A study of customers' direction about designing vehicle's profile.* The Japanese Society for Quality Control, The 26th annual conference, Gifu, Japan.

Nunogaki, N., Shibata, K., Nagaya, A., Ohashi, T., & Amasaka, K. (1996). *A study of customers' direction about designing vehicle's profile: A deployment of "Design SQC" for design business process.* The Japanese Society for Quality Control.

Nunogaki, N., Shibata, K., Nagaya, A., Ohashi, T., & Amasaka, K. (1996). *A study on the "customers' preference" of "Automobile profile design": Development of Design SQC which is useful for Design business process.* The 26th Annual Conference of JSQC, Gifu, Japan.

Ogura, M., Hachiya, T., Masubuchi, K., & Amasaka, K. (2014). Attention-grabbing train car advertisements, *International Journal of Engineering Research and Applications, 4*(1) (Version 2), 167-175.

Ogura, M., Hachiya, T., & Amasaka, K. (2013). A Comprehensive Mixed Media Model for boosting automobile dealer visits. *The China Business Review, 12*(3), 195–203.

Ogura, M., Hachiya, T., & Amasaka, K. (2013). A omprehensive Mixed Media Model for boosting automobile dealer visits. *The China Business Review, 12*(3), 195–203.

Ogura, M., Hachiya, T., & Amasaka, K. (2013b). Attention-grabbing train car advertisements. *The China Business Review, 12*(3), 195–203.

Ogura, M., Hachiya, T., Masubuchi, K., & Amasaka, K. (2014). Attention-grabbing train car advertisements. *International Journal of Engineering Research and Applications, 4*(1), 56–64.

Ohno, T. (1977). *Toyota Production System.* Diamond-sha.

Ohno, T. (1978). *Toyota Production System: Beyond Large-scale production.* I Japanese.

Okabe, Y., Yamaji, M., & Amasaka, K. (2007). Research on the Automobile Package Design Concept Support Methods "CS-APDM": Customer Science approach to achieve CS for vehicle exteriors and package design. *Journal of Japan Society for Production Management, 13*(2), 51–56.

Okada, A., Kijima, M., & Moriguchi, T. (Eds.). (2001). *The mathematical model of marketing.* Asakura-Shoten. (in Japanese)

Okazaki, R., Suzuki, M., & Amasaka, K. (2000). *Study on the sense of values by age using Design SQC.* The Japan Society for Production Management, The 11th Annual Technical Conference, Okayama University, Okayama, Japan.

Okazaki, R., Suzuki, M., & Amasaka, K. (2000). *Study on the sense of values by age using Design SQC.* The Japan Society for Production Management, The 11th Annual Technical Conference, Okayama University, Okayama.

Okutomi, H. (2011). A study on establishment of the structure model on relationship between CS and CL: An example of Japan's auto-dealerships (Bachelor thesis), *School of Science and Engineering, Department of Industrial and System Engineering, Aoyama Gakuin University, Amasaka New JIT Laboratory, Sagamihara,* Kanagawa, Japan. (in Japanese)

Okutomi, H., & Amasaka, K. (2012). Researching customer satisfaction and loyalty to boost marketing effectiveness: A look at Japan's auto dealerships. *Proceedings of the 2nd International Symposium on Operations Management Strategy.* Japanese Operations Management and Strategy Association.

Okutomi, H., & Amasaka, K. (2013). Researching Customer Satisfaction and Loyalty to boost marketing effectiveness: A look at Japan's auto-dealerships. *International Journal of Management & Information Systems*, *17*(4), 193–200. doi:10.19030/ijmis.v17i4.8093

Onodera, T., & Amasaka, K. (2012). Automotive bolts tightening analysis using contact stress simulation: Developing an Optimal CAE Design Approach Model. *Journal of Business & Economics Research*, *10*(7), 435–442. doi:10.19030/jber.v10i7.7148

Parasuraman, A., Zeihaml, V. A., & Berry, L. L. (1988). SERVQUAL: A multiple-item scale for measuring consumer perceptions of service quality. *Journal of Retailing*, *64*(1), 12–40.

Piersma, N., & Jonker, J. J. (2004). Determining the Optimal Direct Mailing Frequency. *European Journal of Operational Research*, *158*(1), 173–182. doi:10.1016/S0377-2217(03)00349-7

Pires, S., & Cardoza, G. (2007). A study of new supply chain management practices in the Brazilian and Spanish auto-industries. *International Journal of Automotive Technology and Management*, *7*(1), 72–87. doi:10.1504/IJATM.2007.013384

Power, J. D., & Associates. (1998). *Vehicle Dependability Study*. JD Power. https://www. jdpower. com/releases/80401car.html/

PowerJD. (2009). UTL: http;//jdpower.com/

Prajogo, D. (2006). Progress of quality management practices in Australian manufacturing firms. *The TQM Magazine*, *18*(5), 501–513. doi:10.1108/09544780610685476

Ramarapu, N. K., Mehra, S., & Frolick, M. N. (1995). A comparative analysis and review of JIT implementation research. *International Journal of Operations & Production Management*, *15*(1), 38–49. doi:10.1108/01443579510077188

Roger, G. S., & Flynn, B. B. (2001). *High performance manufacturing*. John Wiley & Sons, Inc.

Sakai, H., & Amasaka, K. (2003). *Construction of "V-IOS" for promoting intelligence operator: Development and effectiveness for visual manual format*. The Japan Society for Production Management, The 18th Annual Conference, Nagasaki Institute of Applied Science, Nagasaki, Japan.

Sakai, H., & Amasaka, K. (2003). *Construction of "V-IOS" for promoting intelligence operator: Development and effectiveness for visual manual format*. The Japan Society for Production Management.

Sakai, H., & Amasaka, K. (2008). Human-Integrated Assist Systems for intelligence operators, Encyclopedia of Networked and Virtual Iorganizations, 2(G-Pr), 678-687.

Sakai, H., & Amasaka, K. (2008). Human-Integrated Assist Systems for intelligence operators, Encyclopedia of Networked and Virtual Iorganizations, II(G-Pr), 678-687.

Sakai, H., & Amasaka, K. (2008b). Human-Integrated Assist System for Intelligence Operators, Encyclopedia of Networked and Virtual Organization, II(G-Pr), 678-687.

Sakai, H., & Amasaka, K. (2005). V-MICS, Advanced TPS for strategic production administration: Innovative maintenance combining DB and CG. *Journal of Advanced Manufacturing Systems*, *4*(6), 5–20. doi:10.1142/S0219686705000540

Sakai, H., & Amasaka, K. (2006). Strategic HI-POS, Intelligence Production Operating System: Applying Advanced TPS to Toyota's Global Production Strategy, WSEAS (World Scientific and Engineering and Society). *Transactions on Advances in Engineering Education*, *3*(3), 223–230.

Sakai, H., & Amasaka, K. (2006). Strategic *HI-POS*, Intelligence Production Operating System: Applying *Advanced TPS* to Toyota's global production strategy. *WSEAS Transactions on Advances in Engineering Education*, *3*(3), 223–230.

Sakai, H., & Amasaka, K. (2006a). TPS-LAS Model using process layout CAE system at Toyota, Advanced TPS: Key to global production strategy *New JIT*. *Journal of Advanced Manufacturing Systems*, 5(2), 1–14. doi:10.1142/S0219686706000790

Sakai, H., & Amasaka, K. (2006b). Strategic HI-POS, intelligence production operating system: Applying Advanced TPS to Toyota's global production strategy. *WSEAS Transactions on Advances in Engineering Education*, 3(3), 223–230.

Sakai, H., & Amasaka, K. (2007). Human digital pipeline method using total linkage through design to manufacturing. *Journal of Advanced Manufacturing Systems*, 6(2), 101–113. doi:10.1142/S0219686707000929

Sakai, H., & Amasaka, K. (2007a). The robot reliability design and improvement method and Advanced Toyota Production System. *The Industrial Robot*, 34(4), 310–316. doi:10.1108/01439910710749636

Sakai, H., & Amasaka, K. (2007b). Development of a robot control method for curved seal extrusion for high productivity in an advanced Toyota production system. *International Journal of Computer Integrated Manufacturing*, 20(5), 486–496. doi:10.1080/09511920601160262

Sakai, H., & Amasaka, K. (2008). Demonstrative verification study for the next generation production model: Application of the Advanced Toyota Production System. *Journal of Advanced Manufacturing Systems*, 7(2), 195–219. doi:10.1142/S0219686708001577

Sakai, H., & Amasaka, K. (2009). Proposal and demonstration of V-MICS-EM by digital Engineering: Robot operation and maintenance by utilizing Visual Manual. *International Journal of Manufacturing Technology and Management*, 18(4), 344–355. doi:10.1504/IJMTM.2009.027769

Sakalsis, M. M., Hifumi, S., Yamaji, M., & Amasaka, K. (2008). Developing a Strategic Advertisement Method "VUC-MIN" to enhance the desire of customers for visiting dealers. *Proceedings of the International Symposium on Management Engineering*, Waseda University, Kitakyushu, Japan.

Sakalsiz, M. M. (2009). *The proposal of New Turkish Production System utilizing Advanced TPS*. [Master's thesis, Graduate School of Science and Engineering, Aoyama Gakuin University].

Sakatoku, T. (2006). *Partnering performance measurement for assembly maker and suppliers* [A Thesis for the master's degree, Graduate School of Science and Engineering, Aoyama Gakuin University].

Sasaki, S. (1972). Collection and analysis of reliability information in automotive industries. *The 2nd reliability and maintainability symposium*. Union of Japanese Scientist and Engineers.

Sato, Y., Toda, A., Ono, S., & Nakamura, K. (1999). A study of the sealing mechanism of radial lip seal with helical robs–measurement of the lubricant fluid behavior under sealing contact. *SAE Technical Paper Series*.

Seuring, S., Muller, M., Goldbach, M., & Schneidewind, U. (Eds.). (2003). *Strategy and Organization in Supply Chains*. Physica-Verlag Heidelberg.

Shan, H. (2012). *The proposal of China Local Automobile Manufacturer's Production System*, [Thesis, Graduate School of Science and Engineering, Aoyama Gakuin University].

Shan, H., Yeap, Y. S., & Amasaka, K. (2011). *Proposal of a New Malaysia Production Model "NMPM": A new integrated production system of Japan and Malaysia*. International Conference on Business Management 2011, Miyazaki Sangyo-Keiei University, Miyazaki, Japan.

Shan, H., Yeap, Y. S., & Amasaka, K. (2011). *Proposal of a New Malaysia Production Model "NMPM": a new integrated production system of Japan and Malaysia*. Proceedings of International Conference on Business Management 2011, Miyazaki Sangyo-Keiei University, Japan.

Shimakawa, K., Katayama, K., Oshima, K., & Amasaka, K. (2006). *Proposal of Strategic marketing model for customer value maximization.* The Japan Society for Production Management, The 23th Annual Technical Conference, Osaka, Japan.

Shimamura, K., & Kobayashi, T. (1997). *New edition: New Advertisements.* Dentsu.

Shimizu, H., & Amasaka, K. (1975). Quality assurance of low-speed steering force. *Quality Control* (an e*xtra edition), in JUSE (Union of Japanese Scientists and Engineers), 26* (11), 42-46.

Shimizu, K. (2004). *Theory and strategy of advertisement.* Sousei-Sha.

Shimizu, K. (2004). *Theory and Strategy of Advertisement.* Sousei-Sha. (in Japanese)

Shimura, T., & Sakurai, Y. (2015). *The elucidation of the nut loosening mechanism: Combined use of the three-dimensional system experiment by actual vehicle data and CAE analysis* [Bachelor's degree thesis, School of Science & Engineering, Aoyama Gakuin University]

Shinogi, T., Aihara, S., & Amasak, K. (2014). Constructing an Automobile Color Matching Model (ACMM). *IOSR Journal of Business and Management, 16*(7), 7–14. doi:10.9790/487X-16730714

Shinohara, A., Sakamoto, H., & Shimizu, Y. (1996). *Invitation to Kansei Engineering.* Morikita-Shuppan.

Siang, Y. Y., Sakalsiz, M. M., & Amasaka, K. (2010). Proposal of New Turkish Production System (NTPS): Integration and evolution of Japanese and Turkish production system. *Journal of Business Case Study, 6*(6), 69–76. doi:10.19030/jbcs.v6i6.260

Smith, D. A. (2009). Online accessibility concerns in shaping consumer relationships in the automotive industry. *Online Information Review, 33*(1), 77–95. doi:10.1108/14684520910944409

Suzuki, M., Okazaki, R., & Amasaka, K. (2000). *The research studies of the value by the generation by employing Design SQC: The development of Customer Science by utilizing Science SQC.*, The Japan Society for Production Management.

Suzuki, M., Arafune, S., & Wadachi, M. (Eds.). (2005). *Encyclopedia of Physics.* Asakura-Shoten. (in Japanese)

Suzumura, H., Sugimoto, Y., Furusawa, N., Amasaka, K., Eri, Y., Asaji, K., Furugori, N., & Fukumoto, K. (1998). *The development of working conditions for aging worker on assembly line (#1).* The Japan Society for Production Management, The 8th Annual Technical Conference; 1998, Kyushu-Sangyo University, Japan.

Takahashi, T., Ueno, T., Yamaji, M., & Amasaka, K. (2010). Establishment of Highly Precise CAE Analysis Model using automotive bolts. *The International Business & Economics Research Journal, 9*(5), 103–113. doi:10.19030/iber.v9i5.574

Takebuchi, S., Asami, H., Nakamura, T., & Amasaka, K. (2010). *Creation of Automobile Exterior Color Design Approach Model "A-ACAM".* The 40th International Conference on Computers & Industrial Engineering, Awaji Island, Japan.

Takebuchi, S., Asami, H., & Amasaka, K. (2012b). An Automobile Exterior Design Approach Model linking form and color. *China-USA Business Review, 11*(8), 1113–1123.

Takebuchi, S., Asami, H., & Amasaka, K. (2012b). An Automobile Exterior Design Approach Model linking form and color. *Journal of China-USA Business Review, 11*(8), 1113–1123.

Takebuchi, S., Nakamura, T., Asami, H., & Amasaka, K. (2012a). The Automobile Exterior Color Design Approach Model. *Journal of Japan Industrial Management Association, 62*(6E), 303–310.

Taketomi, T., Horiguchi, Y., & Hirabayashi, T. (1997). Evaluation of Intellectual Property Value and Recovery of Invested Resources. *R & D Management, 2*, 32–43.

Takimoto, H., Ando, T., Yamaji, M., & Amasaka, K. (2010). The proposal and validity of the Customer Science Dual System, *China-USA. Business Review (Federal Reserve Bank of Philadelphia)*, *9*(3), 29–38.

Tanaka, M., Miyagawa, H., Asaba, E., & Hongo, K. (1981). Application of the finite element method of bolt-nut joints: Fundamental studies on analysis of bolt-nut joints using the finite element method. *Bulletin of the JSME*, *24*(192), 1064–1071. doi:10.1299/jsme1958.24.1064

Taylor, D., & Brunt, D. (2001). *Manufacturing Operations and Supply Chain Management - Lean Approach*, Boston, USA: Thomson Learning.

Taylor, D., & Brunt, D. (2001). *Manufacturing operations and supply chain management - Lean approach*, Thomson Learning, London.

Taylor, D., & Brunt, D. (2001). *Manufacturing operations and supply chain management–The lean approach*, Thomson Leaning, London.

Taylor, D., & Brunt, D. (2001). *Manufacturing operations and supply chain management–The lean approach–*, Thomson Leaning, London.

Taylor, D., & Brunt, D. (2001). *Manufacturing operations and supply chain management- Lean Approach* (1st ed.). Thomson Learning.

The Japan Machinery Federation and the Japan Society of Industrial Machinery Manufacturers. (1995a). *Research Report: Advanced Technology Introduction in Machinery Industry*. The Japan Machinery Federation and the Japan Society of Industrial Machinery Manufacturers.

The Japan Machinery Federation., & Japan Society of Industrial Machinery Manufacturers. (1995b). Research Report: Production System Model Considering Aged Workers, Tokyo, 1-2.

Toyoda, S., Nishio, Y., & Amasaka, K. (2014). *Matching methods of the car proportion, form, color based on the customer sensitivity approaches and that validity*. Japanese Operations Management and Strategy Association, The 6th Annual Conference on Takushoku University, Tokyo.

Toyoda, S., Koizumi, K., & Amasaka, K. (2015b). Creating a Bicycle Design Approach Model based on fashion styles, *IOSR (International Organization of Scientific Research). Journal of Computational Engineering*, *17*(3), 1–8.

Toyoda, S., Nishio, Y., & Amasaka, K. (2015). Creating a Vehicle Proportion, Form, and Color Matching Model. *International Organization of Scientific Research*, *III*(3), 9–16.

Toyoda, S., Nishio, Y., & Amasaka, K. (2015a). Creating a Vehicle Proportion, Form & Color Matching Model. *Journals International Organization of Scientific Research*, *17*(3), 9–16.

Toyoda, S., Nishio, Y., & Amasaka, K. (2015a). Creating a Vehicle Proportion, Form, and Color Matching Model. *Journals International Organization of Scientific Research*, *17*(3), 9–16.

Toyota Motor Corp. (1987). *Creation Unlimited, 50 Years History of Toyota Motor Corp.* Toyota.

Toyota Motor Corp. (1993). Toyota Technical Review: Special edition for SQC at Toyota. Toyota Motor Corp.

Toyota Motor Corp. (1999). *Toyota's TQM Activities-To creative better cars.* Toyota Motor Corp.

Toyota Motor Corp., & Toyota Motor Kyushu Corp. (1994). Development of a new automobile assembly line. *Business Report Awarded with Okochi Prize*, *1993*(40th), 377–381.

Toyota Motor Corporation. (1987). *Creation is infinite: The history of Toyota-50 years.* Toyota Motor Corporation.

Toyota Motor Corporation. (1993). Toyota Technical Review [hosted by Amasaka, K.]. *SQC at Toyota, 43*(Special issue), 1–172.

Toyota Motor Corporation. (1993). *Toyota Technical Review: SQC at Toyota.* Toyota.

Toyota Motor Corporation. (1996). *The Toyota Production System.* International Public Affairs Division Operations Management Consulting Division.

Toyota Motor Corporation. (1997). *Creation is infinite: The history of the Toyota Motor Corporation 50 year. (in Japanese).* Toyota.

Toyota Motor Corporation. (1997). *Creation is infinite: The history of the Toyota Motor Corporation 50 year.* Toyota.

Tsujitani, M., & Amasaka, K. (1993). Analysis of Enumerated Data (1): Logit Transformation and Graph Analysis [in Japanese]. *Quality Control, 44*(4), 61–66.

Tsunoi, M., Anabuki, K., Yamaji, M., & Amasaka, K. (2009). *A study of Patent Evaluation Method "A-PPM" for Corporate Strategy.* The 11th Annual Conference of Japan Society of Kansei Engineering, Shibaura Institute of Technology, Tokyo..

Tsunoi, M., Yamaji, M., & Amasaka, K. (2010). A study of building an Intellectual Working Value Improvement Model, IWV-IM. *The International Business & Economics Research Journal, 9*(11), 79–84. doi:10.19030/iber.v9i11.33

Uchida, K., Tsunoi, M., & Amasaka, K. (2012). Creating Working Value Evaluation Model, WVEM. *International Journal of Management & Information Systems, 16*(4), 299–306.

Ueno, H. (2003). Formation of a strategic patent group, and a patent portfolio strategy. *Japan Marketing Journal, 80*, 25–35.

Ueno, T., Yamaji, M., Tsubaki, H., & Amasaka, K. (2009). Establishment of bolt tightening simulation system for automotive industry: Application of the Highly Reliable CAE Model. *The International Business & Economics Research Journal, 8*(5), 57–67.

Umezawa, K. (1999). Miscellaneous Impressions of Intellectual Property Rights. *Intellectual Property Management, 49*(3), 353–36.

Umezawa, U., & Amasaka, K. (1999). "Partnering" as the platform of quality management in Toyota Group, *Operation Research . The Operations Research Society of Japan, 44*(10), 24–35.

Umezawa, Y. (1994). Essential studies of re-engineering from organization theory. *Soshiki Kagaku, 28*(1), 4–20.

Umezawa, Y., & Amasaka, K. (1999). Partnering as a platform of Quality Management in Toyota Group. [in Japanese]. *Journal of the Operations Research Society of Japan, 44*(10), 24–35.

Umezawa, Y., & Amasaka, K. (1999). Partnering as platform of quality management in Toyota Group, *Operations Research . Operations Research Society of Japan, 44*(10), 560–571.

Vecchio, D., Sasco, A. Jr, & Cann, I. C. (2003). Occupational risk in health care and research. *American Journal of Industrial Medicine, 43*(4), 369–397. doi:10.1002/ajim.10191 PMID:12645094

Wakaizumi, T. (2005). *Proof theory research on the Japanese quality management consultation model (A-QMDS) construction* [Master's thesis, Graduate school of Science and Engineering, Amasaka New JIT Laboratory].

Whaley, R. C., Petitet, A., & Dongarra, J. J. (2000). *Automated empirical optimization of software and the ATLAS project. Technical report, University of Tennessee.* Department of Computer Science.

Womack, J. P., & Jones, D. (1994). *From lean production to the lean enterprise.* Harvard Business Review.

Womack, J. P., & Jones, D. T. (1994). From Lean Production to the Lean Enterprise. *Harvard Business Review*, (March-April), 93–103.

Womack, J. P., Jones, D. T., & Roos, D. (1991). *The machine that changed the world – The story of Lean Production*. Rawson.

Womack, J. P., Jones, D., & Roos, D. (1990). *The machine that change the world – The story of Lean Production*. Rawson/Harper Perennial.

Yamada, H., & Amasaka, K. (2011). Highly-reliable CAE analysis approach-Application in automotive bolt analysis. *China-USA Business Review*, *10*(3), 199–205.

Yamaji, M., & Amasaka, K. (2006). New Japan Quality Management Model, Hyper-cycle model "QA & TQM Dual System": Implementation of New JIT for strategic Management technology. *Proceedings of the International Manufacturing Leaders Forum*. Springer.

Yamaji, M., & Amasaka, K. (2007). Proposal and validity of Global Intelligence Partnering Model "GIPM-CS" for corporate strategy. *International IFIP TC 5.7 Conference on Advanced in Production Management System*. Springer.

Yamaji, M., & Amasaka, K. (2007a). Proposal and validity of Global Intelligence Partnering Model for Corporate Strategy, GIPM-CS. *International IFIP TC 5.7 Conference on Advanced in Production Management System*. Springer.

Yamaji, M., & Amasaka, K. (2007b). Proposal and validity of Global Intelligence Partnering Model for Corporate Strategy, GIPM-CS. *International IFIP-TC 5.7 Conference on Advanced in Production Management System*. Springer.

Yamaji, M., & Amasaka, K. (2009). Strategic Productivity Improvement Model for White-Collar Workers Employing Science TQM. *The Journal of Japanese Operations Management and Strategy, (JOMS-Special Issue), 1*(1), 30-43.

Yamaji, M., & Amasaka, K. (2008). New Japan Quality Management Model: Implementation of New JIT for strategic management technology. *The International Business & Economics Research Journal*, *7*(3), 107–114.

Yamaji, M., & Amasaka, K. (2009). An Intelligence Design Concept Method utilizing Customer Science. *The Open Industrial and Manufacturing Engineering Journal*, *2*(1), 10–15. doi:10.2174/1874152500902010021

Yamaji, M., & Amasaka, K. (2009). Proposal and validity of Intelligent Customer Information Marketing Model: Strategic development of *Advanced TMS, The Academic Journal of China - USA. Business Review (Federal Reserve Bank of Philadelphia)*, *8*(8), 53–62.

Yamaji, M., & Amasaka, K. (2009). Strategic Productivity Improvement Model for white-collar workers employing Science TQM, *JOMS. The Japanese Operations Management and Strategy Association*, *1*(1), 30–43.

Yamaji, M., & Amasaka, K. (2009). Strategic Productivity Improvement Model for white-collar workers employing Science TQM, *JOMS. The Journal of Japanese Operations Management and Strategy*, *1*(1), 30–46.

Yamaji, M., & Amasaka, K. (2009). Strategic productivity improvement model for white-collar workers employing Science TQM. *The Journal of Japanese Operations Management and Strategy*, *1*(1), 30–46.

Yamaji, M., & Amasaka, K. (2009a). Strategic Productivity Improvement Model for white-collar workers employing Science TQM. *The Journal of Japanese Operations Management and Strategy*, *1*(1), 30–46.

Yamaji, M., & Amasaka, K. (2009b). Proposal and validity of Intelligent Customer Information Marketing Model: Strategic development of *Advanced TMS, China - USA. Business Review (Federal Reserve Bank of Philadelphia)*, *8*(8), 53–62.

Yamaji, M., Hifumi, S., Sakalsiz, M. M., & Amasaka, K. (2010). Developing a strategic advertisement method "VUC-MIN" to enhance the desire of customers for visiting dealers. *Journal of Business Case Studies*, *6*(3), 1–11. doi:10.19030/jbcs.v6i3.871

Yamaji, M., Sakai, H., & Amasaka, K. (2006). Intellectual working Value Improvement Model utilizing Advanced TPS: Applying New JIT, Toyota's global production strategy. *Proceedings of the International Applied Business Research Conference, Cancun, Mexico.*

Yamaji, M., Sakai, H., & Amasaka, K. (2007). Evolution of technology and skill in production workplaces utilizing Advanced TPS. *Journal of Business & Economics Research*, *5*(6), 61–68.

Yamaji, M., Sakai, H., & Amasaka, K. (2007b). Evolution of technology and skills in production workplaces *ss &.* *Economic Research Journal*, *5*(6), 61–68.

Yamaji, M., Sakatoku, T., & Amasaka, K. (2008). Partnering Performance Measurement "PPM-AS" to strengthen corporate management of Japanese automobile assembly makers and suppliers. *International Journal of Electronic Business Management*, *6*(3), 139–145.

Yanagisawa, K., Yamazaki, M., Yoshioka, K., & Amasaka, K. (2013). Comparison of experienced and inexperienced machine workers. *International Journal of Operations and Quantitative Management*, *19*(4), 259–274.

Yazaki, K., Takimoto, H., & Amasaka, K. (2013). Designing vehicle form based on subjective customer impressions. *Journal of China-USA Business Review*, *12*(7), 728–734.

Yazaki, K., Tanitsu, H., Hayashi, H., & Amasaka, K. (2012). A model for design auto- instrumentation to appeal to young male customers. *Journal of Business Case Studies*, *8*(4), 417–426. doi:10.19030/jbcs.v8i4.7035

Yeap, Y. S., Murat, M. S., & Amasaka, K. (2010). Proposal of New Turkish Production System, NTPS: Integration and evolution of Japanese and Turkish Production System. *Journal of Business Case Study*, *6*(6), 69–76.

Yeap, Y. S., Sakalsiz, M. M., & Amasaka, K. (2010). Proposal of New Turkish Production System, NTPS, Integration and Evolution of Japanese and Turkish Production System. *Journal of Business Case Study*, *6*, 69–76.

About the Author

Kakuro Amasaka was born in Aomori Prefecture, Japan, on May 5, 1947. He received a Bachelor of Engineering degree from Hachinohe National College of Technology, Hachinohe, Japan, in 1968, and a Doctor of Engineering degree specializing in Precision Mechanical and System Engineering, Statistics and Quality Control from Hiroshima University, Hiroshima, Japan, in 1997. Since joining Toyota Motor Corp., Japan, in 1968, Dr. Amasaka worked as a quality management consultant for many divisions. He was an engineer and manager of the Production Engineering Div., Quality Assurance Div., Overseas Engineering Div., Manufacturing Div., and TQM Promotion Div. (1968-1997), and the General Manager of the TQM Promotion Div. (1998-2000). Dr. Amasaka became a professor of the College of Science and Engineering, and the Graduate School of Science and Engineering at Aoyama Gakuin University, Tokyo, Japan in April 2000. Now, he is a professor emeritus of Aoyama Gakuin University (2016-present). His specialties include: production engineering (Just in Time and Toyota Production System), statistical science, multivariate analysis, reliability engineering and information processing engineering.

Index

A

Advanced TDS, 96-97, 107, 114, 350, 385, 394, 430

automobile manufacturing 24-25, 30, 114, 155, 159, 164, 220, 339, 365

automobile profile 101, 120, 244, 248, 250, 252, 257-258, 261-262, 269-270

Automobile Sales 106, 169, 172-173, 175, 183-184, 192, 394, 407, 409

Auto-profile design 258, 269, 272, 292

auto-sales marketing 169-170

B

bolt-nut loosening 160, 296-298, 304, 307-308, 311, 313, 315-316

bolt-nut tightening 131, 296-298, 300-306, 309-311, 313-316

business process 7, 16, 18-19, 22-23, 71, 83, 94, 106-107, 114, 117-118, 120, 128, 131, 140, 157, 175, 197, 199, 212, 229, 244-245, 248, 258, 265, 270, 352, 354, 400

C

CAE Analysis 103, 118, 123, 130-133, 217-219, 296, 298, 304, 306, 308-313, 316

Corporate Engineering 113, 119-120, 133

corporate strategy 14, 25, 100, 119, 127, 133, 201, 227-229, 232, 239, 430

E

Engineering Strategy 113, 119-120, 128, 133, 158, 227, 239

Exterior Design 101, 113, 115-117, 119-121, 128-129, 133, 244-248, 250, 257-258, 263-265, 269-272, 274, 383

G

Global Engineering 139, 142, 164, 328-329

global manufacturing 94, 97, 106, 139-142, 145, 147, 157-158, 164, 329, 334, 430

global production 1-2, 4-7, 9, 48, 92-93, 103, 106, 113-114, 122, 139-144, 149, 157, 160, 164, 197-198, 201, 319-320, 323, 329, 340, 345, 349, 351-352, 354, 356-357, 359, 361, 363, 365, 371-372, 430

H

high-cost performance 298, 313, 316

human resources 6, 13-15, 70, 94, 106, 140, 160, 201, 323, 352-353, 355, 359, 361, 364, 366, 369-370

I

intellectual property 227, 231-232, 236, 238-239

J

JIT Laboratory 78, 120-121, 227-228, 231, 247, 258

JIT strategy 47-49, 53, 60, 95, 97, 99-101, 103, 164, 197, 199, 201, 220, 227, 239, 244, 296, 313, 316, 319, 321, 340, 349-350, 377, 394, 406-407

L

loosening mechanism 296-297, 301, 304, 307, 311, 313, 316

M

management platform 70, 88

management technology 1-7, 9, 14-15, 19-20, 23-26, 47-48, 92-95, 99, 101-102, 107, 113-114, 117, 140-141, 145, 147, 158, 169, 197-200, 227-231, 239, 319, 353, 356, 385, 394, 430

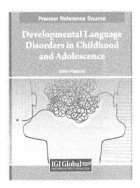

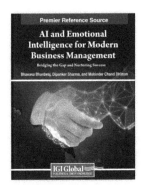

  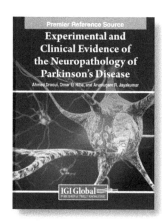 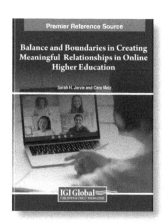

www.igi-global.com

Publishing Tomorrow's Research Today

IGI Global's Open Access Journal Program

Including Nearly 200 Peer-Reviewed, Gold (Full) Open Access Journals across IGI Global's Three Academic Subject Areas:
Business & Management; Scientific, Technical, and Medical (STM); and Education

**Consider Submitting Your Manuscript to One of These Nearly 200
Open Access Journals for to Increase Their Discoverability & Citation Impact**

| Web of Science Impact Factor | 6.5 | Web of Science Impact Factor | 4.7 | Web of Science Impact Factor | 3.2 | Web of Science Impact Factor | 2.6 |

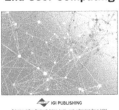

JOURNAL OF
**Organizational and
End User Computing**

JOURNAL OF
**Global Information
Management**

INTERNATIONAL JOURNAL ON
**Semantic Web and
Information Systems**

JOURNAL OF
**Database
Management**

Choosing IGI Global's Open Access Journal Program Can Greatly Increase the Reach of Your Research

Higher Usage
Open access papers are 2-3 times more likely to be read than non-open access papers.

Higher Download Rates
Open access papers benefit from 89% higher download rates than non-open access papers.

Higher Citation Rates
Open access papers are 47% more likely to be cited than non-open access papers.

Submitting an article to a journal offers an invaluable opportunity for you to share your work with the broader academic community, fostering knowledge dissemination and constructive feedback.

Submit an Article and Browse the IGI Global Call for Papers Pages

We can work with you to find the journal most well-suited for your next research manuscript.
For open access publishing support, contact: journaleditor@igi-global.com

Milton Keynes UK
Ingram Content Group UK Ltd.
UKHW050828270524
443319UK00015B/764